Community Ecology

Analytical Methods Using R and Excel®

Mark Gardener

DATA IN THE WILD SERIES

Pelagic Publishing | www.pelagicpublishing.com

Published by Pelagic Publishing
www.pelagicpublishing.com
PO Box 725, Exeter, EX1 9QU

Community Ecology
Analytical Methods Using R and Excel®

ISBN 978–1–907807–61–9 (Pbk)
ISBN 978–1–907807–62–6 (Hbk)
ISBN 978–1–907807–63–3 (ePub)
ISBN 978–1–907807–65–7 (PDF)
ISBN 978–1–907807–64–0 (Mobi)

Windows, Excel and Word and are trademarks of the Microsoft Corporation. For more information visit www. microsoft.com. OpenOffice.org is a trademark of Oracle. For more information visit www.openoffice.org. LibreOffice is a trademark of The Document Foundation. For more information visit www.libreoffice.org. Apple Macintosh is a trademark of Apple Inc. For more information visit www.apple.com.

British Library Cataloguing in Publication Data
A catalogue record for this book is available from the British Library.

Cover image: Over under water picture, showing Fairy Basslets (*Pseudanthias tuka*) amongst Cabbage Coral (*Turbinaria reniformis*) and tropical island in the background. Indo Pacific. © David Fleetham/OceanwideImages.com

Typeset by Swales & Willis Ltd, Exeter, Devon, UK

About the author

Mark Gardener (www.gardenersown.co.uk) is an ecologist, lecturer and writer working in the UK. His primary area of research was in pollination ecology and he has worked in the UK and around the world (principally Australia and the United States). Since his doctorate he has worked in many areas of ecology, often as a teacher and supervisor. He believes that ecological data, especially community data, are the most complicated and ill-behaved and are consequently the most fun to work with. He was introduced to R by a like-minded pedant whilst working in Australia during his doctorate. Learning R was not only fun but opened up a new avenue, making the study of community ecology a whole lot easier. He is currently self-employed and runs courses in ecology, data analysis and R for a variety of organisations. Mark lives in rural Devon with his wife Christine, a biochemist who consequently has little need of statistics.

Acknowledgements

There are so many people to thank that it is hard to know where to begin. I am sure that I will leave some people out, so I apologise in advance. Thanks to Richard Rowe (James Cook University) for inspiring me to use R. Data were contributed from various sources, especially from MSc students doing Biological Recording; thanks especially to Robin Cure, Jessie MacKay, Mark Latham, John Handley and Hing Kin Lee for your hard-won data. The MSc programme helped me to see the potential of 'proper' biological records and I thank Sarah Whild for giving me the opportunity to undertake some teaching on the course. Thanks also to the Field Studies Council in general: many data examples have arisen from field courses I've been involved with.

Software used

Several versions of Microsoft's Excel® spreadsheet were used in the preparation of this book. Most of the examples presented show version 2007 for Microsoft Windows® although other versions may also be illustrated.

The main version of the R program used was 2.12.1 for Macintosh: The R Foundation for Statistical Computing, Vienna, Austria, ISBN 3-900051-07-0, http://www.R-project.org/. Other versions were used in testing code.

Support material

Free support material is available on the Community Ecology companion website, which can be accessed via the book's resources page: http://www.pelagicpublishing.com/community-ecology-resources.html

Reader feedback

We welcome feedback from readers – please email us at info@pelagicpublishing.com and tell us what you thought about this book. Please include the book title in the subject line of your email.

Publish with Pelagic Publishing

We publish scientific books to the highest editorial standards in all life science disciplines, with a particular focus on ecology, conservation and environment. Pelagic Publishing produces books that set new benchmarks, share advances in research methods and encourage and inform wildlife investigation for all.

If you are interested in publishing with Pelagic please contact editor@pelagicpublishing.com with a synopsis of your book, a brief history of your previous written work and a statement describing the impact you would like your book to have on readers.

Contents

Introduction

Interactions between species are of fundamental importance to all living systems and the framework we have for studying these interactions is community ecology. This is important to our understanding of the planet's biological diversity and how species interactions relate to the functioning of ecosystems at all scales. Species do not live in isolation and the study of community ecology is of practical application in a wide range of conservation issues.

The study of ecological community data involves many methods of analysis. In this book you will learn many of the mainstays of community analysis including: diversity, similarity and cluster analysis, ordination and multivariate analyses. This book is for undergraduate and postgraduate students and researchers seeking a step-by-step methodology for analysing plant and animal communities using R and Excel.

Microsoft's Excel spreadsheet is virtually ubiquitous and familiar to most computer users. It is a robust program that makes an excellent storage and manipulation system for many kinds of data, including community data. The R program is a powerful and flexible analytical system able to conduct a huge variety of analytical methods, which means that the user only has to learn one program to address many research questions. Its other advantage is that it is open source and therefore free. Novel analytical methods are being added constantly to the already comprehensive suite of tools available in R.

What you will learn in this book

This book is intended to give you some insights into some of the analytical methods employed by ecologists in the study of communities. The book is not intended to be a mathematical or theoretical treatise but inevitably there is some maths! I've tried to keep this in the background and to focus on how to undertake the appropriate analysis at the right time. There are many published works concerning ecological theory; this book is intended to support them by providing a framework for learning how to analyse your data.

The book does not cover every aspect of community ecology. There are a few minor omissions – I hope to cover some of these in later works.

How this book is arranged

There are four main strands to scientific study: planning, recording, analysis and reporting. The first few chapters deal with the planning and recording aspects of study. You will see how to use the main software tools, Excel and R, to help you arrange and begin to make sense of your data. Later chapters deal more explicitly with the grand themes of community ecology, which are:

- *Diversity* – the study of diversity is split into several chapters covering species richness, diversity indices, *beta* diversity and dominance–diversity models.
- *Similarity and clustering* – this is contained in one chapter covering similarity, hierarchical clustering and clustering by partitioning.
- *Association analysis* – this shows how you can identify which species belong to which community by studying the associations between species. The study of associations leads into the identification of indicator species.
- *Ordination* – there is a wide range of methods of ordination and they all have similar aims; to represent complicated species community data in a more simplified form.

The reporting element is not covered explicitly; however the presentation of results is shown throughout the book. A more dedicated coverage of statistical and scientific reporting can be found in my previous work, *Statistics for Ecologists Using R and Excel*.

Throughout the book you will see example exercises that are intended for you to try out. In fact they are expressly aimed at helping you on a practical level – reading how to do something is fine but you need to do it for yourself to learn it properly. The *Have a Go* exercises are hard to miss.

Have a Go: Learn something by doing it

The Have a Go exercises are intended to give you practical experience at various analytical methods. Many will refer to supplementary data, which you can get from the companion website. Some data are intended to be used in Excel and others are for using with R.

Most of the *Have a Go* exercises utilise data that is available on the companion website. The material on the website includes various spreadsheets, some containing data and some allowing analytical processes. The *CERE.RData* file is the most helpful – this is an R file, which contains data and custom R commands. You can use the data for the exercises (and for practice) and the custom commands to help you carry out a variety of analytical processes. The custom commands are mentioned throughout the book and the website contains a complete directory.

You will also see tips and notes, which will stand out from the main text. These are 'useful' items of detail pertaining to the text but which I felt were important to highlight.

Tips and Notes: Useful additional information

The companion website contains supplementary data, which you can use for the exercises. There are also spreadsheets and useful custom R commands that you can use for your own analyses.

At the end of each chapter there is a summary table to help give you an overview of the material in that chapter. There are also some self-assessment exercises for you to try out. The answers are in Appendix 1.

Support files

The companion website (see resources page: http://www.pelagicpublishing.com/community-ecology-resources.html) contains support material that includes spreadsheet calculations and data in Excel and CSV (comma separated values) format. There is also an R data file, which contains custom R commands and datasets. Instructions on how to load the R data into your copy of R are on the website. In brief you need to use the `load()` command, for Windows or Mac you can type the following:

```
load(file.choose())
```

This will open a browser window and you can select the *CERE.RData* file. On Linux machines you'll need to replace the `file.choose()` part with the exact filename in quotes, see the website for more details.

    I hope that you will find this book helpful, useful and interesting. Above all, I hope that it helps you to discover that analysis of community ecology is not the 'boring maths' at the end of your fieldwork but an enjoyable and enlightening experience.

Mark Gardener, Devon 2013

1. Starting to look at communities

The study of community ecology is complicated and challenging, which makes it all the more fun, of course. Ecology is a science and like all science subjects there is an approach to study that helps to facilitate progress.

1.1   A scientific approach

Science is a way of looking at the natural world. In short, the process goes along the following lines:

- You have an idea about something.
- You come up with a hypothesis.
- You work out a way of testing this hypothesis.
- You collect appropriate data in order to apply a test.
- You test the hypothesis and decide if the original idea is supported or rejected.
- If the hypothesis is rejected, then the original idea is modified to take the new findings into account.
- The process then repeats.

In this way, ideas are continually refined and our knowledge of the natural world is expanded. You can split the scientific process into four parts (more or less): planning, recording, analysing and reporting.

- *Planning*: This is the stage where you work out what you are going to do. Formulate your idea(s), undertake background research, decide what your hypothesis will be and determine a method of collecting the appropriate data and a means by which the hypothesis may be tested.
- *Recording*: The means of data collection is determined at the planning stage although you may undertake a small pilot study to see if it works out. After the pilot stage you may return to the planning stage and refine the methodology. Data are finally collected and arranged in a manner that allows you to begin the analysis.
- *Analysing*: The method of analysis should have been determined at the planning stage. Analytical methods (often involving statistics) are used to test the null hypothesis. If the null hypothesis is rejected then this supports the original idea/hypothesis.

- *Reporting*: Disseminating your work is vitally important. Your results need to be delivered in an appropriate manner so they can be understood by your peers (and often by the public). Part of the reporting process is to determine what the future direction needs to be.

In community ecology the scientific process operates in the same way as in any other branch of science. Generally you are dealing with complicated situations with many species and samples – methods of analysis in community ecology are specialised because of this complexity.

1.2 The topics of community ecology

There are many ways to set about analysing community data. The subject can be split into several broad themes, which can help you to determine the best approach for your situation and requirements.

1.2.1 Diversity

Diversity is concerned with how many different species there are in a given area. Strictly speaking there are two main strands of diversity – in the first you simply count the number of different species in an area. In the second case you take into account the abundance of the species – this leads to the notion of the *diversity index*.

The term *diversity* (or *biodiversity*) is a much used term both in science and in general use. Its meaning in science is not necessarily the same as that understood by the general public. You can think of diversity as being expressed in two forms:

- The number of different species in a given area.
- The number of species and their relative abundance in a given area.

The first form, number of species in an area, is called *species richness* (see Chapter 7). This is an easy measure to understand and you can calculate it from simple species lists. The second form, involving relative abundance of species, is more complicated because of course you have an extra dimension, abundance information (see Chapter 8).

Whichever measure of diversity is under question, the scale of measurement is particularly important. Diversity is usually expressed at three scales (see Chapter 10):

- *Alpha diversity* – this is diversity measured in a single habitat or sampling unit (e.g. a quadrat); it is the smallest unit of measurement.
- *Beta diversity* – this is the diversity between habitats.
- *Gamma diversity* – this is the diversity of a larger sampling unit, such as a landscape that is composed of many habitats.

The three scales of measurement of diversity are linked by a simple relationship:

Alpha × *beta* = *gamma*

In some measures of diversity however, the relationship can be additive rather than multiplicative (see Chapter 10).

The species richness measure of diversity can be used when you do not have abundance information – which can be useful. Species richness can also be used as the response variable in analyses in certain circumstances (see Section 7.1).

When you have abundance information you are able to carry out different analyses, for example:

- Diversity indices.
- Species abundance curves.

A *diversity index* is a way to take into account the evenness of a community – if a single species dominates a community the index is smaller, if the species are all more even in abundance the index is larger (see Chapter 8).

Species abundance curves are another way to look at the evenness of a community – the abundance of each species is plotted on a graph, with the most abundant being plotted first (see Chapter 11).

1.2.2 Similarity and clustering

Similarity and *clustering*: this is where you look to see how similar things are based on their composition (see Section 12.1). In community ecology this tends to be the similarity of sites or habitats based on the species present. The idea of clustering stems from this – you form clusters of things based on how similar they are.

There are two main approaches to clustering:

- *Hierarchical clustering* – in this approach the data are repeatedly split into smaller units until you end up with a kind of 'family tree', which shows the relationship between items (see Section 12.2.1).
- *Clustering by partitioning* – in this approach you take the data and build clusters based on how similar they are; the data are clumped around so-called *medoids*, which are the centres of the various groups (see Section 12.2.2).

You can explore similarity and create clusters of samples even if you do not have species abundance information – simple presence-absence data can be used.

1.2.3 Association analysis

Association analysis is a way to link species together to find out which species tend to be found in the same samples and which ones tend to be found in different samples. This is one way to identify communities – species that tend to be found together will likely be from the same community. You can set about sampling in two main ways:

- By area – in this approach you sample in a geographical area and identify the various associations (which can be positive or negative) and so identify the communities in that area (see Section 13.1).
- By transect – in this approach you sample along a transect, usually because of some underlying environmental gradient. Often this will lead to a succession of communities and your association analysis will help you to identify them (see Section 13.2).

The association analysis gives you values for the 'strength' of the various associations – this can be thought of as akin to the similarity and clustering kind of analyses (Chapter 12). A spin-off from association analysis is the idea of *indicator species* (see Section 13.4). Here you look to see if certain species can be regarded as indicative of a particular community. An ideal indicator species would be one that shows great specificity for a single community.

1.2.4 Ordination

The term *ordination* covers a range of methods that look to simplify a complicated situation and present it in a simpler fashion (see Chapter 14). This sounds appealing! In practice you are looking at communities of species across a range of sites or habitats and the methods of ordination look to present your results in a kind of scatter plot. Things that appear close are more similar to one another than things that are far apart. Think of it as being an extension to the similarity and clustering idea.

There are several methods of ordination (see Chapter 14) but you can split the general idea of ordination into two broad themes:

- *Indirect gradient analysis* – in this approach you analyse the species composition and the patterns you observe allow you to infer environmental gradients that the species may be responding to (see Section 14.2).
- *Direct gradient analysis* – in this approach you already have environmental data which you use to help reorder the samples and species data into meaningful patterns (see Section 14.3). A spin-off from this approach is that you can test hypotheses about the effects of the environmental variable(s) that you measured.

Ordination is a very commonly used analytical approach in community ecology because the main aim of the various methods is to distil the complicated community data into a simpler and more readily understood form.

1.3   Getting data – using a spreadsheet

A spreadsheet is an invaluable tool in science and data analysis. Learning to use one is a good skill to acquire. With a spreadsheet you are able to manipulate data and summarise details in different ways quite easily. You can also use a spreadsheet to prepare data for further analysis in other computer programs. It is important that you formalise the data into a standard format, as you shall see later (in Chapter 3). This will make the analysis run smoothly and allow others to follow what you have done. It also allows you to see what you did later on (it is easy to forget the details).

Your spreadsheet is useful as part of the planning process. You may need to look at old data; these might not be arranged in an appropriate fashion so using the spreadsheet will allow you to organise your data. The spreadsheet will allow you to perform some simple manipulations and run some straightforward analyses, looking at means for example, as well as producing simple summary graphs. This will help you to understand what data you have and what they might show. You will see a variety of ways of manipulating data as you go along (e.g. Section 4.2).

If you do not have past data and are starting from scratch, then your initial site visits and pilot studies will need to be dealt with. The spreadsheet should be the first thing you look to, as this will help you arrange your data into a format that facilitates further study. Once you have some initial data (be it old records or pilot data) you can continue with the planning process.

1.4 Aims and hypotheses

A hypothesis is your idea of what you are trying to determine but phrased in a specific manner. The hypothesis should relate to a single testable item.

In reality you cannot usually 'prove' your hypothesis – it is like a court of law when you do not have to prove your innocence, you are assumed innocent until proven otherwise. In statistics, the equivalent is the *null hypothesis*. This is often written as H0 (or H_0) and you aim to reject your null hypothesis and therefore, by implication, accept the alternative (usually written as H1 or H_1).

The H0 is not simply the opposite of what you thought (called the *alternative hypothesis*, H1) but is written as such to imply that no difference, no pattern, exists (I like to think of it as the *dull* hypothesis).

Getting your hypotheses correct (and also the null hypotheses) is an important step in the planning process as it allows you to decide what data you will need to collect in order to reject the H0. You will examine hypotheses again later (Section 5.2).

Allied to your hypothesis is the analytical method you will use later to help test and support (or otherwise) your hypothesis. Even at this early stage you should have some idea of the statistical test or analytical approach you are going to apply. Certain statistical tests are suitable for certain kinds of data and you can therefore make some early decisions. You may alter your approach, change the method of analysis and even modify your hypothesis as part of your planning process.

Some kinds of analysis do not lend themselves to a hypothesis test – this is particularly so in community ecology. When you have several species and several habitats your analysis may be concerned with looking for patterns in the data to highlight relationships that were not evident from the raw data. These analytical methods are important but you cannot always perform a hypothesis test. However, you still need to plan your approach and decide what method of analysis is best to help you make sense of the ecological situation (see Chapter 5) – if the best approach is to carry out an analysis that does not test a null hypothesis then that is what you go with.

1.5 Summary

Topic	Key Points
The scientific process	Community ecology is like any other science, there is a process: planning, recording, analysis and reporting. Each stage is important and helps with the following processes.
Topics in community ecology	Community ecology is a broad subject but can be split into various topics or themes.
Diversity	*Diversity* (or biodiversity) is concerned with how many different species there are in a given sampling area. In *species richness* you are only concerned with how many different species there are.
	A *diversity index* takes relative abundance into account. Diversity itself can be split into categories according to scale: *alpha*, *beta* and *gamma*. The first is diversity of a single sample, the last is overall diversity and *beta* diversity relates to the changes between samples.

Topic	Key Points
	Species abundance curves can help you determine how even a community is; so if a community is dominated by one or two species the community would not be even.
Similarity	*Similarity* is a way to determine how 'close' samples are in terms of their species composition. You can use species abundance or simple presence-absence data. Allied to this is the idea of dissimilarity where a high value indicates little similarity.
Clustering	*Clustering* involves creating groups of samples based on their species composition. There are two main approaches: *hierarchical* and *partitioning*.
	In hierarchical clustering the samples are split into smaller and smaller groups to make a 'family tree'.
	In partitioning methods, samples are built around cluster centres (medoids).
Association analysis	*Association* analysis is a way to identify which species tend to be found 'together' and which tend to be found separately. This can use used to identify communities by one of two means:
	• *Area analysis* – you sample in a geographical area. • *Transect analysis* – you sample along a transect (usually along an environmental gradient).
	Association analysis is akin to clustering and similarity and you can use the approach for that purpose.
	A spin-off is that you can identify *indicator species*, those that are 'indicative' of a particular community and tend to be found there and nowhere else.
Ordination	*Ordination* is the name given to a variety of analytical methods that seek to reorder samples and species in a meaningful way. The general aim is to represent the complicated community (and sometimes environmental) data in a more easily interpretable form; usually a 2D plot.
	There are two main approaches to ordination:
Indirect gradient analysis	With *indirect gradient analysis* you carry out the analysis using species and sample data. Any environmental information is inferred at the end from the patterns you observe.
Direct gradient analysis	With *direct gradient analysis* you utilise environmental data right at the outset and use it to help form the order of the samples and species.
Data recording	It is important to get your data into a sensible order. A spreadsheet is an invaluable tool and allows you to keep your data in order as well as giving you the chance to carry out a range of analyses.
Aims and hypotheses	A *hypothesis* is a testable statement that you can back-up with a statistical test. The *null hypothesis* is actually what you test. In community ecology it is not always possible to carry out a statistical test but you should still have clear aims to provide a framework for your work.

1.6 Exercises

1.1 What are the main topics in community ecology, as set out in this book?

1.2 Diversity can be measured at various scales, from simple samples to whole landscapes. What are the 'units' of diversity and how are they related?

1.3 What are the main reasons for carrying out association analysis?

1.4 With indirect gradient analysis you can test hypotheses about the relationship between species composition and environment – TRUE or FALSE?

1.5 If you had an idea regarding the number of species and an environmental variable your hypothesis might run along these lines 'there is a positive correlation between species richness and soil moisture'. What would an appropriate null hypothesis be?

The answers to these exercises can be found in Appendix 1.

2. Software tools for community ecology

Learning to use your spreadsheet is time well spent. It is important that you can manipulate data and produce summaries, including graphs. You will see later how the spreadsheet is used for a variety of aspects of data manipulation as well as for the production of graphs. Many statistical tests can be performed using a spreadsheet but there comes a point when it is better to use a dedicated computer program for the job. The more complicated the data analyses are the more cumbersome it is to use a spreadsheet and the more sensible it is to use a dedicated analytical program. There are many on the market, some are cheap (or even free) and others are expensive. Some programs will interface with your spreadsheet and others are totally separate. Some programs are specific to certain types of analysis and others are more general.

In this book you will focus on two programs:

- *Microsoft Excel*: this spreadsheet is common and widely available. There are alternatives and indeed the *Open Office* spreadsheet uses the same set of formulae and can be regarded as equivalent. The *Libre Office* spreadsheet is a derivative of Open Office and similarly equivalent to *Excel*.
- *R*: the R project for statistical computing is a huge open-source undertaking that is fast becoming the *de facto* standard for analysis in many fields of science, engineering and business, to name just a few. It is a powerful and flexible system.

Excel is particularly useful as a data management system, and throughout this book you will see it used mainly in that fashion although it is capable of undertaking some statistical analyses and producing various graphs. The R program is very powerful and flexible, and you will see this used for the majority of the analyses. Once you learn how to use R it is almost as easy to create a complicated community analysis as it is to carry out a simple *t*-test.

2.1 Excel

A spreadsheet in an invaluable tool. The most common is *Microsoft Excel* and it has many uses:

- For data storage.
- As a database.
- For preliminary summary.
- For summary graphs.
- For simple (and not so simple) statistical analyses.

Generally the more complicated the analysis you are going to undertake, the less likely it is that you will use a spreadsheet to do the analysis. However, when you have more complicated data it is really important to manage the data carefully and this is a strength of the spreadsheet. It can act like a database. Part of your planning process should be to determine how you are going to arrange your data – getting the layout correct from the start can save an immense amount of time later on.

2.1.1 Getting Excel

There are many versions of Excel and your computer may already have a version installed when you purchased it. The basic functions that Excel uses have not changed for quite some while so even if your version is older than described here, you should be able to carry out the same manipulations. You will mainly see Excel 2007 for Windows described here. If you have purchased a copy of Excel (possibly as part of the Office suite) then you can install this following the instructions that came with your software. Generally, the defaults that come with the installation are fine although it can be useful to add extra options, especially the *Analysis ToolPak*, which will be described next.

2.1.2 Installing the *Analysis Toolpak*

The *Analysis ToolPak* is an add-in for Excel that allows various statistical analyses to be carried out without the need to use complicated formulae. The add-in is not installed as standard and you will need to set up the tool before you can use it. The add-ins are generally ready for installation once Excel is installed and you usually do not require the original disk.

The statistical methods available via the *Analysis ToolPak* are not very relevant to most community studies and are more likely to be of use for examining hypotheses relating to individual species. However, you may be looking at the number of species in a given area (a measure called *species richness*) and some basic statistical routines could be helpful. You will see more about species richness in Chapter 7.

In order to install the *Analysis ToolPak* (or any other add-in) you need to click the Office button (at the top left of the screen) and select *Excel Options*.

In Figure 2.1 you can see that there are several add-ins already active and some not yet ready. To activate (i.e. install) the add-in, you click the *Go* button at the bottom of the screen. You then select which add-ins you wish to activate (Figure 2.2).

Once you have selected the add-ins to activate, you click the OK button to proceed. The add-ins are usually available to use immediately after this process.

To use the *Analysis ToolPak* you use the *Data* button on the *Ribbon* and select the *Data Analysis* button (Figure 2.3).

Once you have selected this, you are presented with various analysis tools (Figure 2.4).

Each tool requires the data to be set out in a particular manner; help is available using the *Help* button.

2.2   Other spreadsheets

The Excel spreadsheet that comes as part of the Microsoft Office suite is not the only spreadsheet and there are others available – of particular note is the Open Office program. This is available from http://www.openoffice.org and there are versions available for Windows, Mac and Linux. An offshoot of Open Office is Libre Office and this is available at http://www.libreoffice.org.

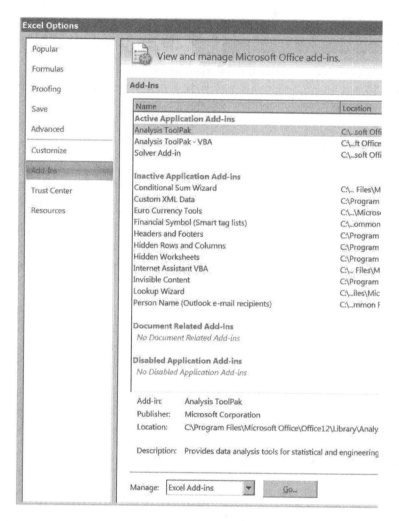

Figure 2.1 Selecting Excel add-ins from the Options menu.

Other spreadsheets generally use the same functions as Excel, so it is possible to use another program to produce the same result. Graphics will almost certainly be produced in a different manner and you will see graphics demonstrated with Excel 2007 for Windows throughout this book.

2.3 The R program

The program called R is a powerful environment for statistical computing. It is available free at the Comprehensive R Archive Network (CRAN) on the Internet at http://www.r-project.org. It is open source and available for all major operating systems.

R was developed from a commercial programming language called S. The original authors were called Robert and Ross so they called their program R as a sort of joke. This is what the R website says about the program:

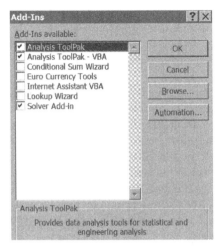

Figure 2.2 Selecting the add-ins for Excel.

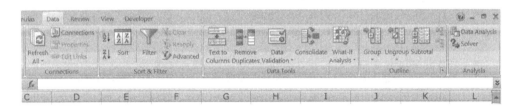

Figure 2.3 The *Analysis ToolPak* is available from the Data Analysis button on the Excel Data Ribbon.

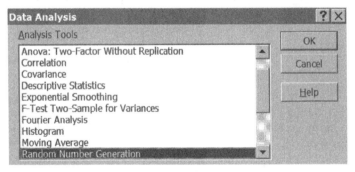

Figure 2.4 The *Analysis ToolPak* provides a range of analytical tools.

R is an open-source (GPL) statistical environment modeled after S and S-Plus. The S language was developed in the late 1980s at AT&T labs. The R project was started by Robert Gentleman and Ross Ihaka (hence the name R) of the Statistics Department of the University of Auckland in 1995. It has quickly gained a widespread audience. It is currently maintained by the R core-development team, a hard-working, international team of volunteer developers. The R project web page is the main site for information on R. At this site are directions for obtaining the software, accompanying packages and other sources of documentation.

R is a powerful statistical program but it is first and foremost a programming language. Many routines have been written for R by people all over the world and made freely available from the R project website as 'packages'. However, the basic installation (for Linux, Windows or Mac) contains a powerful set of tools for most purposes.

Because R is a programming language it can seem a bit daunting; you have to type in commands to get it to work; however, it does have a graphical user interface (GUI) to make things easier and it is not so different from typing formulae into Excel. You can also copy and paste text from other applications (e.g. word processors). So if you have a library of these commands, it is easy to pop in the ones you need for the task at hand.

R will cope with a huge variety of analyses and someone will have written a routine to perform nearly any type of calculation. R comes with a powerful set of routines built in at the start but there are some useful extra 'packages' available on the website. These include routines for more specialised analyses covering many aspects of scientific research as well as other fields (e.g. economics).

There are many advantages in using R:

- It is free, which allows anyone to access it.
- It is open source; this means that many bugs are ironed out.
- It is extremely powerful and will handle very complex analyses as easily as simple ones.
- It will handle a wide variety of analyses. This is one of the most important features: you only need to know how to use R and you can do more or less any type of analysis; there is no need to learn several different (and expensive) programs.
- It uses simple text commands. At first this seems hard but it is actually quite easy. The upshot is that you can build up a library of commands and copy/paste them when you need them.
- Documentation. There is a wealth of help for R. The CRAN site itself hosts a lot of material but there are also other websites that provide examples and documentation. Simply adding CRAN (or R) to a web search command will bring up plenty of options.

2.3.1 Getting R

Getting R is easy via the Internet. The R Project website is a vast enterprise and has local mirror sites in many countries. The first step is to visit the main R Project webpage at http://www.r-project.org.

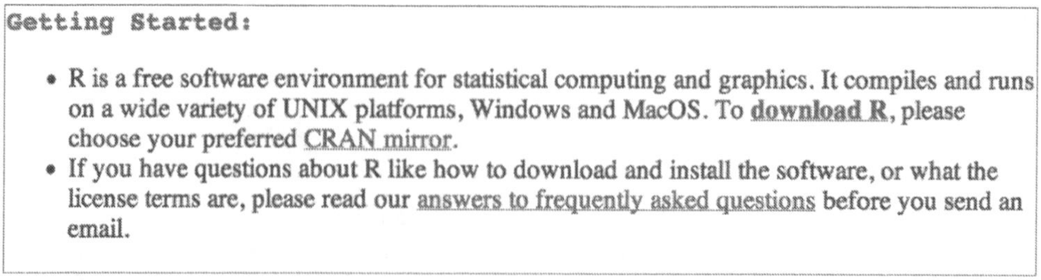

Getting Started:

- R is a free software environment for statistical computing and graphics. It compiles and runs on a wide variety of UNIX platforms, Windows and MacOS. To **download R**, please choose your preferred CRAN mirror.
- If you have questions about R like how to download and install the software, or what the license terms are, please read our answers to frequently asked questions before you send an email.

Figure 2.5 Getting R from the R Project website. Click the download link and select the nearest mirror site.

Once you have clicked the *download R* link (Figure 2.5), you have the chance to select a mirror site. These mirror sites are hosted in servers across the world and using a local one will generally result in a speedier download.

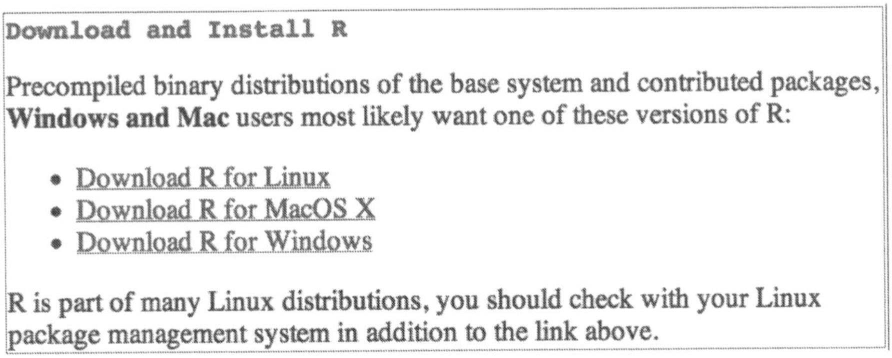

Download and Install R

Precompiled binary distributions of the base system and contributed packages, **Windows and Mac** users most likely want one of these versions of R:

- Download R for Linux
- Download R for MacOS X
- Download R for Windows

R is part of many Linux distributions, you should check with your Linux package management system in addition to the link above.

Figure 2.6 Getting R from the R Project website. Once you have selected the mirror site for your location you can choose the file to download.

Once you have selected a mirror site, you can click the link that relates to your operating system (Figure 2.6). If you use a Mac then you will go to a page where you can select the best option for you (there are versions for various flavours of OSX). If you use Windows then you will go to a Windows-specific page. (Figure 2.7) If you are a Linux user then read the documentation; you can often install R through the terminal and link to a version in a distro-specific repository.

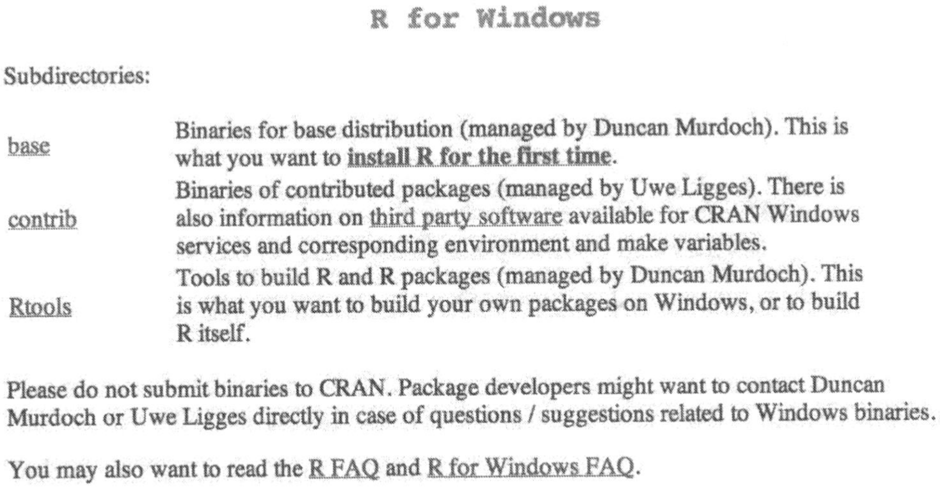

R for Windows

Subdirectories:

base	Binaries for base distribution (managed by Duncan Murdoch). This is what you want to install R for the first time.
contrib	Binaries of contributed packages (managed by Uwe Ligges). There is also information on third party software available for CRAN Windows services and corresponding environment and make variables.
Rtools	Tools to build R and R packages (managed by Duncan Murdoch). This is what you want to build your own packages on Windows, or to build R itself.

Please do not submit binaries to CRAN. Package developers might want to contact Duncan Murdoch or Uwe Ligges directly in case of questions / suggestions related to Windows binaries.

You may also want to read the R FAQ and R for Windows FAQ.

Figure 2.7 Getting R from the R Project website. The Windows-specific page allows you to get the version that is right for your Windows OS.

Assuming you have navigated to the Windows page, you will see something similar to Figure 2.7. Most users will want to select the base link, which will take you to a page where you can (finally) get the latest version of the installer file (Figure 2.8).

R-2.15.1 for Windows (32/64 bit)

Download R 2.15.1 for Windows (47 megabytes, 32/64 bit)

Installation and other instructions
New features in this version

Figure 2.8 Getting R from the R Project website. The final link will download the latest version (the one shown was current as of September 2012).

Now the final step is to click the link and download the installer file. This is an EXE file and it will download in the usual manner according to the setup of your computer.

2.3.2 Installing R

Once you have downloaded the install file, you need to run it to get R onto your computer. The process depends upon your operating system:

- If you use a Mac you need to double-click the disk image file to mount the virtual disk. Then double-click the package file to install R.
- If you use Linux you can simply double click the file you downloaded and instal-lation will proceed. If you install via the terminal then the terminal commands you use will carry out the process of installation.
- If you use Windows then you need to find the EXE file and run it. If you use Vista or later then it is a good idea to right-click the file and run as administrator (Figure 2.9).

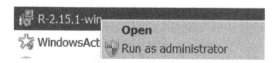

Figure 2.9 Installing R. If you have Windows Vista or later it is a good idea to right-click the install file and run as administrator.

The installation process asks a few basic questions, allowing you to select a language other than English for example. It is usual to accept the default location for the R files (a directory called R). The next screen asks if you wish to use customised startup options. In most cases for installing programs you are strongly suggested to say 'no' and to accept the defaults – this is no different, accept the defaults.

2.4   Summary

Topic	Key Points
Excel spreadsheet	Excel is virtually ubiquitous and is a reliable and easy to use program (at least for basic things). Excel is most useful as data storage software but you can also use it to help manage and manipulate your data, and as a simple database.
	Excel can carry out a range of statistical analyses and is especially useful as a 'first look' and to produce summary graphs and charts.
Other spreadsheets	The Open Office and Libre Office software packages are analogous to Microsoft Office. The spreadsheets use the same formulae/functions and in most senses are equivalent.
Analysis ToolPak	The *Analysis ToolPak* is an add-in for Excel (later version Windows only) that allows you to run a variety of basic statistical analyses more easily that using regular formulae.
The R program for statistical computing	The R program is a powerful statistical and graphical environment. It is a huge Open Source project and is free, with versions running on all computer systems (Windows, Mac and Linux).
	The R program is very flexible and there are packages available for it that will carry out all the analyses you can imagine (and many you cannot).
	The program uses basic text commands to 'drive' it. This permits great flexibility as you can save commands and reuse them.
	The program is available from the Internet at www.r-project.org where you can also obtain much documentation.

2.5   Exercises

2.1 What are the main uses for Excel (or any spreadsheet) in community ecology?

2.2 If you install the *Analysis ToolPak* for Excel – to help you carry out a range of statistical operations – where in Excel can you access the ToolPak?

2.3 The R program for statistical computing is available for a nominal fee from the Internet – TRUE or FALSE?

2.4 The R program is only useful for complicated statistical procedures – TRUE or FALSE?

The answers to these exercises can be found in Appendix 1.

3. Recording your data

The data you write down are of fundamental importance to your ability to make sense of your research at a later stage. If you are collecting new data then you are able to work out the recording of the data as part of your initial planning. If you have past data then you may have to spend some time rearranging before you can do anything useful.

3.1 Biological data

It is easy to write down a string of numbers in a notebook. You might even be able to do a variety of analyses on the spot; however, if you simply record a string of numbers and nothing else you will soon forget what the numbers represented. Worse still, nobody else will have a clue what the numbers mean and your carefully collected data will become useless.

All recorded data need to conform to certain standards in order to be useful at a later stage. The minimum you ought to record is:

- *Who*: the name of the person that recorded the data.
- *What*: the species you are dealing with.
- *Where*: the location that the data were collected from.
- *When*: the date that the data were recorded.

There are other items that may be added, depending upon your purpose, as you shall see later.

3.1.1 Biological data and science

Your data are important. In fact they are the most important part of your research. It is therefore essential that you record and store your data in a format that can be used in the future. There are some elements of your data that may not seem immediately important but which nevertheless are essential if future researchers need to make sense of them.

You need to write down our data in a way that makes sense to you at the time and also will make sense to future scientists looking to repeat or verify your work. Table 3.1 shows some biological data in an appropriate format. Not all the data are shown here (the table would be too big).

Every record (e.g. a row in Table 3.1) always has who, what, where and when. This is important for several reasons:

- It allows the data to be used for multiple purposes.
- It ensures that the data you collect can be checked for accuracy.

Table 3.1 An example of biological data: bat species abundance at various sites around Milton Keynes (only part of the data is shown).

Species	Recorder	Date	Site	Grid Ref	Quantity
Pipistrellus pipistrellus	Atherton, M	5-Aug-06	Ouzel Valley	SP880375	21
Myotis daubentonii	Atherton, M	5-Aug-06	Ouzel Valley	SP880375	23
Nyctalus noctula	Atherton, M	5-Aug-06	Ouzel Valley	SP880375	26
Plecotus auritus	Atherton, M	5-Aug-06	Ouzel Valley	SP880375	54
Myotis natteri	Atherton, M	5-Aug-06	Ouzel Valley	SP880375	54
Pipistrellus pipistrellus	Atherton, M	5-Aug-06	Newport	SP874445	43
Myotis daubentonii	Atherton, M	5-Aug-06	Newport	SP874445	11
Nyctalus noctula	Atherton, M	5-Aug-06	Newport	SP874445	9

- It means that you won't forget some important aspect of the data.
- It allows someone else to repeat the exercise exactly.

In the example above, you can see that someone (M Atherton) is trying to ascertain the abundance of various species of bat at sites around Milton Keynes in the UK. It would be easy for him to forget the date because it doesn't seem to matter that much. But if someone tries to repeat his experiment, they need to know what time of year he was surveying at. Alternatively, if environmental conditions change, it will be essential to know what year he did the work.

If you fail to collect complete biological data, or fail to retain and communicate all the details in full, then your work may be rendered unrepeatable and therefore useless as a contribution to science.

Once your biological data are compiled in this format, you can sort them by the various columns, export the grid references to mapping programs, and convert the data into tables for further calculations using a spreadsheet. They can also be imported into databases and other computer programs for statistical analysis.

Data collection in the field

When you are in the field and using your field notebook, you may well use shortcuts to record the information required. There seems little point in writing the site name and grid reference more than once for example. You may decide to use separate recording sheets to write down the information. These can be prepared in advance and printed as required. Once again there will be items that do not need to be repeated, a single date at the top of every sheet would be sufficient for example; however, when you transfer the data onto a computer it is a simple matter to copy the date or your name in a column.

In general, you should aim to create a column for each item of data that you collect. If you were looking at species abundance at several sites for example, then you would need at least two columns, one for the abundance data and one for the site. In your field notebook or recording sheet you may keep separate pages for each site and end up with a column of figures for each site. When you return to base and transfer the data to the spreadsheet, you should write our data in the 'standard format', i.e. one column for each thing (as in Table 3.1).

Having this strict *biological recording format* allows great flexibility, especially if you end up with a lot of data. Your data are now in the form of a database and your spreadsheet will be able to extract and summarise your data easily. You will see this kind of operation in Section 4.2.

Supporting information

As part of your planning process (including maybe a pilot study), you should decide what data you are going to collect. Just because you can collect information on 25 different environmental variables does not mean that you should. The date, location and the name of the person collecting the data are basic items that you always need but there may also be additional information that will help you to understand the biological situation as you process the data later. These things include field sketches and site photographs.

A field sketch can be very helpful because you can record details that may be hard to represent in any other manner. A sketch can also help you to remember where you placed your quadrats; a grid reference is fine but meaningless without a map! Photographs may also be helpful and digital photography enables lots of images to be captured with minimum fuss; however, it is also easy to get carried away and forget what you were there for in the first place. Any supporting information should be just that – support for the main event: your data.

3.2   Arranging your data

As in the example in Table 3.1, it is important to have data arranged in an appropriate format. When you enter data into your spreadsheet you ought to start with a few basics which correspond to the *who, what, where* and *when*. There are extra items that may be entered depending on the level of study. These will largely correspond to your needs and the level of detail required. If you are collecting data for analysis then it is also important to set out your data in a similar fashion. This makes manipulating the data more straightforward and also maintains the multi-purpose nature of your work. You need to move from planning to recording and on to analysis in a seamless fashion. Having your data organised is really important!

When you collect biological data, enter each record on a separate line and set out your spreadsheet so that each column represents a factor. For example, Table 3.2 shows a small part of a complex dataset. Here you have recorded the abundance of several butterfly species. You could have recorded the species in several columns, one for each; however, you also have different locations. These locations are themselves further subdivided by management. If you wrote down the information separately you would end up with several

Table 3.2 Data table layout. Complex data are best set out in separate columns. Here butterfly abundance is recorded for four different factors.

Transect	Year	ssp	Man	Count
N	1996	pbf	no	1.15
N	1997	pbf	no	1.54
N	1998	pbf	no	0
N	1996	pbf	yes	1.54
N	1997	pbf	yes	4.62
N	1998	pbf	yes	0
N	1996	pbf	yes	2.78
N	1997	pbf	yes	1.67
N	1998	pbf	yes	0
S	1996	swf	yes	7.11
S	1997	swf	yes	25.53
S	1998	swf	yes	2.37

smaller tables of data and it would be difficult to carry out any actual analyses. By recording the information in separate columns you can carry out analyses more easily.

The data in Table 3.2 can be split into various subsections using your spreadsheet and the *filter* command (Section 4.2.2). You can also use the *Pivot Table* function to review the data (Section 4.2.7).

Now you have gone through the planning process. Ideally, you would have worked out a hypothesis and know what data you need to collect to support your hypothesis (or to reject it). You ought to know at this stage what type of analysis you are going to run on your data (Chapter 5).

3.3   Summary

Topic	Key Points
Importance of biological records	Your data are important. You need to keep them in good order so that they can be used effectively. Good data can be checked easily for accuracy and the project can be repeated.
Elements of biological records	There are some basic elements of biological records that should always be included:
	Who – the person recording the data.
	What – the item of data, e.g. the species name.
	Where – the location where the data were recorded.
	When – the time the data were collected; usually just the date but more precise information may be important for some studies.
Record structure and arrangement	All data should have a rigid structure. Each column should represent a single variable. Each row should be a single 'record', a unique combination of the columns.
	This structure allows the data to be used like a database and allows you to manipulate the data later using sorting, filters or Pivot Tables.
Recording in the field	There are some elements that you need not record in the field. It is usually only necessary to record the date once, for example. However, when you return home you should ensure that the date is transferred to all lines of the records on the computer.
Supporting information	Additional elements that are not strictly data can be useful. Field sketches and photographs can help later, especially if it is some while before you carry out the analyses or another researcher conducts the analyses. However, only include useful information – it is easy to get carried away with a digital camera.

3.4   Exercises

3.1   What are the basic elements of a biological record?

3.2   What sort of items should make up the columns of your data?

3.3   It is important to write down the date because you will need to show your supervisor when you were out recording data – TRUE or FALSE?

The answers to these exercises can be found in Appendix 1.

4. Beginning data exploration: using software tools

In order to make sense of your data you will need to use some of the tools you have come across already, your spreadsheet and the R program. Excel is able to carry out a range of statistical tasks but it was never designed as a tool for community analyses. It therefore makes sense to use something that was designed as a statistical environment – this is where R comes in. R is very powerful and flexible – you can carry out a simple correlation as easily as a complicated community analysis.

In this chapter you will find out a bit more about using R (Section 4.1) and Excel (Section 4.2) – the two mainstays of your analytical world. The next section will give you a quick tour of R – this will form the mainstay of most of the analytical routines.

4.1   Beginning to use R

Once you have installed R, run it using the regular methods: you may have a shortcut on the desktop or use the *Start* button. Once you have run the program, you will see the main input window and a welcome text message. This will look something like Figure 4.1 if you are using Windows. There is a > and cursor | to show that you can type at that point. In the examples, you will see the > to indicate where you have typed a command, and lines beginning with anything else are the results of your typing.

The program appearance (GUI) is somewhat sparse compared to most Windows programs (Figure 4.1). You are expected to type commands in the window. This sounds a bit daunting but is actually not that hard. Once you know a few basics, you can start to explore more and more powerful commands because R has an extensive help system. There are many resources available on the R website and it is worth looking at some of the recommended documents (most are PDF) and working through those. Of course this book itself will provide a good starting point! You might also look at the companion book *Statistics for Ecologists Using R and Excel* (Gardener 2012).

After a while you can start to build up a library of commands in a basic text editor; it is easy to copy and paste commands into R. It is also easy to save a snapshot of the work you have been doing for someone else to look over.

4.1 1 Getting help

Everyone needs a bit of help from time to time and you are bound to want help with using R. You can get help in various ways:

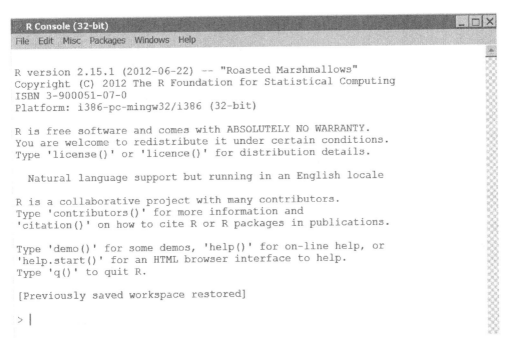

Figure 4.1 The R program interface is a bit sparse compared to most Windows programs.

- The internal help system.
- Online help.
- A book (like the one you are reading).

Help within R

R has extensive help. If you know the name of a command and want to find out more (there are often additional options), then type one of the following:

```
> help(topic)
> ?topic
```

You replace the word *topic* with the name of the command you want to find out about. Newer versions of R do not use the Windows style of help but open the help system in your default web browser. You can access the main index by typing:

```
> help.start()
```

Note that you do need the brackets at the end. These commands are listed in the opening welcome message (Figure 4.1).

The way the help system is displayed depends upon the operating system you are using:

- In Windows the help entries always open in your default web browser.
- If you use a Mac then the help commands open in a browser-like window but typing help.start() will always open the HTML index in your default browser.
- If you use Linux then the 'regular' help appears as plain text but help.start() will open the HTML help system in the default browser.

Online help

Lots of people use R and many of them write about it on the Internet. The R website is a good start but just adding CRAN to a web search may well get you what you need. CRAN is the acronym for the 'Comprehensive R Archive Network'. Adding R to your search also works well. The R website itself has links to many help articles.

4.1.2 Starting with R

You have to start somewhere so begin gently with some really simple arithmetic.

Simple maths

R can function like a regular calculator. Start by typing some maths:

```
> 23 + 7
> 14 + 11 + (23*2)
> sqrt(17-1)
> pi*2
```

Notice that R does not mind if you have spaces or not. Your answers look something like the following:

```
[1] 30
[1] 70
[1] 4
[1] 6.283185
```

The [1] indicates that the value displayed is the first in a series (and in this case the only value). If you have several lines of answers, then each line begins with a [n] label where n is a number relating to 'how far along' the list the first in that line is. This maths is very useful but you really want to store some of the results so that you can get them back later and use them in longer calculations. To do this you create variable names. For example:

```
> answer1 = 23 + 21 + 17 + (19/2)
> answer2 = sqrt(23)
> answer3 = answer1 + (answer1 * answer2)
```

This time you will notice that you do not see the result of your calculations. R does not assume that just because you created these objects you necessarily want to see them. It is easy enough to view the results; you need to type in the name of the thing you created:

```
> answer1
[1] 70.5
> answer2
[1] 4.795832
> answer3
[1] 408.6061
```

In older versions of R, a sort of arrow was used instead of the = sign:

```
> answer4 <- 93 - 21 + sqrt(41)
> 77/5 + 9 -> answer5
```

This is actually a bit more flexible than the = sign (try replacing the arrows above with = and see what happens) but for most practical purposes, = is quicker to type.

Data names

Everything in R needs a name. In a spreadsheet, each cell has a row and column reference, e.g. B12, which allows you to call up data as you like. R does not set out data like this so you must assign labels to everything so you (and R) can keep track of it. You need to give your data a name; R expects everything to have a name and this allows great flexibility. Think of these names as like the memory on a calculator; with R you can have loads of different things stored at the same time.

Names can be more or less any combination of letters and numbers. Any name you create must start with a letter, but other than that you can use numbers and letters freely. The only other character allowed is a full stop. R is case sensitive so the following are all different:

```
data1
Data1
DATA1
data.1
```

It is a good idea to make your names meaningful but short!

4.1.3 Inputting data

The first thing to do is to get some data into R so you can start doing some basic things. The most basic method of inputting data is to type it directly.

The c() command (short for combine, or concatenate) reads data from the keyboard so:

```
> data1 = c(24, 16, 23, 17)
```

Here you make an object called *data1* and combine the four numbers in the brackets to make it. Each value has to be followed by a comma. The spaces are optional, R ignores them, but they can be useful to aid clarity and avoid mistakes when typing.

You may easily add more items to your data using the c() command; for example:

```
> data2 = c(data1, 23, 25, 19, 20, 18, 21, 20)
```

Now you have a new item called *data2*. You could also have overwritten our original item and effectively added to it without creating a new object:

```
> data1 = c(data1, 23, 25, 19, 20, 18, 21, 20)
```

Here you put the old data at the beginning because you wanted to add the new values after, but you could easily place the old data at the other end or even in the middle.

Many of the data sets that you will have will be too large to input directly from the keyboard. You'll see a method to import CSV files later (Section 4.3).

4.1.4 Seeing what data items you have

Once you have a few objects in the memory, you will naturally want to see what you have. Do this by simply typing the name of a variable. For small datasets this is not a problem but if you have lots of data then this might be a bit tedious (R does its best to wrap the text to fit the screen).

```
[1] 24 16 23 17
[1] 24 16 23 17 23 25 19 20 18 21 20
```

The examples above show your *data1* and *data2* items.

List items `ls ()` command

The `ls()` command will show you what objects reside in the memory of R. This will usually consist of the samples that you entered but you may also see other objects that arise from calculations that R performs.

```
> ls()
[1] "beetle.cca"    "biol"    "biol.cca"    "env"    "op"
```

The example above shows five objects, some are data typed in directly and others are results from various calculations.

Remove items `rm ()` command

You might wish to remove some clutter and using the `rm()` command allows you to delete objects from R.

```
> rm(data1)
```

This will now get rid of the *data1* object that you created earlier. There is no warning – it just gets deleted!

4.1.5 Summary statistics in R

Now that you have a few objects in R, you will want to explore them. R provides a lot of basic commands that will do this.

The `summary ()` command

The `summary()` command will provide basic summary statistics about your chosen object. Some statistical commands create many sub-objects that do not display by default, so typing `summary()` will generally provide a lot of extra information. The basic command displays something like this:

```
> summary(data1)
   Min. 1st Qu.   Median   Mean 3rd Qu.    Max.
  16.00   16.75    20.00   20.00   23.25   24.00
```

Here you see that the basic stats are the middle values (mean and median) as well as the ends (max and min) and something in between, the interquartiles. If you perform a `summary()` on other types of data you may get something different.

Statistical summary functions

There are a number of useful statistical functions that you can select; here are some examples:

```
length(data1)
mean(data1)
median(data1)
max(data1)
min(data1)
sd(data1)
```

The first shows you how many items there are in your data (the number of observations, also called replicates). The next determines the mean (a kind of average). The median() command determines another kind of average. You can determine the largest and smallest values using the max() and min() commands. The final sd() command works out the standard deviation (a measure of variability). Try them out on some data and become familiar with them; you will meet some of them again shortly.

4.1.6 Previous commands and R history

Now you have a few basic commands at your disposal. R stores your commands in an internal list, which you can access using the up arrow. The up and down arrows cycle through recent commands; allowing you to edit a previous command. The left and right arrows move through the command line. Alternatively, you can click the command you want to edit. When you exit R, the previous commands will usually be stored automatically.

Saving history of commands

You can save (or load) a list of previous commands using the GUI. In Windows, the *File* menu gives the option to load or save a history file (Figure 4.2). This may be useful if you have been working on a specific project and wish to recall the commands later.

Saving the workspace

You can also save the workspace; this includes all the objects currently in the memory. Imagine you are working on several projects: one might be a bird survey, another a plant database and the third an invertebrate study. It might be useful to keep them all separate. R allows you to save the workspace as a file that you can call up later. You can also send the file to a colleague (or tutor) who can access the objects you were working on (Figure 4.2).

Saving a snapshot of your R session

If you have been working on a series of analyses it may be useful to save an overview of what you have been doing. You can do this using the *File > Save To File* menu option (Figure 4.2). R will save the input window as a plain text file, which you can open in a text editor. You can use this to keep a note of what you did or send it to a colleague (or tutor). You can also use the text file as the basis for your own library of commands; you can copy and paste into R and tweak them as required – R is very flexible.

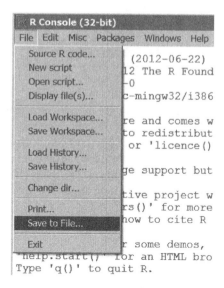

Figure 4.2 The File menu in R allows you to save a snapshot of the R working area to a plain text file.

Annotating work

R accepts input as plain text. This means you can paste text from other applications. It is useful to have a library of commands at your disposal. You can annotate these so that you know what they are. R uses the hash character # to indicate that the following text is an annotation and is therefore not executed as a command.

```
> vegemite(vegdata, use= "Domin") # from the vegan package
```

Here the note reminds the user that the *vegan* library needs to be loaded because this command is not part of the original R program.

4.1.7 Additional libraries of commands

The basic distribution of R appears like a single program but it is in fact a modular system and you can see the various elements by using the `search()` command like so:

```
> search()
 [1]  ".GlobalEnv"         "tools:RGUI"          "package:stats"
 [4]  "package:graphics"   "package:grDevices"   "package:utils"
 [7]  "package:datasets"   "package:methods"     "Autoloads"
[10]  "package:base"
```

You can see that R is built from several *packages*. Now R is a very capable program but it cannot do everything 'out of the box'. At various times people have written libraries of specialised R commands that are bundled into new packages. These are available to download and install. You will need some additional packages in order to carry out some community ecology – you'll see exactly what you need at the appropriate juncture but the following text will give you some general guidelines.

Finding R packages

You can see the variety of packages available on the R website (simply follow the links) – at time of writing there were over 4000 available! Packages are listed alphabetically and also by date. The *Task Views* section is useful as it lists the packages available by topic. You can also do a web search and will often find the name of the package that you need for a particular analysis.

Installing R packages

Once you know the name of a package that you need, you will want to get it installed in your copy of R. There are several ways to do this.

If you are using Windows, the GUI has a dedicated *Packages* menu (Figure 4.3). Mac users have a similar menu.

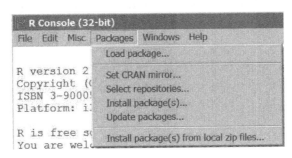

Figure 4.3 The Windows version of R has a menu item to handle packages of R commands.

The *Install package(s)* menu item is what you want to download and install packages. If you are not connected to the Internet you can use the *Install package(s) from local zip files* menu item to install a package from a zip file – this assumes you have already downloaded the zip file of course (get zip files from the R website).

You can also install packages by using a command typed into R:

```
> install.packages("package_name")
```

You simply replace the package_name with the name of the package you wish to download and install – you need quote marks around the name. Generally it is best to install packages using the command, as this will also install any additional packages that your selected package needs to operate.

Running R packages

Once you have the additional packages you needed the R commands contained within them need to be 'activated'. The packages are in your system but not ready for use until you use the library() command (you could also use the *Package* menu in Windows):

```
> library("vegan")
Loading required package: permute
This is vegan 2.0-0
```

In this case the *vegan* package is made ready. You can see that the command has also opened the *permute* package. If a package requires some other routines in order to work correctly, it will open the appropriate package.

Now the commands within the *vegan* package are ready for use (as are those in `permute` in this case). You can see that the packages are ready by using the `search()` command again:

```
> search()
 [1] ".GlobalEnv"          "package:vegan"   "package:permute"
 [4] "tools:RGUI"          "package:stats"   "package:graphics"
 [7] "package:grDevices"   "package:utils"   "package:datasets"
[10] "package:methods"     "Autoloads"       "package:base"
```

You can 'close' a package and make the commands unavailable by using the `detach()` command:

```
> detach(package:vegan)
```

This would detach the *vegan* package – see that quotes are not needed this time. Note that the package is not deleted from your system, you can still open it again using the `library()` command.

4.1.8 Exiting R

When you are ready to end your session in R, you can quit using several methods. You can use the buttons in the GUI like a regular program or use the `quit()` or `q()` commands (they are the same).

```
> q()
```

R will now ask if you want to save the workspace (Figure 4.4). It is a good idea to say *yes*. This will save the history of commands, which is useful. It will also save any items in memory. When you run R later, you can type `ls()` to list the objects that R has saved.

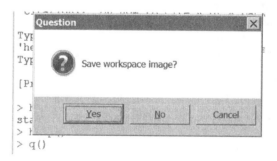

Figure 4.4 When quitting R you are prompted to save the workspace, thus preserving the data in memory.

Now you have been introduced to the R program, it is time to look at things you can do in Excel. This is the subject of the following section.

4.2 Manipulating data in a spreadsheet

Usually you collect data in a field notebook (or recording sheets) and then transfer them to a spreadsheet. This allows you to tidy up and order the data in a way that will allow you to explore it more easily. It is also a convenient way to store and transfer data.

In Chapter 3 you looked at how to arrange the data in a logical manner using a biological recording format.

This is only the start – once you have your data into a spreadsheet you are able to rearrange and sort them in a variety of ways that are potentially useful. Your spreadsheet can act as a database: the more complicated your data are the more important it is to be able to manipulate them easily. You can use various functions in your spreadsheet to help manage your data:

- *Sorting*: Being able to rearrange data into a new order is a simple but surprisingly useful way to explore your data. You can arrange in order of abundance for example and spot rare species at a glance.
- *Data filtering*: A filter allows you to select a block of data that matches certain criteria. Perhaps you need to extract a subset of your data for a separate analysis.
- *Paste Special*: Everyone is familiar with copy and paste operations as a way to move blocks of text or numbers from one place to another. The *Paste Special* command gives you more control over this – a useful operation is being able to switch rows and columns, essentially rotating a table.
- *Saving in different formats*: Your spreadsheet can save data in a variety of formats. This allows you to transfer data to a dedicated program for analysis. The comma separated value (CSV) format is the most widely used and many analytical programs (including R) can read it.
- *Lookup tables*: You can use a table of references values to act as replacements for other data. This can be used for many purposes, to translate a DAFOR value (a non-numerical abundance scale, where A = abundant, O = occasional, etc.) into a number, for example.
- *Pivot Tables*: Your spreadsheet can work like a database and the Pivot Table is the key. This allows you to rearrange and divide your data into meaningful chunks. You can use a Pivot Table as the basis for a summary of results or simply to rearrange your data into the correct layout for analysis.

These functions and operations are the subject of the following sections.

4.2.1 Sorting

Simply reordering your data into a sensible order can be useful – you can look at your data in various rearrangements that can help to make sense of the data. You can access sorting via the *Data > Sort* menu on the Ribbon in Excel 2007 (Figure 4.5).

The menu box now allows you to select the columns you wish to sort (and whether you want an ascending or descending order). It is essential to have column names so you recall what the data represent. Ensure the button is ticked where it says *My data has headers*.

4.2.2 Data filtering

Often you need to examine part of your data. For example you may have collected data on the abundance of plant communities at several sites. If you wish to look at a single site or a single species (or indeed both) then you can use the filtering ability of Excel. The exact command varies a little according to the version of the program you are using; Figure 4.6 shows Excel 2007 and the filtering option using the *Data* menu.

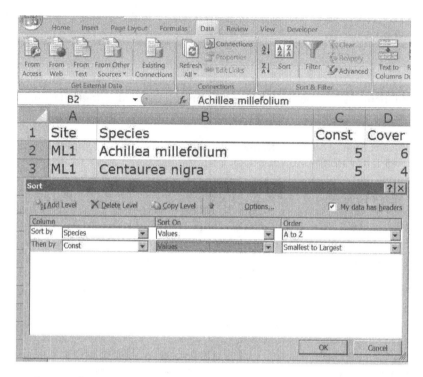

Figure 4.5 Sorting data in Excel, which provides a range of sorting options.

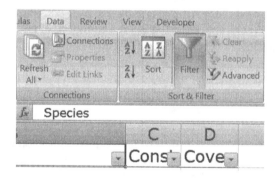

Figure 4.6 Using the filter option (in the *Data* menu) to select parts of the data. Once the filter options is selected drop-down arrows appear in the column headings.

You can highlight the columns of data you wish to filter. Excel will select all the columns by default so you do not make an active selection. Once the filter is applied, you see a little arrow in each column heading. It is important to give each column a heading. You can then click the arrow to bring up a list.

Figure 4.7 shows the filter in action. You can select to display a variety of options from the list available. Here you have site names so could select all, several or just one site to display. By adding a filter to subsequent columns you can narrow the focus.

Figure 4.7 Filtering in Excel. Once a filter is applied, it can be invoked using the drop-down menu.

You can see which columns have an active filter because the drop-down arrow button looks like a funnel (Figure 4.8).

You will think of many uses for filters once you get started – once you have a set of filtered data you can copy it as a block to another location. You cannot use summary statistics on filtered data so be careful. If you select cells and carry out some function you'll see small green triangles in the corner of the cells. When you click on a cell with a green triangle an icon appears with an exclamation mark. Select this and you will see something like Figure 4.9.

If you want to carry out functions on filtered data then copy and paste the data to a new location (a new tab or sheet) before anything else.

You can of course simply use the filter as a way to explore a large dataset – by clicking on the filter buttons for each column you can see quickly the data, a quick way to see a species list for example or find out how many sites were used.

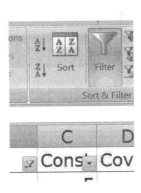

Figure 4.8 When a column is filtered the badge changes to a funnel icon.

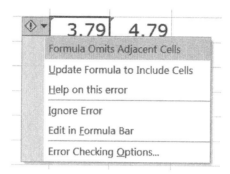

Figure 4.9 Filtered cells are non-adjacent, which can lead to errors if functions are applied. Copy the cells to a new location to avoid this.

4.2.3 Paste Special

The copy and paste commands are so general that it is easy to take them for granted. However, you can fine-tune the paste command by using *Paste Special*. This allows you to preserve certain attributes but not others.

It is possible to rotate the rows and columns using *Paste Special*. Some programs require the species to be the rows and the samples to be the columns but other programs need the samples to be the rows. It is simple enough to rotate the data. First of all you highlight the data you want and copy to the clipboard. Next you place the cursor where we want the data to appear (make a new file or worksheet before copying if you want) and click *Edit > Paste Special*. If you are using a later version of Excel then you can select a variety of paste options in the *Home* menu (Figure 4.10) – right-clicking also works.

You now have a variety of options (Figure 4.11). If your data were created using formulae then it is advisable to select *Values* from the options at the top. This will not

Figure 4.10 The paste options are found in the *Home* menu of later versions of Excel.

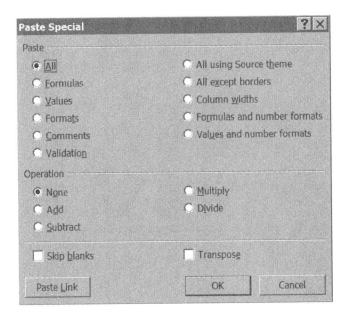

Figure 4.11 Using *Paste Special* in Excel. There are a variety of options for moving blocks of data from one place to another.

keep the formulae but because they would not be recognised correctly (since you are moving the items) then this is desirable. If you want to preserve formatting (e.g. italic species names) then you can select this option. In this case you also wish to rotate the data so that rows become columns and columns become rows so you select the *Transpose* button. The samples (sites) are now represented as rows and the species are columns.

4.2.4 File formats

You may need to save your data in a format that is not an Excel workbook. Some programs require plain text files, whilst others will accept text where the columns are separated by commas or tab characters. To do this you need to select the worksheet you want to save and click *File > Save As*. In Office 2007 you access this via the *Office* button as shown in Figure 4.12. The comma separated variables (CSV) format in particular is useful as many programs (including R) can read this. You can access the CSV format via the *Other Formats* option.

At the bottom of the menu box there will be a drop-down menu that allows you to select the file type required (Figure 4.13). CSV is a common format but tab delimited text and space delimited text are also used at times.

The CSV format is plain text resulting in the loss of some formatting and you will only be able to save the active tab; multiple worksheets are not allowed. Because of these reasons, Excel will give you a warning message to ensure you know what you are doing (Figure 4.14).

The resulting file is plain text and, when you view it in a word processor, the data are separated by commas. It is a good idea to keep a master XLS file; you can then save the bits you want as CSV to separate files as required. This means that you keep your original data file intact and use portions for separate purposes.

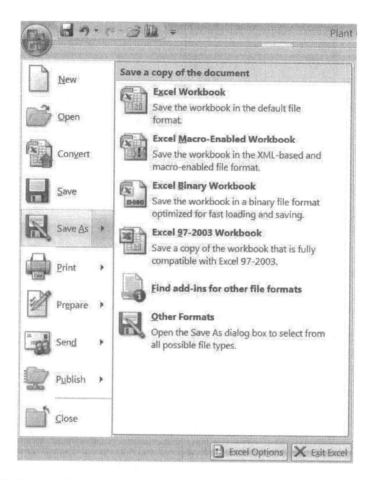

Figure 4.12 The Save As menu in Excel 2007.

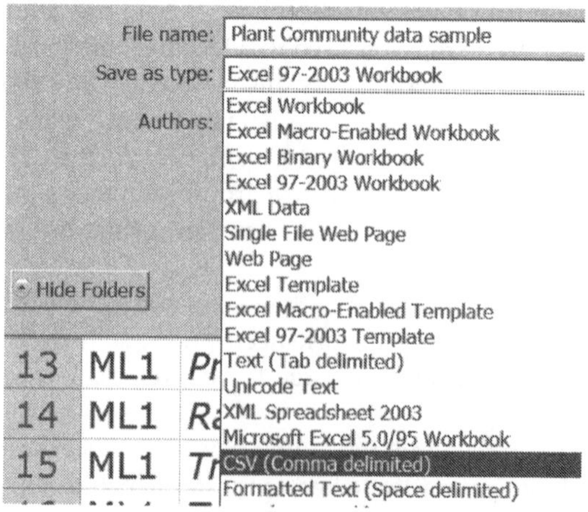

Figure 4.13 The CSV or comma delimited format is used by many other programs.

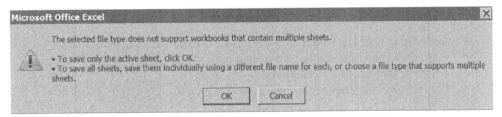

Figure 4.14 Excel will give you a warning if you try to save a file in a non-Excel format.

In Figure 4.15 you can see what a CSV file looks like in a plain text editor (Notepad). As the name 'comma separated variables' suggests, you see the data separated by commas. When the CSV data are read by a spreadsheet, a comma causes a jump to a new column.

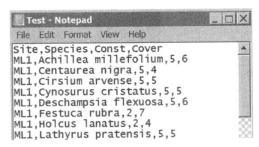

Figure 4.15 CSV format. In this format, commas separate data and when Excel reads the file it will place each item into a separate column when it comes across a comma. Many analytical programs use the CSV format.

4.2.5 Opening a CSV file

When you open a CSV in a spreadsheet, the program will convert the commas into column breaks and you see what you expect from a regular spreadsheet. You can do all the things you might wish in a spreadsheet, such as creating graphs; however, if you want the spreadsheet elements to remain, you must save the file as an Excel file. When you hit the *Save* button you will get a warning message to remind you that the opened file was not a native Excel file; you then have the choice of choosing the format.

You can also open a CSV file in other programs – usually this will be a dedicated analytical program. You will see how to open a CSV file using R in Section 4.3. If you double-click on a CSV file it will usually open in your spreadsheet, but the CSV file might be associated with another program so try it and see.

4.2.6 Lookup tables

A lookup table is usually used to replace one item with another – actually the replacement value is made as a copy so the original is intact. Your lookup table should contain at least two columns (or rows) – the first column contains a list of items that you want replaced; the next column contains the items to replace them with. You then create a simple formula that searches your original data – the lookup table then carries out all the replacements and places the results in a new location.

There are many uses for lookup tables – here are a few ideas:

- Converting DAFOR values into numerical values.
- Converting site names into grid references.
- Creating abbreviated species names.
- Converting frequency scores into Roman numerals.

Once you start to use lookup tables you will undoubtedly find other uses. Some of the ideas in this list could be achieved in a different manner but the lookup table is often the simplest.

The lookup table itself is a passive object – you use one of two functions in Excel to carry out the actual replacement job: VLOOKUP and HLOOKUP. The first function looks at columns of data, whilst the second function looks at rows of data. You will see both of these in action in the examples that follow.

Using VLOOKUP

For this example you'll see lookup tables used with the British NVC system. In plant community studies involving the NVC system each community at a site is assessed using five quadrats. Every plant can then be given a *constancy* score, which relates to how many of the five quadrats the species was present within. Species with scores of 4–5 are therefore widespread, whilst species with scores of 1–2 are less widespread.

The NVC system also uses the Domin scale as a measure of abundance (a *cover* score). It is easier (and faster) to assess an approximate percentage cover using the Domin scale than to measure (or estimate) percentage cover exactly. The NVC system uses the constancy score more heavily than the cover score. When you write out tables of communities using the NVC system you express the cover score (Domin) as a numerical value but the constancy is usually written as a Roman numeral. This makes it easier to spot when you are looking at tables of results.

You can use a lookup table to convert regular numbers to Roman numerals. You can also go the other way of course and convert Roman numerals to regular (Arabic) numbers. There is a function that will convert from Arabic to Roman (it is called ROMAN), but there is not one for the other, so a lookup table is useful.

You start by making a lookup table in your spreadsheet. Usually it is best to put the table in a separate worksheet but in this example it is placed alongside the original data so you can see it more easily. Figure 4.16 shows some community data from an NVC survey.

	A	B	C	D	E	F	G	H
			Const	Cover	CS		Dom	Rom
1	Site	Species						
2	ML1	Achillea millefolium	5	6			1	I
3	ML1	Centaurea nigra	5	4			2	II
4	ML1	Lathyrus pratensis	5	5			3	III
5	ML1	Leucanthemum vulgare	5	5			4	IV
6	ML1	Lotus corniculatus	5	7			5	V
7	ML1	Plantago lanceolata	5	5				

Figure 4.16 Lookup tables are a useful way to create new fields in your data. Here you want to replace constancy values with Roman numerals. Note that the lookup table must be sorted in ascending order.

The values in column C are *constancy* values and you want to replace them with Roman numerals. The new values will be placed in column E.

The lookup table is created in columns G and H in Figure 4.16 – note that **the first column of the lookup table must be sorted in ascending order.** This column contains the items that you will search the original data for. It is worthwhile checking the sorting before you proceed – the simplest way is to highlight the complete lookup table (including the headings) and to click the *Sort* button in the *Data* menu – the AZ↓ button will do the job (Figure 4.17).

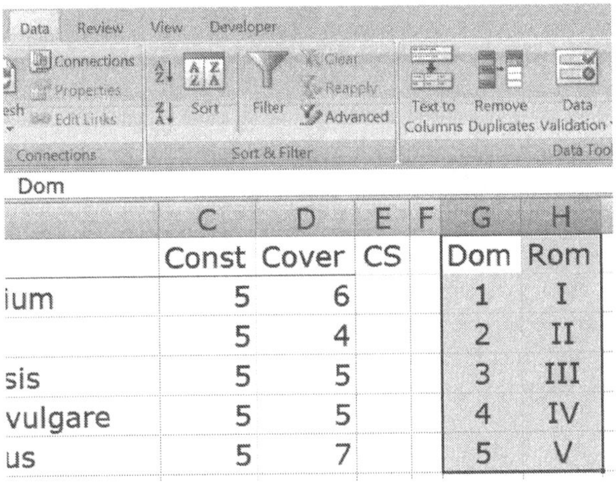

Figure 4.17 Your lookup tables need to be sorted in ascending order. Use the *Sort* button in the *Data* menu to achieve this.

To replace the values you use the VLOOKUP function in Excel. All functions start by typing the = sign. This formula has three parts: the first is the item you are looking for. In this case it is C2; that is 5 in the *Const* column C. The second is the reference location of the table. In this case it starts at G2 and carries over to H6. The best way to enter the location is to highlight it with the mouse (Figure 4.18).

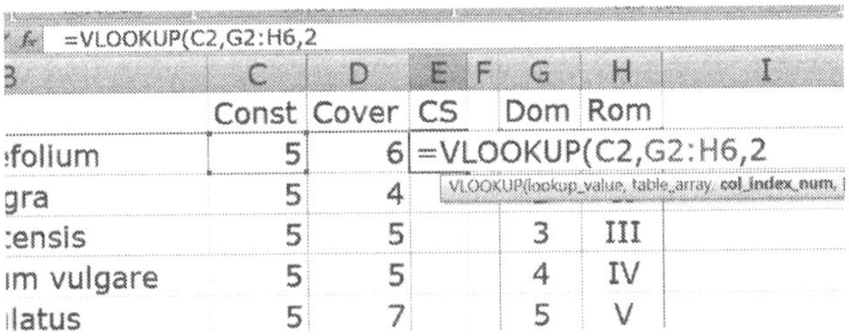

Figure 4.18 The VLOOKUP command is used to replace on value with another.

The table could be in a different sheet (this is in fact often preferable) in which case the name of the sheet appears before the reference e.g. Sheet3!G2:H6 (this will happen automatically if you select the range of cells with your mouse). The third part of the formula tells Excel to look in the second column of the table (where you have the Roman numerals) and that is what will end up in the cell (Figure 4.19).

	=VLOOKUP(C2,G2:H6,2)						
	C	D	E	F	G	H	
	Const	Cover	CS		Dom	Rom	
folium	5	6	V		1	I	
gra	5	4			2	II	
ensis	5	5			3	III	
m vulgare	5	5			4	IV	
latus	5	7			5	V	

Figure 4.19 The lookup table has been used with the VLOOKUP function to create the first entry in the new column.

Now you have the Roman numeral V in the cell as a replacement for the regular 5. The advantage over search and replace is that you now have both values and have not lost the original. The next step is to copy the formula down the rest of the column; however, if you do that there will be a problem: Excel will replace the C2 on the next line with C3. That is fine because that is what you want; however, it will also replace the location of your table with G3:H7 and that it **not** what you want. You want the row and column references to remain fixed. You need to edit the formula to tell Excel that the G2:H6 reference is not to change. You do this using the dollar sign (Figure 4.20).

	=VLOOKUP(C2,$G$2:$H$6,2)						
	C	D	E	F	G	H	I
	Const	Cover	CS		Dom	Rom	
folium	5	6		=VLOOKUP(C2,$G$2:$H$6,2)			
gra	5	4			2	II	
ensis	5	5			3	III	
m vulgare	5	5			4	IV	
latus	5	7			5	V	

Figure 4.20 The lookup formula needs some tweaking before it can be copied down the rest of the cells.

You click in the formula bar and add dollar signs in front to give $G$2:$H$6, everything else can stay the same. Press ENTER and the result looks the same as before. Now you can copy the cell to the clipboard and paste into the rest of the column (Figure 4.21).

C	D	E	F	G	H
Const	Cover	CS		Dom	Rom
5	6	V		1	I
5	4	V		2	II
5	5	V		3	III
5	5	V		4	IV
5	7	V		5	V
5	5	V			
3	1	III			
3	4	III			
5	8	V			
5	7	V			

Figure 4.21 The completed table. The lookup table used to change Arabic to roman numerals.

The formula has been copied down the column correctly and you have successfully replaced all the Arabic numerals with Roman numerals.

If your data are arranged in a slightly different manner – rows rather than columns – then you can still use a lookup table, but you need to use the HLOOKUP command rather than VLOOKUP.

Using HLOOKUP

Your data may sometimes be arranged in rows. Figure 4.22 shows some data on seashore seaweed communities: the data represent abundance of various seaweeds at heights above low tide level.

A	B	C	D	E	F	G	H	I	J	K	L	M	N
1							Transect Station						
2 Species	1	2	3	4	5	6	7	8	9	10	11	12	13
3 Fucus vesiculosus					F		F	F	A	O	C	O	C
4 Ulva latuca						F		F	F	O	F	O	F
5 Enteromorpha spp.	F	C				C	A						
6 Chondrus crispus								R			R		R
7 Coralina officinalis								O		A	F	F	

Figure 4.22 Seaweed community data. A lookup table will be used to convert DAFOR data to a numerical value.

The data in Figure 4.22 are recorded using a DAFOR type of scale where ACFOR is used to represent abundance (Abundant, Common, Frequent, Occasional, Rare) – thus you have an ordinal scale. You can use a lookup table to convert the values from ACFOR into numeric values. In this case you could use a lookup table that was in columns but you'll see the HLOOKUP function used instead.

Here there is an added complication – some cells in the table are blank. The blank cells represent zero abundance – it is easier to record values for species you observe and to ignore missing ones. However, if you try to use a formula on a blank cell you'll probably get an error (most likely the result be #NA). There are two sensible ways to proceed:

- Replace all blank cells with something definite, like a dash -.
- Use a modified command to deal with missing values (use the IF function).

The two options are considered in the following examples.

Replace blank cells

You could highlight all your original data and use the *Home > Find & Select* button to replace all the blank cells with a dash -. You must highlight the data you want to replace otherwise Excel will decide where the replacements should go and they may not be where you expect! You would then add the dash into your lookup table – the replacement value might as well be a dash too. You need to make sure the table is sorted into ascending order for the lookup table to work correctly. You can alter the sort settings to order from left to right like so: Open the *Data > Sort* menu item, from the Sort box select *Options*. You will now see that you can alter the sort order (Figure 4.23).

Now you have the blank cells replaced with – characters you can set about using the lookup table (Figure 4.24).

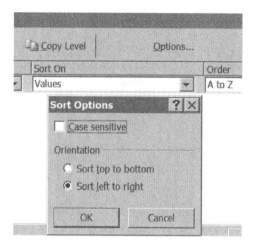

Figure 4.23 Sort order options can reorder from left to right for horizontal lookup tables.

A	B	C	D	E	F	G	H	I	J	K	L	M	N
1							Transect Station						
2 Species	1	2	3	4	5	6	7	8	9	10	11	12	13
3 Fucus vesiculosus	-	-	-	-	F	-	F	F	A	O	C	O	C
4 Ulva latuca	-	-	-	-	-	F	-	F	F	O	F	O	F
5 Enteromorpha spp.	-	F	C	-	-	C	A	-	-	-	-	-	-
6 Chondrus crispus	-	-	-	-	-	-	-	R	-	-	R	-	R
7 Coralina officinalis	-	-	-	-	-	-	-	O	-	A	F	F	-
8													
9		-	A	C	F	O	R	< This row is what you look for					
10		-	5	4	3	2	1	< This row is what you replace with					

Figure 4.24 Blank cells are replaced by dashes, which also appear in the horizontal lookup table. This avoids errors due to formulae applied to blank cells.

The next step is to create a template to hold the replacement values. The simplest way is to copy and paste the whole lot and then delete the data values – leaving the headings intact (Figure 4.25). You'll probably want to make a new worksheet for this; Figure 4.26 shows the template in the same worksheet so you can see it more clearly.

Now you can use the HLOOKUP function to look for a data value. The first value in the function is the original data item you want to replace. The next value is the range of cells that make up the lookup table. The final value is the row that corresponds to the row of the table you want to use as the replacement (Figure 4.26).

You will need to copy and paste the formula over all the blank cells in your replacement. However, as before, Excel will modify the contents of the formula. You need to 'fix' the location of the lookup table by using dollar signs in the formula. You want to end up with =HLOOKUP(B3,$B$9:$G$10,2) in this case. Remember that you only need to add $ to the bit that corresponds to lookup table $B$9:$G$10.

Once you have edited the formula you can carry on and copy/paste to the rest of your template. Now you will have replaced the ACFOR values (and dashes) with numerical

Name Box A	B	C	D	E	F	G	H	I	J	K	L	M	N
1					Transect Station								
2 Species	1	2	3	4	5	6	7	8	9	10	11	12	13
3 Fucus vesiculosus	-	-	-	-	F	-	F	F	A	O	C	O	C
4 Ulva latuca	-	-	-	-	-	F	-	F	F	O	F	O	F
5 Enteromorpha spp.	-	F	C	-	-	C	A	-	-	-	-	-	-
6 Chondrus crispus	-	-	-	-	-	-	R	-	-	-	R	-	R
7 Coralina officinalis	-	-	-	-	-	-	O	-	-	A	F	F	-
8													
9	-	A	C	F	O	R	< This row is what you look for						
10	-	5	4	3	2	1	< This row is what you replace with						
11													
12					Transect Station								
13 Species	1	2	3	4	5	6	7	8	9	10	11	12	13
14 Fucus vesiculosus													
15 Ulva latuca													
16 Enteromorpha spp.													
17 Chondrus crispus													
18 Coralina officinalis													

Figure 4.25 A blank template created ready to hold the replacement values from the original data.

SUM   ✕ ✓ ƒ =HLOOKUP(B3,B9:G10,2)

A	B	C	D	E	F	G
1						Transec
2 Species	1	2	3	4	5	6
3 Fucus vesiculosus	-	-	-	-	F	-
4 Ulva latuca	-	-	-	-	-	F
5 Enteromorpha spp.	-	F	C	-	-	C
6 Chondrus crispus	-	-	-	-	-	-
7 Coralina officinalis	-	-	-	-	-	-
8						
9	-	A	C	F	O	R
10	-	5	4	3	2	1
11						
12						Transec
13 Species	1	2	3	4	5	6
14 Fucus \ =HLOOKUP(B3,B9:G10,2)						

Figure 4.26 The HLOOKUP function is used to replace one value with another using a horizontal lookup table.

values (with dashes representing zero values – that is, originally empty). You can see the result in Figure 4.27.

Using the dash character is the simplest replacement for a blank cell as it is small! It will also be ignored if you create a graph – essentially treated as zero. The disadvantage is that you have to modify your original data and you may want to retain the empty cells. The alternative is to leave the blank cells intact but to account for them in the HLOOKUP formula.

	-	A	C	F	O	R		< This row is what you look for
	-	5	4	3	2	1		< This row is what you replace with

							Transect Station						
Species	1	2	3	4	5	6	7	8	9	10	11	12	13
Fucus vesiculosus	-	-	-	-	3	-	3	3	5	2	4	2	4
Ulva latuca	-	-	-	-	-	3	-	3	3	2	3	2	3
Enteromorpha spp.	-	3	4	-	-	4	5	-	-	-	-	-	-
Chondrus crispus	-	-	-	-	-	-	-	1	-	-	1	-	1
Coralina officinalis	-	-	-	-	-	-	-	2	-	5	3	3	-

Figure 4.27 The result of replacing ACFOR text with numerical values using a horizontal lookup table.

Use the IF function to ignore blank cells

If your original data contains blank cells then any formulae would give an error. If you use a lookup table to replace values using the HLOOKUP function (or indeed the VLOOKUP function) then this could be a problem. You saw earlier how to use search and replace to alter the blank cells to something else. However, this is not always ideal as you might wish to keep the original data as they were.

The way to overcome this is to account for the possibility of a blank cell in the formula. You need to use the IF command. The general form of the function is:

IF(*compare with something, what to do if the comparison is TRUE, otherwise do this*)

There are three parts: the first part allows you to decide something, then you use a comma and insert what to do if this decision was TRUE, finally you add another comma and write down what to do if the result of the decision was not true (i.e. FALSE).

To look for a blank cell you use a pair of double quotes. This is the first part of the formula IF(*cell* = ""...) meaning that you look to see if the cell is blank. If it is blank then you want to keep it that way so you add a pair of double quotes, which will force the cell to remain blank. Finally you insert the HLOOKUP part. In practice you would select a non-empty cell and create your HLOOKUP formula. Then you could add the IF part at the beginning. Once you are happy the formula works on one cell, you can copy and paste over all the data. Doing it this way means you are less likely to get confused and make a mistake.

In Figure 4.28 you can see the IF function in tandem with HLOOKUP for the first cell in the data (it happens to be a blank one).

The result is of course a blank in this case as the original cell was blank. Before the formula can be copied over the rest of the data you need to 'fix' the reference to the lookup table as before by adding dollar signs: $C$9:$G$10 is what you are aiming for in this case (Figure 4.29).

After you have fixed the cell reference to the lookup table you can finish off by copying the formula across all the cells. Your original data remain intact and you now have a copy that you can use for whatever purpose you like.

4.2.7 Pivot Tables – use Excel like a database

Your spreadsheet can act like a database. In fact, if you look at a database file you will see that it is arranged in rows and columns just like a spreadsheet. The key to using your

SUM — ✗ ✓ ƒ =IF(B3="","",HLOOKUP(B3,C9:G10,2))

	Transect Station								
Species	1	2	3	4	5	6	7	8	9
Fucus vesiculosus					F		F	F	A
Ulva latuca						F		F	F
Enteromorpha spp.	F	C				C	A		
Chondrus crispus							R		
Coralina officinalis							O		
		A	C	F	O	R				< This row is
		5	4	3	2	1				< This row is

	Transect Station								
Species	1	2	3	4	5	6	7	8	9
Fucus vesiculosus	=IF(								

Figure 4.28 The IF function is used to account for blank cells. If the target cell is not blank the HLOOKUP function is evaluated.

SUM — ✗ ✓ ƒ =IF(B3="","",HLOOKUP(B3,$C$9:$G$10,2))

	Transect Station												
Species	1	2	3	4	5	6	7	8	9	10	11	12	13
Fucus vesiculosus					F		F	F	A	O	C	O	C
Ulva latuca						F		F	F	O	F	O	F
Enteromorpha spp.	F	C				C	A						
Chondrus crispus							R				R		R
Coralina officinalis							O				A	F	F
		A	C	F	O	R								< This row is what you look for
		5	4	3	2	1								< This row is what you replace with

	Transect Station												
Species	1	2	3	4	5	6	7	8	9	10	11	12	13
Fucus vesiculosus (2))					3		3	3	5	2	4	2	4
Ulva latuca						3		3	3	2	3	2	3
Enteromorpha spp.	3	4				4	5						
Chondrus crispus							1				1		1
Coralina officinalis							2				5	3	3

Figure 4.29 The reference to the lookup table must be fixed using $ signs before the formula is copied to other cells.

spreadsheet as a database is in setting out the data in a meaningful manner – this is where the biological recording format comes in (Chapter 3). Each row of your spreadsheet should be unique – each row becomes a *biological record*. Each column should represent a separate variable – think of the columns as *fields* in the database.

The Pivot Table function of Excel is the key to using your data as a database. The column headings will be used to help you rearrange and organise your data. It is important to think about the columns you use right at the planning stage – getting the recording correct at the beginning can save you a lot of time later on.

There are many uses for Pivot Tables and the following examples will give you a flavour for what you can do.

Make a contingency table from frequency data

In Table 4.1 you can see some data on bird communities. The recorder has spent time in various locations that are different habitat types. At each location every bird spotted has been recorded. If you look carefully you will see that every row is a unique combination of variables – ignore the *Qty* variable for now.

At the moment each bird *Species* in Table 4.1 is represented in several rows. Each *Habitat* is also represented by several rows. It would be useful to create a new table where the species were in rows and the habitats were the columns – then you could see the relationships between the species and the habitats more clearly. This is where a Pivot Table will come in.

To start the process you need your data in Excel of course. The *Pivot Table* button is found in the *Insert* menu in Excel 2007 (Figure 4.30). In other versions you may find it under the *Data* menu. There is no need to highlight the data but do make sure that you click once somewhere in the block of data.

Table 4.1 Bird community data arranged in *Biological Recording* format. Each row is a separate and unique record. Each column is a separate variable and will act as a field in a database, which can be manipulated using a *Pivot Table*.

Species	Site	GR	Date	Recorder	Qty	Habitat
Blackbird	Springfield	SP873385	04-Jun-07	Starling, C	47	Garden
Chaffinch	Springfield	SP873385	04-Jun-07	Starling, C	19	Garden
Great Tit	Springfield	SP873385	04-Jun-07	Starling, C	50	Garden
House Sparrow	Springfield	SP873385	04-Jun-07	Starling, C	46	Garden
Robin	Springfield	SP873385	04-Jun-07	Starling, C	9	Garden
Song Thrush	Springfield	SP873385	04-Jun-07	Starling, C	4	Garden
Blackbird	Campbell	SP865395	04-Jun-07	Starling, C	40	Parkland
Chaffinch	Campbell	SP865395	04-Jun-07	Starling, C	5	Parkland
Great Tit	Campbell	SP865395	04-Jun-07	Starling, C	10	Parkland
House Sparrow	Campbell	SP865395	04-Jun-07	Starling, C	8	Parkland
Song Thrush	Campbell	SP865395	04-Jun-07	Starling, C	6	Parkland
Blackbird	Ouzel Valley	SP880375	04-Jun-07	Starling, C	10	Hedgerow
Chaffinch	Ouzel Valley	SP880375	04-Jun-07	Starling, C	3	Hedgerow
House Sparrow	Ouzel Valley	SP880375	04-Jun-07	Starling, C	16	Hedgerow
Robin	Ouzel Valley	SP880375	04-Jun-07	Starling, C	3	Hedgerow
Blackbird	Linford	SP847403	04-Jun-07	Starling, C	2	Woodland
Chaffinch	Linford	SP847403	04-Jun-07	Starling, C	2	Woodland
Robin	Linford	SP847403	04-Jun-07	Starling, C	2	Woodland
Blackbird	Kingston Br	SP915385	04-Jun-07	Starling, C	2	Pasture
Great Tit	Kingston Br	SP915385	04-Jun-07	Starling, C	7	Pasture
House Sparrow	Kingston Br	SP915385	04-Jun-07	Starling, C	4	Pasture

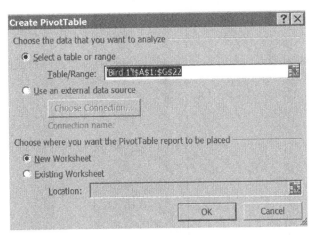

Figure 4.30 The *Pivot Table* button is found in the *Insert* menu in Excel 2007.

Once you click this you will see a dialogue box – this will allow you to check the range of data (Excel will automatically highlight it for you) and decide where you wish to place the result (Figure 4.31).

Create PivotTable

Choose the data that you want to analyze

⦿ Select a table or range

    Table/Range: [Bird 1'!$A$1:$G$22]

◯ Use an external data source

    [Choose Connection...]

    Connection name:

Choose where you want the PivotTable report to be placed

⦿ New Worksheet

◯ Existing Worksheet

    Location: [   ]

            [OK]   [Cancel]

Figure 4.31 Once the *Pivot Table* has been initiated you can check the input data and decide where to place the resulting table.

It makes sense to place your result in a separate worksheet from your original data. Once you click the OK button you will be ready to build your table (Figure 4.32).

You can now begin to assemble the fields into the Pivot Table to create the result you want. In older versions of Excel you could drag the fields into a skeleton outline of the table. In newer versions you drag the fields from the list to the sections at the bottom of the *Pivot Table Field List* box (Figure 4.32).

You want to end up with a table that has the bird species forming the rows, the habitats forming the columns, and the quantity making up the bulk of the table. Start by dragging the *Species* item in the box at the top into the *Row Labels* box at the bottom. The table begins to form and immediately you will see the bird species names appear in a table (Figure 4.33).

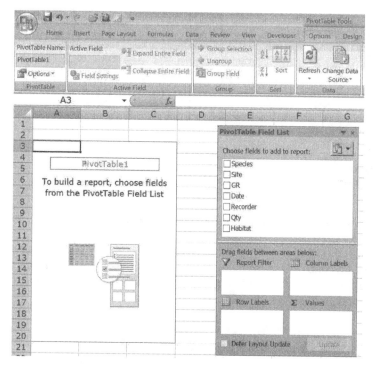

Figure 4.32 Once the data for the *Pivot Table* has been selected, the report can be generated. To assemble your *Pivot Table* you must drag fields from the list at the top to the appropriate section at the bottom.

Figure 4.33 As new fields are dragged into place, the *Pivot Table* forms.

You will also see a *Grand Total* item appear at the bottom – don't worry about this at the moment as you will refine the table shortly. Now you need to drag the *Habitat* item from the top box to the *Column Labels* box at the bottom. The habitat names now appear in the table, which will not have any actual data yet (Figure 4.34).

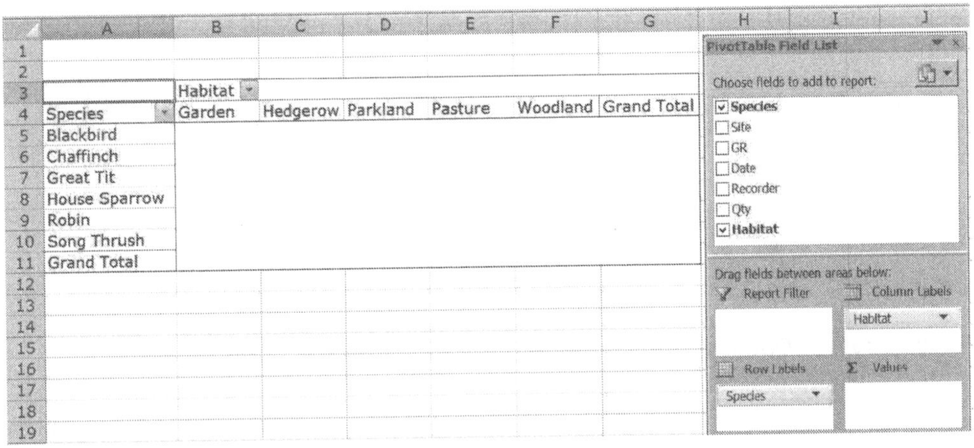

	Habitat					
Species	Garden	Hedgerow	Parkland	Pasture	Woodland	Grand Total
Blackbird						
Chaffinch						
Great Tit						
House Sparrow						
Robin						
Song Thrush						
Grand Total						

Figure 4.34 The *Pivot Table* has row and column headings but no data – yet.

Now you can add the data by dragging the *Qty* item from the box at the top into the *Values* box at the bottom. The table updates automatically once more and now you see the final result (Figure 4.35).

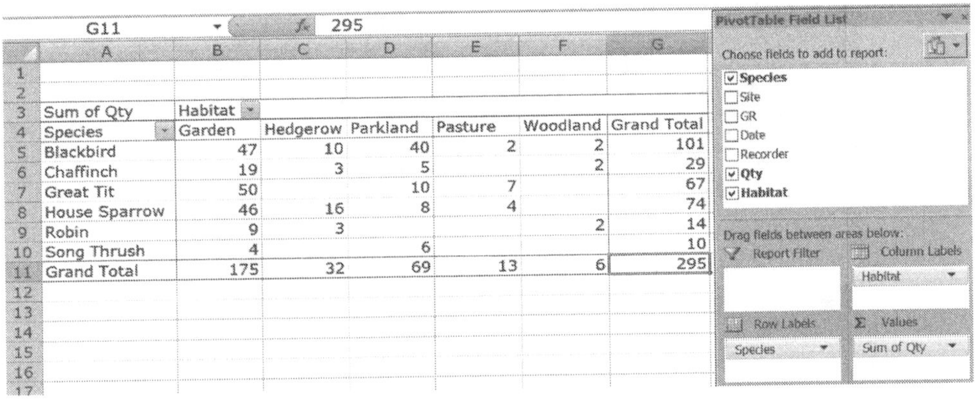

Sum of Qty	Habitat					
Species	Garden	Hedgerow	Parkland	Pasture	Woodland	Grand Total
Blackbird	47	10	40	2	2	101
Chaffinch	19	3	5		2	29
Great Tit	50		10	7		67
House Sparrow	46	16	8	4		74
Robin	9	3			2	14
Song Thrush	4		6			10
Grand Total	175	32	69	13	6	295

Figure 4.35 The *Pivot Table* completed with data.

The result contains empty cells – this makes it easy to see the values but you may need to alter empty cells to zero values before carrying out any mathematical operations.

Replace empty cells with zeroes

Once you click on a *Pivot Table* the *Ribbon* will show a *Pivot Table Tools* menu item. There are two menus within the *Tools*: *Options* and *Design*. These provide many ways to alter the appearance of your table. In the current table (Figure 4.35) you can see that there are empty cells – these relate to habitats where particular species were not observed. If you want to do any mathematical operations on the table then you ought to have zeroes rather than blanks. You can get the Pivot Table to replace blank cells with more or less anything (including 0) by using the *Options* button under the *Pivot Table Tools > Options* menu (Figure 4.36).

Figure 4.36 The Options button provides many ways to alter your *Pivot Table*.

Once you click the *Options* button you will see a dialogue box; in the *Layout & Format* tab you can see a box entitled *For empty cells show:* with a blank box (Figure 4.37).

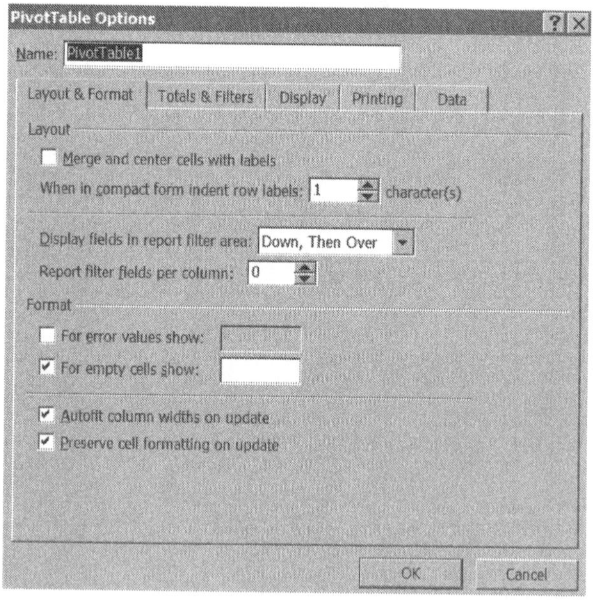

Figure 4.37 The *Pivot Table Options* dialogue provides many ways to alter your *Pivot Table*.

If you type a 0 (zero) into the 'empty cells' box you can replace all the empty cells with a 0; press the OK button to finish. Your final table will appear like Figure 4.38.

Notice that there are arrow buttons in the *Species* and *Habitat* title cells. These are *Filter* buttons and you can use them to refine the result (Figure 4.39).

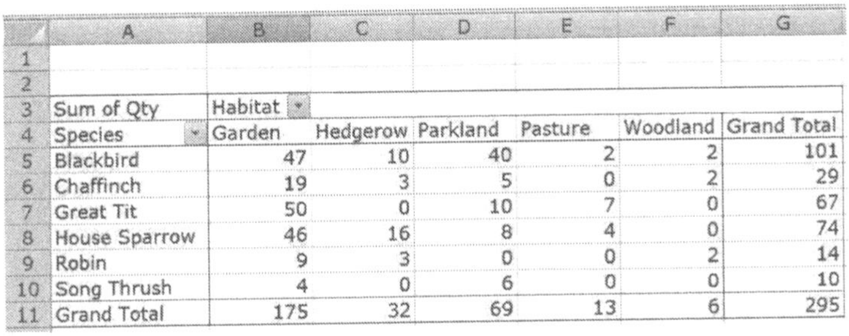

	A	B	C	D	E	F	G
1							
2							
3	Sum of Qty	Habitat					
4	Species	Garden	Hedgerow	Parkland	Pasture	Woodland	Grand Total
5	Blackbird	47	10	40	2	2	101
6	Chaffinch	19	3	5	0	2	29
7	Great Tit	50	0	10	7	0	67
8	House Sparrow	46	16	8	4	0	74
9	Robin	9	3	0	0	2	14
10	Song Thrush	4	0	6	0	0	10
11	Grand Total	175	32	69	13	6	295

Figure 4.38 Empty cells replaced by zero values – this is important if you need to undertake further maths with your data.

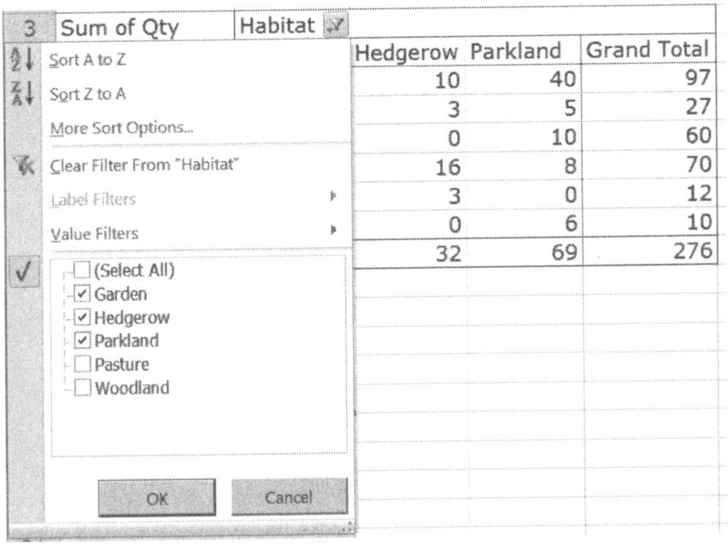

Figure 4.39 Additional filters can be applied to headings to refine your results in a *Pivot Table*.

Use filters to refine results

You can use filters in Pivot Tables in much the same way as filters in regular columns (see Section 4.2.2). You simply click on the filter button in the Pivot Table and select your data (Figure 4.39). The table updates once you have clicked the OK button.

There are many other ways to create Pivot Tables – you can use several data fields, you aren't restricted to one. In the following example you'll see how to use multiple data fields to create a *constancy table* for an NVC survey as well as using filters.

Use multiple fields to make a constancy table

You can use a *Pivot Table* to take *Biological Records* and create a report in a completely different layout. You aren't restricted to using a single data field in your tables. In this example you'll see how to take plant data from an NVC survey and construct a *constancy table* that shows the species with their maximum abundance and constancy (that is a value from 0 to 5, a frequency), for several sites.

A portion of the data is shown in Table 4.2, where you can see four columns. The first column is the species name, the second is a quadrat number – this varies from 1 to 5 as the NVC methodology uses five quadrats for each community. There is a column entitled *Domin*, which contains the abundance (using the Domin scale). The last column contains the site name – in this case they are called simply *upper* and *lower* and relate to two different areas of a site (called Tarn Foot).

Table 4.2 Plant community data from an NVC survey in *Biological Recording* format. Only the first few rows are shown (there are 98 in total).

Species	Quadrat	Domin	Site
Carex flacca	1	4	Upper
Carex nigra	1	6	Upper
Festuca ovina	1	7	Upper
Galium saxatile	1	2	Upper
Juncus effusus	1	1	Upper
Juncus squarrosus	1	4	Upper
Luzula campestris	1	2	Upper
Pleurozium schreberi	1	1	Upper
Rhytidiadelphus squarrosus	1	8	Upper
Vaccinium myrtillus	1	1	Upper
Carex flacca	2	2	Upper
Carex nigra	2	4	Upper
Festuca ovina	2	7	Upper
Galium saxatile	2	3	Upper

The first step is to get the Pivot Table started by clicking once in the block of data and then going to *Insert > Pivot Table* on the *Ribbon*. You should see the 'marching ants' around the data that Excel has highlighted for you – select the option that places the Pivot Table in a new worksheet and click the OK button.

You want the *Species* field to go in the *Row Labels* box. Drag the *Site* field into the *Column Labels* box. You will now have a table with headings but no data (Figure 4.40).

At this point you only have a species list and so the next step is to add the data. You want to end up with the maximum Domin value for each species – each species will have up to five values as there are five quadrats for each site. You also want to know how many of the five quadrats each plant was found in – this is a frequency from 0 to 5 and is called *constancy* in NVC survey parlance.

What you must do is to drag the *Domin* field into the *Values* box at the bottom. Do this twice! Now you have two lots of data the same – for the time being (Figure 4.41).

At the moment both the fields say the same thing *Sum of Domin* – they are only differentiated because the second one has a 2 at the end to read *Sum of Domin2* (you may have to make the *Field List* box wider to see). The point is that the values are recorded using the *Sum* of the data. The triangle by the name gives a clue that you can do something else.

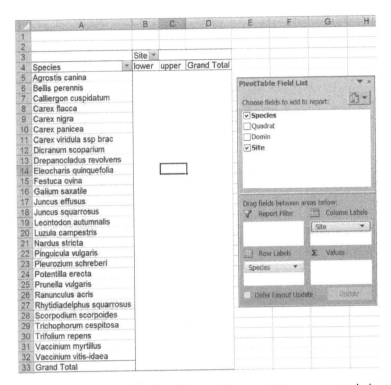

Figure 4.40 The *Pivot Table* has headings but no data – yet. However, you do have an overall species list.

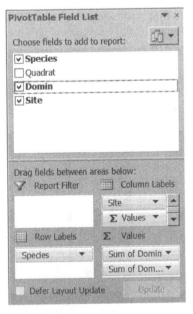

Figure 4.41 Two identical fields in the Value box. You will modify these to create a *Constancy Table*.

Alter value field settings

If you click on the field item that says *Sum of Domin* you see a new pop-up menu (Figure 4.42). This allows you to alter the *Value Field Settings*.

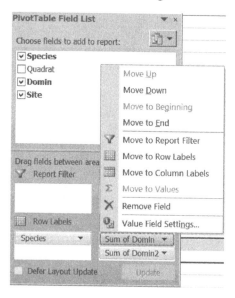

Figure 4.42 A data field can be customised using the *Value Field Settings* option.

Once you click the *Value Field Settings* option you see a new dialogue box that allows you to alter various field settings. You can change the *Summarize by* option and set it to *Count* rather than *Sum* (Figure 4.43). Once you click OK the Pivot Table updates to reflect the changes. You now have a value that represents *constancy* – that is, how many (out of five, since you have five quadrats for each site) quadrats each species was found in. You only recorded species that were found so the count reflects the frequency.

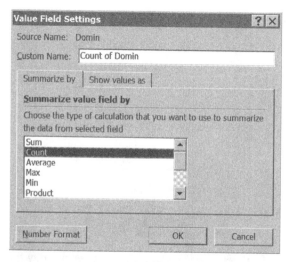

Figure 4.43 A value data field can be customised to show various summary statistics other than the default *Sum*.

Now you want to modify the other value to show the maximum *Domin* value that each species was recorded at for each site. Click the field name that says *Sum of Domin2* and select the *Value Field Settings* option. You can now change the field to display *Max* rather than *Sum*. The table updates once you have clicked OK.

Your table now shows a constancy table for the two sites (upper and lower), displaying *Count of Domin* and *Max of Domin2* as the column headings. There is more you can do to improve this – you do not really want the Total columns or rows for example.

Alter row and column totals settings

Your constancy table does not really need all the totals parts – they simply get in the way. You can turn off the *Pivot Table Field List* box using the buttons at the top right of that box or use the *Field List* button from the *Pivot Table Tools > Options* menu on the *Ribbon*. This gets the field list out of the way.

You can now go to the *Grand Totals* button, which you will find on the *Pivot Table Tools > Design* menu in the *Ribbon* (Figure 4.44).

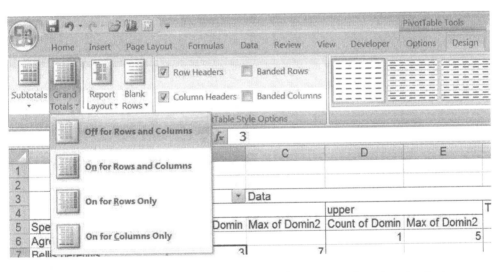

Figure 4.44 Options for dealing with totals in the Pivot Table are found on the *Design* menu.

There are four options for *Grand Totals* and the icons show what the results look like. Once your *Grant Totals* are all off the constancy table looks useful (Figure 4.45).

Here you only have two sites but you can imagine that with many sites the table could become somewhat unwieldy. You can use the filter settings to alter what you display.

Alter filter settings

The *Species* and *Site* headings contain a filter button (Figure 4.45), which you can use to select one or several items to display – you saw this earlier (Figure 4.39) with the bird contingency table. When you display all the sites you inevitably have gaps in the data – in this case hardly any species are common to both sites. If you want to produce a constancy table for each site you have to display each site in turn using the filter.

Species	Site	Data		
	lower		upper	
	Count of Domin	Max of Domin2	Count of Domin	Max of Domin2
Agrostis canina			1	5
Bellis perennis	3	7		
Calliergon cuspidatum	5	7		
Carex flacca			4	4
Carex nigra	3	7	4	6
Carex panicea	5	6		
Carex viridula ssp brac	5	7		
Dicranum scoparium	1	1		
Drepanocladus revolvens	5	7		
Eleocharis quinquefolia	5	5		
Festuca ovina			5	9
Galium saxatile			5	3
Juncus effusus			1	1
Juncus squarrosus			3	5
Leontodon autumnalis	2	1		
Luzula campestris			4	4
Nardus stricta			4	5
Pinguicula vulgaris	2	3		
Pleurozium schreberi			3	2
Potentilla erecta			2	3
Prunella vulgaris	2	1		
Ranunculus acris	2	3		
Rhytidiadelphus squarrosus			5	8
Scorpodium scorpoides	5	7		
Trichophorum cespitosa			4	5
Trifolium repens	1	1		
Vaccinium myrtillus			5	3
Vaccinium vitis-idaea			1	1

Figure 4.45 The final constancy table created using the *Pivot Table* function.

You can move the *Site* field to the *Report Filter* box – this will merge all the data together but allows you to select the site you wish to display from that filter (Figure 4.46).

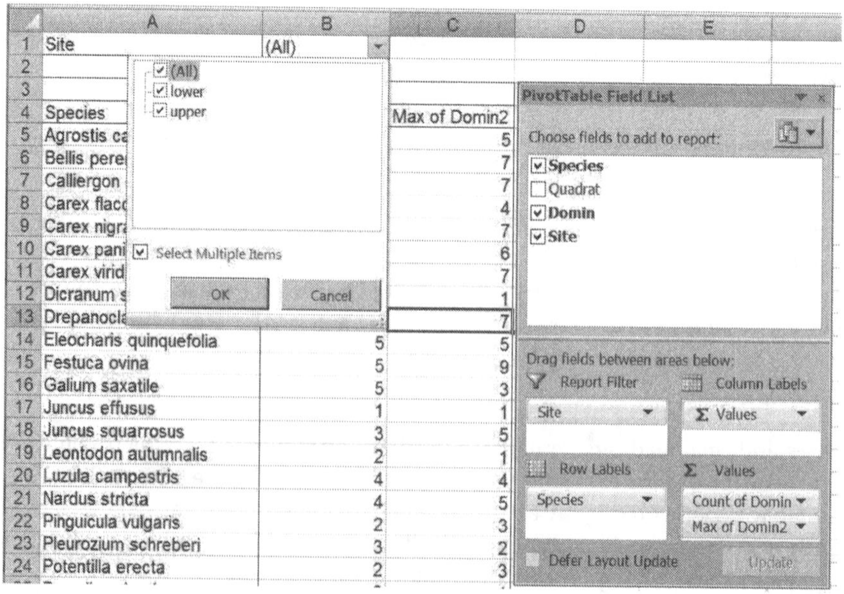

Figure 4.46 The *Report Filter* provides an upper level of filter control.

This overall filter is no more useful for the current situation but imagine if you had surveys for different years – you could use the year as an overall filter and rapidly be able to switch between years and sites.

Multiple row fields

Rather than use the filter option to create individual constancy tables you can use multiple fields in your rows. If you reset the filter on the constancy table and drag the *Site* field into the *Row Labels* box you can produce a series of constancy tables. The position of the fields in the box is important so make sure that *Site* is above *Species* (Figure 4.47).

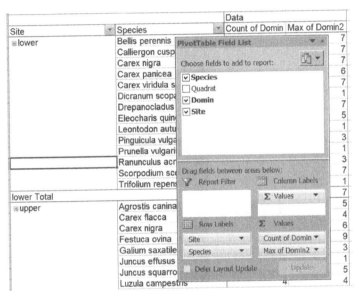

Figure 4.47 Multiple fields can be placed in the *Row Labels* section to create a series of stacked tables.

In Figure 4.47 you can see that you now have subtotals, which are not very useful here. You can turn these off using the *Subtotals* button, which is found in the *Pivot Table Tools > Design* menu on the *Ribbon*. The final result appears like Figure 4.48.

You can use multiple fields in any part of a Pivot Table, but order matters so experiment until you get the layout that's best.

Pivot table design

There are plenty of other things that you can do with Pivot Tables – you will see them used again. They key thing to remember is that you are using columns as indices to help you sort and rearrange your data. When you are designing your data collection and recording methods think carefully about how you will use the data – it is easier to create an index column right from the start than to wade through hundreds of lines of data and add one later.

The *Design* menu in the *Pivot Table Tools* section of the *Ribbon* contains a few other buttons that you can use to make your tables look different. In general 'pretty' tables are nothing more than window dressing and usually obscure the data rather than clarify

Site	Species	Data	
		Count of Domin	Max of Domin2
⊟ lower	Bellis perennis	3	7
	Calliergon cuspidatum	5	7
	Carex nigra	3	7
	Carex panicea	5	6
	Carex viridula ssp brac	5	7
	Dicranum scoparium	1	1
	Drepanocladus revolvens	5	7
	Eleocharis quinquefolia	5	5
	Leontodon autumnalis	2	1
	Pinguicula vulgaris	2	3
	Prunella vulgaris	2	1
	Ranunculus acris	2	3
	Scorpodium scorpoides	5	7
	Trifolium repens	1	1
⊟ upper	Agrostis canina	1	5
	Carex flacca	4	4
	Carex nigra	4	6
	Festuca ovina	5	9
	Galium saxatile	5	3
	Juncus effusus	1	1
	Juncus squarrosus	3	5
	Luzula campestris	4	4
	Nardus stricta	4	5
	Pleurozium schreberi	3	2
	Potentilla erecta	2	3
	Rhytidiadelphus squarrosus	5	8
	Trichophorum cespitosa	4	5
	Vaccinium myrtillus	5	3
	Vaccinium vitis-idaea	1	1

Figure 4.48 The final stacked series of constancy tables.

matters! However, it is worth looking at a couple of ways of presenting tables that can be useful.

Colour your table

Adding colour is generally something to be avoided but if you do want to make a Pivot Table for a presentation then the *Pivot Table Styles* section of the *Design* menu can be useful (Figure 4.49).

You can also click on the *New Pivot Table Style* button and create a custom style. Find this button at the bottom of the *Pivot Table Styles* item (Figure 4.49).

The *Report Layout* button is on the left of the *Design* menu (Figure 4.50). You can use this to alter how the field headings are set out.

In Figure 4.50 you can see the effect of applying the compact layout to a Pivot Table and using banded rows. This can help the reader to pick out items – lots of rows of data can be hard to read. In the end how you display your table is up to you of course but subtle tends to have more positive impact!

Copy results to a new location

Once you have made a Pivot Table you may well want to keep it for some other purpose. The *Move Pivot Table* item in the *Options* menu will allow you to shift an entire table to another worksheet. This is of limited use! It is generally more helpful to copy the table to the clipboard and paste it to a new worksheet or workbook.

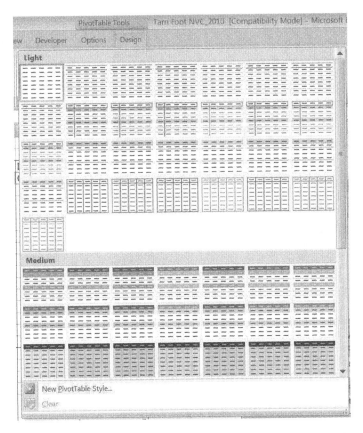

Figure 4.49 The *Pivot Table Styles* section has a range of built-in layouts that can help (or hinder) presentation.

Start by highlighting the appropriate cells of your table. Now use the *Copy* button, this is in the *Clipboard* section of the *Home* menu (it looks like two sheets of paper and is under the pair of scissors). Now decide where you want to put the table – a new workbook or worksheet. Navigate to the new location and use the *Paste* button, which is on the *Home* menu (Figure 4.51). Do not just hit the *Paste* button but use the small triangle at the bottom to bring up a mini-menu (Figure 4.51).

If you simply use the basic *Paste* option you would get a new Pivot Table – you want to avoid this because it contains formatting that could interfere with any other mathematical operations you want to perform. You could use *Paste Special* like you saw in Section 4.2.3, but a quick method is to hit the *Paste Values* option. This transfers the data and strips out any formatting. You can alter the heading names – *constancy* and *cover* would seem appropriate for the current NVC constancy table example. Now if you carry out any additional data manipulation you do not need to worry that some hidden formatting will interfere with the process.

Pivot Tables in other programs

So far you have only seen the Pivot Table used in Excel for Windows 2007. However, the Pivot Table is available in other spreadsheets too.

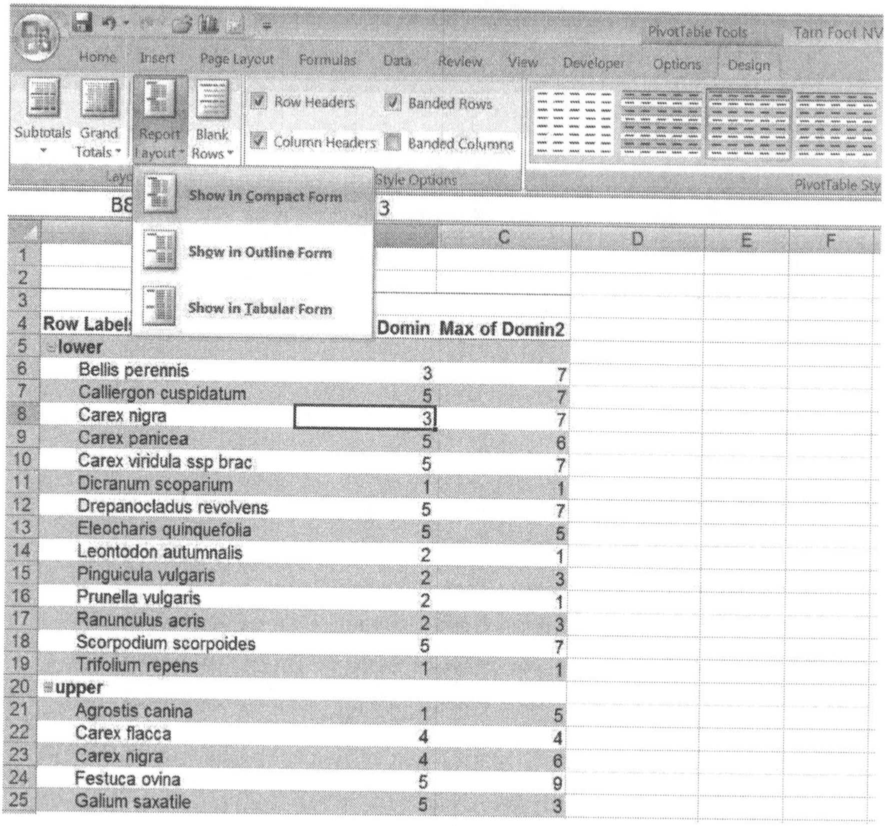

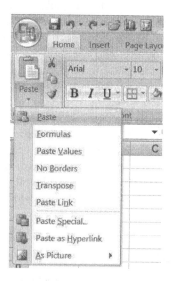

Figure 4.50 The *Report Layout* button can alter the appearance of your *Pivot Table*.

Figure 4.51 The *Paste* button provides a range of paste options.

Versions of Excel

If you have an older version of Excel (or a Mac) you can still use a Pivot Table – it is likely to be found under the *Data* menu. Older versions use a slightly different way of constructing the table – you drag the fields into the skeleton table rather than the four boxes in the *Pivot Table Field List* window. The menus for customising the Pivot Table and fields are also slightly different from the *Ribbon* set-up. Generally speaking you can use the right-click to bring up context-sensitive menus that allow you to customise your Pivot Table. The right-click also works in Excel 2007.

Open Office

Open Office is the most widely used alternative to Excel and is bundled into Linux operating systems. Libre Office is a derivative of Open Office and newer versions of Linux use this. The *Pivot Table* option is found under the *Data* menu. In older versions of Open Office the *Pivot Table* was called *Data Pilot*.

The actual building process is more like the older versions of Excel – you are presented with a dummy table, into which you drag the fields (Figure 4.52).

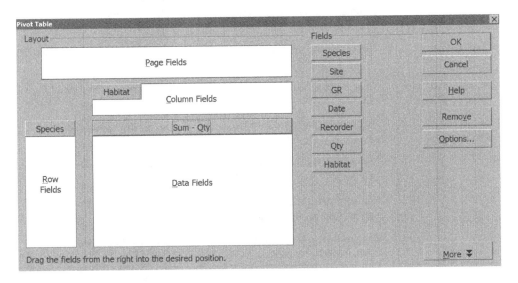

Figure 4.52 The Pivot Table building process in Open Office.

You can set options for fields (like turning off subtotals) by selecting the field with a mouse click and clicking the *Options* button. The *More* button brings up more options – for the basic Pivot Table you can determine where to place the final table and turn off row and column totals for example.

The Pivot Table function is a bit more limited than in Excel:

- There are fewer formatting and style options.
- You cannot set a character for blank cells.
- You cannot add a field to the same place twice (as you did for the constancy table example).

You can still use multiple fields in any one zone of the table, as long as they are different. These limitations are not insurmountable but you have to put in a bit more work.

Cross-classifying in R

There are several commands in R that carry out similar process to the Pivot Table. There are commands that will use cross-classifying factors to produce summary tables and others that can use summary functions across grouping variables, Table 4.3 shows a few of the commands with a brief explanation of their operation.

Table 4.3 R commands for producing Pivot Tables and related summaries.

Command	Explanation
table	This command uses cross-classifying factors to build a contingency table of the counts at each combination of factor levels. If you use more than two factors the result will be a table with more than two dimensions.
ftable	This command produces contingency tables, unlike the table() command the factors are 'collapsed' so that the resulting table is two-dimensional.
xtabs	This command produces contingency tables but can also use a variable of counts to build the table.
rowSums, colSums	These commands produce the sum of values for rows or columns of a table.
rowMeans, colMeans	These commands produce the mean of values for rows or columns of a table.
apply	This command allows you to carry out a function over the rows or columns of a table.
tapply	This command allows you to apply a function to a variable using another variable as a grouping factor.
aggregate	This command is similar to the tapply() command but produces the result in a different form.

You will find more details about these commands in Section 6.3.3.

4.3   Getting data from Excel into R

Excel is an invaluable data management tool and useful for initial 'data mining', that is, having a look at your data in various ways in order to get a feel for what you are dealing with. This is what causes statisticians to break out in a cold sweat – in traditional hypothesis-driven analysis you always set a hypothesis first. In community studies you cannot always be so narrow and you are often looking for possible patterns and associations. The analytical approaches are not always driven by classic hypothesis testing. This doesn't mean that you do not need to plan your approach though – it means that you know from the start that you aren't using a classic hypothesis test.

Once you have gained an overview of your data you will need to carry out some analyses. In community ecology many of the analytical methods are rather too heavy-duty for Excel and at some point you will need to prepare your data for use in another program. For most purposes this means using R since it will carry out virtually all the analysis we are likely to need.

4.3.1 Saving data using Excel

Preparing data items usually involves saving them (or some of them) to a disk file in a non-Excel format. The CSV format is common (see Section 4.2.4) and is accepted by many analytical programs, including R. You saw how to save Excel data as CSV in Section 4.2.4 by using the *Save As* option from the *Office* button.

You may not want or need all of your data so you can use filtering or Pivot Tables as necessary to prepare the data you need. Once you have prepared the data you can save it as a CSV file. This leaves your original data intact. As part of your data management you could keep all the files in one folder, with the original data as Excel format and other items as CSV files.

4.3.2 Reading data into R

The `c()` and `scan()` commands are useful for entering small samples of data direct from the keyboard but you will usually have a lot more – especially so in community studies. The `read.csv()` command allows you to read CSV files. These may be prepared easily from spreadsheets (see Section 4.2.4). By default R expects the first row to contain variable names.

If you use Windows or Mac operating system then you can use the command like so:

```
> my.data = read.csv(file.choose())
```

As usual when using R you assign a name to store the data. In the above example you call this *my.data*. The `read.csv` part tells R to look for a CSV file. The `file.choose()` part is and instruction that tells R to open an explorer-type window so you can choose the file you want from your computer (Figure 4.53), otherwise you must specify the exact filename yourself (this is what you have to do if you are running R in Linux). There are several variations on this command but since the CSV format is so readily produced it seems unnecessary to learn anything other than the basic form of the command (try looking at the help for this command in R).

Once the data are read into R you can check they are in place by typing the name you gave it. You can also use the `ls()` command to get a list of data stored in R (see Section 4.1.4). If you exit R you will be asked if you want to save the workspace – if you click on the *Yes* option your data will be saved in R so that it will be ready next time you run the program (see Section 4.1.8).

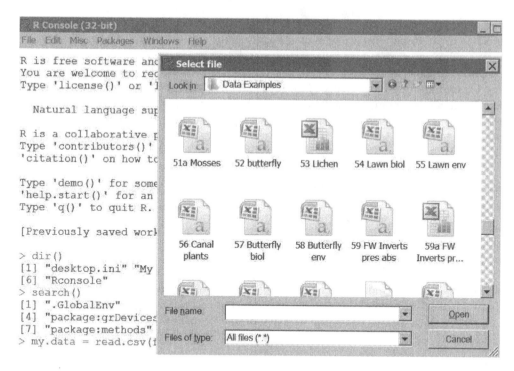

Figure 4.53 Importing data to R using the *read.csv()* command. The *file.choose()* instruction opens an explorer-type window.

4.4 Summary

Topic	Key Points
Getting help in R	Get help on a topic by typing `help(topic)`.
	Use `help.start()` to open the help system in your web browser.
	There is a lot of help on the Internet, try searching online.
Data names and R	R is an object-oriented program, meaning that everything needs a name. You can use upper and lower case letters and the numbers 0–9. The only punctuation allowed is the full stop and underscore.
	Data names are case sensitive and must start with a letter.
Additional command packages	There are hundreds of additional packages of R commands available for R. You can get these from the R website. They can be downloaded directly from R if you know the name of the package via the `install.packages()` command.
	These packages are usually written with a specific topic in mind, like community ecology for example!
	One a package has been installed you can 'load' it using the `library()` command.

Sorting data in Excel	Data can be rearranged easily using the *Data > Sort* button. You can sort using more than one column.
Filtering data in Excel	Use the *Data > Filter* button to select parts of your data. You can use more than one column as the base for a filter.
Paste Special	The *Edit > Paste > Paste Special* button allows finer control of the paste operation. You can copy just the formatting or just the values as well as transposing data (i.e. making rows into columns and vice versa).
Data formats	You can save data in a variety of formats in addition to the standard XLS or XLSX format. Many computer programs use comma separated values (CSV), which you can produce easily from the File menu (or the Office button).
Lookup tables	Lookup tables are useful as references. You can use the HLOOKUP and VLOOKUP functions in Excel to match a value in the data and return a value from the lookup table.
Pivot Tables	The Pivot Table is one of the main strengths of Excel. It allows you to use your spreadsheet like a database and permits you to select and rearrange your data quickly and easily. The Pivot Table can also summarise data by mean or sum and even produce summary graphs.
Cross-tabulation in R	The R `xtabs()` command is analogous to the Pivot Table in Excel. Use it for cross-tabulation, particularly for transferring data in biological recording format to a community dataset arrangement.
Get data from Excel to R	It is possible to get R to read XLS files but it is 'safer' to save your data in CSV format from Excel. Then you can use the `read.csv()` command to get the data into R.

4.5 Exercises

4.1 Which of the following is **not** a valid data name in R?

A) `data.1`
B) `DaTa_1`
C) `1_data`
D) `da.ta.1`

4.2 To get additional R command packages you must get ZIP files from the R website – TRUE or FALSE?

4.3 If you want to filter data in Excel you must apply a filter for each column – TRUE or FALSE?

4.4 When you build a Pivot Table you need to drag fields from a list into one (or more) of four boxes, what do these four boxes do?

4.5 Summarise the main steps required to get data from Excel into R.

The answers to these exercises can be found in Appendix 1.

5. Exploring data: choosing your analytical method

Looking at communities of organisms is generally going to be rather more messy and complicated than studying a single species. The world is an uncertain place and ecological data are not as 'neat and tidy' as in some branches of study. Some might argue that this makes ecology more fun.

You will need to find a way to make sense of your data. You can use Pivot Tables and graphs to help you gain some insights into what you are dealing with, as you will see in Chapter 6. This is an important step because it can help you to refine your approach and make sure that you are 'on track'. Overviews and summary reports can be useful for many reasons, for example:

- Helping you see what you are dealing with.
- Refining your analytical approach.
- Showing others what you are up to (e.g. supervisor, teacher or employer).

Sooner or later though, you are going to need something more. Ideally you should have worked out the analytical approach right at the planning stage. However, things do not always work out so neatly! In this chapter you will learn some of the ways that you can approach the study of communities – this will help you in your own planning process.

You can split the analytical approaches into different chunks – this can help you to decide what it is you want to study and help you plan how to go about doing it. Planning your analytical approach right at the outset is the ideal of course. If you left the planning side of things out, and happen to have data already, then knowing what you can do with it (if anything!) is helpful.

5.1   Categories of study

If it helpful to split the areas of study into broad blocks so that you can determine what areas of study you want to look at. You could for example think of a basic split as being like so:

- Single species studies.
- Multiple species (communities) studies.

Each of these rather broad areas could then be subdivided (Figure 5.1).

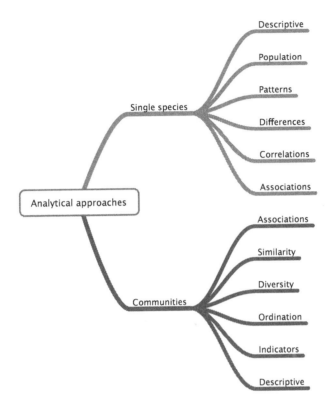

Figure 5.1 Broad areas of analytical approach.

The single species approach contains most of the 'classic' hypothesis tests. Some of these can be pressed into service for use with community studies. The general strands of the single species theme are:

- *Descriptive* – in terms of a single species descriptive studies can involve looking at life-history information such as flowering time or reproductive strategy. If you are measuring something, such as abundance, then you will use averages and some measure of dispersion (e.g. standard deviation) to summarise the values you collected.
- *Populations* – population studies involve attempts to assess how many individuals of a species there are in a given area. Such studies often attempt to compare population size between years.
- *Patterns* – there are two broad kinds of pattern that you might look at: one-dimensional or two-dimensional. Single dimensional patterns involve looking at 'runs', that is sequences. Two-dimensional patterns look for evidence of clusters or of regular patterns.
- *Differences* – this is where you are looking to compare something – this could be a measure of abundance (e.g. plants in a quadrat) or something else, like size or weight. In the simplest cases you have a single variable to compare and two samples. However, in more complicated cases you might have multiple samples.

- *Correlations* – here you are looking to find links between things. Generally this will be a biological measurement and some environmental variable. As things become more complicated you may have several variables to assess.
- *Associations* – this kind of analysis involves 'count' data. You are looking to compare categories of things. You can also compare one set of categories to a 'standard', this kind of analysis is called a *Goodness of Fit* test. You can think of association analysis as being the foundation of many community studies.

When it comes to multi-species or community studies the general strands are as follows:

- *Associations* – this kind of analysis can be extended to include several species – you may have species and habitats for example. If you look at co-occurrence of species you have the beginnings of various community analyses such as indicator species and similarity.
- *Similarity and clustering* – here you are looking to see how similar your samples are in terms of their species composition. You can then go on to organise samples into clusters, either as an hierarchical dendrogram or by partitioning into 'clumps'.
- *Diversity* – there are two main areas of diversity study – the simplest involves counting how many different species are in a set area. This *Species Richness*, can be looked at using some of the 'classic' hypothesis tests that you also use in single species studies. The relative abundance of the species in a sample can also be used as the basis for looking at diversity in a more complex way.
- *Ordination* – this branch of study aims to take complicated data, multiple species and multiple sites, and find patterns in the data that can be represented in a simpler two-dimensional form.
- *Indicators* – here you are trying to assess whether certain species can be considered as indicative of a certain habitat. You can think of this as a kind of association analysis and indeed one way of looking at indicator species is to use chi squared tests.
- *Descriptive* – community data can be described in many ways and descriptive studies aim to create useful definitions to help classify your samples. One example would be the NVC system for looking at plant communities in the UK.

In the following sections you will see how these areas of study can be used to examine your community data. Not all of the topics are considered in detail in this book but it is useful to get an overview of the methods available to you.

5.2 How 'classic' hypothesis testing can be used in community studies

You can split the study of single species into various broad categories, Figure 5.2 shows one way you can visualise these categories.

These branches of study lead to some form of hypothesis test – all, that is, except the descriptive branch. In community studies the 'classic' hypothesis tests are generally less useful but there are occasions when such tests can be pressed into service, as you will see next.

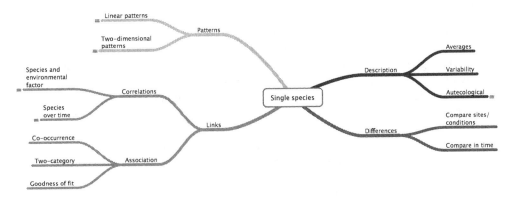

Figure 5.2 Single-species studies can be split into broad categories.

5.2.1 Differences hypothesis tests

In differences tests you are usually looking at some single measure, such as abundance, weight or size. You have several samples and look to determine differences between them. These different sample can come from different sites for example. In more complicated cases you may have additional factors to take into account. The sorts of tests that you would carry out depend on the situation you have as well as the distribution of the data (i.e. normally distributed or not). Commonly used tests include:

- Student's *t*-test.
- Wilcoxon rank sum test (also called the Mann–Whitney *U*-test).
- Analysis of variance (ANOVA).
- Kruskal–Wallis test.

When you have community data these tests become less useful. There are two main exceptions to this:

- Species richness – the number of different species in a given area.
- The different species are a variable in the analysis.

If you are studying *species richness* then you are essentially 'converting' the community into a single value – the number of species in a given area. This means that you could use species richness as the *response variable* in a differences analysis such as:

- Comparing species richness at different sites.
- Comparing species richness at different times.

The species themselves may be the *predictor variable* in an analysis. For example, you may have recorded the time spent foraging at a feeding location for several different species. The foraging time is the response variable, the predictor variable is the species. You could carry out a test to see if there were differences between species in the time spent foraging.

5.2.2 Correlation hypothesis tests

Correlation is a way to link two variables – often a biological factor and an environmental factor. In simple correlation you have a single response variable and a single predictor variable. In regression analysis the correlation is linked to a mathematical relationship. The idea of correlation can be 'extended' to include several predictor variables in a method generally called multiple regression. There are several commonly used methods you could encounter in the general theme of 'correlation':

- *Simple correlation* – here you determine the strength of the relationship between two variables. A commonly used method of correlation is the Spearman rank test.
- *Linear regression* – in this case there is an assumed mathematical relationship between two variables. The regression method determines the strength of the relationship as well as the coefficients of the relationship (e.g. $y = mx + c$).
- *Multiple regression* – this extends the idea of regression to incorporate several predictor variables. For example you could use this to determine which habitat variable was most important in determining the abundance of a species.
- *Logistic regression* – this is similar to linear or multiple regression with one important difference, the response variable is *binomial*. This usually means that you have the abundance of a species recorded as present or absent (i.e. 1 or 0).

Since the underlying idea of correlation and regression is to look at a continuous variable as the predictor variable, it is not going to be a method you encounter in community studies except in two main cases:

- *Species richness* – you can correlate the number of species in a community with some other variable, this could be an environmental variable or time.
- *Ordination* – the results of most ordination methods produce *axis scores*, which are used to draw the community data in a 2-D plot. You can use correlation to examine the link between an axis of an ordination and another variable, such as an environmental factor.

If your community data comprise only species lists (or presence-absence data) then species richness is what you have. You can use species richness as the response variable in correlation-like tests such as:

- Comparing species richness over time.
- Comparing species richness in response to an environmental factor.

There is a method of analysis called *analysis of covariance* (ANCOVA), which you can think of as a mixture of ANOVA and regression. This could allow you to use the different species as a predictor variable in an analysis.

5.2.3 Association hypothesis tests

In association analysis you are using count data that are split into categories. There are two main forms of association test:

- Two-category analysis – this kind of test compares two sets of categories and is what you would think of as a classic *chi squared* test.
- Single-category analysis – this kind of test compares a single set of categorical data against a 'standard'; this is usually known as a *goodness of fit* test.

If you have several species, that is, a community, you might use association analysis to examine the association between the species and some other category. This could be anything that can be placed into categories, habitat type for example.

You can also look at species co-occurrence – you record species present in a quadrat and count how many co-occurrences you get for each pair of species. This approach can identify groups of species that tend to live together and those that tend to live apart. In other words, you can begin to sort out various communities. Using a similar approach you can also build up a picture of the relationships between species in a kind of family tree. This can give you a pictorial representation of the communities.

Another use for association is in the area of indicator species – if your test shows a strong positive association for one particular species with a particular habitat then that species could be taken as an indicator of that habitat. Of course the species would also have to show a strong lack of association with other habitats.

The goodness of fit approach is probably less useful for community studies but there are occasions where you might be able to press it into service. For example – you can compare a community survey with previously obtained survey data. The goodness of fit analysis will tell you if the two surveys are different.

You can think of association analysis as being the 'classic' hypothesis test that is most suited to community studies. The method is flexible and can be used in a variety of ways.

5.2.4 Population studies

Population studies are not really hypothesis based. Usually you are attempting to estimate the population size of a single species. The estimate might be compared to previous measures and it is possible to use a hypothesis test to compare two estimates (since they have confidence intervals) or to compare populations over time.

For communities these population estimates are not terribly useful but if you had population estimates for all the species in a community then of course you could compare them in various ways.

5.2.5 Classic hypothesis tests that are not useful for community studies

There are some hypothesis-based tests that are hard to press into service for community studies. They are mentioned here simply so you can get a more complete understanding of the analytical methods available.

One-dimensional patterns

The *Runs test* is used to look for one-dimensional patterns where there are two options (i.e. you have binary data). An example might be where you are looking at a bee species visiting flowers of two different colours. You note which colour the bee visits and your data will show one of three patterns: the bee could visit colours alternately, the visits could be to all of one colour then all of the other colour, and the last option is that the visits are random.

Two-dimensional patterns

Nearest neighbour analysis looks for two-dimensional patterns. For example, you look at a colony of nesting birds and measure the distance of each nest to its nearest neighbour. Your analysis can detect if the pattern is regular, clumped or random.

5.3 Analytical methods for community studies

When it comes to studying communities you are not always so reliant on using a 'classic' hypothesis test. It is usually more important to see how your communities relate to one another and to spot patterns in the data. There are various approaches that you can take to looking at community data (Figure 5.3).

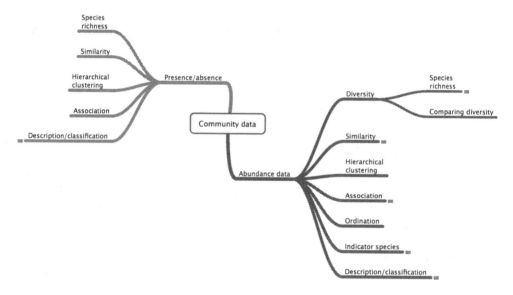

Figure 5.3 The study of community data by theme and by kind of data.

As you can see from Figure 5.3, you can think of your community data in two main ways:

- Your data are species lists – that is, presence-absence.
- Your data are counts or abundances.

If you only have lists of species present there is a range of approaches available to you. If your data are more 'detailed' and you have some kind of measure of abundance then you are able to carry out a wider range of analyses. Not only can you carry more tests but your results will be more 'sensitive' – the more information you have to input the better the output is likely to be.

5.3.1 Diversity

The term 'diversity' is often misunderstood, especially by the general public. In the simplest sense diversity is the number of different species in a given area of study – species richness. You can treat species richness like any other measurement and use it in many analytical methods for example:

- Differences tests – such as comparing species richness between habitats or in relation to some factor.
- Correlation tests – such as linking species richness to some environmental variable or comparing changes over time.

This simple measure of diversity can be useful – especially when all you have to work with is lists of species (i.e. you only have presence-absence data). However, it does not tell you anything about the relative abundance of the species in the communities. A more useful approach to the study of diversity is to look at the relative abundance of the species in a community. This leads to the notion of *diversity indices*.

In a broad sense, the higher a diversity index the more species there are. However, if one species is very dominant then the index will be reduced slightly. Communities with high diversity have many species but also the abundance is evenly spread between the species present – no one species is overwhelmingly dominant. You will learn more about diversity in Chapters 7–11.

Another element to the study of diversity is the scale of measurement. The diversity of a single sample is called *alpha* diversity. The diversity of many samples, taken as a whole, is called *gamma* diversity (you can think of it as landscape diversity). The change in diversity from sample to sample is called *beta* diversity diversity. So, landscapes with many different communities have high *beta* diversity. The study of *beta* diversity is covered in Chapter 10 and there are many analytical methods allied to that.

5.3.2 Similarity and clustering

If you have lists of species from two communities then you can look to see how many species are common to both. This is a measure of *similarity*. If your data contain abundance information then your measure of similarity can be more 'accurate'. The study of similarity becomes more meaningful when you have many samples to compare as you can look at the similarity of all the various pairwise combinations. You can now begin to see which groups of samples resemble one another.

Once you have a matrix of pairwise similarity measures you can assemble the samples into groups with closely similar samples forming clusters. You can represent the situation graphically using a diagram called a *dendrogram* – think of it as like a family tree. The dendrogram shows the relationships between the various samples and you can see at a glance how some samples (communities) are more closely 'related' to some samples than others. This arrangement is called *hierarchical cluster analysis* (see Section 12.2.1). Another way to set about clustering is to form your data into 'clumps'; methods that do this are usually referred to as *partitioning* methods (see Section 12.2.2).

There are various measures of similarity and they are sometimes known as *dissimilarity indices* if they measure how 'far apart' samples are. Most methods of ordination (see Chapter 14) use a dissimilarity measure as the starting point for the analysis. You will learn more about the study of similarity in Chapter 12.

5.3.3 Associations

Tests for association come in two basic forms:

- Two sets of categories – chi squared test.
- Single set of categories – goodness of fit test.

These two forms of association test are 'classic' hypothesis tests. When it comes to looking at community data association tests can be pressed into service in a variety of ways (see Section 5.2.3). In summary these are:

- Identification of communities via species co-occurrence.
- Similarity and hierarchical clustering.
- Indicator species analysis.

You can also use the goodness of fit approach to compare two communities – the test looks to see if two samples are 'the same' (i.e. how good the fit between them is).

In many ways you can think of association analysis as the basis for the study of communities; you will learn more about association analysis in Chapter 13.

5.3.4 Ordination

Ordination is a general term for methods that look to represent multi-dimensional data in a simpler form – usually the simpler form is two-dimensional.

If you had information about a single species and an environmental factor you might be able to show the data in a scatter plot. If you had two environmental variables you could try using a 3-D plot. When you have more variables it becomes impossible to show the data as you do not have enough spatial dimensions. The same goes for community data as you have information about many species. The various methods of ordination look to find patterns in the data and to represent them in a simpler form. There are two main approaches:

- *Indirect gradient analysis* – with *indirect* gradient analysis (Section 14.2) you are starting with just the species composition in various samples. Any environmental gradients must be inferred after the main analysis. Usually the data are passed through a mathematical algorithm to create a distance matrix (similar to that used in similarity analysis). This simplifies the data and allows you to 'project' the results in two (or more) axes that can be drawn as a 2-D scatter plot.
- *Direct gradient analysis* – with *direct* gradient analysis (Section 14.3) you already have environmental data and the methods of analysis use these data to help sort out the patterns in the communities. The spin-off is that you can test hypotheses about species composition and environmental variables directly.

The results of ordination show the data in a way that allows you to spot patterns that you would not have been able to see before. Sites and species that are close together in the final results plot are more closely similar than those further apart.

You can take the results of ordination as data in their own right and use them in further analyses, such as correlation. One method of ordination – canonical correspondence analysis – uses environmental variables to constrain the manipulation of the data, thus you can see how the environmental variables 'map' onto the biological data. You will learn more about ordination in Chapter 14.

5.3.5 Indicator species

The basic premise of indicator species analysis is to determine if a species can be considered to be indicative of a particular habitat or group. If you have an 'indicator species' it would be useful, as you can use the presence of such a species to indicate that you had a particular habitat. There are three main ways that you can set about the analysis of indicators:

- Association analysis – chi squared tests.
- Explicit measures of 'indicativeness' – for example the Dufrene Legendre approach to indicator values.
- Methods of ordination – the TWINSPAN program is designed to split your community data into groups, as part of this some species can be considered as indicators of certain groups – either groups of species or groups of habitats.

In this book you'll see indicator species analysis covered by the association analysis approach (Chapter 13).

5.4 Summary

Topic	Key Points
Single species studies	You can split the study of single species into various categories: descriptive, population, patterns, differences, correlations and associations. Some of these approaches can be pressed into service for community data.
Community descriptions • Community components • Location • Physiognomy • Community properties	Communities can be described in various ways, with the key species components being most important. You can also describe where the community is found in terms of geography as well as 'habitat'. The physiognomy of the community can be important as well as descriptions of any inherent properties of the community.

Diversity
 	
• Species richness
 	
  • Comparing richness
  • Correlating richness
 	
• Diversity indices
	The simplest form of diversity is species richness, the number of species in a given area. This measure can be used as the response variable in a variety of analyses such as comparing richness to some environmental factor. Using relative abundance gives rise to more sensitive measures of diversity, diversity indices, which give information about dominance.

Diversity and scale	Diversity can be measured at different scales, from a single sample to a whole landscape. The smallest scale (a single sample) has *alpha* diversity and many samples together have *gamma* diversity. The study of *beta* diversity is the study of changes in communities across samples.
Similarity and dissimilarity • Similarity indices • Matrices of similarity	Two communities can be compared in terms of the species present. This gives a similarity index or distance measure. The opposite is a measure of dissimilarity.
Clustering • Hierarchical clustering • Clustering by partitioning	This can be used to create a hierarchical clustering dendrogram to visualise the relationship between samples. Clustering by partitioning creates 'clumps' of data around group centres (called medoids).

Topic	Key Points
Ordination • Indirect gradient analysis • Direct gradient analysis	Methods of ordination aim to represent community data in a 2-D form. This gives rise to axis scores that can be used to create scatter plots or to correlate to other variables. The 2-D plots can show patterns in the data because the closer together points are the more similar they are. *Indirect gradient analysis* uses community data and any environmental gradients are inferred afterwards. *Direct gradient analysis* uses the environmental information directly to help form the order of the species and samples.
Association • Community ID • Goodness of fit • Indicators	Methods of association analysis can be used in a variety of ways and the chi squared approach can be regarded as a foundation of community analysis. Communities can be 'discovered' by using a co-occurrence analysis. The chi squared statistic can be used as a distance measure in hierarchical clustering. Goodness of fit can be used to compare communities. Association can lead to analysis of indicator species.
Indicator species	If a species is associated positively with a single habitat but not with others then it can be regarded as an indicator species. Tests of association can be used to look for indicators.

5.5 Exercises

5.1 What broad kinds of analysis can you carry out using species richness?

5.2 Which of these analytical approaches is **least** likely to be useful in community studies?

A) Correlation
B) Populations
C) Differences
D) Regression analysis
E) Analysis of covariance

5.3 You can use species composition to cluster samples – what are the main approaches to clustering?

5.4 What are two main strands of ordination analysis?

5.5 You can use association analysis as the basis for a clustering analysis – TRUE or FALSE?

The answers to these exercises can be found in Appendix 1.

6. Exploring data: getting insights

You should always spend some time to look over your data critically before you launch into any complicated analyses. This is an important phase in your investigation because you have the opportunity to carry out a range of tasks:

- Check for errors.
- Check the data are arranged appropriately.
- Add additional information.
- Summarise your data numerically.
- Produce summary graphs.

Checking for errors is obviously important; in community analyses you are dealing with species names and misspellings are common. You may wish to add additional information – this might be an index or grouping variable to help visualise the data or perhaps abbreviated species names.

Summarising your data numerically in tables is helpful to gain an overview of the situation. You might spot patterns in the data that you had not considered before; the overview might also form the basis of an interim report. Graphical summaries are often really helpful in visualising complicated data – especially if you are presenting information to others who are not so familiar with the data as you are.

At the checking stage your spreadsheet is still the most useful software tool available. You'll find using filters, lookup tables and Pivot Tables will help to knock your data into shape. You'll see how to use Excel for error checking (Section 6.1) and adding additional information (Section 6.2). Later in the chapter you'll see how to get an overview of your data (Section 6.3). Excel will be used as the starting point but at some stage you will need to use R, and you'll find Section 6.3.3 will give you some insights into how you can use R to your advantage.

6.1 Error checking

Whenever you have a lot of data you are bound to get a few mistakes creeping in. It is easy to misspell a name when you are typing species names. Another issue is that of additional spaces – this can happen if you use copy and paste, for example. What happens is that extra spaces are added to the end of the name. The name will look perfectly all right to you but Excel (and other programs) will regard the extra space as a separate name.

Errant data present a different problem and one that is harder to solve. If you have simply missed an entry then there will be a blank – you can spot this. If data are incorrectly entered then you have a more difficult time sorting things out.

6.1.1 Checking for spelling mistakes

In the following exercise you can have a go at spotting and correcting some errors in a simple set of plant community data. You will use a Pivot Table to help spot the errors and filters to help correct the mistakes.

Have a Go: Check data for spelling errors

For this exercise you will need the data file *Plant species lists with errors.xls*.

1. Open the file in your spreadsheet. There are two columns – one for the site names and one for the species names.

2. Click once anywhere in the data (you are going to make a Pivot Table).

3. Now click on the *Insert* button in the *Ribbon* and then click on the *PivotTable* button.

4. The Pivot Table wizard will start and ask you to select the data and the place for the result. The data should already be selected (look for the marching ants). If it is not then you will have to highlight it yourself using the mouse. Choose *New Worksheet* as the location for the result and then click the OK button.

5. The *Pivot Table Field List* box will now appear and you can start to create a Pivot Table. Begin by dragging the *Site* field item into the *Column Labels* box.

6. Now drag the *Species* field item into the *Row Labels* box. You will now have a skeleton Pivot Table, which contains species names and site names but no data (Figure 6.1).

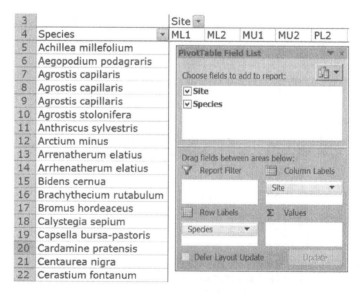

Figure 6.1 An 'empty' Pivot Table can be used to help spot errors in species names.

7. Look down the species list and see if you can spot any names that appear similar. Near the top you can see *Agrostis capillaris*. There are three entries that appear similar. The first one looks like a misspelling with a missing 'l'. Deal with that first.

8. Click on the *Data* tab and return to the main data sheet.

9. Click once anywhere in the block of data then go to the *Home* menu in the *Ribbon*. Click the *Sort & Filter* button and select *Filter*.

10. You will see filter icons appear in the heading row. Click the one in the *Species* column and bring up a dialogue box. You can now turn off the *(Select All)* option.

11. Near the top of the list you will see the misspelt *Agrostis capilaris*. Click this and then the OK button. Now you will see only the entry that is misspelt. You can correct it.

12. After correcting the entry to read *Agrostis capillaris* you can reset the filter (the simplest way is to turn filtering off using the *Home > Sort & Filter > Filter* button). Now return to the Pivot Table.

13. The original (incorrect) entry is still visible. You need to refresh the data. You want the *Refresh* button, which is found under the *Pivot Table Tools > Options* menu (you need to click anywhere in the Pivot Table to make the *Pivot Table Tools* menu visible). Once you click the button the old entry disappears. Now you can deal with the apparent duplication.

14. Return to the main data sheet. Turn on the *Filter* option (via *Home > Sort & Filter*) and select to display the *Agrostis capillaris* items. Now you can check these individually, some have an additional space after the name. You can edit each one separately or try using a replacement option.

15. You can use a search and replace approach to look for extra spaces at the end of *Agrostis capillaris*. You need to go to the *Replace* button via the *Home > Find & Select* menu. Search for 'aris' (you do not need the quotes) and replace with 'aris' (quotes not needed). The *Find All* button will show you which items were found and the *Replace All* button will carry out the replacement.

16. Return to the Pivot Table and click the *Refresh* button (as you did in step 13).

17. Now look for other errors and deal with them in a similar manner. You will see *Arrenatherum elatius* and *Arrhenatherum elatius* for example. The first is a spelling error (the 'h' is missing).

18. In fact there are no further mistakes (I hope) so you can now save the data with its new Pivot Table (you will use this later). You might want to alter the filename so that the original is intact.

You can use the same approach to check the site names and any other data field. In this case there are no other data.

Checking for errors is not just something you can do at the end of your data collection – look over the data routinely to check your progress.

6.2 Adding extra information

Sometimes you will want to add extra information to your data. This might relate to environmental data that you have collected or something as simple as adding abbreviated species names. Additional columns of data can be useful for splitting up your data into different groupings. Ideally you will have thought of this right at the start of the planning process but the course of true research does not always run smoothly.

Adding data might seem like a simple typing process but usually the sorts of things that you want to add can be dealt with using a lookup table in conjunction with a simple Excel formula.

In the following exercise you can have a go at adding some extra information to some data on ground beetle communities. You'll make abbreviated species names (useful for displaying in summary tables and graphs) and add some environmental data that can be used as an index to help group and display the data.

Have a Go: Add grouping variables and abbreviated species names

For this exercise you will need the data file *Ground beetles.xls*.

1. Open the spreadsheet file. There are two tabs – one for the main data and one for additional information to be incorporated into the main data. The *Beetles* tab contains the main data. There are three columns; species name, the number captured and a sample name. Each row shows the number of individuals captured from a set of pitfall traps (a block of ten traps) placed in one of three habitat types (*Edge, Grass* and *Wood*). Each of the habitat types has been sampled six times, which results in labels E1, E2 ... E6, and so on.

2. Look at the *Lookup* tab in the spreadsheet. There are two separate blocks of data. The first shows the sample names along with the maximum height of the vegetation at that sampling area. The third column shows the habitat type. The next block of data shows the species names and corresponding abbreviations. The abbreviations are formed by taking three letters from the genus, adding a period, then three letters from the trivial name. In most cases this '3.3' system is adequate to differentiate the species. Here a fourth letter had to be added to the end of some of the names.

3. Now return to the *Beetles* tab, you are going to use a Lookup table to add the abbreviated species names to the fourth column. Start by making a title for the column – type 'Abbr' into the first cell of column four (you do not need the quotes).

4. You will need the VLOOKUP formula. Start the process by selecting the D2 cell (the first data cell of the new *Abbr* column). The formula can be entered in two ways: you can use the function wizard or type it directly. To use the function wizard you click the f_x button in the formula bar. If you don't use the wizard you simply press the 'equals' key (=). Now type the name of the formula (lower case is fine).

5. Now type a left bracket '(' and you will see that you are invited to enter various arguments. The *lookup_value* is the cell you want to search for – the species name – in this case the value is in cell A2. You can type 'A2' or click with the mouse. If you are using the keyboard you can now type a comma. If you are using the wizard you can click in the next box.

6. The next part you require is the *table_array*. This is the range of cells that contains the data you are looking to replace. This is in the *Lookup* tab and is the block of cells with the species names and abbreviations. The function is going to match the *lookup_value* with an item in the first column of the *table_array*. Use the mouse to highlight the table but do not include the header items (these are for your convenience). Once you have highlighted the cells you can enter a comma or click in the next box of the wizard. You'll see the name of the sheet as well as the actual cell reference (in this case it is *Lookup!E2F49*).

7. Now you need to tell Excel what data to take from the table, the *col_index_num*. This is simply a column number. You type '2' here – note that this means the second column of the *table_array*, it is actually column F in the spreadsheet.

8. If you are using the keyboard you can now type in a closing bracket ')' and press ENTER. If you are using the wizard you click the OK button.

9. The abbreviated name has now been copied from the lookup table. You now need to copy the cell down the entire column. However, if you do this now the formula will also alter the *table_array* references, which will be incorrect. You need to alter the formula to 'fix' the position of the *table_array*. Edit the formula and add dollar signs '$' to the cell references of the *table_array* part. The reference should now read $E$2:$F$49. Notice that you need a $ in front of the column (the letter) and the row (the number) reference.

10. Now you can copy the formula down the rest of the column. You can copy it to the clipboard or use drag and drop, if enabled on your computer. The abbreviation should now appear for all the species in the list (Figure 6.2).

D2			f_x	=VLOOKUP(A2,Lookup!$E$2:$F$49,2)	
	A		B	C	D
1	Species		Quantity	Sample	Abbr
2	Pterostichus melanarius		4	E5	Pte.mel
3	Leistus spinibarbis		2	E6	Lei.spi
4	Carabus violaceus		1	E6	Car.vio
5	Platynus assimilis		4	E5	Pla.ass
6	Agonum muelleri		1	E5	Ago.mue

Figure 6.2 A lookup table used to add abbreviated species names to data.

11. You now need to use a similar process to add *Max Ht* and *Habitat* to the data. Don't forget that the *table_array* should not include the lookup table headings (these are for your benefit). The *Max Ht* item will be column 2 in the *col_index_num* and the *Habitat* will be column 3. In both cases you will use the *Sample* column as your *lookup_value*. The first formula will be: VLOOKUP(C2,Lookup!$A$2:$C$19,2) for the *Max Ht* and VLOOKUP(C2,Lookup!$A$2:$C$19,3) for the *Habitat*.

12. Once you have copied the formula down the columns you can save the file – you might want to use a different filename to keep the old and new versions separate.

Additional columns can be useful to provide index columns for arranging your data in different ways.

Think carefully about how you might want to group and arrange your data. The more familiar you are with Pivot Tables the more you'll see what you can do, and this will give you new ideas for adding index columns.

Tip: Use an Excel formula for abbreviated species names

You can use a function in Excel to create abbreviated names:

= LEFT(cr,3)&"."&MID(cr,FIND(" ",cr)+1,3)

Simply replace the cr part with the cell reference that contains the 'long' name of your species. You'll end up with three letters of the genus, a dot then three letters of the trivial name. You can of course edit the formula to produce other combinations.

If your species names contain layer information, e.g. *Quercus robur* (s), you can use a longer form of the function:

=IF(RIGHT(cr,1)=")",LEFT(cr,3)&MID(cr,FIND(" ",cr)+1,3)&"."&MID(cr,FIND("(",cr)+1,1),LEFT(cr,3)&"."&MID(cr,FIND(" ",cr)+1,3))

The genus and trivial name are now joined together, and the layer information follows a period. The function would convert the example above to Querob.s.

6.3 Getting an overview of your data

It is really important to gain some sort of overview of your data before you launch into detailed analyses. You have already seen how to carry out some basic error checking (Section 6.1) and how to add extra indexing information to your data (Section 6.2). In this section you will see how to get overviews of your data – there are two main ways:

- *Numerically* – this usually means tables of averages, with your data split into meaningful groupings.
- *Graphically* – it is generally easier to see patterns in the data using a graph than with a table of figures. You can split your data into meaningful chunks using grouping variables.

Both ways have their merits – it largely depends on your purpose which one you use. Simple site species lists can be very useful for example but graphs are generally more understandable, especially to someone who is not familiar with the data.

You will not be surprised to find out that Pivot Tables are very useful in producing overviews of your data. The great advantage is that you can produce a range of different summaries very rapidly and do not need to alter your original data – this saves time and effort. Of course, this does suppose that you have your data arranged in a sensible layout to begin with (Chapter 3). Later in the chapter you'll see how to use R for data overview (Section 6.3.3) but first you will see how to use Excel.

6.3.1 Numerical overview

The exact nature of the overview you go for will depend on what kind of data you have and your exact purpose. In general you want to get an overview that allows you to make some sense of the data – you are trying to simplify the original data to make things more digestible. Your digest may be used to present a report or simply to help you see what you are dealing with. You can probably simplify the kinds of data you have into three main sorts:

- Simple species lists.
- Communities at different sites (possibly with environmental data).
- Communities at different times.

In any event you can use filters and Pivot Tables to help you prepare summaries with minimal fuss.

Usually a graphical summary is more easily digestible – this is the subject of the next section.

6.3.2 Graphical overview

Graphs are generally more easily understood than raw numbers. This is especially true if the reader is not familiar with the data concerned. You can think of graphs as being in three broad groups, depending on the purpose for which they are required:

- *Differences* – bar charts and box-whisker plots. For example, changes in abundance between sites.
- *Correlations* – scatter plots. For example, links between abundance and some environmental factor.
- *Time series* – line plots. For example, changes in abundance over time.

You can easily create bar charts, scatter plots and line plots in Excel but box-whisker plots are harder to produce (but it is possible). The difficulty is that when you have community data you have a lot of different things, and your graphs can become so congested that it is hard to read them. Since the purpose of a graph is to make a summary very clear with minimal effort on the part of the reader, having a complicated graph is a poor idea.

The 'trick' here is to produce selective graphs and not to crowd each plot with too much information. You can use the Pivot Table function to produce the summary data from which you will make your graphs. Later versions of Excel also allow *Pivot Charts* to be produced. These are linked to the Pivot Table closely and change when you alter the Pivot Table. Since 2010 Excel has also been able to produce *Sparklines*, which are 'micro-graphs', these can be extremely useful.

Use charts with Pivot Table data

You can use a Pivot Table as the basis for a chart. Once created, the Pivot Table acts like any other data in Excel. Of course, if you later change the Pivot Table the graph you make will alter. To prevent your graph from changing when you alter the Pivot Table you have several options:

- Make a new Pivot Table for subsequent graphs.
- Copy the Pivot Table data to a new worksheet using *Home > Paste > Paste Special,* and then make the graph.
- Copy or save the graph to a separate file – make sure that you use a picture format otherwise the graph might still be linked to the original Pivot Table data.

Which of these options suits you best will depend what you are going to do with the graph in the long term.

Tip: Save a graph as PDF from Excel

PDF is a high-quality format that most publishers use to make publication quality images. You can save graphs in PDF format by getting an Office Add-In: *SaveAsPDF.exe* from the Microsoft website (search for Microsoft Office PDF in a search engine). Once you have the Add-In simply click on your graph then use the Office button and select *Save As > PDF*. One of the Options when saving is to save only the selected chart.

In the following exercise you will get a chance to create a summary graph of species abundance at different sites.

Have a Go: Make a bar chart comparing species abundance with site

For this exercise you will look at the same plant data as you did earlier but this time you have abundance information. The data are in the file: *Plant species and abundance.xls*.

The file contains three columns: one called *Site* for the site name, one called *Species* for the species names, and one called *Qty* for the abundance information. Values for *Qty* range from 0.2 to 10 and are based on Domin scores in five sample quadrats.

1. Start by making a Pivot Table. Click once in the block of data then use the *Pivot Table* button via *Insert > Pivot Table* on the Excel *Ribbon*. The data should be selected automatically – place the new Pivot Table in a fresh worksheet.

2. Construct the Pivot Table by dragging the fields from the *Pivot Table List* box into the appropriate sections: *Species* for the *Row Labels*, *Site* for the *Column Labels*, and *Qty* for the *Values*. You should now have a completed Pivot Table showing the abundance of the various species at the sites.

3. Prepare the table so that you can make a bar chart. The *Values* field item should read *Sum of Qty*. If it does not then click on the field in the *Values* box and alter the *Value Field Settings* so that it does. Turn off the *Grand Total* for rows and columns – click once in the Pivot Table then use the *Grand Totals* button via the *Pivot Table Tools > Design* menu.

4. Click outside of the Pivot Table data – ensure that you are not adjacent to the data either. When you begin the chart-making process Excel will search around the currently selected cursor for data and will select it automatically. It usually makes incorrect decisions so it is best to populate the graph yourself. Now start the chart-making process by using *Insert > Column > 2-D Column*. Choose the top-most option (*Clustered Column*).

5. Move and resize the blank chart frame as you like – a place to the right of the main Pivot Table data is sensible.

6. Click once in the empty chart frame, then use the *Select Data* button via the *Chart Tools > Design* menu. Use the *Add* button in the *Legend Entries (Series)* section to add some data.

7. You should see the *Edit Series* dialogue box. Click in the *Series name* section and then click on a species name in the Pivot Table to use as the label – choose *Agrostis capillaris*. Now click in the *Series values* box – delete anything that appears there.

Now use the mouse to select the cells relating to the abundance of *Agrostis capillaris* (including the blank cells where abundance is zero). Click the OK button to return to the *Select Data Source* dialogue box.

8. Now click the *Edit* button in the *Horizontal (Category) Axis Labels* section. You want to select the site names when the *Axis Labels* box appears. Click OK and return to the *Select Data Source* dialogue box. Click the OK button and your graph is prepared.

9. The graph needs some work to make it completely acceptable. Click once on the graph and then use the buttons in the *Chart Tools > Layout* menu to do some editing. You can delete the main title and add axis titles at the very least.

10. So far you have a graph that shows the abundance of a single species across the various sites. Add a second species for a comparison by clicking once on the graph and then using the *Select Data* button via the *Chart Tools > Design* menu.

11. Click the *Add* button. Now click on the *Cirsium arvense* species name to select that as the *Series name* label. Click in the *Series values* box and delete anything that is in there. Select the cells relating to the abundance of *Cirsium arvense* (include the blank cells). Now click the OK button to return to the *Select Data Source* dialogue box. There is no need to edit the *Horizontal (Category) Axis Labels* since you already set these for the previous data. Simply click the OK button and return the graph.

12. Your graph now should show two species compared across the various sites (Figure 6.3).

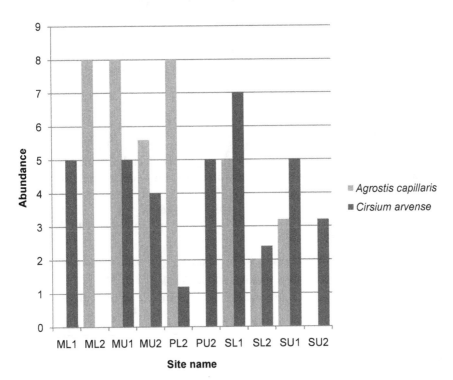

Figure 6.3 Bar chart to show species abundance at various sites using Pivot Table data.

13. You can change the data that are displayed; click once in the graph then use the *Select Data* button via the *Chart Tools > Design* menu. When the *Select Data Source* dialogue box appears select the *Agrostis capillaris* item from the list of series. Now click the *Edit* button.

14. When the *Edit Series* window appears change the *Series name* by clicking on a new species name from the Pivot Table. Choose the *Ranunculus repens* entry. Now move to the *Series values* box. Make sure the contents are highlighted or deleted before you select the new data. Select the data corresponding to the *Ranunculus repens* abundance from the Pivot Table. Click the OK button to return to the *Select Data Source* window. Click OK again and return to the newly modified graph.

You can add more species and alter the ones you choose to display. If you alter the Pivot Table however, the graph will change so to 'fix' the graph you should save it to a graphics file when you are happy with it.

The graphs you have made so far do not really show much detail about the entire communities, as you have focused on just a few species. It is possible to show more detail but you risk creating a graph that is too hard to read easily. The compromise would be to show all the species but for only one or two sites at a time.

Tip: Saving charts as PDF

If you save a graph as a PDF you should alter the fonts so that they are readable by PDF programs such as Adobe Illustrator. The Microsoft Office fonts do not always display properly so use 'standard' fonts such as Arial, Times or Verdana.

It would be useful if you could link your graph to the Pivot Table so that it updated automatically when you changed the table. Later versions of Excel use something called a Pivot Chart, which does exactly this.

Use Pivot Charts

A Pivot Chart is a graph that maintains close links to your Pivot Table data. This means that if you alter the Pivot Table the graph is updated to reflect the changes automatically. This is potentially very useful but there is a limitation – you cannot create scatter plots. You can make bar charts and line graphs.

Editing and manipulating the finished graph is the same as for any other kind of graph – the difference comes with the way you create the graph to begin with. There are two ways to link a graph to a Pivot Table:

- Make a Pivot Chart from scratch – simply use the *Pivot Chart* button to make a chart in a similar fashion to making a Pivot Table.
- Link a graph to a Pivot Table – create a Pivot Table and then make a chart using the Pivot Table data.

Tip: Rotate axis labels

It can be helpful to re-orient axis labels to make them fit better or simply to be more readable. To do this you can right-click on the axis and select *Format Axis*. Then go to the *Alignment* section. There are presets for *Text direction* and you can also specify a *Custom angle*.

6.3.3 Using R for data exploration

The R program is able to produce results analogous to Pivot Tables that Excel produces. There are commands that can use cross-classifying factors to produce summary tables in various ways (look back to Table 4.3 for a list). R has powerful graphical capabilities too and you can use a range of commands to visualise your data.

In general it is a good idea to use Excel as your starting point and to carry out your 'first look' at the data using Excel if you can. Error checking and adding extra 'useful' information is best conducted using Excel since this is the program that you'll use as your repository for the data and what you will use to add more data should you acquire it. Once you have your data 'prepared' you can use R to carry out the various community analyses.

However, it is still useful to know how to carry out some of these summary operations using R. The starting point will usually be a CSV file, since this is the basic data format that R will read most easily. Excel can save CSV files quite easily (Section 4.2.4) and they can be read into R using the `read.csv()` command (Section 4.3.2).

Pivot Tables and data summary using R

R can make contingency tables that are similar to the Pivot Table of Excel. You can also use various commands to summarise the tables that you produce. Other commands can use grouping factor to apply a summary function to a data variable.

Cross-classifying

There are three main commands that produce contingency tables:

- `table()` – this is the simplest and produces a table of counts.
- `ftable()` – this produces a table of counts but 'collapses' multi-dimensional tables into a more convenient two-dimensional form.
- `xtabs()` – this command can use a numerical variable as the count data.

These commands are very similar but it is the subtle differences that make them useful. In the following exercise you can have a go at using the three commands to create tables for some data on ground beetle communities that you met previously.

Have a Go: Use cross-classifying commands in R to make Pivot Tables

In this exercise you will need the beetle community data that you saw earlier. If you saved the modified file then use that. If you did not, you can use a version with the habitat data incorporated, the file is called: *Ground beetles and habitat.xls*.

The data represent ground beetle communities from three types of habitat: *Grass, Edge* and *Wood*. Each habitat has data from six sampling locations – they are labelled *G1, G2, G3* and so on. Each sampling location (18 in total) has also recorded the maximum vegetation height, called *Max Ht* in the spreadsheet.

1. Open the data in your spreadsheet and make sure that you can see the main data (not the *Lookup* worksheet). Save the data as a CSV, use the *Save As* button from the *Office* button and choose the CSV (Comma delimited) option.

2. Open the R program. Read in the CSV data you just saved and make a new object called *gb* to contain the data like so:

```
> gb = read.csv(file.choose())
```

3. Now get a preview of the data by looking at the top few lines of the data:

```
> head(gb)
                    Species Quantity Sample   Abbr Max.Ht Habitat
1  Pterostichus melanarius        4     E5 Pte.mel    2.5    Edge
2     Leistus spinibarbis        2     E6 Lei.spi    2.0    Edge
3         Carabus violaceus        1     E6 Car.vio    2.0    Edge
4        Platynus assimilis        4     E5 Pla.ass    2.5    Edge
5          Agonum muelleri        1     E5 Ago.mue    2.5    Edge
6  Calathus rotundicollis       94     E5 Cal.rot    2.5    Edge
```

4. Use the `table()` command to make a simple contingency table. The first instruction will form the rows of the table and the second will form the columns like so:

```
> table(gb$Abbr, gb$Sample)

        E1 E2 E3 E4 E5 E6 G1 G2 G3 G4 G5 G6 W1 W2 W3 W4 W5 W6
Aba.par  1  1  1  1  1  1  1  1  1  1  1  1  1  1  1  1  1  1
Acu.dub  0  0  0  0  0  0  1  0  0  1  1  0  0  0  0  0  0  0
Ago.afr  0  0  0  0  0  0  1  1  1  1  1  1  0  0  0  0  0  0
Ago.ful  0  0  0  0  0  0  1  0  0  0  1  0  0  0  0  0  0  0
...
```

5. You now have the counts of species at the different samples. Add another variable to the command to make a three-dimensional table:

```
> table(gb$Abbr, gb$Sample, gb$Habitat)
, , = Edge

        E1 E2 E3 E4 E5 E6 G1 G2 G3 G4 G5 G6 W1 W2 W3 W4 W5 W6
Aba.par  1  1  1  1  1  1  0  0  0  0  0  0  0  0  0  0  0  0
Acu.dub  0  0  0  0  0  0  0  0  0  0  0  0  0  0  0  0  0  0
...
```

6. Now you have three dimensions – effectively three separate tables that are linked together. It would be convenient to show the results in a single table. Use the `ftable()` command to 'collapse' the table to two dimensions like so:

```
> ftable(Sample ~ Abbr + Habitat, data = gb)
         Sample E1 E2 E3 E4 E5 E6 G1 G2 G3 G4 G5 G6 W1 W2 W3 W4 W5 W6
Abbr    Habitat
Aba.par Edge     1  1  1  1  1  1  0  0  0  0  0  0  0  0  0  0  0  0
        Grass    0  0  0  0  0  0  1  1  1  1  1  1  0  0  0  0  0  0
        Wood     0  0  0  0  0  0  0  0  0  0  0  0  1  1  1  1  1  1
Acu.dub Edge     0  0  0  0  0  0  0  0  0  0  0  0  0  0  0  0  0  0
        Grass    0  0  0  0  0  0  1  0  0  1  1  0  0  0  0  0  0  0
        Wood     0  0  0  0  0  0  0  0  0  0  0  0  0  0  0  0  0  0
...
```

7. In this case you use a formula to specify how the table should be constructed. Variables to the left of the ~ form the columns. Variables to the right of the ~ form the rows of the table. Try using different combinations and see what happens. The result does not always fit very neatly onto screen but you can make the window larger.

8. The `ftable()` command can make a table similar to the `table()` command if you

only use two variables. The main difference is that there are additional headings. Try replicating the original table like so:

```
> ftable(Sample ~ Abbr, data = gb)
          Sample E1 E2 E3 E4 E5 E6 G1 G2 G3 G4 G5 G6 W1 W2 W3 W4 W5 W6
Abbr
Aba.par          1  1  1  1  1  1  1  1  1  1  1  1  1  1  1  1  1  1
Acu.dub          0  0  0  0  0  0  1  0  0  1  1  0  0  0  0  0  0  0
Ago.afr          0  0  0  0  0  0  1  1  1  1  1  1  0  0  0  0  0  0
...
```

9. The `xtabs()` command works much like the `table()` command but is able to use an additional variable for the count data. The input is in the form of a formula but this time the variable to the left of the ~ forms the count data. Variables to the right of the ~ are the rows, columns and additional dimensions of the table itself. Start by omitting the count data and creating a presence-absence table like so:

```
> xtabs(~ Abbr + Sample, data = gb)
           Sample
Abbr        E1 E2 E3 E4 E5 E6 G1 G2 G3 G4 G5 G6 W1 W2 W3 W4 W5 W6
  Aba.par    1  1  1  1  1  1  1  1  1  1  1  1  1  1  1  1  1  1
  Acu.dub    0  0  0  0  0  0  1  0  0  1  1  0  0  0  0  0  0  0
  Ago.afr    0  0  0  0  0  0  1  1  1  1  1  1  0  0  0  0  0  0
...
```

10. Now add the abundance data to the command to form a table that shows the abundance of the species at each sampling location:

```
> xtabs(Quantity ~ Abbr + Sample, data = gb)
           Sample
Abbr        E1  E2  E3  E4  E5  E6  G1  G2 G3 G4  G5  G6  W1  W2  W3  W4  W5 W6
  Aba.par  388 325 295 350 407 381 178 114 80 99 133 105 426 368 338 253 298 330
  Acu.dub    0   0   0   0   0   0   1   0  0  1   1   0   0   0   0   0   0   0
  Ago.afr    0   0   0   0   0   0   1   1  1  4   6   7   0   0   0   0   0   0
...
```

If you save the results of a cross-classifying command, you will have a new data object that can be used for further analyses.

The `xtabs()` command is especially useful, as it allows you to take data in biological recording format and reassemble it in other ways, ready for analysis.

Summarising tables

There are various summary commands you can use on tables of data, such as a contingency table:

- `rowSums()`, `colSums()` – these commands simply return the sums of the rows or columns of your data table.
- `rowMeans()`, `colMeans()` – these commands simply return the means of the rows or columns of your data table.
- `apply()` – this command allows you to apply a function to the rows or columns of a data table.

These commands work on any data objects that have rows and columns – in the following exercise you can have a go at using the commands to summarise species richness and frequency for the ground beetle community data.

Have a Go: Use R table summary commands to explore species richness and frequency

In this exercise you will need the ground beetle community data that you used previously. The data represent samples of ground beetles from 18 sampling locations that come from three different habitat types. If you already have the data in R (you called it *gb*) you can skip straight to step 3. The data are in an Excel file and also a CSV that is ready to import to R, the file is called *Ground beetles and habitat.csv*.

1. Open R and read the CSV file into the program using the `read.csv()` command. Assign the data to an object called *gb* like so:

```
> gb = read.csv(file.choose())
```

2. You can use the browser window that opens to locate the *Ground beetles and habitat. csv* file, which will be read into the *gb* object.

3. Make a simple contingency table of presence-absence by using the `table()` command. Assign the result to a named object like so:

```
> gb.pa = table(gb$Abbr, gb$Sample)
```

4. The first instruction formed the rows of the table (*Abbr*), whilst the second instruction formed the columns (*Sample*). Have a look at the top few lines of the table you just made using the `head()` command:

```
> head(gb.pa)

          E1 E2 E3 E4 E5 E6 G1 G2 G3 G4 G5 G6 W1 W2 W3 W4 W5 W6
  Aba.par  1  1  1  1  1  1  1  1  1  1  1  1  1  1  1  1  1  1
  Acu.dub  0  0  0  0  0  0  1  0  0  1  1  0  0  0  0  0  0  0
  Ago.afr  0  0  0  0  0  0  1  1  1  1  1  1  0  0  0  0  0  0
  Ago.ful  0  0  0  0  0  0  1  0  0  0  1  0  0  0  0  0  0  0
  Ago.mue  0  0  0  1  1  0  0  0  0  0  0  0  0  0  0  0  0  0
  Ago.vid  0  0  0  0  0  0  1  0  0  1  0  0  0  0  0  0  0  0
```

5. You now see each species shown as present (1) or absent (0) in the table. The sum of the columns will be the species richness for each sample. The sum of the rows will be the frequency of the species across all the samples.

6. Use the `colSums()` command to obtain the species richness for the samples:

```
> colSums(gb.pa)
 E1 E2 E3 E4 E5 E6 G1 G2 G3 G4 G5 G6 W1 W2 W3 W4 W5 W6
 17 14 15 25 21 17 28 22 18 28 26 24 12 11 11 12 12 10
```

7. Now look at the species frequency using the `rowSums()` command:

```
> rowSums(gb.pa)
 Aba.par Acu.dub Ago.afr Ago.ful Ago.mue Ago.vid Ama.aen Ama.com
      18       3       6       2       2       2       1       1
 Ama.fam Ama.ple Ama.sim Bad.bul Bem.big Bem.gut Bem.lam Bem.man
       1       4       1       3       7       2      10      12
 Bem.obt Bra.har Bra.sha Cal.fus Cal.rot Car.vio Cli.fos Cur.aul
       3       1       4       6      18      12       6       1
 Cyc.car Har.ruf Lei.fer Lei.ful Lei.ruf Lei.spi Lor.pil Neb.bre
       6       1       2       5       9       5       2      18
 Not.big Not.ruf Ocy.har Pat.atr Pla.ass Poe.cup Pte.mad Pte.mel
       9       4       3       1      17      12      18      11
```

```
   Pte.nige Pte.nigr Pte.obl  Pte.str  Pte.ver  Sto.pum  Syn.niv  Tre.qua
       18       12        4       17        4       11        4        4
```

8. You can also use the `apply()` command to achieve the same results:

```
> apply(gb.pa, MARGIN = 2, FUN = sum)
> apply(gb.pa, MARGIN = 1, FUN = sum)
```

In the first command the `MARGIN = 1` instruction applies the `sum()` function to the columns. Setting `MARGIN = 2` applies the function to the rows.

Tip: Create abbreviated names using R

The `abbreviate()` command can be used to create abbreviations of text labels, such as species names. The basic form of the command is:

`abbreviate(names.arg, minlength = 4, method = "left.kept")`

You can also use `method = "both.sides"` to produce a slightly different abbreviation.

It is useful to be able to apply summary functions to rows or columns of simple tables but often your data are arranged in a different way with response variables and predictors. In this case you'll need to use the predictor variables as groupings to help summarise the response variables – you'll see how to do this next.

Applying summary functions using grouping variables

There are some commands that can take a variable and apply a summary command to it using another variable as a grouping factor. The main commands that you will encounter are:

- `tapply()` – This command takes a column of data and splits it into groups using another variable. A summary function is then applied to each group. The result is placed in a `matrix`.
- `aggregate()` – This command takes a column of data and splits it into groups using another variable. A summary function is then applied to each group. The result is placed in a `data.frame`.

The main difference between the two commands is the form of the output. The `tapply()` command produces a `matrix` as its result, whilst `aggregate()` makes a `data.frame`. These are two different kinds of R object that have slightly different properties. In the following example you can have a go at using these commands to summarise the ground beetle data that you saw previously.

Have a Go: Use grouping variables to summarise data

In this exercise you will need the ground beetle community data that you used previously. The data represent samples of ground beetles from 18 sampling locations that come from 3 different habitat types. If you already have the data in R (you called it *gb*) you can skip straight to step 3. The data are in an Excel file and also a CSV that is ready to import to R, the file is called *Ground beetles and habitat.csv*.

1. Open R and read the CSV file into the program using the read.csv() command. Assign the data to an object called *gb* like so:

    ```
    > gb = read.csv(file.choose())
    ```

2. You can use the browser window that opens to locate the *Ground beetles and habitat. csv* file, which will be read into the *gb* object.

3. Remind yourself of the data headings by using the `names()` command:

    ```
    > names(gb)
    [1] "Species" "Quantity" "Sample" "Abbr" "Max.Ht" "Habitat"
    ```

4. Use the `tapply()` command to get the *Max.Ht* for each *Sample*. You will need to tell the command to use the `max()` summary function like so:

    ```
    > tapply(gb$Max.Ht, INDEX = gb$Sample, FUN = max)
      E1   E2   E3   E4   E5   E6   G1   G2   G3   G4   G5   G6   W1
    5.50 15.00 15.00 3.00 2.50 2.00 1.80 1.80 2.00 1.50 1.75 1.75 17.00
      W2   W3   W4   W5   W6
    16.00 18.00 18.0017.0018.00
    ```

5. The result of the `tapply()` command is a `matrix` but it might be more convenient as a `data.frame`. Look at the difference by using the `aggregate()` command. In this case you can use a formula in the command:

    ```
    > aggregate(Max.Ht ~ Sample, data = gb, FUN = max)
        Sample Max.Ht
    1       E1   5.50
    2       E2  15.00
    3       E3  15.00
    ...
    ```

6. Now split up the data by *Habitat* – obtain the mean of the *Quantity* for each species in the three habitats using the aggregate() command:

    ```
    > aggregate(Quantity ~ Abbr + Habitat, data = gb, FUN = mean)
         Abbr Habitat    Quantity
    1  Aba.par    Edge  357.666667
    2  Ago.mue    Edge    1.500000
    3  Bem.big    Edge    1.000000
    4  Bem.gut    Edge    1.000000
    5  Bem.lam    Edge    1.250000
    ...
    ```

7. Notice that you used two grouping variables, both on the right of the ~ in the formula. The result is fine but it might be more useful if each *Habitat* had its own column. Use the `tapply()` command to do this like so:

    ```
    >tapply(gb$Quantity, INDEX=list(gb$Abbr, gb$Habitat), FUN=mean)
                   Edge        Grass          Wood
    Aba.par   357.666667   118.166667    335.500000
    Acu.dub          NA     1.000000            NA
    Ago.afr          NA     3.333333            NA
    Ago.ful          NA     1.500000            NA
    Ago.mue    1.500000           NA            NA
    ...
    ```

8. Note that you had to bundle the two grouping variables together in a `list()` command. The `tapply()` command does not accept the formula input. Note also that there are no 0 values but `NA` is shown instead. This is similar to the blank cell of an Excel Pivot Table and you can think of it as 'missing'.

The `NA` items can be 'taken care of' if you carry out any further analysis because most commands can use the instruction `na.rm = TRUE,` which will eliminate them before any calculation is performed.

So, the two commands `tapply()` and `aggregate()` perform useful summary grouping functions but their output is slightly different – you can choose the one that is most suited to your requirements.

You generally want to avoid `NA` items in your community data as `NA` is treated as 'missing' rather than as zero. When you make your Excel Pivot Tables you can use the *Options* to alter blank cells to 0 but in R you have to alter the `NA` items afterwards. Of course there are some kinds of data where you may really have missing values so take care.

Replacing NA items with 0

If you are using cross-classifying commands such as `table()` or `xtabs()`, you will generally not get `NA` items. They might, however, appear when you use `tapply()` or `aggregate()` because you have attempted to obtain a mean of zero for example. In such cases you might wish to alter the `NA` items to 0.

The key is the `is.na()` command, which allows you to test if a datum is `NA` or not. If you get a `TRUE` result you can alter the datum to 0, if you get `FALSE` the datum is a real value and can remain as it is.

You need to set up a loop using the `for()` command and test each datum to see if it is `NA` using the `is.na()` command. In the following exercise you can have a go at replacing `NA` with 0 in some data.

Have a Go: Replace NA items with 0 (zero)

You will need the *gb* data that you met previously. If you already have this in R then you can go straight to step 3. If not then you will need to get the data, called *Ground beetles and habitat.csv*, into R first.

1. Use the `read.csv()` command to get the ground beetle data and assign them to a named object called *gb*:

    ```
    > gb = read.csv(file.choose())
    ```

2. You can select the *Ground beetles and habitat.csv* file from the browser window that appears. The data are abundances of beetles at three habitat types, each of which was sampled six times.

3. Remind yourself of the layout of the data by looking at the top few lines:

```
> head(gb)
                   Species Quantity Sample    Abbr Max.Ht Habitat
1 Pterostichus melanarius        4     E5 Pte.mel    2.5    Edge
2     Leistus spinibarbis        2     E6 Lei.spi    2.0    Edge
3        Carabus violaceus        1     E6 Car.vio    2.0    Edge
4      Platynus assimilis        4     E5 Pla.ass    2.5    Edge
5         Agonum muelleri        1     E5 Ago.mue    2.5    Edge
6  Calathus rotundicollis       94     E5 Cal.rot    2.5    Edge
```

4. Use the `tapply()` command to summarise the abundance of beetles at the three habitats:

```
> gb.habitat = tapply(gb$Quantity, INDEX = list(gb$Abbr, gb$Habitat),
FUN = mean)
> head(gb.habitat)
              Edge         Grass     Wood
Aba.par   357.6667    118.166667    335.5
Acu.dub         NA      1.000000       NA
Ago.afr         NA      3.333333       NA
Ago.ful         NA      1.500000       NA
Ago.mue     1.5000            NA       NA
Ago.vid         NA      1.000000       NA
```

5. Look at the number of beetles for each habitat using the `colSums()` command:

```
> colSums(gb.habitat)
 Edge    Grass    Wood
   NA       NA      NA
```

6. Use the `na.rm = TRUE` instruction to remove `NA` items before carrying out the calculation:

```
> colSums(gb.habitat, na.rm = TRUE)
    Edge      Grass      Wood
823.3500   331.8167   880.0000
```

7. The *gb.habitat* object is a matrix, which means it is one single block of data, split into rows and columns. Use the `length()` command to see how many items are in the entire dataset:

```
> length(gb.habitat)
[1] 144
```

8. Because the matrix is one single data entity you can change the `NA` items in one go:

```
> for(i in 1:length(gb.habitat)) {
  if(is.na(gb.habitat[i]) == TRUE) gb.habitat[i] = 0}
```

9. You start by creating an index, *i*, which will be used to step through the data. This begins at 1 and carries on for the length of the data matrix (144). After you type the

curly bracket { you can carry on a new line and R will not evaluate what you typed until the closing }. You now use the `if()` command to test if each element of the data is NA – if something is NA then it is replaced by 0, if it is not then it stays as it is (there is no command that needs to be typed).

10. Look now at the data and see that the NA items are replaced with 0s:

```
> head(gb.habitat)
                Edge          Grass        Wood
Aba.par       357.6667     118.166667     335.5
Acu.dub         0.0000       1.000000       0.0
Ago.afr         0.0000       3.333333       0.0
Ago.ful         0.0000       1.500000       0.0
Ago.mue         1.5000       0.000000       0.0
Ago.vid         0.0000       1.000000       0.0
```

If your data were in a `data.frame` then each column would need to be dealt with separately.

Tip: Replace NA with any value using the *gdata* package

The *gdata* package contains various useful utility commands. One is called `NAToUnknown()`, which allows you to quickly replace NA items with the value of your choice.

Graphical summaries using R

R has very powerful graphical capabilities and it is possible to produce a huge variety of graphs using it. There are three main sorts of graph that you can produce that are useful:

- *Bar charts* – used to compare categories of things – `barplot()` command.
- *Box-whisker plots* – used to compare things across categories but provides additional information about each group – `boxplot()` command.
- *Scatter plots* – used to compare two variables – `plot()` command.

There are other graphs you can produce of course but these are the mainstays of your arsenal.

Making bar charts using R

The `barplot()` command creates bar charts. The general form of the command is like so:

```
barplot(height, names.arg = NULL, legend.text = NULL,
        beside = FALSE, horiz = FALSE)
```

There are various other instructions that you can use but Table 6.1 shows the common ones and what effect they have on the command.

Table 6.1 Commonly used instructions for the `barplot()` command.

Instruction	Explanation
`height`	The data to be plotted. This must be either a single vector of numeric values or a matrix. If the data are a single vector then each item value is plotted as a separate category. If the data are a matrix then the columns form the categories and the rows are the grouping variables.
`names.arg = NULL`	An optional vector of names to place under the bars. If this is `NULL` (the default) the names are taken from the names attribute of the data.
`legend.text = NULL`	An optional vector of labels to use in a legend. If `legend.text = TRUE` then the legend is constructed using the names attribute of the data.
`beside = FALSE`	If multiple rows are plotted then by default a stacked bar chart is drawn (each column of the data matrix being split by row). To create a chart with grouped bars use `beside = TRUE`.
`horiz = FALSE`	If `horiz = TRUE` the chart is drawn with horizontal bars.

You can specify other graphical instructions that are common to many other graphs; these include altering the colour of the bars for example.

In the following examples you can have a go at making bar charts.

Have a Go: Make a bar chart using R

For this example you will need to use the ground beetle data you met earlier. If you already have the data in R (called *gb*) then you can go directly to step 3.

1. Open R and prepare to load the data using the `read.csv()` command. Assign a named object to hold the data:

    ```
    > gb = read.csv(file.choose())
    ```

2. You will need to select the file *Ground beetles and habitat.csv* when the browser window opens. If you are using Linux then you'll need to specify the filename (in quotes) exactly.

3. Remind yourself what the data look like by viewing the first few lines:

    ```
    > head(gb)
                       Species Quantity Sample    Abbr Max.Ht Habitat
    1 Pterostichus melanarius         4     E5 Pte.mel    2.5    Edge
    2    Leistus spinibarbis          2     E6 Lei.spi    2.0    Edge
    3        Carabus violaceus        1     E6 Car.vio    2.0    Edge
    4       Platynus assimilis        4     E5 Pla.ass    2.5    Edge
    5         Agonum muelleri         1     E5 Ago.mue    2.5    Edge
    6  Calathus rotundicollis        94     E5 Cal.rot    2.5    Edge
    ```

4. The data show ground beetle abundance at three habitats; each habitat was sampled six times. Each sampling location also has the maximum height of the vegetation at that location.

5. You will need to determine the species richness at each sampling location. You could make a table showing presence-absence and then use the `colSums()` com-

mand. However, you can also combine the commands into one and get the result 'directly' like so:

```
> gb.sr = colSums(table(gb$Abbr, gb$Sample))
> gb.sr
E1 E2 E3 E4 E5 E6 G1 G2 G3 G4 G5 G6 W1 W2 W3 W4 W5 W6
17 14 15 25 21 17 28 22 18 28 26 24 12 11 11 12 12 10
```

6. The *gb.sr* result is a vector with a names attribute. This means that if you use the barplot() command the bars will be named automatically:

```
> barplot(gb.sr, las = 1)
> title(xlab = "Sample", ylab = "Species richness")
```

7. The graph should resemble Figure 6..4. In the command you used las = 1, which makes all the axis labels horizontal. The title() command adds titles to axes.

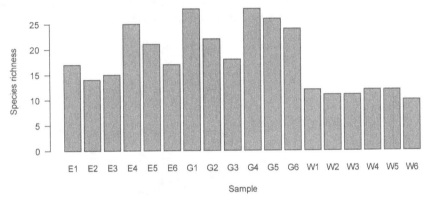

Figure 6.4 A bar chart of species richness created using the barplot() command.

Try using a few other instructions to see how you can alter the appearance of the graph. Try making the bars horizontal for example, or altering the colour of the bars.

When you have more complicated data you can choose to graph only some of the results. A matrix, for example, has rows and columns – you can subset the matrix using the square brackets and giving the parts you want to use as [row, column]. In the following exercise you can have a go at making bar charts from a larger dataset.

Have a Go: Make a grouped bar chart using R

For this exercise you will need to use some butterfly data. The data are in a CSV file called *Butterfly and year data.csv*.

1. Open R and prepare to read in the data using the read.csv() command. Make an object called *bf* to contain the data like so:

```
> bf = read.csv(file.choose())
```

2. You can select the CSV file *Butterfly and year data.csv* once the browser window opens (if you are using Linux you will need to type the filename in quotes in full instead of the `file.choose()` part).

3. The data show abundance of various butterfly species at a site over several years. Remind yourself of the data layout by viewing the first few lines:

```
> head(bf)
      Spp     Yr   Qty
1 Lg.skip   1996     3
2  Lg.wht   1996     4
3  Sm.wht   1996     2
4  GV.wht   1996   119
5  Or.tip   1996    90
6  Sm.cop   1996     9
```

4. Reform the data into a contingency table using the `xtabs()` command. You want to see the abundance for each species and each year like so:

```
> bf.xt = xtabs(Qty ~ Spp +Yr, data = bf)
> bf.xt
      Spp
Yr    Com.blu DkG.frit Gn.hair GV.wht Lg.skip Lg.wht M.bro Or.tip Paint.l
1996      3.0      0.0     0.0  119.0     3.0    4.0  88.0   90.0    49.5
1997      3.0      0.0     1.0   95.0    12.0    1.0  47.0   14.0     0.0
1998      0.0      0.0     0.0   88.0     2.0    0.0  12.5   36.0     0.0
1999      0.0      1.0     1.0  157.5     1.0    1.0  33.0   24.0     0.0
...
```

5. You can select a single row by using the square bracket syntax. Show the first row of the data (corresponding to the 1996):

```
> bf.xt[,1]
1996 1997 1998 1999 2000 2001 2002 2003 2004 2005
   3    3    0    0    4    3    0    1    7    4
```

6. Use the square bracket syntax to show the first column of the data (corresponding to the Common blue):

```
> bf.xt[,1]
Com.blu  DkG.frit  Gn.hair  GV.wht  Lg.skip   Lg.wht    M.bro  Or.tip
    3.0       0.0      0.0   119.0      3.0      4.0     88.0    90.0
Paint.l       PBF      Pea  Red.ad     Ring  S.heath Scot.arg  Sm.cop
   49.5       3.0     48.0     6.0    190.0     12.0      0.0     9.0
 Sm.pbf    Sm.tor   Sm.wht    Wall
   12.0      26.0      2.0     0.0
```

7. Use the `barplot()` command to make a horizontal bar chart of the butterfly abundance for 1996 like so:

```
> barplot(bf.xt[1,], horiz = TRUE, las = 1, cex.names = 0.8)
> title(xlab = "Species abundance", main = "Data for 1996")
```

8. The `horiz = TRUE` instruction makes the bars horizontal, the `las = 1` instruction makes all the axis labels horizontal, `cex.names = 0.8` makes the category names smaller than 'standard' (so they fit in the margin). The `title()` command adds titles to the graph – note that the `xlab` instruction refers to the bottom axis even though it is the response variable. The graph should resemble Figure 6.5.

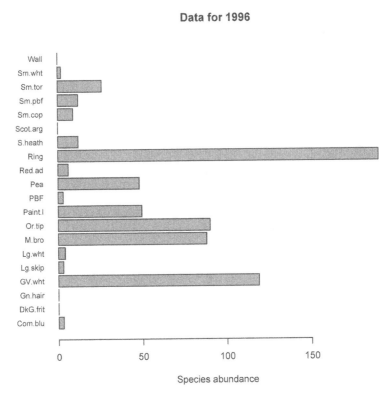

Data for 1996

Species abundance

Figure 6.5 Abundance of butterfly species for 1996 using the `barplot()` command.

9. You can include a range of columns (and/or rows) in the square brackets using x : y, where x and y are the starting and ending values respectively. Compare the years 1996 and 1997 for all species as a horizontal, grouped bar chart like so:

```
> barplot(bf.xt[1:2,], horiz = TRUE, las = 1, cex.names = 0.8,
          legend.text = TRUE, beside = TRUE)
> title(xlab = "Butterfly abundance")
```

10. This time you have grouped the data by year and the `beside = TRUE` instruction keeps the bars in the groups (rather than stacking them). The `legend.text = TRUE` part adds a legend and takes the text from the names of the species. Your graph should resemble Figure 6.6.

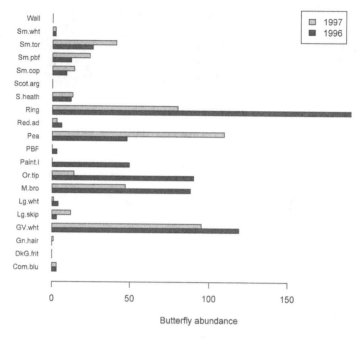

Figure 6.6 Butterfly abundance for two years.

You can specify various combinations of rows and columns, simply by altering the values in the square brackets.

The square brackets allow you to specify [rows, columns] of a 2-D data object. If you leave one blank then it is assumed that you want all items, so [, 1] produces all the rows but only column 1. To specify a continuous range you use a colon so, [1:3, 4:6] would select rows 1–3 and columns 4–6. If you want a discontinuous range you must specify the exact values using a c() command, so [c(1,3), c(4, 7)] would select rows 1 and 3, columns 4 and 7.

Tip: Rotating a data table

You can quickly rotate a data table using the t() command (the result is a matrix object). This is useful if you want to make a bar chart that is grouped by column, rather than row for example.

Making box-whisker plots using R

The box-whisker plot is like a bar chart in that it is designed to show items in various categories. However, the plot itself shows more information (median, interquartiles and range) and is therefore potentially more useful. The command to create box-whisker plots in R is boxplot() and the general form of the command is like so:

```
boxplot(formula, data = NULL, range = 1.5, horizontal = FALSE)
```

The command can accept input in several ways, see Table 6.2. There are other instructions that can be given to the command but Table 6.2 shows the most commonly used ones.

Table 6.2 Commonly used instructions for the `boxplot()` command.

Instruction	Explanation
`formula`	The data to be used for the plot can be specified in three main ways:
	• As separate vectors, separated by commas.
	• As a data object that contains columns that are the samples.
	• As a formula of the form `response ~ predictor`. Multiple predictors can be used.
`data = NULL`	If the data are given as a formula you can specify where the variables are to be found by naming the data object that contains them.
`range = 1.5`	The whiskers extend to 1.5 times the inter-quartile range by default. Any points outside this are shown as plain points (outliers). To show the max–min use `range = 0`.
`horizontal = FALSE`	If TRUE, the boxplot is drawn horizontally.

You can specify other graphical instructions that are common to many other graphs; these include altering the colour of the plot for example.

In the following exercise you can have a go at making a box-whisker plot.

Have a Go: Make a box-whisker plot using R

For this example you will need to use the ground beetle data you met earlier. If you already have the data in R (called *gb*) then you can go directly to step 3.

1. Open R and prepare to load the data using the `read.csv()` command. Assign a named object to hold the data:

   ```
   > gb = read.csv(file.choose())
   ```

2. You will need to select the file *Ground beetles and habitat.csv* when the browser window opens. If you are using Linux then you'll need to specify the filename (in quotes) exactly.

3. Remind yourself what the data look like by viewing the first few lines:

   ```
   > head(gb)
                     Species Quantity Sample   Abbr Max.Ht Habitat
   1 Pterostichus melanarius        4     E5 Pte.mel    2.5    Edge
   2     Leistus spinibarbis        2     E6 Lei.spi    2.0    Edge
   3        Carabus violaceus        1     E6 Car.vio    2.0    Edge
   4      Platynus assimilis        4     E5 Pla.ass    2.5    Edge
   5         Agonum muelleri        1     E5 Ago.mue    2.5    Edge
   6  Calathus rotundicollis       94     E5 Cal.rot    2.5    Edge
   ```

4. The data show ground beetle abundance at three habitats; each habitat was sampled six times. Each sampling location also has the maximum height of the vegetation at that location.

5. Look at the *Abbr* variable. This contains the names of the species – they look like plain text but they are a special R object called a factor. The `levels()` command will show the different items in the variable:

```
> levels(gb$Abbr)
 [1] "Aba.par" "Acu.dub" "Ago.afr" "Ago.ful" "Ago.mue"  "Ago.vid"
 [7] "Ama.aen" "Ama.com" "Ama.fam" "Ama.ple" "Ama.sim"  "Bad.bul"
[13] "Bem.big" "Bem.gut" "Bem.lam" "Bem.man" "Bem.obt"  "Bra.har"
[19] "Bra.sha" "Cal.fus" "Cal.rot" "Car.vio" "Cli.fos"  "Cur.aul"
[25] "Cyc.car" "Har.ruf" "Lei.fer" "Lei.ful" "Lei.ruf"  "Lei.spi"
[31] "Lor.pil" "Neb.bre" "Not.big" "Not.ruf" "Ocy.har"  "Pat.atr"
[37] "Pla.ass" "Poe.cup" "Pte.mad" "Pte.mel" "Pte.nige" "Pte.nigr"
[43] "Pte.obl" "Pte.str" "Pte.ver" "Sto.pum" "Pte.nige" "Pte.nigr"
```

6. Use the `aggregate()` command to summarise the data, include the `subset` instruction to display data for the *Aba.par* species like so:

```
> aggregate(Quantity ~ Abbr + Sample + Habitat, data = gb, FUN = max,
            subset = Abbr %in% "Aba.par")
      Abbr Sample Habitat Quantity
1  Aba.par     E1    Edge      388
2  Aba.par     E2    Edge      325
3  Aba.par     E3    Edge      295
...
```

7. The `subset` instruction can be used as part of the `boxplot()` command to get data for a single species and plot it as a box-whisker plot:

```
> boxplot(Quantity ~ Habitat, data = gb, subset = Abbr %in% "Aba.par")
> title(xlab = "Habitat", ylab = "Abundance")
```

8. You entered the data as a formula with Habitat as the single predictor variable. Using the `subset` instruction ensures that you only plot data for the single species. Your plot should resemble Figure 6.7.

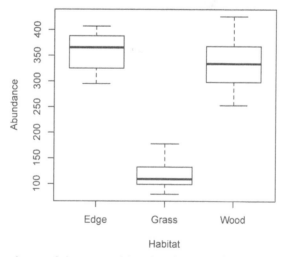

Figure 6.7 Abundance of the ground beetle *Abax parallelepipedus* in three habitat types. Stripes show median abundance, boxes show interquartile range (IQR) and whiskers show max-min.

To alter the species plotted you can simply edit the name in the `subset` instruction.

Making scatter plots using R

Scatter plots are created using the generic `plot ()` command using R. The general form of the command is like so:

```
plot(x, y = NULL, type = "p", pch, col, xlim, ylim)
```

This command is very flexible and many R objects have their own dedicated plotting routines Table 6.3 shows some of the most common instructions used with the `plot ()` command.

Some of the instructions shown in Table 6.3 are common to many other graphs (look at the `par ()` command in R).

In the following exercise you can have a go at making a scatter plot.

Table 6.3 Commonly used instructions for the plot() command.

Instruction	Explanation
`x, y = NULL`	The co-ordinates for the points can be specified in several ways: • As *x*, and *y* values. • As a single object that has a plotting structure (the simplest being a data frame or matrix with two columns). • As a formula of the form response ~ predictor.
`data = NULL`	If the data are given as a formula you can specify where the variables are to be found by naming the data object that contains them.
`type = "p"`	By default points are drawn. Other options include: • '1' – plots lines only that join the co-ordinates. • 'b' – both points and lines, the lines form segments between the points. • 'o' – points and lines overplotted (i.e. there are no gaps in the line. • 'n' – nothing is plotted.
`pch`	The plotting symbol to be used. The default is taken from the currently set graphical parameters but is usually 1, an open circle. Available values are 1–25 (R symbols) and 32–127 (ASCII symbols). You can also specify a symbol by typing it in quotes.
`col`	The colour to be used for the plot. If a numerical value is used the colour is taken from the current colour `palette ()`, otherwise named colours can be used (in quotes). See `colours ()` command for a list.
`Xlim ylim`	The limits of the *x* and *y* axes can be specified by giving the starting and ending values like so: `c(start, end)`.

Have a Go: Make a scatter plot using R

In this exercise you will use the butterfly data that you saw previously. The data are in a CSV file ready for import to R, the file is called *Butterfly and year data.csv*.

1. Open R and use the `read.csv()` command to import the data; create an object to hold the data like so:

```
> bf = read.csv(file.choose())
```

2. If you are using Windows or Mac then you can now choose the file *Butterfly and year data.csv* from the browser window that opens. If are using Linux you will have to replace the `file.choose()` part with the explicit filename (in quotes).

3. Look at the first few lines of the data using the `head()` command:

```
> head(bf)
      Spp   Yr Qty
1 Lg.skip 1996   3
2  Lg.wht 1996   4
3  Sm.wht 1996   2
4  GV.wht 1996 119
5  Or.tip 1996  90
6  Sm.cop 1996   9
```

4. The data are the abundance of various butterfly species, sampled at a site over several years. Look at the total numbers of butterflies per year by using the `aggregate()` command:

```
> (bf.tot = aggregate(Qty ~ Yr, data = bf, FUN = sum))
      Yr   Qty
1   1996 664.5
2   1997 459.5
3   1998 356.0
4   1999 479.0
5   2000 691.0
6   2001 911.5
7   2002 852.0
8   2003 872.0
9   2004 604.5
10  2005 487.0
```

5. Use the `plot()` command to visualise the relationship between total abundance and year:

```
> plot(Qty ~ Yr, data = bf.tot)
```

6. In this case you used a formula to specify the co-ordinates but since the data are in two columns (*x*, *y*) you could use the following command to produce Figure 6.8:

```
> plot(bf.tot)
```

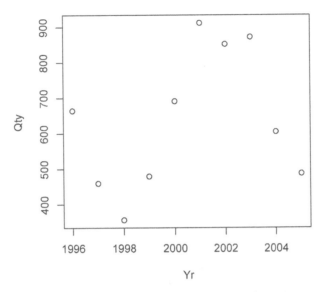

Figure 6.8 Total abundance of all butterfly species and year for a site.

7. Try specifying the x and y values independently like so:

```
> plot(bf.tot$Yr, bf.tot$Qty)
```

Notice how the axis names reflect what you typed in the last step – you could overcome this by using `xlab` and `ylab` instructions as part of the `plot()` command. Try altering some of the graphical instructions, to alter colour or plotting symbol for example.

The `plot()` command is something of a workhorse command and many analytical routines include their own method of plotting.

Tip: Show a result of a calculation without typing the object name

When you make a named object R does not display the result – the object is just created. You can view the object by typing its name. However, if you enclose the entire command in parentheses `()`, the result is presented immediately.

This has been a brief tour of the capabilities of R (and earlier of Excel), the exercises that you'll encounter in subsequent chapters will give you more practice as well as introducing a few extra tricks.

6.4   Summary

Topic	Key Points
Error checking • Using filters • Using Pivot Tables	Checking for errors such as missing values, incorrect values and misspellings is important. Filters and Pivot Tables can help you check that you have accurate data. A common error in community data is to add extra spaces at the end of species names, they are invisible to you but the computer regards the name as a new species.
Abbreviations	Species names (and site names too) can be rather long and it is therefore useful to have abbreviations in addition to the full names. You can use lookup tables in Excel to help you add abbreviations. The `abbreviate()` command in R can make unique abbreviations.
Overview of your data • Pivot Tables • Pivot Charts	Pivot Tables are especially useful in helping you explore your data before starting with the 'main analyses'. You can take data in biological recording format and rearrange it in many helpful ways. Pivot Tables can provide summary statistics as well as graphics (Pivot Charts).
Summary graphs	Use bar charts or box-whisker plots to look at differences. Use scatter plots to explore the relationships between variables (correlations). Use line graphs to look at time-series data.
Tabulation in R	Tabulation and cross-classification can be carried out using R commands: `table()`, `ftable()` and `xtabs()`. The latter is especially useful in allowing you to convert biological recording data into new forms.
Summary statistics in R	You can use various commands to get summary statistics from data tables, e.g. `rowSums()`, `colMeans()`. The `apply()` command allows you to use any command over the rows or columns of a data table. The `tapply()` and `aggregate()` commands allow you to use a summary command using a grouping variable.
Replacing 'missing' values	Missing data items are assigned NA in R. You'll usually need to replace these NA items with 0 for community analyses. But beware, some kinds of data really do have missing values. The *gdata* package contains a useful command, `NAToUnknown()`, that can easily replace NA with other values.
Graphics in R	R has powerful graphical capabilities. Use `barplot()` to make bar charts. Use `boxplot()` to make box-whisker plots. Use `plot()` to make scatter plots.

6.5   Exercises

6.1 What kind of graph would you prepare to summarise the number of species (species richness) in several different habitats?

6.2 You can use Excel to make an abbreviated name, so why bother with a lookup table?

6.3 What are the main ways to go about looking for spelling mistakes?

6.4 Look at the *psa* data in R, which is part of the *CERE.RData* file. These data are in biological recording format; there are columns for *Site*, *Species* and *Qty*. How would you rearrange the data into a community dataset – with species forming the rows and site names forming the columns?

6.5 Look at the *psa* data in R once again. How would you go about making a bar chart of the species richness for each site?

The answers to these exercises can be found in Appendix 1.

7. Diversity: species richness

The term species richness refers simply to how many different kinds of species there are in a given area. Species richness only requires that you have a list of species in order to calculate it. This makes it attractive because you only have to differentiate species (they do not need to be identified) and produce a list. Site lists are commonly produced in conservation and are readily obtained for sites by examining biological records. The trade-off is that you can only carry out a restricted set of analyses when you have simple species lists.

Species richness can be used as the response variable in various sorts of analysis. However, in order to do that you have to ensure that the sampling effort is the same for all your replicates. If you sample plant species for example, a larger quadrat would generally result in more species being found. There are methods for standardising species richness to take into account different sample sizes (see Section 7.3). Once you have taken care of the sampling effort issue there are two main approaches you might take:

- Compare difference in species richness with some variable (see Section 7.1).
- Correlate species richness with a variable (see Section 7.2).

The first step is to get your data into the appropriate format. You can calculate species richness using Excel or R – in the following example you can have a go at working out species richness using R.

Have a Go: Calculate species richness using R

In this exercise you will look at species richness for plants at several sites. You will use the plant data that you met previously when you checked data for errors. The corrected data are called *Plant species errors fixed.xls*. In this exercise you will use R but first you will need to prepare the data so it is in CSV form.

1. Open the spreadsheet *Plant species errors fixed.xls*.

2. Make sure you are on the *Data* tab then click the *Office* button, select the *Save As* option.

3. From the *Save as type* section select the *CSV (Comma delimited)*option, you can see the available options by clicking on the triangle icon to drop down a list. You can keep the same filename since *.csv* will be appended instead of *.xls* so that you (and the computer) know it is a CSV file.

4. You will be warned that multiple sheets are not supported in CSV format – simply click OK. You will also be warned that there are features in the file that are not supported by CSV format – click Yes to save the file. You can now close Excel; if you are asked if you want to save the file once more you can just click No.

5. Now open the R program. You may have a shortcut on the desktop, a quick-launch icon or simply find it via the *Start* button.

6. Read in the CSV data and assign a name:

```
> pd = read.csv(file.choose())
```

7. You can now select the data file. If you are using Linux you will have to specify the filename explicitly (in quotes) as part of the command instead of the `file.choose()` part.

8. You will need to rearrange the data in a similar way to Excel's Pivot Table (see Section 6.3.3 for details of R commands). A command to do this in R is `ftable()`. Assign a variable name and make a table like so:

```
> pd.t = ftable(Site ~ Species, data = pd)
> pd.t
                      Site  ML1 ML2 MU1 MU2 PL2 PU2 SL1 SL2 SU1 SU2
Species
Achillea millefolium          1   1   1   1   0   1   0   0   0   0
Aegopodium podagraris         0   0   0   0   0   0   0   1   0   0
Agrostis capillaris           0   1   1   1   1   0   1   1   1   0
Agrostis stolonifera          0   0   0   0   0   1   0   1   0   0
Anthriscus sylvestris         0   0   0   0   0   0   0   0   1   1
```

9. You can see that R has created a result similar to the Pivot Table in Excel. The column sums will be the species richness for each site; you can get columns sums like so:

```
> colSums(pd.t)
 [1] 15 16 21 14 13 11 16 24 27 26
```

10. You have the species richness but the site names are not shown. To do this you need to extract the names from the original data and assign them to the result. Get the site names from the original data:

```
> levels(pd$Site)
 [1] "ML1" "ML2" "MU1" "MU2" "PL2" "PU2" "SL1" "SL2" "SU1" "SU2"
```

11. Make a named object to 'hold' the species richness result:

```
> pd.sr = colSums(pd.t)
```

12. Use the `names` command to apply the site names to the richness result:

```
> names(pd.sr) = levels(pd$Site)
```

13. Now simply type the name of your result object to view the final species richness along with the site names:

```
> pd.sr
ML1 ML2 MU1 MU2 PL2 PU2 SL1 SL2 SU1 SU2
 15  16  21  14  13  11  16  24  27  26
```

The `ftable()` command is one of several R commands that can be used in a similar way to the Pivot Table of Excel (see Section 6.3.3).

Now that you have seen how to determine species richness it is time to look at some basic analytical approaches such as comparing sites (Section 7.1) and correlating richness with an environmental variable (Section 7.2). You will see how to take into account differing sampling effort in Section 7.3.

7.1 Comparing species richness

Simple comparison of species richness only possible when the data are 'equivalent', that is the they are based upon the same sampling effort. Species richness then becomes a simple response variable that can be used like any other. As long as you have replicated data you can use regular differences tests for example (Section 5.2.1).

You might compare differences in species richness using any grouping factor; in the following exercise you can have a go at comparing species richness between two sites.

Have a Go: Compare species richness between sites

In this exercise you will look at some plant community data. The data are in the file *Plant quadrat data.xls*. The file contains columns for the species name as the scientific binomial and the common name. There is a column entitled *quadrat*, which contains a value from 1 to 5. There are two sites surveyed (*upper* and *lower*) and each survey used five quadrats. At each site the plants found were listed along with an estimate of the cover in the Domin scale. The two sites are named in the column headed *site*. There is also a column for the species abbreviated name. If you look at this you will see that this is constructed using a formula.

You could create a Pivot Table in Excel and calculate the species richness – these values could then be copied and used for a test using the *Analysis ToolPak*. Unfortunately this does not allow you to carry out a U test, which you will need, so you will use R.

1. Open the spreadsheet and click once in the data. Then make a Pivot Table using the *Insert > Pivot Table* button. You want the species richness to end up in the column of totals so arrange the Pivot Table fields like so: Drag the *abbr* field to the *Column Labels* box, drag the *site* field to the *Row Labels* box, drag the *quadrat* field to the *Row Labels* box (put it underneath the *site* field), drag the *abbr* field to the *Values* box.

2. Now the *Grand Total* column on the right contains the species richness. There are subtotals too, which you do not need. The subtotals can remain as you are going to copy and paste the data for each site separately.

3. Open the R program. You need to make two samples, one for each site. Start by making an object to hold the *lower* data:

   ```
   > lower = scan()
   1:
   ```

4. R now displays a 1: and waits for you to enter some data. You could type the values in as simple numbers, separated with spaces but instead you will use the clipboard.

5. Switch back to Excel and highlight the species richness for the *lower* site. Copy the values to the clipboard. Return to R and paste the data in – the cursor will now show a 6: and wait for further input. Press Enter to complete the entry process:

```
> lower = scan()
1: 19
2: 21
3: 18
4: 22
5: 23
6:
Read 5 items
```

6. Now repeat the process for the *upper* site. Make an object called upper and paste the species richness as before:

```
> upper = scan()
1: 12
2: 9
3: 10
4: 14
5: 10
6:
Read 5 items
```

7. You now have two named objects, *lower* and *upper*, which represent species richness for the two sites. The first step in the analysis should be to visualise the data so make a boxplot like so:

```
> boxplot(lower, upper, names = c("Lower", "Upper"))
> title(xlab = "Site", ylab = "Species richness")
```

8. The first command creates a basic box-whisker plot and the second adds the axis titles. The final plot should resemble Figure 7.1.

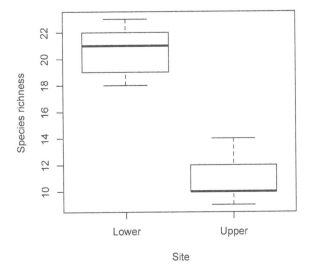

Figure 7.1 Boxplot of species richness at two sites. The stripe shows median richness, the boxes are interquartile range and whiskers max–min.

9. With only five replicates you probably ought to carry out a non-parametric test. The Wilcoxon rank sum test can be undertaken quite simply:

```
> wilcox.test(lower, upper)

    Wilcoxon rank sum test with continuity correction

data:  lower and upper
W = 25, p-value = 0.01193
alternative hypothesis: true location shift is not equal to 0
```

10. You may get a warning about tied ranks, but the correction is small and so it will not affect your result in this case. You can see that there is a significant different in species richness between the two sites.

This method – using copy and paste with the scan() and font Courier New command – works well when you have only two sites but when your data are more complicated you need a different approach.

In the preceding exercise you only compared two samples so the *t*-test or *U*-test was most appropriate. When you have more samples you will need to use analysis of variance (ANOVA) or the Kruskal–Wallis test. If your data are not normally distributed then you'd use the Kruskal–Wallis test; in R you can use the kruskal.test() command:

```
kruskal.test(response ~ predictor, data)
kruskal.test(response, groups)
```

You can give the data as a formula or as two separate variables. If your data are normally distributed then the aov() command can be used for ANOVA:

```
aov(response ~ predictor, data)
```

The command will only accept a formula. With ANOVA you can also run a post-hoc test using the TukeyHSD() command. You supply the aov() result to the command.

Previously you saw how to use the ftable() command to make a summary contingency table in R that resembled a Pivot Table (see Table 4.3 and Section 6.3.3). You used the levels() command to make names for the colSums() result that gave the species richness. This approach is fine for fairly simple comparisons but when you have many sites and different numbers of replicates, things can get messy and a better way is required. The key is to have your data arranged appropriately from the outset.

You want to end up with a Pivot Table that contains species names as columns and sample names as rows – these sample names should all be unique. The data can be the abundance or simple presence-absence (that is 1 for presence and 0 for absence). Any information relating to environmental variables or other data can go in a separate data file, with the rows representing the samples (sites). So, you end up with two files of data, one for the biological data and one for everything else. You can prepare the data using Excel and transfer to R for the main analyses.

In the following example you can have a go at examining species richness in comparison to site using a more generic approach by creating two data files – one for the community data and one for the overall site names.

Have a Go: Create a generic community data file

For this exercise you will use the *Plant quadrat data.xls* file that you met previously. The data show species abundance for two sites, each of which has five quadrats.

The quadrats are numbered 1–5 for each of the sites; this is useful but it is better to have a unique label for each sample. You will make unique labels and create two files – one for the community (presence-absence) data and one for the site names.

1. Open the spreadsheet and make a column heading (in column G) for the unique sample labels – call it *obs* (think of it as observation).

2. In column G make labels for the samples by adding the quadrat number to the site name. Use a function to do this most easily by simply joining the site name and the quadrat name. The first label will be made so: =E2&C2 (Figure 7.2).

	G2				f_x	=E2&C2	
	C	D	E		F		G
1	quadrat	domin	site		abbr		obs
2	1	8	upper		Poa.ann		upper1
3	1	5	upper		Fes.ovi		upper1

Figure 7.2 Using a simple concatenation to make a unique sample label from site name and quadrat number.

3. The & symbol tells Excel to simply join (concatenate) items. Copy the formula down the column to create labels for all the rows.

4. Now make a Pivot Table – click once in the data and then use the *Insert > Pivot Table* button. The data should be highlighted automatically – select to place the table in a new worksheet.

5. You are going to make a data file for the samples and site names first, so drag the *obs* field to the *Row Labels* box. You will see the sample names appear. Now drag the *site* field to the same box, making sure that the *obs* field is above the *site* field.

6. You have the sample names and the sites paired up but you also have subtotals. Turn these off by clicking on the *obs* field button in the *Row Labels* box of the *Pivot Table Field List* window. Select *Field Settings* and set *Subtotals* to *None*. Click OK to finish the process.

7. Make a new worksheet (click on the tab at the bottom) and return to the Pivot Table. Click once anywhere in the Pivot Table then select all the data using Ctrl+A (you can also use the mouse). Copy to the clipboard. Now go to the new blank worksheet and click in cell A1. Do not use a simple paste operation as you will simply end up with a copy of the Pivot Table.

8. Go to the *Home* menu and click the bottom of the *Paste* button and select *Paste Special*. Select *Values* and click OK. You will now see the data. There is also a Grand Total row. Right-click in the margin of this row (over the 12 label) and *Delete* the entire row. Just for tidiness you can rename the worksheet by right-clicking on the tab at the bottom of the screen and using the *Rename* option. Call the worksheet *sites*.

9. Return to the Pivot Table – you are now going to make the community presence/absence data. Get rid of the *site* field from the *Row Labels* box – you can either drag it out using the mouse or click on the tick in the list of fields at the top. You now have simple sample names.

10. Drag the *abbr* field into the *Row Labels* box. Then drag the *abbr* field from the top list into the *Values* box at the bottom. You end up with the *abbr* field in the Pivot Table twice – once as a row label and once as data. The field in the *Values* box should read *Count of abbr*. If it does not then click on it and select *Value Field Settings* and alter the summary method to *Count*.

11. The Grand Total items are not required so click once in the Pivot Table then turn them off via the *Pivot Table Tools > Design > Grand Totals* button.

12. At the moment the Pivot Table shows 1s where a species is present and a blank where it is not. You need to make the blanks into 0s because R will interpret a blank cell as a missing data rather than a 0. Click once in the Pivot Table then use the *Pivot Table Tools > Options > Options* button. On the *Layout & Format* tab you will see a box entitled *For empty cells show*. Type a 0 into the box and make sure the tick is enabled (Figure 7.3).

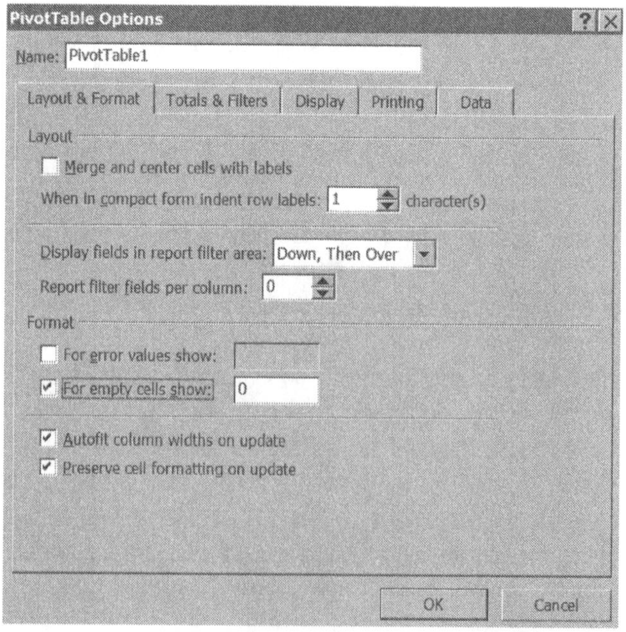

Figure 7.3 The Pivot Table Options menu box allows you to fill empty cells with zeroes.

13. Click the OK button to return to the Pivot Table and now the empty cells should be filled with zeroes. Make a new worksheet and copy the community Pivot Table data into it using *Paste Special* like you did in steps 7 and 8. You can rename the new worksheet *community*. There are no Grand Total items but the first row is not required – cell A1 will read *Count of abbr*. Simple delete the entire row.

14. You will save the data as CSV in a moment but if you wish to keep the Pivot Table and your modifications, now would be a good time to click the *Save* button.

15. Click on the *community* tab so that you have your presence-absence community data on screen. Save the file as a CSV using the *Office* button and *Save As > Other Formats*. Select *CSV (Comma delimited)* as the format and use the name *PQDbiol*. The *.csv* should be appended automatically. You will be asked if you are sure that you want this format – twice.

16. Now click on the *sites* tab to display the sample names and sites that you created earlier. Save this as a CSV and call it *PQDsites*. Notice that the tabs are renamed to reflect the file names you used.

You now have the community data saved as a CSV file with rows representing the various samples. The second file contains rows with the same names but containing the site names. Keeping the two separate is useful – R can easily read information from the second file and use it in analysis of the first. At this point your spreadsheet is still acting like a spreadsheet and you have simply made copies of data to disk (as CSV). If you want to save the file as an Excel file you can use *Save As* once more and select *Excel Workbook* as the format.

It is usually best to keep your biological data separate from any other when you are dealing with community analyses. Any environmental data can go in a separate file with row names identical to the biological data. Using R it is easy enough to link the biological and environmental data. In the following exercise you can have a go at comparing species richness using this method. Unlike the earlier exercise this method 'scales up' more easily.

Have a Go: Use a generic method for comparing species richness

In this exercise you will use the plant data you saw previously. You have plant species abundance for two sites – each site has five quadrats of data. In the previous exercise you created two CSV files – one for the community data and one for the site names. The data files are called *PQDbiol.csv* and *PQDsites.csv*. These will open in Excel if you want to look at them.

1. Open the R program. You need to get the data in the CSV files into R using the `read.csv()` command. The first column of each file contains the sample names. You want these to be treated as labels by R so you will need to add an extra instruction to the command. Start with the community data – assign a name to the data and read the file like so:

```
> pqd = read.csv(file.choose(), row.names = 1)
```

2. The `file.choose()` part will open a browser that allows you to select the file (*PQDbiol.csv*). If you use Linux then you have to replace this with the filename in quotes. Now repeat the process for the site names file – you can use the up arrow to recall the previous command, which you can edit so:

```
> pqd.site = read.csv(file.choose(), row.names = 1)
```

3. The file you want is *PQDsites.csv*. You should now have the data as two named objects, *pqd* and *pqd.site*. You can check this by listing all the objects like so:

```
> ls()
[1] "pqd"    "pqd.site"
```

4. Your listing might include other items if you have been using R previously. You can see the data by typing the name but they will not easily fit on the screen. You can remind yourself of the species names and the sample names by accessing the names of the objects like so:

```
> names(pqd)
 [1] "Ach.mil" "Agr.cap" "Alc.vul" "Ant.odo" "Bel.per" "Bot.lun" "Cam.rot"
 [8] "Car.car" "Car.fla" "Cer.fon" "Cir.vul" "Dac.glo" "Des.ces" "Fes.ovi"
[15] "Gal.sax" "Lol.per" "Lot.cor" "Luz.cam" "Nar.str" "Pla.lan" "Pla.maj"
[22] "Poa.ann" "Poa.tri" "Pot.ere" "Pru.vul" "Ran.rep" "Rhy.squ" "Rum.ace"
[29] "Tar.off" "Tri.rep" "Ver.ser" "Vio.riv"
> row.names(pqd)
[1] "lower1" "lower2" "lower3" "lower4" "lower5" "upper1" "upper2" "upper3"
[9] "upper4" "upper5"
```

5. The site names data will fit more easily since there is only a single column aside from the sample names:

```
> pqd.site
        site
lower1 lower
lower2 lower
lower3 lower
lower4 lower
lower5 lower
upper1 upper
upper2 upper
upper3 upper
upper4 upper
upper5 upper
```

6. The species richness is obtained from the sums of each row since you have presence-absence data. Make a named object to hold the result and use the `rowSums` command like so:

```
> pqd.sr = rowSums(pqd)
> pqd.sr
lower1 lower2 lower3 lower4 lower5 upper1 upper2 upper3 upper4 upper5
    19     21     18     22     23     12      9     10     14     10
```

7. You should represent the data graphically and a boxplot is easily created like so:

```
> boxplot(pqd.sr ~ pqd.site$site)
> title(xlab = "Site", ylab = "Species richness")
```

8. The plot will be very similar to the one you made previously (Figure 7.1). Notice that you used a formula this time – this is in the form `response ~ group`. You can use this for the statistical test too. This time use a *t*-test as the data are normally distributed. You can use the up arrow to recall the `boxplot()` command, which you can edit like so:

```
> t.test(pqd.sr ~ pqd.site$site)

   Welch Two Sample t-test

data:  pqd.sr by pqd.site$site
t = 7.451, df = 7.99, p-value = 7.307e-05
alternative hypothesis: true difference in means is not equal to 0
95 percent confidence interval:
  6.628246 12.571754
sample estimates:
mean in group lower mean in group upper
  20.6                11.0
```

9. You can see that there is a significantly greater species richness for the *lower* site compared to the *upper*.

You can use this method to carry out analyses using any grouping variable; here you looked at a simple site comparison.

As you wish to undertake more complex analyses it becomes more useful to have a dedicated set of commands to use, as you will see next.

Tip: Naming R objects

All R objects need a name – this can be a mixture of letters and numbers as long as the name starts with a letter (note that R is case sensitive). You can use a period or underscore as well and these can be useful to 'group' items together. If your data are called *plants*, for example, then use the period to add a 'name extension' for results of analyses carried out on the data, for example: *plants.sr*, *plants.t* for results of species richness calculations and a *t*-test respectively. This helps you to see what you have done with the data and allows you to save 'projects' more easily.

7.1.1 R packages for community analyses

So far you have used only commands that are built into the basic R program. However, there are many additional packages of commands available (see Section 4.1.7).

The vegan package

A particularly useful package is called *vegan* (*vegetation analysis*), which you will use extensively in community analyses. You can get the *vegan* package using the `install.packages()` command:

```
> install.packages("vegan")
```

This assumes that you are connected to the Internet. You will be asked for a local mirror site: choose one near to you (this minimises download times), and then the download and installation will proceed. You may notice that additional packages are downloaded since some require routines in these other packages to work correctly. You still need to 'activate' the package to access the commands within it – you need the library() command like so:

```
> library(vegan)
```

Now the commands in the *vegan* package are available for you to use. You may also see additional packages loaded since some require routines in other packages and these are loaded as required. See Section 4.1.7 for a reminder of how to manipulate packages.

You will use the *vegan* package often in your study of community data as most of the analytical methods you need are not contained in the base distribution of R. In the following exercise you can have a go at summarising species richness for groups of samples. This is simple using the *vegan* package and much more difficult without it.

Have a Go: Use the *vegan* R-package to summarise species richness for groups of sites

In this exercise you will use the plant quadrat data that you met earlier. The data are in two separate files: *PQDbiol.csv* contains the community data as presence-absence, *PQDsites.csv* contains the site names. Both files contain rows representing the samples, all with unique labels. There are two sites and each was sampled five times.

1. Open R and read in the two data files. If you already have these data you can skip to step 2. If you use Linux then you'll need to specify the filenames explicitly (in quotes), otherwise the file.choose() part will allow you to select the files from your hard drive:

```
> pqd.biol = read.csv(file.choose(), row.names = 1)
> pqd.site = read.csv(file.choose(), row.names = 1)
```

2. Now you have the data you will need to prepare the *vegan* package:

```
> library(vegan)
```

3. Get the species richness for all samples using the specnumber() command:

```
> specnumber(pqd.biol)
lower1 lower2 lower3 lower4 lower5 upper1 upper2 upper3 upper4 upper5
    19     21     18     22     23     12      9     10     14     10
```

4. The specnumber() command can also accept additional instructions, such as MARGIN. The default setting is MARGIN = 1, which gives the number of species per row; use the MARGIN = 2 instruction to get the frequency for all the species (that is species per column):

```
> specnumber(pqd.biol, MARGIN = 2)
Ach.mil Agr.cap Alc.vul Ant.odo Bel.per Bot.lun Cam.rot Car.car Car.fla Cer.fon
      7       4       1       5      10       1       1       5       5       6
Cir.vul Dac.glo Des.ces Fes.ovi Gal.sax Lol.per Lot.cor Luz.cam Nar.str Pla.lan
      5       3       4       9       4       4       2       4       4       4
Pla.maj Poa.ann Poa.tri Pot.ere Pru.vul Ran.rep Rhy.squ Rum.ace Tar.off Tri.rep
      9       5       5       5       7      10       5       4       3      10
```

```
    Ver.ser Vio.riv
        6      1
```

5. The `specnumber()` command can also use a grouping variable via the `groups` instruction; use this to get the species richness for the sites:

```
> specnumber(pqd.biol, groups = pqd.site$site)
lower upper
   28    15
```

6. You can also get the mean species richness by using the `aggregate()` command:

```
> aggregate(specnumber(pqd.biol), by = pqd.site$site, FUN = mean)
   site    x
1 lower 20.6
2 upper 11.0
```

7. The `tapply()` command gives a similar output:

```
> tapply(specnumber(pqd.biol), INDEX = pqd.site$site, FUN = mean)
lower upper
 20.6  11.0
```

The `aggregate()` and `tapply()` commands are not part of the *vegan* package and are useful commands for applying summary functions using grouping variables.

The *vegan* package provides a host of other useful routines but it is not the only package available, as you will see next.

The BiodiversityR package

The *BiodiversityR* package contains some useful routines and utilities for examination of community data. Some of the routines use commands from the *vegan* package but allow the use of grouping variables directly so you do not have to use the basic R commands like `aggregate()` or `tapply()` as you did in the preceding exercise.

To get the package you can use the `install.packages()` command (see Section 4.1.7):

```
> install.packages("BiodiversityR")
```

Once you have the package downloaded and installed you can make it ready using the `library()` command:

```
> library(BiodiversityR)
```

In the following exercise you can have a go at using the *BiodiversityR* package to look at species richness using grouping variables.

Have a Go: Use the *BiodiversityR* package to summarise species richness using grouping variables

In this exercise you will use the plant quadrat data that you met earlier. The data are in two separate files: *PQDbiol.csv* contains the community data as presence-absence, *PQDsites.csv* contains the site names. Both files contain rows representing the samples, all with unique labels. There are two sites and each was sampled five times.

1. Open R and read in the two data files. If you already have these data you can skip to step 2. If you use Linux then you'll need to specify the filenames explicitly (in quotes), otherwise the file.choose() part will allow you to select the files from your hard drive:

```
> pqd.biol = read.csv(file.choose(), row.names = 1)
> pqd.site = read.csv(file.choose(), row.names = 1)
```

2. Now you have the data you will need to prepare the *BiodiversityR* package:

```
> library(BiodiversityR)
```

3. First, use the diversityresult() command to view the species richness of all the individual samples:

```
> diversityresult(pqd.biol, index = "richness", method = "s")
        richness
lower1      19
lower2      21
lower3      18
lower4      22
lower5      23
upper1      12
upper2       9
upper3      10
upper4      14
upper5      10
```

4. Now get an overall species richness:

```
> diversityresult(pqd.biol, index = "richness", method = "all")
        richness
all          32
```

5. Use the diversitycomp() command to compare species richness between the two sites. The command calls routines from the *vegan* library to do this like so:

```
> diversitycomp(pqd.biol, y = pqd.site, factor1 = "site",
              index = "richness", method = "all")

site    n richness
  lower 5      28
  upper 5      15
```

6. Now look at mean species richness, calculated from the replicates for each site:

```
> diversitycomp(pqd.biol, y = pqd.site, factor1 = "site",
              index = "richness", method = "mean")

site    n richness
  lower  5     20.6
  upper  5     11.0
```

7. Calculate the standard deviation of the species richness:

```
> diversitycomp(pqd.biol, y = pqd.site, factor1 = "site",
              index = "richness", method = "sd")
```

```
site      n  richness
   lower 5 2.0736441
   upper 5 2.0000000
```

8. Now use the `diversityresult()` command to look at species richness for a single site:

```
> diversityresult(pqd.biol, y = pqd.site, factor = "site",
          index = "richness", level = "lower", method = "s")
         richness
lower1       19
lower2       21
lower3       18
lower4       22
lower5       23
```

The *BiodiversityR* package allows you to gain some insights without having to use the `aggregate()` or `tapply()` commands. However it can do more than this, as you will see later.

So far you have seen how species richness can be compared using simple grouping variables (such as site) using differences tests. You can use similar methods to prepare your data for analysis of species richness in correlation, as you will see in the next section.

7.2 Correlating species richness over time or against an environmental variable

Species richness can be used like any response variable as long as the sampling effort is the same for all observations. You saw previously how to use species richness as a variable in differences tests (comparing sites, Section 7.1). You can use a similar approach to look at correlations between species richness and other variables.

If you have a simple correlation you could create a Pivot Table in Excel and simply copy and paste the data – either into R or into a new spreadsheet, where you could carry out the correlation. If you have a more complicated situation then it is generally better to create two files – one for the community data and one for the environmental data – like you saw previously.

If you are dealing with time then you could consider the year (or sampling interval) as a variable to correlate to species richness for example. In the following exercise you can have a go at a simple correlation using a Pivot Table and copy/paste.

Have a Go: Correlate species richness and time

For this exercise you will need the butterfly and year data that you met previously; the data are in the file *Butterfly and year data.csv*. The first column contains species names in abbreviated form. The second column contains the year of the survey and the third column contains the abundance.

1. Open the file in your spreadsheet and make a Pivot Table (click on the data then use *Insert > Pivot Table*). You want *Yr* as the *Row Labels*, *Spp* as the *Column Labels* and *Spp*

also in the *Values* box (this will end up as *Count of Spp*). You will see the years as a column on the left and the Grand Total on the right will be the species richness for each year.

2. Use the mouse to highlight the year data and copy to the clipboard.

3. Open the R program. Use the `scan()` command to make an object to hold the data that you will get via the clipboard:

```
> year = scan()
1:
```

4. Once you press Enter, R will display 1: and wait for you to enter some data. Use paste to enter the data from the clipboard. Press Enter and finish the data entry process.

```
> year = scan()
1: 1996
2: 1997
3: 1998
4: 1999
5: 2000
6: 2001
7: 2002
8: 2003
9: 2004
10: 2005
11:
Read 10 items
```

5. Return To Excel and use the mouse to highlight the species richness data – this will be in the *Grand Total* column at the right of the table.

6. Use the `scan()` command again to enter the species richness data:

```
> bfr = scan()
1: 16
2: 15
3: 12
4: 14
5: 15
6: 12
7: 13
8: 14
9: 14
10: 15
11:
Read 10 items
```

7. You now have the two variables to correlate: *year* and *bfr*. You should visualise the relationship with a graph (Figure 7.4):

```
> plot(year, bfr)
```

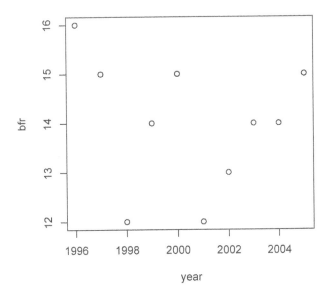

Figure 7.4 Simple scatter plot of species richness and time.

8. Now carry out a correlation test; the Spearman Rho (rank correlation) test is probably most appropriate here. Use the cor.test() command like so:

```
> cor.test(year, bfr, method = "spearman")

  Spearman's rank correlation rho

data:  year and bfr
S = 199.967, p-value = 0.5567
alternative hypothesis: true rho is not equal to 0
sample estimates:
       rho
-0.2119213
```

9. You may get a note about tied ranks – the correction is small and not important here as there is no evident link between species richness and year.

The graph is a little sparse but it is adequate to visualise the relationship (or lack of) between species richness and year.

When you have environmental data for various sites it is more efficient to prepare two data files – one for the community data and the other for the environmental – than to try to copy and paste. You saw this approach in the previous section (Section 7.1).

Tip: Matching biological and environmental row names

Make sure that the row names match between the biological and environmental data files. If your data were prepared with Pivot Tables they should be sorted in the same manner. You can always check this by applying a sort to the CSV files before you enter them into R.

In the following example you can have a go at a correlation between species richness and height of vegetation for some data on ground beetle communities.

Have a Go: Explore the link between species richness and an environmental factor

For this exercise you will need the data on ground beetles that you met earlier. The data are in the file *Ground beetles and habitat.xls*. They show the abundances of ground beetles at three habitat types (*Grass*, *Edge* and *Wood*). Each habitat was sampled six times and at each sampling location the height of the vegetation was recorded.

1. The spreadsheet contains two worksheets: one is the basic data and the other is a lookup table that contains the vegetation height and habitat name matched up to the samples (as well as other data). Start by going to the lookup worksheet and save this as a CSV file via the *Save As* button. Call the file *GBsite.csv*.

2. Now return to the main data and click once in the data and make a Pivot Table using *Insert > Pivot Table*. You want to drag the *Abbr* field to the *Column Labels* box, the *Sample* field to the *Row Labels* box and the *Abbr* field to the *Values* box. The field in the *Value* box should read *Count of Abbr*; if it doesn't then click it and use the *Value Field Settings* option to change it.

3. Turn off the Grand Total fields by clicking once in the Pivot Table and then using the *Pivot Table Tools > Design > Grand Totals* button.

4. You want 0s in place of empty cells so use *Pivot Table Tools > Options > Options* and enter a 0 in the *For empty cells show* box. Click OK to complete the operation.

5. From the *Office* button use *Save As* to save the Pivot Table worksheet as a CSV file. Call it *Gbbiol.csv*.

6. You have now made copies of the two worksheets to disk as CSV. They will need editing so now close the current spreadsheet – say No when asked if you want to save changes.

7. Open the *GBbiol.csv* file. Right-click in the margin of the top row and delete the entire row. Now save the file (it will default to CSV), which is now ready for use with R.

8. Open the *GBsite.csv* file. Delete the unwanted columns D, E and F. Now save the file (it will default to CSV), which is now ready for use with R.

9. Open R and use the `read.csv()` command to read in the community data file *GBbiol.csv*:

    ```
    > gb = read.csv(file.choose(), row.names = 1)
    ```

10. Open R and use the `read.csv()` command to read in the environmental data file *GBsite.csv*:

    ```
    > gb.site = read.csv(file.choose(), row.names = 1)
    ```

11. Calculate the species richness for the samples using the `rowSums()` command:

    ```
    > gb.sr = rowSums(gb)
    > gb.sr
    ```

```
E1 E2 E3 E4 E5 E6 G1 G2 G3 G4 G5 G6 W1 W2 W3 W4 W5 W6
17 14 15 25 21 17 28 22 18 28 26 24 12 11 11 12 12 10
```

12. Visualise the relationship between species richness and height using a scatter plot (Figure 7.5):

```
> plot(gb.sr ~ gb.site$Max.Ht, xlab = "Max Height",
ylab = "Species Richness")
```

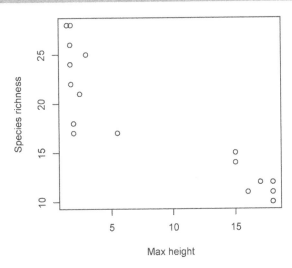

Figure 7.5 Species richness and vegetation height for ground beetles.

13. Now carry out a correlation test using the Spearman rho test via the cor.test() command:

```
> cor.test(~ gb.sr + gb.site$Max.Ht, method = "spearman")
    Spearman's rank correlation rho
data:   gb.sr and gb.site$Max.Ht
S = 1858.343, p-value = 8.083e-08
alternative hypothesis: true rho is not equal to 0
sample estimates:
       rho
-0.9177945
```

You can see that there is a strong correlation between ground beetle species richness and vegetation height in these data.

7.3 Species richness and sampling effort

So far in the analysis of species richness you have assumed that you've used an equal sampling effort for all your samples or sites. One example of this is the idea of species and area sampled: essentially, the bigger the area you sample the more species you will find.

You can overcome the difficulty of sampling area by standardising your sampling areas among sites. However, there are other potential difficulties. If one sample has a large number of individuals then the species richness tends to be greater than samples with few individuals. You can use the idea of *rarefaction* to overcome this and estimate species richness for samples with differing numbers of individuals.

It is generally unlikely that you will find all the species present in the habitat in your samples and you can estimate the expected species richness by looking at the number of new species found in subsequent samples.

The accumulation of new species in subsequent samples can be visualised in *species accumulation curves*, these curves can be compared for different communities and may give insights into patterns of diversity.

7.3.1 Species and area

Simply put, the larger the area you sample the more species you will find. If you plot the number of species against sample area you will usually end up with a graph similar in appearance to Figure 7.6.

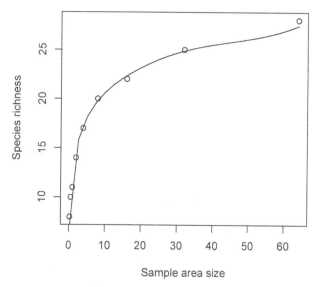

Figure 7.6 Species richness and area sampled. Richness increases rapidly with area but the rate of increase slows down.

The relationship looks like a logarithmic one and if you plot the log of species richness against the log of area you end up with a straight line (Figure 7.7).

This logarithmic relationship is potentially useful as you could use it to estimate species richness for areas other than the one you actually sampled. You'll see more of this in the sections on species accumulation curves (Section 7.3.3), and estimating total species richness (Section 7.3.4).

7.3.2 Rarefaction

If you have samples where there are large differences in abundances then the estimates of species richness will be skewed in favour of the samples with larger abundances. The idea

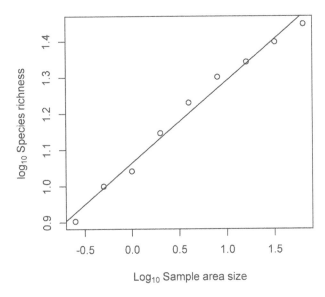

Figure 7.7 Species richness and area are related logarithmically.

of *rarefaction* is to standardise the sample sizes so that you are comparing species richness at equivalent abundance. Rarefaction is calculated by creating new samples of equal size by randomly selecting from your original data. The species richness is calculated for each new sample and then averaged. You end up with a new estimate for all your original samples that is based on a particular sample size. You select the appropriate sample size by looking at your original data. The method of rarefaction requires that you have abundance data in the form of counts, so is not really suitable for plant data. This also means that presence-absence data cannot be used.

In the following example you can have a go at rarefaction using data on counts of ground beetles that you've met previously.

Have a Go: Calculate rarefied species richness

For this exercise you will need the ground beetle data that you have met before. However, this time you will need the complete abundance data rather than simply presence-absence information. The data are in the file called *Ground beetles and abundance.csv*, a CSV file that is ready to import into R. The data were constructed using a Pivot Table. As an additional exercise you might want to take the original Excel file and make a Pivot Table yourself.

1. Start by opening R and using the `read.csv()` command to get the data from the *Beetles and abundance.csv* file:

   ```
   > gb.biol = read.csv(file.choose(), row.names = 1)
   ```

2. The columns are the species abundances and the rows are the sites; you can remind yourself of the species names and site names by using the following commands:

```
> names(gb.biol)
> row.names(gb.biol)
```

3. Prepare the *vegan* package using the `library()` command like so:

```
> library(vegan)
```

4. Use the `specnumber()` command to get the species richness:

```
> specnumber(gb.biol)
E1 E2 E3 E4 E5 E6 G1 G2 G3 G4 G5 G6 W1 W2 W3 W4 W5 W6
17 14 15 25 21 17 28 22 18 28 26 24 12 11 11 12 12 10
```

5. You need to use the `rarefy()` command to get rarefied species richness at equivalently sized samples:

```
> rarefy(gb.biol)
Error in rarefy(gb.biol):
  The size of 'sample' must be given --
Hint: Smallest site maximum 205
```

6. Note that you need to specify the appropriate sample size. The command gives you a hint but look at the total numbers of individuals per site by using the `rowSums()` command like so:

```
> rowSums(gb.biol)
 E1   E2   E3   E4   E5   E6   G1   G2   G3   G4   G5   G6    W1    W2
715  734  646  867  971  901  365  287  205  336  359  331  1092  912
 W3   W4   W5   W6
802  755  886  814
```

7. Simplify the display and get the range of values like so:

```
> range(rowSums(gb.biol))
[1]  205 1092
```

8. Now calculate the rarefied species richness for a sample size that equals the minimum abundance:

```
> rarefy(gb.biol, sample = 205)
        E1         E2         E3         E4         E5         E6
10.5506699 10.5925407 10.4655012 15.7981348 13.2306451 11.9053383
        G1         G2         G3         G4         G5         G6
21.8411432 19.9080559 18.0000000 23.8209662 21.6576426 21.1285529
        W1         W2         W3         W4         W5         W6
 7.9910002  8.7463204  8.8480663  9.1556382  8.6262030  7.6315870
attr(,"Subsample")
[1] 205
```

9. Use a value for sample size that is equal to (or larger than) the maximum to get the simple (non-rarefied) species richness:

```
> rarefy(gb.biol, sample = 1092)
E1 E2 E3 E4 E5 E6 G1 G2 G3 G4 G5 G6 W1 W2 W3 W4 W5 W6
17 14 15 25 21 17 28 22 18 28 26 24 12 11 11 12 12 10
attr(,"Subsample")
[1] 1092
```

Try using other values as the sample.

Rarefaction allows you to estimate species richness when your samples contain differing numbers of individuals. Thus, you can use rarefaction to calculate species richness before comparing richness using a grouping variable or in correlating richness to an environmental factor.

Standardising communities using rarefaction

You can use rarefaction on your community data to 'adjust' for differences in abundance of the various species. The `rrarefy()` command in the *vegan* package will carry out the necessary computations to produce a new community dataset that is based on random resampling of the original.

The command requires two instructions, the name of your community data and the sample size at which to calculate the new community:

```
rrarefy(x, sample)
```

Generally you will likely want to standardise using the same sample size for all of your sites but it is possible to specify a different sample size for each. In the following exercise you can have a go at creating a rarefied community using the ground beetle data that you have been using.

Have a Go: Standardise a community dataset using rarefaction

For this exercise you will need the ground beetle data that you have met before. The data are in the file called *Ground beetles and abundance.csv*, a CSV file that is ready to import into R. The file contains the abundance of ground beetles at 18 sites in three different habitat types. If you already have these data in R then you can go straight to step 2.

1. Start by opening R and using the `read.csv()` command to get the data from the *Beetles and habitat.csv* file:

    ```
    > gb.biol = read.csv(file.choose(), row.names = 1)
    ```

2. The columns are the species abundances and the rows are the sites; you can remind yourself of the species names and site names by using the following commands:

    ```
    > names(gb.biol)
    > row.names(gb.biol)
    ```

3. Prepare the *vegan* package using the `library()` command like so:

    ```
    > library(vegan)
    ```

4. Look at the number of individuals in each sample using the `rowSums()` command:

    ```
    > range(rowSums(gb.biol))
    [1]  205 1092
    ```

5. You can see that the lowest value is 205 so this will be the largest sample size that you can use in the `rrarefy()` command. Make a rarefied community using a sample size of 205 like so:

```
> rrarefy(gb.biol, sample = 205)
```

6. The sample size can be given as a value for each site – use `rowSums()` to determine the largest size for each site in the data:

```
> rowSums(gb.biol)
  E1   E2   E3   E4   E5   E6   G1   G2   G3   G4   G5   G6   W1   W2   W3
 715  734  646  867  971  901  365  287  205  336  359  331 1092  912  802
  W4   W5   W6
 755  886  814
```

7. Now use the `rowSums()` result and set this as the sampling size in the `rrarefy()` command like so:

```
> rrarefy(gb.biol, sample = rowSums(gb.biol))
```

8. Using the `rowSums()` result gives the same community as the original data! You can use a subset to visualise the effects of rarefaction. Try the following to get a subset of the community:

```
> gb.biol[1:3, c("Aba.par", "Neb.bre", "Pte.mad", "Pte.str")]
     Aba.par  Neb.bre  Pte.mad  Pte.str
E1      388       59      210       21
E2      325      125      218       14
E3      295       99      192        7
```

9. Now look at the same subset for a rarefied community using a sample size of 200:

```
> rrarefy(gb.biol, sample = 200)[1:3,
         c("Aba.par", "Neb.bre", "Pte.mad", "Pte.str")]
     Aba.par  Neb.bre  Pte.mad  Pte.str
E1      111       15       57        7
E2       84       31       65        4
E3       89       32       61        3
```

Try out other values for sample size (these will need to be 205 or less) and see the effects. Try the same value more than once – the randomisation will produce slightly different values.

Using rarefaction to predict likelihood of occurrence

You can use the idea of rarefaction to give an idea of the likelihood of a species occurring in a site. The `drarefy()` command in the *vegan* package does this job:

```
> drarefy(x, sample)
```

You simply need to specify the community data and the sample size much like you did for the `rrarefy()` command. Usually you will want the sample size to be consistent for all the sites in your dataset but you can specify a separate value for each site (the values must be integers). The following example shows the `drarefy()` command in use on the ground beetle data you've met before.

Have a Go: Use rarefaction to predict probability of species occurrence

For this exercise you will need the ground beetle data that you have met before. The data are in the file called *Ground beetles and abundance.csv*, a CSV file that is ready to import into R. The file contains the abundance of ground beetles at 18 sites in three different habitat types. If you already have these data in R then you can go straight to step 2.

1. Start by opening R and using the `read.csv()` command to get the data from the *Beetles and habitat.csv* file:

    ```
    > gb.biol = read.csv(file.choose(), row.names = 1)
    ```

2. The columns are the species abundances and the rows are the sites; you can remind yourself of the species names and site names by using the following commands:

    ```
    > names(gb.biol)
    > row.names(gb.biol)
    ```

3. Prepare the *vegan* package using the `library()` command like so:

    ```
    > library(vegan)
    ```

4. Look at the number of individuals in each sample using the `rowSums()` command:

    ```
    > range(rowSums(gb.biol))
    [1]  205 1092
    ```

5. You can see that the lowest value is 205 so this will be the largest sample size that you can use in the `drarefy()` command. Make a rarefied community using a sample size of 205 like so:

    ```
    > drarefy(gb.biol, sample = 205)
    ```

6. View a portion of the original data like so:

    ```
    > gb.biol[1:3, 37:40]
       Pla.ass  Poe.cup  Pte.mad  Pte.mel
    E1       2        2      210        4
    E2       4        3      218       11
    E3       2       13      192        4
    ```

7. Now view the probabilities of occurrence on this portion of the data like so:

    ```
    > drarefy(gb.biol, sample = 200)[1:3, 37:40]
          Pla.ass      Poe.cup   Pte.mad    Pte.mel
    E1  0.4814793    0.4814793         1  0.7317227
    E2  0.7207160    0.6155242         1  0.9706256
    E3  0.5236758    0.9923363         1  0.7737480
    ```

8. You can specify separate sample values for each site but they need to be integers. Try scaling each site to 50% of its original overall abundance like so:

    ```
    > drarefy(gb.biol, sample = floor(rowSums(gb.biol)*0.5))[1:3, 37:40]
          Pla.ass    Poe.cup  Pte.mad     Pte.mel
    E1  0.7496503  0.7496503        1   0.9376759
    E2  0.9380112  0.8755116        1   0.9995475
    E3  0.7503876  0.9998921        1   0.9380809
    ```

The `floor()` command rounds down values to an integer.

Tip: Rounding to make integer values

The `floor()` command makes a number into an integer by knocking off any decimal portion – effectively rounding down. The corresponding upward rounding is carried out by the `ceiling()` command.

7.3.3 Species accumulation curves

Species accumulation curves can be used to compare the diversity properties of communities. The simplest form of species accumulation curve is derived by plotting the species richness against the number of samples. However, most commonly you use a randomisation method where you repeatedly sub-sample your data and then find the mean species area curve. There are several methods for carrying out the randomisation process and the *vegan* package will allow you to calculate species area curves using a variety of these methods.

In the following example you can have a go at generating species area curves using the ground beetle data you met previously.

Have a Go: Generate species area curves

For this exercise you will need the ground beetle data with abundance information that you met earlier. The data are in a CSV file, ready to import to R, called *Ground beetles and abundance.csv*. If you already have the data in R you can go directly to step 2.

1. Start by opening R and using the `read.csv()` command to get the data from the *Beetles and habitat.csv* file:

   ```
   > gb.biol = read.csv(file.choose(), row.names = 1)
   ```

2. The columns are the species abundances and the rows are the sites; you can remind yourself of the species names and site names by using the following commands:

   ```
   > names(gb.biol)
   > row.names(gb.biol)
   ```

3. Prepare the *vegan* package using the `library()` command like so:

   ```
   > library(vegan)
   ```

4. Start by looking at the simplest form of species accumulation by taking sites in the order they are in the dataset. Use the `specaccum()` command with the `method = "collector"` instruction like so:

   ```
   > gb.sa = specaccum(gb.biol, method = "collector")
   > gb.sa
   Species Accumulation Curve
   Accumulation method: collector
   Call: specaccum(comm = gb.biol, method = "collector")

   Sites      1  2  3  4  5  6  7  8  9 10 11 12 13 14 15 16 17 18
   Richness  17 17 17 26 26 27 41 43 44 46 47 47 48 48 48 48 48 48
   ```

5. The result of the calculation contains several items, which you can access individually. Look at the items available by using the names() command:

```
> names(gb.sa)
[1] "call"  "method"  "sites" "richness" "sd" "perm"
```

6. Not all items in the result will necessarily contain information. Access the species richness explicitly like so:

```
> gb.sa$richness
E1 E2 E3 E4 E5 E6 G1 G2 G3 G4 G5 G6 W1 W2 W3 W4 W5 W6
17 17 17 26 26 27 41 43 44 46 47 47 48 48 48 48 48 48
```

7. Plot the species accumulation curve using the plot() command. Your result should resemble Figure 7.8 (part a):

```
> plot(gb.sa)
```

8. Now use the "exact" method and plot a new accumulation curve, your plot should resemble Figure 7.8 (part b):

```
> gb.sa = specaccum(gb.biol, method = "exact")
> plot(gb.sa)
```

9. Notice that the "exact" method produced error bars. You can see the standard deviation from the result object like so:

```
> gb.sa$sd
 [1] 6.1053003 6.5802637 6.5421229 6.2829013 5.9278892 5.5294688 5.1125998
 [8] 4.6930636 4.2750450 3.8632318 3.4573788 3.0580040 2.6616127 2.2629269
[15] 1.8439555 1.4034329 0.8314794 0.0000000
```

10. Use the "random" method to produce another species accumulation curve, your plot should resemble Figure 7.8 (part c):

```
> gb.sa = specaccum(gb.biol, method = "random")
> plot(gb.sa)
```

11. The "random" method has a separate summary() and boxplot() commands. Use the boxplot() command to produce a species accumulation curve that should resemble Figure 7.8 (part d):

```
> boxplot(gb.sa)
```

12. Use an individual-based accumulation model (rather than a site-based one) via the method = "rarefaction" instruction:

```
> gb.sa = specaccum(gb.biol, method = "rarefaction")
```

13. The plot() command can display the species accumulation curve by individuals rather than sites if you use the xvar = "individuals" instruction (the default is xvar = "sites"). Note that this only works for results from the specaccum() command. Make an individual-based plot; yours should resemble Figure 7.9:

```
> plot(gb.sa, xvar = "individuals")
```

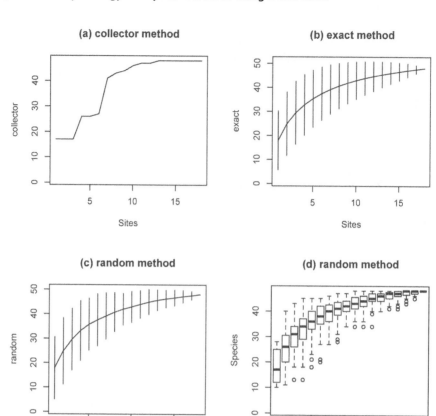

Figure 7.8 Species accumulation curves using various accumulator functions: a = "collector" method, b = "exact" method, c = "random" method, d = "random" method and boxplot output.

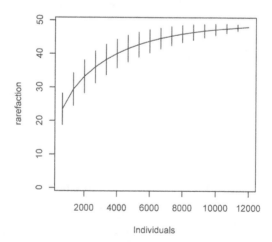

Figure 7.9 Species accumulation curve based on an individual accumulation model (rarefaction).

14. You can use other instructions with the `plot()` command. For example, the `ci` instruction can suppress the confidence intervals if you use `ci = 0` (the default is `ci = 2`). You can also produce a confidence 'envelope' or a filled polygon by using the `ci.type` instruction, `ci.type = "line"`, produces the former whilst `ci.type = "polygon"` produces the latter. Try the following commands to view the options: your figure should resemble Figure 7.10:

```
> par(mfrow = c(2, 2)) # sets the graph window into 2 rows and 2
                         columns
> plot(gb.sa, xvar = "individuals", main = "Individual-based accumulator")
> plot(gb.sa, main = "Site based, no CI", ci = 0)
> plot(gb.sa, ci.type = "line", main = "Confidence envelope")
> plot(gb.sa, ci.type = "polygon", main = "Confidence polygon",
        ci.col = "gray50")
> par(mfrow = c(1, 1)) # resets the graph window to a single
                         row/column
```

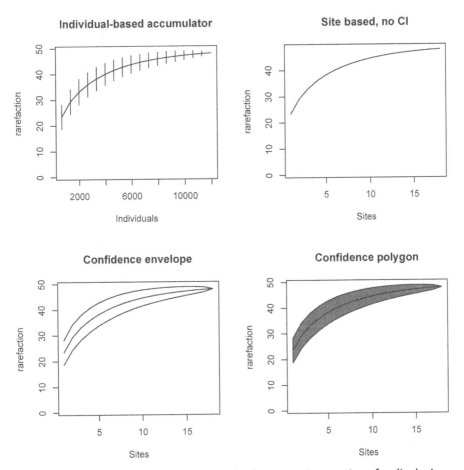

Figure 7.10 Species accumulation curves displaying various options for displaying confidence intervals.

For other options regarding the `plot()` command and species accumulation curves, take a look at the help entry for the `specaccum()` command in the *vegan* package.

Earlier, in Section 7.3.1 'Species and area' it was suggested that the relationship between species richness and area was logarithmic. In fact several models have been suggested for the species and area relationship; these are generally non-linear models such as the Michaelis–Menten model, which is also found in connection with enzyme kinetics. The *vegan* package allows you to fit a range of non-linear models to species accumulation curves using the `fitspecaccum()` command.

In the following exercise you can have a go at fitting some non-linear models to a species accumulation curve. You'll use the ground beetle data once more.

Have a Go: Fit non-linear models to species accumulation curves

For this exercise you will need the ground beetle data with abundance information that you met earlier. The data are in a CSV file, ready to import to R, called *Ground beetles and abundance.csv*. If you already have the data in R you can go directly to step 2.

1. Start by opening R and using the `read.csv()` command to get the data from the *Beetles and habitat.csv* file:

    ```
    > gb.biol = read.csv(file.choose(), row.names = 1)
    ```

2. The columns are the species abundances and the rows are the sites; you can remind yourself of the species names and site names by using the following commands:

    ```
    > names(gb.biol)
    > row.names(gb.biol)
    ```

3. Prepare the *vegan* package using the `library()` command like so:

    ```
    > library(vegan)
    ```

4. Start by making a species accumulation model:

    ```
    > gb.sa = specaccum(gb.biol, method = "random")
    ```

5. Now use the species accumulation result as the basis for fitting a species-area model:

    ```
    > gb.nlm = fitspecaccum(gb.sa, model = "arrhenius")
    ```

6. Use the `names()` command to see the various components of the result:

    ```
    > names(gb.nlm)
     [1] "call"    "method" "sites"  "richness"  "sd"
     [6] "perm"    "fitted" "residuals" "coefficients" "models"
    ```

7. Use the `plot()` command to view the model. Your graph should resemble Figure 7.11:

    ```
    > plot(gb.nlm)
    ```

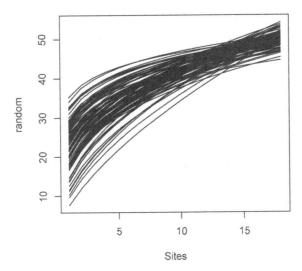

Figure 7.11 Modelling a species–area relationship using an Arrhenius non-linear model.

8. Redraw the plot but this time make the lines paler – use the up-arrow to recall the previous command and edit like so:

   ```
   > plot(gb.nlm, col = "gray90")
   ```

9. Now overlay a boxplot of the species accumulation to make a graph similar to Figure 7.12. You will need the add = TRUE instruction like so:

   ```
   > boxplot(gb.sa, add = TRUE, pch = "+", col = "gray80")
   ```

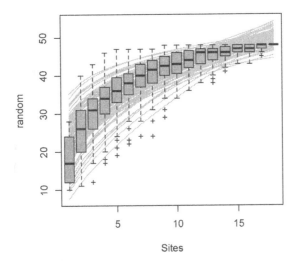

Figure 7.12 Species accumulation curve and an Arrhenius model for ground beetle community.

The `pch` instruction sets the character for the outliers in a `boxplot()` command (points that lie greater than 1.5 times the IQR from the median). The `col` command alters plot colours. Using the `add = TRUE` command ensures that the boxplot is placed on the existing graph.

You can use a variety of models. Options are: `"arrhenius"`, `"gleason"`, `"gitay"`, `"lomolino"`, `"asymp"`, `"gompertz"`, `"michaelis-menten"`, `"logis"` and `"weibull"`. Not all of these models will 'succeed'. Experiment with different models and look at the differences in the plots.

The `fitspecaccum()` command can accept community data as its starting point, in which case the species accumulation curve is produced using the `"random"` method (by default, you can alter the `method` instruction) and the non-linear model fitted to that.

The commands in the *vegan* package allow you to investigate species accumulation but it is not always particularly easy to obtain subsets of the data or to compare these subsets. The *BiodiversityR* package accesses many of the commands in *vegan* but adds routines that allow you to subset and compare more easily. In the following example you can have a go at subsetting and comparing species accumulation curves using the ground beetle data that you've seen already.

Have a Go: Compare species area curves for subsets of community data

For this exercise you will need the ground beetle data with abundance information that you met earlier. The data are in a CSV file, ready to import to R, called *Ground beetles and abundance.csv*. You will also need the data that contains the site information, *GBsite.csv*. If you already have the data in R you can go directly to step 2.

1. Start by opening R and using the `read.csv()` command to get the data from the *Beetles and habitat.csv* and *Gbsite.csv* files:

    ```
    > gb.biol = read.csv(file.choose(), row.names = 1)
    > gb.site = read.csv(file.choose(), row.names = 1)
    ```

2. For the abundance data the columns are the species abundances and the rows are the sites; you can remind yourself of the species names and site names by using the following commands:

    ```
    > names(gb.biol)
    > row.names(gb.biol)
    ```

3. The *gb.site* data contain rows corresponding to the sites but the columns contain information about the maximum height of the vegetation and the habitat type. Simply type the name of the object to view the data:

    ```
    > gb.site
    ```

4. Prepare the *BiodiversityR* package using the `library()` command like so:

    ```
    > library(BiodiversityR)
    ```

5. Use the `specaccum()` command in the *vegan* package to get a species accumulation curve for the *Grass* habitat like so:

```
> specaccum(gb.biol[gb.site$Habitat == "Grass",])
Species Accumulation Curve
Accumulation method: exact
Call: specaccum(comm = gb.biol[gb.site$Habitat == "Grass", ])

Sites       1.000000   2.000000   3.000000   4.000000   5.000000   6
Richness 24.333333  29.466667  32.350000  34.200000  35.666667  37
sd          3.543382   2.927873   2.261418   1.645013   1.105542   0
```

6. Now try the same thing but using the `accumresult()` command from the *BiodiversityR* package. You can use some of the same instructions as for the `specaccum()` command, to set the `method` for example; here you will accept the default (`method = "exact"`):

```
> accumresult(gb.biol, y = gb.site, factor = "Habitat", level = "Grass")
Species Accumulation Curve
Accumulation method: exact
Call: specaccum(comm = x, method = method, permutations = permutations,
conditioned = conditioned, gamma = gamma)

Sites       1.000000  2.000000  3.000000  4.000000  5.000000  6
Richness 24.333333 29.466667 32.350000 34.200000 35.666667 37
sd          3.543382  2.927873  2.261418  1.645013  1.105542  0
```

7. The `accumcomp()` command creates several species accumulation curves, each corresponding to levels of a factor you specify. Try comparing habitats like so:

```
> accumcomp(gb.biol, y = gb.site, factor = "Habitat", plotit = F)

, ,  = Sites

           obs
Habitat 1 2 3 4 5 6
  Edge  1 2 3 4 5 6
  Grass 1 2 3 4 5 6
  Wood  1 2 3 4 5 6

, ,  = Richness

             obs
Habitat           1         2       3         4          5  6
  Edge  18.16667 21.26667 23.30 24.80000 26.00000 27
  Grass 24.33333 29.46667 32.35 34.20000 35.66667 37
  Wood  11.33333 12.20000 12.60 12.86667 13.00000 13

, ,  = sd

           obs
Habitat          1          2          3          4         5 6
  Edge  3.760171 3.7601496 3.214135 2.5688721 1.825742 0
  Grass 3.543382 2.9278728 2.261418 1.6450135 1.105542 0
  Wood  0.745356 0.4898979 0.400000 0.2494438 0.000000 0
```

8. Use the `accumcomp()` command to plot the result and compare the species area curves graphically. Your graph should resemble Figure 7.13:

```
> accumcomp(gb.biol, y = gb.site, factor = "Habitat", xlim = c(0,7),
         plotit = T, rainbow = F, legend = F)
```

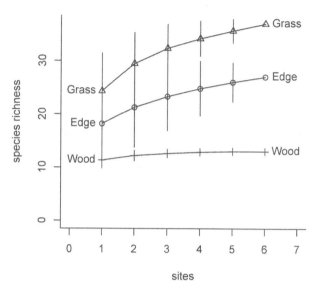

Figure 7.13 Species accumulation curves for ground beetle communities in three adjacent habitats.

9. Incorporate the total abundance for each site to the *gb.site* data using the rowSums() command like so:

```
> gb.site$tot.abund = rowSums(gb.biol)
```

10. Now use the scale instruction to scale the *x*-axis of the species abundance curve by number of individuals rather than by site:

```
> accumcomp(gb.biol, y = gb.site, factor = "Habitat",
            scale = "tot.abund",
    plotit = T, rainbow = F, legend = F)

,  , = Sites

  obs
Habitat        1          2       3         4         5      6
   Edge  805.6667 1611.3333 2417.0 3222.667 4028.333 4834
  Grass  313.8333  627.6667  941.5 1255.333 1569.167 1883
   Wood  876.8333 1753.6667 2630.5 3507.333 4384.167 5261
...
```

You may need to alter the limits of the *x*-axis to get the plot (not shown) to fit nicely. There are various additional instructions that you can give to the accumcomp() command. Some are generally used in graphical commands and others are specific to this one command. For example you used xlim to alter the *x*-axis – this is a general instruction. The rainbow = F instruction, however, is specific to accumcomp() and turns off the colours. Try setting rainbow = T and see the difference for yourself. If you set legend = T you can add a legend – you will need to click with the mouse on the graph to set the position of the top left of the legend box.

7.3.4 Estimating total species richness

Whenever you sample from a community you are likely to miss some species, generally the rarer ones. There are various methods used to estimate the total number of species in samples. These methods can be divided into two main groups:

- Incidence-based estimates – these methods use the frequencies of species in collections of sites.
- Abundance-based estimates – these methods use counts of species in a single site.

Incidence-based estimates can be calculated by the `specpool()` command in the *vegan* package. The various estimates are shown in Table 7.1.

Table 7.1 Incidence-based species richness estimators used by the `specpool()` command. S_p = estimated species pool, S_o = observed # species, a_1 and a_2 = observed # species occurring in one or two sites only, p_i = frequency of species i, N = number of sites.

Estimate	Formula
Chao	$S_p = S_o + a_1{}^2/(2\,a_2)$
Jackknife (1st order)	$S_p = S_o + a_1\,(N-1)/N$
Jackknife (2nd order)	$S_p = S_o + a_1\,(2N-3)/N - a_2\,(N-2)^2/N\,(N-1)$
Bootstrap	$S_p = S_o + \Sigma(1-p_i)\,N$

Abundance-based estimates can be calculated using the `estimateR()` command in the *vegan* package. The various estimates are shown in Table 7.2.

Table 7.2 Abundance-based species richness estimators used by the `estimateR()` command. S_p = estimated species pool, S_o = observed # species, a_1 and a_2 = abundance of species occurring in one or two sites, S_{rare} = number of rare species (threshold = 10), S_{abund} = number of abundant species, N_{rare} = # individuals in rare species.

Estimate	Formula
Chao	$S_p = S_o + a_1\,(a_1 - 1)/[2(a_2 + 1)]$
ACE	$S_p = S_{abund} + S_{rare}/C_{ACE} + a_1/C_{ACE} \times \gamma^2$
	where $C_{ACE} = 1 - a_1/N_{rare}$
	$\gamma^2 = \max\{S_{rare}/C_{ACE}\ [\Sigma_{i=1..10}\ i\,(i-1) \times a_i]/N_{rare}\ /(N_{rare} - 1) -1, 0\}$

All these estimators except the second order jackknife also have formulae for determining standard error.

Incidence-based estimators of species richness

The incidence-based estimators are useful in that they allow you to estimate species richness when you only have presence-absence data (that is, site lists), although they will of course work when you also have abundance data. In the following exercise you can have a go at using some commands in the *vegan* package to calculate estimated species richness using incidence-based estimators.

Have a Go: Use incidence-based estimators of species richness

For this exercise you will need the ground beetle data with abundance information that you met earlier. The data are in a CSV file, ready to import to R, called *Ground beetles and abundance.csv*. You will also need the data that contains the site information, *GBsite.csv*. If you already have the data in R you can go directly to step 2.

1. Start by opening R and using the `read.csv()` command to get the data from the *Beetles and habitat.csv* and *Gbsite.csv* files:

```
> gb.biol = read.csv(file.choose(), row.names = 1)
> gb.site = read.csv(file.choose(), row.names = 1)
```

2. For the abundance data the columns are the species abundances and the rows are the sites; you can remind yourself of the species names and site names by using the following commands:

```
> names(gb.biol)
> row.names(gb.biol)
```

3. The *gb.site* data contain rows corresponding to the sites but the columns contain information about the maximum height of the vegetation and the habitat type. Simply type the name of the object to view the data:

```
> gb.site
```

4. Prepare the *vegan* package using the `library()` command like so:

```
> library(vegan)
```

5. Use the `specpool()` command to estimate total species richness for all the sites combined:

```
> specpool(gb.biol)
     Species    chao chao.se   jack1 jack1.se    jack2     boot boot.se  n
All       48 53.33333 4.929127 55.55556 3.832931 57.64706 51.81422 3.021546 18
```

6. The pool instruction can be used to create separate pools of species – a grouping variable. Use this to obtain estimates for species richness for the three habitat types:

```
> specpool(gb.biol, pool = gb.site$Habitat)
       Species chao  chao.se   jack1 jack1.se    jack2     boot  boot.se n
Edge        27   33 6.480741 32.00000 4.564355 34.40000 29.30678 2.7193315 6
Grass       37   53 16.492423 43.66667 3.726780 47.93333 39.94114 2.1077498 6
Wood        13   13 0.000000 13.00000           NA 11.93333 13.17563 0.3842571 6
```

7. The result of the `specpool()` command can be subsetted, allowing you to obtain a single estimate. Start by creating an object to hold the result then show the estimates for the *Grass* habitat:

```
> gb.sp["Grass",]
      Species chao  chao.se   jack1 jack1.se    jack2     boot  boot.se n
Grass      37   53 16.492423 43.66667 3.726780 47.93333 39.94114 2.1077498 6
```

8. Now use the square brackets to subset the result to show the `"chao"` estimate and its standard error:

```
> gb.sp[,c("chao", "chao.se")]
      chao    chao.se
Edge    33  6.4807407
Grass   53 16.4924225
Wood    13  0.0000000
```

9. Use the `specpool2vect()` command to make a vector result that shows the estimated number of species for each site. The command makes a result that contains the same number of elements as there are sites:

```
> specpool2vect(gb.sp, index = "chao")
 [1] 33 33 33 33 33 33 53 53 53 53 53 53 13 13 13 13 13 13
```

10. The `poolaccum()` command generates species accumulation curves for estimates of the total species richness based on cumulative sampling – use it like so:

```
> (gb.sp = poolaccum(gb.biol))
    N     S       Chao Jackknife 1 Jackknife 2 Bootstrap
 3 28.54 44.978501   37.120000   40.195000 32.623333
 4 32.03 53.032813   41.645000   45.611667 36.561914
 5 34.78 51.287744   45.116000   49.704500 39.611629
 6 36.95 53.949481   47.341667   51.799000 41.841447
 7 38.83 53.421486   49.030000   53.018095 43.699444
 8 40.81 54.778723   50.951250   54.629464 45.702782
 9 42.07 52.502363   52.007778   55.408611 46.906038
10 43.12 51.157712   52.534000   55.255778 47.796421
11 44.32 51.835305   53.465455   56.070545 48.893167
12 45.04 52.556143   53.867500   56.400076 49.470025
13 45.72 53.122142   54.147692   56.466859 49.985233
14 46.25 53.954262   54.570000   57.043846 50.440354
15 46.57 54.576942   54.708667   57.308667 50.650076
16 47.01 54.303145   54.931875   57.319458 50.998691
17 47.71 53.760595   55.446471   57.670000 51.616699
18 48.00 53.333333   55.555556   57.647059 51.814221
```

11. The `poolaccum()` command produces a complicated result but there is a dedicated `summary()` command – use this to access just the "`jack2`" results:

```
> summary(gb.sp)$jack2
        N Jackknife 2       2.5%       97.5%     Std.Dev
 [1,]   3   40.195000 17.762500 61.133333 11.6567792
 [2,]   4   45.611667 22.168750 65.427083 12.4265475
 [3,]   5   49.704500 26.211250 67.933750  9.7544476
 [4,]   6   51.799000 29.556667 68.865833  9.2481981
 [5,]   7   53.018095 36.922619 67.446429  8.1038880
 [6,]   8   54.629464 42.153125 68.821875  7.3986802
 [7,]   9   55.408611 42.486458 67.642708  6.2979471
 [8,]  10   55.255778 43.056111 64.731944  5.4609041
 [9,]  11   56.070545 44.913409 64.550227  4.8664600
[10,]  12   56.400076 44.920644 66.749621  4.8674851
[11,]  13   56.466859 49.959135 66.053686  3.9405637
[12,]  14   57.043846 51.434341 65.373352  3.4872903
[13,]  15   57.308667 52.352381 65.380952  3.0488949
[14,]  16   57.319458 51.003750 62.041667  2.7493656
```

```
[15,] 17   57.670000 54.447518 62.099265  1.7875324
[16,] 18   57.647059 57.647059 57.647059  0.0000000
```

12. The `summary()` command for `poolaccum()` results is also able to display one (or more) results using the display instruction, use this to view the `"boot"` result:

```
> summary(gb.sp, display = "boot")
$boot
         N Bootstrap      2.5%      97.5%      Std.Dev
[1,]  3 32.623333 16.004630 46.366667 8.31130239
...
```

13. Use the `plot()` command to shows the results graphically. The command uses the *lattice* package to present the results, which should appear similar to Figure 7.14:

```
> plot(gb.sp)
```

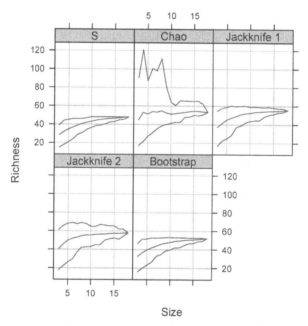

Figure 7.14 Species richness and 95% confidence intervals for ground beetle communities using four incidence-based estimators.

The `plot()` routine for results of the `poolaccum()` command uses the *lattice* package to present the curves in separate windows. You can alter the appearance using various additional instructions:

- `display` – choose to display one or more of the results, the options are: `"s"`, `"chao"`, `"jack1"`, `"jack2"`, `"boot"`. If you want more than one use the `c()` command, e.g. `display = c("jack1", "jack2")`.
- `type` – choose what is plotted, the default is `type = c("l", "g")`, which shows the background grid and shows the plotted values as a line. Other options include `"p"` for points and `"b"` for both lines and points.

- pch – the plotting character as a number or a symbol (in quotes).
- col – the colour of the line(s), usually as two values, the first for the main line and the second for the confidence envelope. Set col = "black" to make both black.
- xlab, ylab – text labels for the *x* and *y* axes.

Altering the background colour for the strips above each min-window is slightly more complicated, the following plot() command will alter the colour to "gray90" for example:

```
> plot(gb.sp, strip = function(..., bg)
            strip.default(..., bg = "gray90"))
```

Try out a few options and experiment.

Abundance-based estimators of species richness

When you have abundance data (as opposed to presence-absence) you can use a different approach and estimate the total species richness for each site.

Have a Go: Use abundance-based estimators of species richness

For this exercise you will need the ground beetle data with abundance information that you met earlier. The data are in a CSV file, ready to import to R, called *Ground beetles and abundance.csv*. You will also need the data that contains the site information, *GBsite.csv*. If you already have the data in R you can go directly to step 2.

1. Start by opening R and using the read.csv() command to get the data from the *Beetles and habitat.csv* and *Gbsite.csv* files:

   ```
   > gb.biol = read.csv(file.choose(), row.names = 1)
   > gb.site = read.csv(file.choose(), row.names = 1)
   ```

2. For the abundance data the columns are the species abundances and the rows are the sites; you can remind yourself of the species names and site names by using the following commands:

   ```
   > names(gb.biol)
   > row.names(gb.biol)
   ```

3. The *gb.site* data contain rows corresponding to the sites but the columns contain information about the maximum height of the vegetation and the habitat type. Simply type the name of the object to view the data:

   ```
   > gb.site
   ```

4. Prepare the *vegan* package using the library() command like so:

   ```
   > library(vegan)
   ```

5. Use the estimateR() command to get species richness using two abundance-based estimators:

```
> gb.sp = estimateR(gb.biol)
> gb.sp
              E1          E2          E3          E4          E5          E6
S.obs      17.000000  14.0000000  15.000000  25.000000  21.000000  17.000000
S.chao1    19.500000  14.0000000  20.000000  26.428571  26.000000  18.000000
se.chao1    4.882888   0.5357584  17.139137   2.505780  10.173495   3.395769
S.ACE      20.157895  14.5358796  21.650546  28.382593  26.007191  18.923316
se.ACE      2.122604   1.8637332   2.300889   2.386429   2.503418   2.093945
              G1          G2          G3          G4          G5          G6
S.obs      28.000000  22.000000  18.000000  28.000000  26.000000  24.000000
S.chao1    44.500000  25.000000  25.500000  35.200000  33.000000  27.750000
se.chao1   20.199010   4.802343  23.622024   9.017561  10.270418   6.480741
S.ACE      42.769264  28.040429  25.233017  36.903809  33.524960  28.855866
se.ACE      3.184495   2.660935   2.697379   2.986485   2.796557   2.564232
              W1          W2          W3          W4          W5          W6
S.obs      12.000000  11.000000  11.000000  12.000000  12.000000  10.000000
S.chao1    13.500000  12.000000  11.500000  12.000000  13.500000  10.500000
se.chao1    7.193747        NaN   3.741657   0.728869   7.193747   3.741657
S.ACE      15.936298  13.355556  14.461818  12.428571  15.600000  12.222222
se.ACE      2.195112   1.498649   1.743736   1.682506   1.735174   1.425282
```

6. The result of the `estimateR()` command is a matrix – use the square brackets syntax to view elements of the result. Start by viewing the ACE estimates for the *Edge* samples:

```
> gb.sp["S.ACE", 1:6]
       E1        E2        E3        E4        E5        E6
20.15789  14.53588  21.65055  28.38259  26.00719  18.92332
```

7. Now use the `estaccumR()` command to obtain a species accumulation curve for the entire ground beetle dataset:

```
> gb.sp = estaccumR(gb.biol)
> gb.sp
    N     S     Chao      ACE
 1 18.02 21.21400 22.50692
 2 24.58 30.09254 31.18091
 3 29.37 35.42203 36.63966
 4 32.83 38.85835 40.55689
 5 35.63 41.13349 42.98770
 6 37.97 42.83917 44.95058
 7 39.41 43.82074 45.94125
 8 40.64 44.46734 46.67243
 9 42.13 45.61553 47.64325
10 43.14 46.02172 48.06785
11 44.06 46.39754 48.23952
12 44.84 46.87283 48.32690
13 45.53 47.30027 48.47995
14 46.29 47.74560 48.75451
15 46.71 48.07083 48.86870
16 47.16 48.35298 49.02973
17 47.60 48.69179 49.28060
18 48.00 49.00000 49.53934
```

8. The result of the `estaccumR()` command is a complicated object (a list). Look at the elements using the `names()` command:

```
> names(gb.sp)
[1] "S"     "chao" "ace"   "N"     "means"
```

9. Each element (except `"means"`) is a matrix containing columns representing the permutations (the default is 100) and rows corresponding to the accumulated sites. Get the mean values for the ACE estimator like so:

```
> rowMeans(gb.sp$ace)
 [1] 22.50692 31.18091 36.63966 40.55689 42.98770 44.95058 45.94125 46.67243
 [9] 47.64325 48.06785 48.23952 48.32690 48.47995 48.75451 48.86870 49.02973
[17] 49.28060 49.53934
```

10. Try a different way to get the ACE mean values:

```
> gb.sp$means[,4]
```

11. Use the dedicated `summary()` command to view the estimates along with 95% confidence intervals and standard deviation:

```
> summary(gb.sp)
```

12. Since the regular output from `summary()` is long, try focusing on a single model:

```
> summary(gb.sp)$chao
       N     Chao      2.5%     97.5%    Std.Dev
 [1,]  1 21.21400 10.50000 44.50000  9.2007966
 [2,]  2 30.09254 12.00000 53.06562 11.0404550
 [3,]  3 35.42203 17.63333 53.62500 10.0163159
 [4,]  4 38.85835 20.71250 58.68000  9.8546444
 [5,]  5 41.13349 22.25000 57.07750  8.4833904
 [6,]  6 42.83917 27.06786 56.50000  7.2662471
 [7,]  7 43.82074 29.02250 53.17500  6.1834665
 [8,]  8 44.46734 30.50000 51.00000  4.9692635
 [9,]  9 45.61553 38.45938 50.76250  3.4237123
[10,] 10 46.02172 40.42857 50.12778  2.7746110
[11,] 11 46.39754 39.90875 50.46607  2.7100192
[12,] 12 46.87283 41.44792 50.00000  2.1938013
[13,] 13 47.30027 43.73750 49.84167  1.5905613
[14,] 14 47.74560 44.41562 49.52500  1.3355357
[15,] 15 48.07083 45.42857 49.00000  1.0189719
[16,] 16 48.35298 45.86875 49.00000  0.8234388
[17,] 17 48.69179 47.50000 49.00000  0.4968028
[18,] 18 49.00000 49.00000 49.00000  0.0000000
```

13. The `summary()` command allows you to focus on one (or more) models using the display instruction – get the same result as step 12 using this:

```
> summary(gb.sp, display = "chao")
```

14. The `plot()` command uses the *lattice* graphics package – use it to compare the Chao and ACE estimates, your graph should resemble Figure 7.15:

```
> plot(gb.sp, col = "black",
    strip = function(..., bg) strip.default(..., bg = "gray90"),
    display = c("chao", "ace"))
```

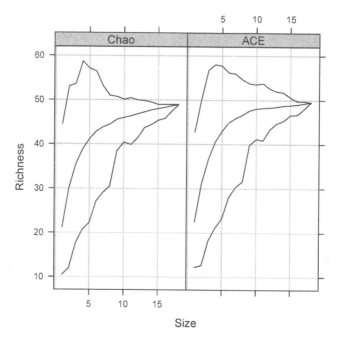

Figure 7.15 Species richness accumulation curves and 95% confidence intervals for ground beetle communities using two abundance-based estimators.

There are various options that you can alter for the `plot()` command – see the previous exercise on incidence-based models for details.

Comparing subsets of your data for estimated species richness

The `specpool()` and `estimateR()` commands operate on individual samples so it is relatively easy to obtain subsets of the data. The `poolaccum()` and `estaccumR()` commands, however, use all the sites to create an accumulative result. It would be useful to be able to extract subsets of the data so that you can focus on certain groups of samples. The *BiodiversityR* package provides some routines to permit subsetting but it doesn't permit a full investigation of the results for the species accumulation commands. This means that you need to be a bit more creative and find alternative ways to subset your data.

The ground beetle data that you've been investigating, for example, contain 18 samples from three different habitat types. It might be sensible to look at the species accumulation for each of the three habitats rather than lumping them altogether. In the following exercise you can have a go at splitting the data into subsets, which you can investigate separately.

Have a Go: Compare estimates of species richness in subsets of your data

For this exercise you will need the ground beetle data with abundance information that you met earlier. The data are in a CSV file, ready to import to R, called *Ground beetles and abundance.csv*. You will also need the data that contains the site information, *GBsite.csv*. If you already have the data in R you can go directly to step 2.

1. Start by opening R and using the `read.csv()` command to get the data from the *Beetles and habitat.csv* and *Gbsite.csv* files:

```
> gb.biol = read.csv(file.choose(), row.names = 1)
> gb.site = read.csv(file.choose(), row.names = 1)
```

2. For the abundance data the columns are the species abundances and the rows are the sites; you can remind yourself of the species names and site names by using the following commands:

```
> names(gb.biol)
> row.names(gb.biol)
```

3. The *gb.site* data contain rows corresponding to the sites but the columns contain information about the maximum height of the vegetation and the habitat type. Simply type the name of the object to view the data:

```
> gb.site
```

4. Prepare the *vegan* package using the `library()` command like so:

```
> library(vegan)
```

5. Start by making a simple index and adding it to the *gb.site* data:

```
> gb.site$index = 1:length(gb.site$Habitat)
```

6. Now you have a simple index number that corresponds to the site names and habitats. Make a new object that contains the index values in columns corresponding to the habitat types. You will need to use the `unstack()` command and force the result to become a list (using as.list) like so:

```
> gb.index = as.list(unstack(gb.site, form = index ~ Habitat))
> gb.index
$Edge
[1] 1 2 3 4 5 6

$Grass
[1] 7 8 9 10 11 12

$Wood
[1] 13 14 15 16 17 18
```

7. You are going to use the index values to pick out rows of the original data to use as subsets in the accumulation routines. Start by making a simple function:

```
> pacc = function(x, data,...) {poolaccum(data[x,])}
```

8. The *pacc* function you made will allow you to use the underlying `poolaccum()` command as an instruction in the `lapply()` command. This takes values from a list and applies a function to each item. Use this to get species accumulation results like so:

```
> gb.sp = lapply(gb.index, FUN = pacc, data = gb.biol)
> gb.sp
$Edge
  N       S       Chao  Jackknife 1  Jackknife 2   Bootstrap
  3   23.78  36.107333    28.033333    29.600000   25.794815
  4   25.19  35.383667    29.945000    32.058333   27.402773
  5   26.32  35.090000    31.344000    33.762000   28.635558
  6   27.00  33.000000    32.000000    34.400000   29.306777
```

```
$Grass
  N       S        Chao    Jackknife 1   Jackknife 2    Bootstrap
  3    32.66    39.817936    38.753333     40.720000    35.608148
  4    34.41    39.779075    40.177500     41.965833    37.247344
  5    35.94    43.575333    41.972000     44.763500    38.778349
  6    37.00    53.000000    43.666667     47.933333    39.941144

$Wood
  N       S        Chao    Jackknife 1   Jackknife 2    Bootstrap
  3    12.58    13.2375      13.333333     13.443333    12.974074
  4    12.90    13.2000      13.725000     14.008333    13.303750
  5    13.00    13.3450      13.552000     13.376500    13.328518
  6    13.00    13.0000      13.000000     11.933333    13.175626
```

The result of your calculation is a list that contains one element for each of the habitats. Each of these elements is a separate `poolaccum()` result.

Try creating a complimentary function for `estaccumR()` by using the following:

```
> eacc = function(x, data,...) {estaccumR(data[x,])}
```

You can explore the results in the same manner as you did earlier but to get an individual habitat result you will need to append the name of the habitat to the result using the dollar sign, e.g. gb.sp$Edge.

Tip: Combine rows using a grouping variable

The `rowSum()` command will use a grouping variable to add together the rows of a dataset. For example: rowSum(gb.biol, group = gb.site$Habitat) will make new data item by combining the data for each of the three habitats in the *gb* data. Note however that each habitat contains amalgamated data (the sum) so you lose information about the replicates.

7.4 Summary

Topic	Key Points
Calculating richness	Species richness is simply the number of different species in any sample. The `specnumber()` command in the *vegan* package can calculate richness. You could also use a Pivot Table in Excel. If your data are presence-absence you can simply sum the rows of your dataset.
Comparing richness	You can use species richness as the response variable in hypotheses tests regarding differences between samples. For simple *t*-test or *U*-test you can use the `t.test()` or `wilcox.test()` commands in R.
	Differences tests are best summarised graphically with box-whisker plots; use the boxplot() command in R.
	If you have multiple samples you can use ANOVA or the Kruskal–Wallis test using `aov()` or `kruskal.test()` commands. A post-hoc test can be carried out (for ANOVA) using the `TukeyHSD()` command.

Topic	Key Points
Correlating richness	You can use species richness as the response variable in correlations. For simple correlation tests you can use the `cor.test()` command in R for example.
	Correlations are best summarised graphically using scatter plots; use the plot() command in R to do this.
	It is also possible to use multiple regression, via the `lm()` command in R or with the *Analysis ToolPak* in Excel.
R packages for community ecology	There are many R command packages available that can undertake the kinds of analysis required for community ecology. A useful package is *vegan* but *BiodiversityR* is also helpful.
	Get and install packages from R using the `install.packages()` command.
	Make a package 'ready' with the `library()` command.
Richness and area	Species richness increases with the area studied. The relationship is often logarithmic. So, log(richness) vs. log(area) gives a straight line.
Rarefaction	Samples with higher abundances tend to have more species. Rarefaction is a way to 'standardise' on a set sample size.
	Use the `rarefy()` command in the *vegan* package to generate the expected species richness for a community dataset.
	Use the `rrarefy()` command to make a new community dataset 'standardised' to a maximum abundance.
	Use the `drarefy()` command to predict the probability of occurrence for species in a rarefied community.
Species accumulation curves	Species accumulation curves (SACs) provide a picture of accumulating species across samples. The SAC can provide information about patterns of diversity.
	Use the `specaccum()` command in the *vegan* package to carry out SAC calculations using various randomisation methods.
	The `fitspecaccum()` command uses non-linear modelling to find the 'shape' of the SAC.
Estimating total richness	You can estimate the total number of species (i.e. including species you missed during sampling) for samples using various models.
Incidence-based richness estimation	Incidence-based models (e.g. Chao, Jackknife, Bootstrap) look at frequencies of occurrence in collections of sites.
	Use the `specpool()` command in the *vegan* package.
Abundance-based richness estimation	Abundance-based models (e.g. Chao, ACE) look at counts of individuals in single sites.
	Use the `estimateR()` command in the *vegan* package.

7.5   Exercises

7.1 The *gb.biol* dataset contains abundances for ground beetles at three different habitats. Each habitat has six replicate samples. Compare richness between the three habitats (the *gb.site* data item contains the *Habitat* variable).

7.2 Species richness increases with area in a linear fashion – TRUE or FALSE?

7.3 Species richness can be affected by the abundance in sample. Look at the *DeVries* data item, which shows tropical forest butterfly abundance for two samples (canopy and understorey). What is the change in species richness for each sample when rarefaction is undertaken?

7.4 Which of the following are not models available for use in Species Accumulation Curves via the `fitspecaccum()` command?

A) `"arrhenius"`
B) `"michaelis-menten"`
C) `"logit"`
D) `"lomolino"`
E) `none – they are all models`

7.5 Look at the *DeVries* data again – which species richness estimator provides the closest estimate to the observed values?

The answers to these exercises can be found in Appendix 1.

8. Diversity: indices

If two sites have the same number of species they obviously have the same species richness but there can be large differences in the relative abundance of the species (Figure 8.1).

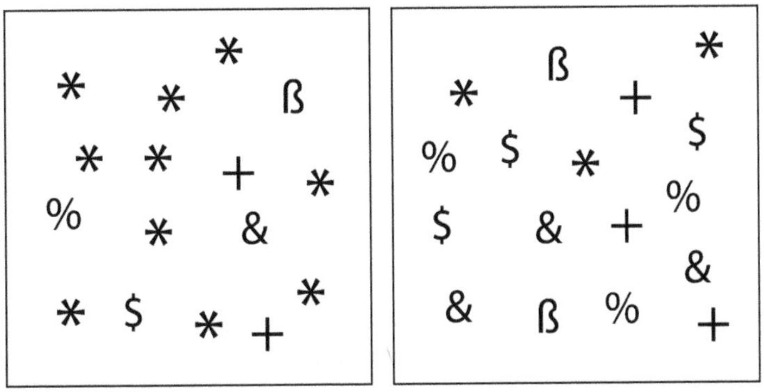

Figure 8.1 Differences in diversity. Both sites have species richness of five but the site on the left is dominated by one species '*' whilst the species in the right site are more evenly abundant.

If one species is dominant it seems obvious that the diversity at that site is lower than where the species are more evenly abundant. Various diversity indices attempt to take into account the relative abundance of species. The result of a diversity index is a value that increases not only as the number of species increases but as the relative abundance becomes more even. The two most commonly used indices are Simpson's and Shannon, which you will meet shortly. For the sites shown in Figure 8.1 the species richness is 5 for both sites, Simpson's index gives 0.5 and 0.8, whilst the Shannon index is 1.08 and 1.61.

8.1   Simpson's index

Simpson's index uses the abundance of each species in a sample to determine the proportion that each species contributes to the total. In Table 8.1 you can see how the index is calculated using one of the imaginary communities in Figure 8.1.

Table 8.1 Calculating Simpson's original diversity index (Simpson's S).

Abundance	P	P²
10	0.667	0.444
2	0.133	0.018
1	0.067	0.004
1	0.067	0.004
1	0.067	0.004
$\Sigma = 15$	$\Sigma = 1.000$	$\Sigma = 0.476$

In the first column you can see the abundance of each 'species' in the community. The total (15) is given at the bottom. The second column shows the proportion that each species contributes towards the total (i.e. the individual abundance, n, divided by the total abundance of everything, N). The sum of these proportions is of course unity. The third column shows the proportions squared, a common 'trick' in statistics. The sum of the squared proportions (0.476) is the index, sometimes called Simpson's S index. The formal representation of the index is shown in Figure 8.2.

$$S = \Sigma \left(\frac{n}{N} \right)^2$$

Figure 8.2 The formula for Simpson's S index. The probability that two randomly selected individuals from a community will be the same species.

The figure represents the likelihood of finding a species of the same type when you've already taken a species at random from a community. This means that the larger the value the less diverse a community must be. If you only had a single species in your community the index would be 1, as you would be bound to find the same species each time you sampled at random. This is not perhaps the most intuitive of diversity indices as you would probably wish to have your index show a large value for a more diverse community. The answer is simple enough, simply subtract Simpson's S index from 1 (Figure 8.3).

$$D = 1 - \Sigma \left(\frac{n}{N} \right)^2$$

Figure 8.3 The 'classic' Simpson's diversity index. The probability that two individuals sampled at random from a community will be different species.

This index has several advantageous properties: a larger value indicates a more diverse community; the index is bounded 0–1 and is probabilistic. The index should be used for genuine counts of individuals but it is often applied to other situations, such as percentage cover of plants. You can calculate Simpson's diversity readily using a spreadsheet. In the following exercise you can have a go at making such a spreadsheet.

Have a Go: Make a spreadsheet to calculate Simpson's diversity

For this exercise you'll be creating a spreadsheet from new so all you will need is your spreadsheet.

1. Start by opening Excel – this should start with a new blank workbook but if it does not then you can create one by clicking the *Office* button and then *New > Blank Workbook*.

2. Click in cell B1 and type 'Count' as a heading label (you do not need the quotes).

3. Place appropriate headings in columns C and D for the proportions and proportion-squared (P and P^2 will do quite well). To get the superscript you will need to first type 'P2' in the cell. Press Enter and then click once on the cell. Now click in the formula bar and you will see the cursor at the end of the line. Highlight the 2 in the formula bar and then use the *Home > Format > Format Cells* button. Click on the Superscript box and then OK.

4. Now fill in some values for species counts in column B – you can use the values in Table 8.1 for now.

5. Leave a few lines blank so that you can add extra data. Click in cell A12 and type 'Sum' as a row label.

6. In cell B12 enter a formula to add up the count values for all the species. Use =SUM(B2: B11) – you can type this directly or use the mouse to help you select the cells.

7. Now click in cell C2 and enter a formula to work out the proportion that the species in row 2 contributes to the total – this will be =B2/B12. You will need to edit this formula to 'fix' the B12 part – this is the total abundance and you want the cell reference to remain the same when you copy the formula down the column. Add a dollar sign before the B and the 12 so that the final formula reads =B2/$B$12.

8. Copy the proportion formula down the column (to cell C11) – you can copy the cell to the clipboard and then highlight the 'target' cells and use a paste operation.

9. Check that the sum of the proportions is unity – copy the formula for total abundance in cell B12 and paste it into C12.

10. Now calculate the square of the proportions – click in cell D2 and enter a formula to square the contents of cell C2 like so: =C2^2.

11. Copy the cell D2 and paste down the column as far as D11 – you now have the squared proportions in column D.

12. Simpson's *S* index is the sum of the squared proportions so copy the cell C12 into D12 or make a new formula =SUM(D2:D11).

13. Tidy up the formatting of the values by reducing the number of decimal places. Highlight the cells C2:D11, which are the calculated values. Now from the *Home* menu use *Format > Format Cells* and alter the formatting to a number with three decimal places. You might also alter the appearance of row 12 to make these sums stand out from the other calculations – make them bold and add a border; these options are on the *Home* menu.

14. So far you have worked out Simpson's *S* index. Add some other summary statistics, in cell A14 type 'N', then under that type 'No. Spp', 'S' and 'D' so that you have labels in cells A14:A17.

15. In cell B14 type a formula to represent the total abundance – you can simply refer to cell B12 like so: =B12, or create the original =SUM(B2:B11) formula.

16. In cell B15 type a formula to work out species richness – this will be the number of items in column B, use =COUNT(B2:B11).

17. In cell B16 type a formula to show Simpson's S index, the simplest way is to refer to cell D12, were you already have the result ' =D12'.

18. In cell B17 type a formula to show Simpson's D index. This is 1 – S so use: = 1-D12 to get the result.

19. Now save the spreadsheet – call it what you like but *Diversity calculator.xls* would seem sensible. You will use this spreadsheet later.

If you need to add extra rows then this is achieved easily enough, remember though that the newly inserted rows will not contain any formulae so you will need to copy down into the new cells.

It is fairly easy to take your recorded data and get it into a format to allow you to work out the diversity – you can use a Pivot Table to arrange your data and then use copy and paste to explore diversity using a separate spreadsheet. If you saved your Pivot Table data you could easily create the formulae to calculate diversity for each sample. If you had several samples this could be a bit tedious so you may prefer to use R instead.

You can use R to examine Simpson's diversity in your data quite easily. There are packages, such as *vegan* or *BiodiversityR*, that will calculate Simpson's index for you and allow you to use grouping variables to combine sites. You can also calculate Simpson's index without any special packages, in the following exercise you can have a go at working out Simpson's diversity using various methods with R.

Have a Go: Use R to calculate Simpson's diversity

For this exercise you will need the ground beetle data with abundance information that you met earlier. The data are in a CSV file, ready to import to R, called *Ground beetles and abundance.csv*. You will also need the data that contains the site information, *Gbsite.csv*. However, you will start by making some simpler data that are based on Figure 8.1.

1. Start by opening R. Make two samples to represent the community data shown in Figure 8.1; call the samples *a* and *b* and use the following to create a data matrix called *divc*:

```
> a = c(10, 2, 1, 1, 1)
> b = c(3, 3, 3, 3, 3)
> divc = rbind(a, b) # joins items together in rows
> divc
    [,1] [,2] [,3] [,4] [,5]
a    10    2    1    1    1
b     3    3    3    3    3
```

2. You now have a matrix object that contains the two samples bundled together; the columns are not labelled but this does not matter. Use the rm() command to remove the *a* and *b* objects:

```
> rm(a, b)
```

3. The `prop.table()` command calculates the proportion of each item of a matrix (a form of table) – you can get the proportions for rows, columns or for the entire matrix by using the `margin` instruction. Use the command to get proportions for the rows like so:

```
> prop.table(divc, margin = 1)
      [,1]       [,2]       [,3]        [,4]        [,5]
a 0.66666667 0.13333333 0.066666667 0.066666667 0.066666667
b 0.20000000 0.20000000 0.200000000 0.200000000 0.200000000
```

4. The `rowSums()` command will add up rows of a table (or matrix) so you can use it to determine Simpson's S index like so:

```
> rowSums(prop.table(divc, margin = 1)^2)
        a          b
0.47555556 0.20000000
```

5. Determine Simpson's D index by subtracting the previous result from 1 like so:

```
> 1 - rowSums(prop.table(divc, margin = 1)^2)
        a          b
0.52444444 0.80000000
```

6. Now get the ground beetle data from the file named called *Ground beetles and abundance.csv*. You will also need the data that contains the site information, *GBsite.csv*. If you already have these data in R then go to step 8:

7. Get the data from the *Beetles and habitat.csv* and *Gbsite.csv* files into R:

```
> gb.biol = read.csv(file.choose(), row.names = 1)
> gb.site = read.csv(file.choose(), row.names = 1)
```

8. For the abundance data the columns are the species abundances and the rows are the sites; you can remind yourself of the species names and site names by using the following commands:

```
> names(gb.biol)
> row.names(gb.biol)
```

9. The *gb.site* data contain rows corresponding to the sites but the columns contain information about the maximum height of the vegetation and the habitat type. Simply type the name of the object to view the data:

```
> gb.site
```

10. Prepare the *vegan* and *BiodiversityR* packages using the `library()` command. The *BiodiversityR* package requires *vegan* and will load it for you if not already available:

```
> library(BiodiversityR)
```

11. Use the `diversity()` command in *vegan* to get Simpson's D index for the ground beetle data:

```
> diversity(gb.biol, index = "simpson")
        E1         E2         E3         E4         E5         E6
0.61113991 0.68546429 0.67810963 0.74167109 0.72862551 0.71002376
        G1         G2         G3         G4         G5         G6
0.72592982 0.77485462 0.77282570 0.84628330 0.80159217 0.82136892
        W1         W2         W3         W4         W5         W6
0.69905339 0.73755627 0.72230272 0.75207579 0.75986884 0.73750823
```

12. Since Simpson's S index is 1 − D you can modify the previous command to get this like so:

```
> 1- diversity(gb.biol, index = "simpson")
```

13. The `diversitycomp()` command in the *BiodiversityR* package allows you to compare diversity using a grouping variable – use the *Habitat* variable in the *gb.site* data to compare Simpson's D index like so:

```
> diversitycomp(gb.biol, y = gb.site, factor1 = "Habitat",
                index = "Simpson")

Habitat n   Simpson
  Edge  6 0.70153378
  Grass 6 0.79758941
  Wood  6 0.73918472
```

14. The result of the `diversitycomp()` command is a `matrix` that contains the number of sites as well as the indices. The diversity index is in the second column so to determine the Simpson's S index you must use the following:

```
> 1 - diversitycomp(gb.biol, y = gb.site, factor1 = "Habitat",
                    index = "Simpson")[,2]
      Edge       Grass       Wood
0.29846622 0.20241059 0.26081528
```

15. The `diversityresult()` command allows you to view results for one level of a grouping factor – use it to see the Simpson's D index for the *Wood* habitats:

```
> diversityresult(gb.biol, y = gb.site, factor = "Habitat", level = "Wood",
                  index = "Simpson", method = "s")
      Simpson
W1 0.69905339
W2 0.73755627
W3 0.72230272
W4 0.75207579
W5 0.75986884
W6 0.73750823
```

16. By using the `method = "all"` instruction you can combine all the sites for a given grouping variable:

```
> diversityresult(gb.biol, y = gb.site, factor = "Habitat", level = "Wood",
                  index = "Simpson", method = "all")
       Simpson
all 0.73918472
```

The *BiodiversityR* package is useful because it makes it easy to combine sites by a grouping variable. You can do this using commands from the base R packages but it is a bit more involved.

Tip: Convert a `data.frame` into a matrix

If you import your data using the `read.csv()` command it will end up as a `data.frame` object in R. Sometimes you need your data to be a matrix – you can convert an object to a matrix using the `as.matrix()` command.

Note: Hurlbert's unbiased Simpson's index

You can calculate Hurlbert's unbiased Simpson's index using the `rarefy()` command in the *vegan* package. You need to set the sample size to 2 and then subtract unity from the result like so: `rarefy(x, sample = 2) -1`, where x is your community sample.

The Simpson's index has one more 'trick' up its sleeve – there is an interesting relationship between the species richness and Simpson's index, as you will see next.

8.1.1 Simpson's index: 'effective species' and evenness

Ideally you would like your diversity index to increase in value each time you added a new species to the list. Even better would be if the index grew larger if the new species was equal in abundance to other species present.

Simpson's index can be modified so that when all species are equally abundant the value is equal to the species richness. The modification is simple:

Effective species = $1/S$

This can also be written as $1/(1 - D)$. The index is often called 'inverse Simpson's'. As the species become more uneven in their abundances the value decreases from a maximum that equals the species richness.

If you divide the 'effective species' value by the actual number of species present you get a value representing 'evenness' of the community. If the evenness is unity then the species present are exactly equal in abundance. You can look at effective species quite easily using Excel. In the following exercise you can have a go at modifying the diversity calculator spreadsheet to show effective species and evenness.

Have a Go: Use Excel to determine Simpson's inverse index (effective species) and evenness

For this exercise you'll need the spreadsheet you made earlier. There is a completed version available with the download files, called *Diversity Simpson D.xls*. You can use this to view the various formulae and can explore the effects of the various indices by altering the data.

1. Open the spreadsheet you created earlier. You should have a column containing counts for various species, a column for the proportions of each species to the total and a column of squared proportions. You should also have some rows showing various results such as total abundance, species richness, Simpson's S index and Simpson's D index. If you do not then you can open the completed file *Diversity Simpson D.xls* or follow the earlier exercise.

2. Click in cell A18 and type a label '1/S' for the inverse Simpson's result (effective species). Now in cell B18 type a formula to compute the result from the Simpson's S index already present. The formula will be: =1/B16.

3. Now go to cell A19 and type a label for evenness: 'S-Even'. Then in cell B19 type a formula to divide the 1/S result by the species richness: =B18/B15.

4. Use the spreadsheet to look at the effects of altering the species abundances. Start by altering all the current values to 3 (you should end up with five lots of 3 to give a total

abundance of 15). Note that $1/S$ is 5, which is the same as the species richness. The evenness is 1, which is not surprising as all the species have the same abundance.

5. Now add another species, use the same abundance (3) as the others. The species richness and $1/S$ increase (to 6) but the evenness remains at 1.

6. Save the new version of the spreadsheet – you will add more in later exercises.

7. Explore the effects of altering the abundances – try adding more species too.

Note that the evenness decreases whether you increase or decrease an individual abundance value. However, it goes down less when you make abundance larger.

You can also use R to look at $1/S$, the effective species, as well as evenness. It is possible to do this without special packages but it is a lot more difficult. In the following exercise you can have a go at using R to look at effective species and evenness using the *vegan* and *BiodiversityR* packages.

Have a Go: Use R to explore Simpson's inverse index (effective species) and evenness

For this exercise you will need the ground beetle data with abundance information that you met earlier. The data are in a CSV file, ready to import to R, called *Ground beetles and abundance.csv*. You will also need the data that contains the site information, *GBsite.csv*. If you already have the data in R you can go directly to step 2.

1. Start by opening R and using the read.csv() command to get the data from the *Beetles and habitat.csv* and *Gbsite.csv* files:

```
> gb.biol = read.csv(file.choose(), row.names = 1)
> gb.site = read.csv(file.choose(), row.names = 1)
```

2. Prepare the *BiodiversityR* and *vegan* packages. The former requires the latter so you only need to load one to get the other:

```
> library(BiodiversityR)
```

3. The diversity() command in *vegan* allows the inverse Simpson's index to be calculated like so:

```
> diversity(gb.biol, index = "invsimpson")
       E1        E2        E3        E4        E5        E6
2.5716190 3.1792893 3.1066478 3.8710340 3.6849447 3.4485584
       G1        G2        G3        G4        G5        G6
3.6487005 6.5054742 5.0401236 5.5981299 3.3228485 3.8103406
       W1        W2        W3        W4        W5        W6
4.4415745 4.4019064 3.6010436 4.0334907 4.1643908 3.8096432
```

4. The species richness can be determined using the specnumber() command. Calculate evenness by dividing the inverse Simpson's index by the species richness:

```
> diversity(gb.biol, index = "invsimpson")/specnumber(gb.biol)
        E1         E2         E3         E4         E5         E6
0.15127171 0.22709209 0.20710985 0.15484136 0.17547356 0.20285638
```

```
        G1         G2         G3         G4         G5         G6
0.13031073 0.20188975 0.24455035 0.23233837 0.19385091 0.23325541
        W1         W2         W3         W4         W5         W6
0.27690404 0.34639460 0.32736760 0.33612422 0.34703257 0.38096432
```

5. Compare the effective species using the *Habitat* grouping variable like so:

```
> diversitycomp(gb.biol, y = gb.site, factor1 = "Habitat",
                index = "inverseSimpson", method = "all")


Habitat n inverseSimpson
   Edge  6      3.3504630
   Grass 6      4.9404530
   Wood  6      3.8341311
```

6. Getting evenness is more tricky – the result of the `diversitycomp()` command is a matrix so you can get the 1/*S* results like so:

```
> diversitycomp(gb.biol, y = gb.site, factor1 = "Habitat",
                index = "inverseSimpson", method = "all")[,2]
    Edge     Grass     Wood
3.3504630 4.9404530 3.8341311
```

7. The `specnumber()` command can get species richness using a grouping variable like so:

```
> specnumber(gb.biol, groups = gb.site$Habitat)
 Edge Grass  Wood
   27    37    13
```

8. Get the evenness by combining steps 6 and 7 like so:

```
> diversitycomp(gb.biol, y = gb.site, factor1 = "Habitat",
  index = "inverseSimpson", method = "all")[,2]/specnumber(gb.biol,
  groups = gb.site$Habitat)
       Edge       Grass       Wood
0.12409122 0.13352576 0.29493316
```

9. Use the `diversityresult()` command to view a single level of the grouping factor `Habitat`:

```
> diversityresult(gb.biol, y = gb.site, factor = "Habitat",
                  level = "Grass", method = "s",
                  index = "inverseSimpson")
   inverseSimpson
G1      3.6487005
G2      4.4415746
G3      4.4019064
G4      6.5054742
G5      5.0401236
G6      5.5981299
```

10. Now combine the sites corresponding to the *Grass Habitat*:

```
> diversityresult(gb.biol, y = gb.site, factor = "Habitat",
                  level = "Grass", method = "all",
                  index = "inverseSimpson")
     inverseSimpson
all       4.940453
```

11. The `diversityresult()` command also uses an `index = "richness"` to determine species richness. You can use this to get evenness but you need to split the process into steps like so:

```
> iS = diversityresult(gb.biol, y = gb.site, factor = "Habitat",
                       level = "Grass", method = "s",
                       index = "inverseSimpson")
> SR = diversityresult(gb.biol, y = gb.site, factor = "Habitat",
                       level = "Grass", method = "s",
                       index = "richness")
> gb.even = iS/SR
> names(gb.even) = "S.Even"
> gb.even
       S.Even
G1 0.13031073
G2 0.20188975
G3 0.24455035
G4 0.23233837
G5 0.19385091
G6 0.23325541
```

In step 11 you split the process into chunks to make it easier to follow. The *iS* result object contains the inverse Simpson values. The *SR* result object contains the species richness. The *gb.even* result is the evenness but the *iS* object contained the name 'inverseSimpson' so you have to use the `names()` command to change this.

Try using a similar process to step 11 to calculate the evenness for the all the sites in the *Grass Habitat* combined (hint: just alter `method = "s"` to method = "all").

Tip: Retyping long commands

If you have to modify a long command you can use the up arrow to cycle back through previous commands, which you can then edit.

8.2   Shannon index

The Shannon index uses the proportions of the various species but in a slightly different way – using a logarithm. The index is sometimes called the Shannon–Wiener or Shannon–Weaver index (more properly it should be termed an entropy). The proportion of each species is not squared but multiplied by the log of the proportion. In Table 8.2 you can see the index calculated for one of the communities shown in Figure 8.1.

Table 8.2 Calculating the Shannon diversity index.

Abundance	P	P.ln(P)
10	0.667	−0.270
2	0.133	−0.269
1	0.067	−0.181
1	0.067	−0.181
1	0.067	−0.181
$\Sigma = 15$	$\Sigma = 1.000$	$\Sigma = -1.081$

In the first column you can see the abundance of each 'species' in the community, the second columns shows the proportion that each makes towards the total. The third column show the proportions multiplied by their natural logarithm. The final index is simply the absolute magnitude of the sum of this third column (the values are always negative). Formally this is written as shown in Figure 8.4.

$$H = -\sum\left(\frac{n}{N}\right)\text{Ln}\left(\frac{n}{N}\right)$$

Figure 8.4 The formula for calculation of the Shannon diversity index, H.

If you calculate the index for the other community in Figure 8.4 you get a value of 1.609. You can see readily that the index is not bounded 0–1 like Simpson's S or D indices – it is not probabilistic in the same manner either and this makes it harder to interpret readily. One reason perhaps for its success is that the difference between the index for two communities is larger than for Simpson's index and so quite small differences can be enumerated more easily.

Any logarithm could be used and it would be perhaps sensible to use base 2 but generally the natural log is used (base e). If you use base 2 then the result is equivalent to how many yes/no questions you'd have to ask in order to identify an individual.

It is easy to calculate the Shannon index using a spreadsheet and in the following exercise you can have a go at making a spreadsheet for yourself.

Have a Go: Use Excel to calculate the Shannon diversity index

For this exercise you'll need the spreadsheet you made earlier. There is a completed version available with the download files, called *Diversity Simpson SEven.xls*. You can use this to view the various formula and can explore the effects of the various indices by altering the data.

1. Open the spreadsheet you created earlier. You should have a column containing counts for various species, a column for the proportions of each species to the total and a column of squared proportions. You should also have some rows showing various results such as total abundance, species richness, Simpson's S index and Simpson's D index and evenness. If you did not then you can open the completed file *Diversity Simpson SEven.xls* or follow the earlier exercises.

2. Click in cell E1 and type a heading: 'P.Ln(P)' will do. You are going to use this column for the main calculations.

3. Now click in cell E2 and type a formula that will determine the natural log of the proportion multiplied by the proportion: =C2*LN(C2).

4. Copy the formula down the column – notice that the results are all negative. There is a problem however, where the abundance is left blank you see an error message #NUM!

5. You will need to edit the formula in column E so that blank cells are ignored – use an IF statement to place a blank if the abundance is blank by editing the formula in cell E2 like so: =IF(B2=" "," ",C2*LN(C2)).

6. The IF part looks to see if the abundance is a blank (that is the pair of double quotes
    ""); if it is then a blank is returned but if not then the value in C2 (the proportion) is
    multiplied by its natural log. Copy cell E2 down the rest of the column (to cell E11).
    Copy the formula for the sum of the column from that in D12 or simply retype it as
    =SUM(E2:E11).

7. Make a label for the result in cell A20, 'H' or 'H$_e$' will be fine. Then in cell B20 type
    a formula to show the Shannon index: =-D12, that is the negative of the sum in cell
    D12.

8. Now click in cell F1 and type a new heading 'P.Log2(P)' – you are going to calculate
    the Shannon index for log base 2 rather than the natural log.

9. Type a formula in cell F2 that will multiply the proportion in cell C2 by the log2 of
    itself. Use the LOG and an IF functions like so: =IF(B2="","",C2*LOG(C2,2)).

10. Copy the formula down the column and make a sum result in cell F12.

11. In cell A21 type a heading for the result: 'H$_2$' would be best.

12. In cell B21 type a formula to show the Shannon index for log base 2: =-F12.

Save the spreadsheet and have a go at altering the values for abundance in the *Count*
column. Look at the differences between the Shannon and Simpson's indices. If you
type just two values for *Count*, the H$_2$ index will be 1 as long as the two values are the
same. This represents how many yes/no questions you'd have to ask before you could
identify a species selected at random. Add more values until H$_2$ = 2 (you need four all
the same). How many will you need to get H$_2$ = 3?

Using log base 2 is interesting but most often the natural log is used. If you need to compare a Shannon diversity index that was calculated using a different base then you will
have to convert the bases so they are equivalent. The 'simplest' way is to 'unlog' the Shannon index and reapply the logarithm using the base of your choice. For example:

```
> divc # Some data
   [,1] [,2] [,3] [,4] [,5]
a   10    2    1    1    1
b    3    3    3    3    3

> (be = diversity(divc, base = 10)) # Shannon using base10
        a          b
0.46928726 0.69897000

> 10^be # Unlog the values (10^ because of base10)
        a          b
2.9463698 5.0000000

> log(10^be) # Take natural log of "unlogged" values
        a          b
1.0805738 1.6094379

> diversity(divc) # Shannon index calculated using natural log
        a          b
1.0805738 1.6094379
```

In the preceding example you see some data where the Shannon index is calculated using log base 10. To 'unlog' the values you use 10 to the power of the values you want to unlog, so $10^{0.469} = 2.946$. You now take the natural log of this result and obtain the same as if you'd used the natural log right from the start.

The natural log is used in the majority of cases so it is sensible to use this for all your calculations so that your results can be compared to others more easily. If you want to explore other logarithms then do this in parallel with the natural log so that you report both in your reports.

In the preceding example you see that the `diversity()` command was used in R. In the following exercise you can have a go at using R to look at the Shannon index for yourself.

Have a Go: Use R to look at the Shannon diversity index

For this exercise you will need the ground beetle data with abundance information that you met earlier. The data are in a CSV file, ready to import to R, called *Ground beetles and abundance.csv*. You will also need the data that contains the site information, *GBsite.csv*. If you already have the data in R you can go directly to step 2.

1. Start by opening R and using the `read.csv()` command to get the data from the *Beetles and habitat.csv* and *Gbsite.csv* files:

```
> gb.biol = read.csv(file.choose(), row.names = 1)
> gb.site = read.csv(file.choose(), row.names = 1)
```

2. Prepare the *BiodiversityR* and *vegan* packages. The former requires the latter so you only need to load one to get the other:

```
> library(BiodiversityR)
```

3. The packages make it easier to work out the Shannon index but it is possible to do this without them. Have a go at making some data and using basic commands to determine the Shannon index. Start by making some data:

```
> a = c(10, 2, 1, 1, 1)    # a data sample
> b = c(3, 3, 3, 3, 3)     # another data sample
> divc = rbind(a, b)       # join the data in rows
> rm(a, b)                 # remove the original items
> divc                     # look at the new data
  [,1] [,2] [,3] [,4] [,5]
a   10    2    1    1    1
b    3    3    3    3    3
```

4. Because the data are in a matrix you can use the `prop.table()` command to work out the proportions of each 'species' in its row (site or sample). You need to tell the command to use the rows in the calculation by using the `margin` instruction:

```
> prop.table(divc, margin = 1)
        [,1]       [,2]        [,3]        [,4]        [,5]
a 0.66666667 0.13333333 0.066666667 0.066666667 0.066666667
b 0.20000000 0.20000000 0.200000000 0.200000000 0.200000000
```

5. Now determine the Shannon index for the data like so:

```
> abs(rowSums(prop.table(divc, margin = 1)*log(prop.table(divc,
     margin = 1))))
        a         b
1.0805738 1.6094379
```

6. The `diversity()` command in *vegan* will calculate the Shannon index by default but specify the `index` instruction in full:

```
> diversity(gb.biol, index = "shannon")
        E1        E2        E3        E4        E5        E6
1.2670962 1.4245054 1.4121788 1.7371456 1.6369227 1.5574768
        G1        G2        G3        G4        G5        G6
1.9351634 2.0002724 1.9294264 2.3602503 2.1395667 2.2051749
        W1        W2        W3        W4        W5        W6
1.3972508 1.5821021 1.5395110 1.6010734 1.6163330 1.5524702
```

7. The `diversitycomp()` command in the *BiodiversityR* package allows you to compare diversity using a grouping variable – use the *Habitat* variable in the *gb.site* data to compare the Shannon index like so:

```
> diversitycomp(gb.biol, y = gb.site, factor1 = "Habitat",
                index = "Shannon")

Habitat n   Shannon
  Edge  6 1.5588424
  Grass 6 2.1624256
  Wood  6 1.5667275
```

8. The `diversityresult()` command allows you to view results for one level of a grouping factor – use it to see the Shannon index for the *Edge* habitats:

```
> diversityresult(gb.biol, y = gb.site, factor = "Habitat",
                  level = "Edge", index = "Shannon", method = "s")
        Shannon
E1 1.2670962
E2 1.4245054
E3 1.4121788
E4 1.7371456
E5 1.6369227
E6 1.5574768
```

9. By using the `method = "all"` instruction you can combine all the sites for a given grouping variable:

```
> diversityresult(gb.biol, y = gb.site, factor = "Habitat",
                  level = "Edge", index = "Shannon", method = "all")
        Shannon
all 1.5588424
```

The *BiodiversityR* package is useful because it makes it easy to combine sites by a grouping variable. You can do this using commands from the base R packages but it is a bit more involved.

Tip: Copy and paste

You can use copy and paste to help transfer commands to and from R. However, beware of smart quotes, which R does not usually recognise. It is usually better to use a text editor like `Notepad.exe`, and stick to plain text if you want to keep a 'library' of useful commands.

8.2.1 Shannon index: 'effective species' and evenness

The Shannon index can be modified in a similar way to Simpson's index to produce measures of effective species and evenness. The calculation depends upon the base of the logarithm used to generate the Shannon index, for the natural log you get:

Effective species $= \exp(H)$

For log to the base 2 you would use 2^H and for base 10 you would use 10^H. If you divide the effective species by the actual species richness you get a value that ranges from 0 to 1 and shows the evenness:

Evenness = effective species ÷ species richness

The closer the value is to unity the more even the community is. There is another way to determine evenness without having to calculate the effective species – divide the Shannon index by the logarithm of the species richness. The logarithm used should match the one used for calculating the Shannon index. This version of evenness is sometimes called *J-evenness*, with the alternative being called *E-evenness*.

It is easy enough to determine effective species and evenness using Excel and in the following exercise you can have a go at this.

Have a Go: Use Excel to calculate Shannon evenness and effective species

For this exercise you'll need the spreadsheet you made earlier. There is a completed version available with the download files, called *Diversity Shannon.xls*. You can use this to view the various formula and can explore the effects of the various indices by altering the data.

1. Open the spreadsheet you created earlier. You should have a column containing counts for various species, a column for the proportions of each species to the total and other columns for square proportions and the logged values. You should also have some rows showing various results such as total abundance, species richness, Simpson's indices and the Shannon index. If you did not then you can open the completed file *Diversity Shannon.xls* or follow the earlier exercises.

2. Click on cell A22 and type a label for the Shannon effective species result, 'exp(H)' will do. Now go to cell B22 and type in a formula to calculate this: =EXP(B20).

3. Now click in cell A23 and type a label for effective species using log base 2, '2H' will suffice. Then click in cell B23 and type a formula to calculate the effective species: =2^B21.

4. Next you want to determine evenness; start with cell A24 and type a label, 'E-even'. Then move to cell B24 and type a formula for the calculation: =B22/B15.

5. Go to cell A25 and type a label, 'J-even'. In cell B25 type a formula for this version of evenness: =B20/LN(B15).

6. In cell A26 type a label for evenness based on log base 2, 'E_2-even' will do. Then in cell B26 type a formula to calculate the evenness: =B23/B15.

7. Now go to cell A27 and type a label for J-evenness based on log base 2, 'J_2-even' for example. In cell B27 type a formula to calculate the value: =B21/LOG(B15,2).

Save the spreadsheet and explore the effects of altering the values in the *Count* column. When the values for abundance are all the same the effective species values are also the same and the measures of evenness are all 1. However, as you alter the abundances and move away from a completely even community the values alter.

Using the spreadsheet to calculate diversity indices and evenness is easy enough for a single sample but when you have several samples or simply a lot of species then using R is probably easier. R also allows you to readily compare the Shannon index using grouping variables as you have seen already with Simpson's index. In the following exercise you can have a go at exploring effective species and evenness using R.

Have a Go: Use R to explore effective species and evenness

For this exercise you will need the ground beetle data with abundance information that you met earlier. The data are in a CSV file, ready to import to R, called *Ground beetles and abundance.csv*. You will also need the data that contains the site information, *GBsite.csv*. If you already have the data in R you can go directly to step 2.

1. Start by opening R and using the `read.csv()` command to get the data from the *Beetles and habitat.csv* and *Gbsite.csv* files:

```
> gb.biol = read.csv(file.choose(), row.names = 1)
> gb.site = read.csv(file.choose(), row.names = 1)
```

2. Prepare the *BiodiversityR* and *vegan* packages. The former requires the latter so you only need to load one to get the other:

```
> library(BiodiversityR)
```

3. Use the `diversity()` command in *vegan* to look at effective species, there is no special instruction to do this so you will have to use the `exp()` command:

```
> exp(diversity(gb.biol, index = "shannon"))
        E1         E2         E3         E4         E5         E6
 3.5505275  4.1558019  4.1048895  5.6811040  5.1393300  4.7468291
        G1         G2         G3         G4         G5         G6
 6.9251752  7.3910692  6.8855595 10.5936032  8.4957560  9.0718377
        W1         W2         W3         W4         W5         W6
 4.0440668  4.8651720  4.6623099  4.9583517  5.0345946  4.7231230
```

4. To get the effective species using a different base you can use the `base` instruction in the `diversity()` command. However, you'll need to match the `base` in the final calculation like so (the result is the same as before):

```
> 2^(diversity(gb.biol, index = "shannon", base = 2))
```

5. To get E-evenness you can use the `specnumber()` command to get species richness:

```
> exp(diversity(gb.biol, index = "shannon")) / specnumber(gb.biol)
         E1         E2         E3         E4         E5         E6
0.20885456 0.29684299 0.27365930 0.22724416 0.24473000 0.27922524
         G1         G2         G3         G4         G5         G6
0.24732769 0.33595769 0.38253108 0.37834297 0.32675985 0.37799324
         W1         W2         W3         W4         W5         W6
0.33700557 0.44228836 0.42384635 0.41319597 0.41954955 0.47231230
```

6. Now use the Shannon index and the log of species richness to get J-evenness:

```
> diversity(gb.biol, index = "shannon") / log(specnumber(gb.biol))
         E1         E2         E3         E4         E5         E6
0.44722936 0.53977812 0.52147439 0.53967462 0.53766157 0.54972099
         G1         G2         G3         G4         G5         G6
0.58074567 0.64711904 0.66753572 0.70831497 0.65669225 0.69387587
         W1         W2         W3         W4         W5         W6
0.56229510 0.65978781 0.64202596 0.64431932 0.65046025 0.67422926
```

7. The *BiodiversityR* package allows you to calculate E-evenness and J-evenness directly, but only for the natural log:

```
> diversityresult(gb.biol, index = "Jevenness", method = "s")
> diversityresult(gb.biol, index = "Eevenness", method = "s")
```

8. Use the `diversitycomp()` command to compare E-evenness for the three habitat types:

```
> diversitycomp(gb.biol, y = gb.site, factor1 = "Habitat",
                index = "Eevenness")

Habitat  n  Eevenness
   Edge  6  0.17604872
  Grass  6  0.23492420
   Wood  6  0.36853415
```

9. Look at J-evenness for the *Edge* habitats:

```
> diversityresult(gb.biol, y = gb.site, factor = "Habitat",
                  level = "Edge", index = "Jevenness", method = "s")
      Jevenness
E1 0.44722936
E2 0.53977812
E3 0.52147439
E4 0.53967462
E5 0.53766157
E6 0.54972099
```

10. If you want to look at effective species of evenness using a logarithmic base other than *e*, the natural log, then you have to create your basic Shannon index first and then convert the result into a new base. Do this for log base 2 like so:

```
> gb.H = diversitycomp(gb.biol, y = gb.site, factor1 = "Habitat",
                 index = "Shannon")[,2]
> gb.H
     Edge      Grass       Wood
1.5588424 2.1624256 1.5667275

> gb.H = log(exp(gb.H), base = 2)
     Edge      Grass       Wood
2.2489341 3.1197206 2.2603099
```

11. Now you can determine J-evenness for example by using the `specnumber()` command to obtain species richness for the *Habitat* groupings:

```
> gb.H / log2(specnumber(gb.biol, groups = gb.site$Habitat,
                 MARGIN = 1))
      Edge       Grass        Wood
0.47297315 0.59885758 0.61082199
```

The *BiodiversityR* package is useful because it makes it easy to combine sites by a grouping variable. You can do this with commands from the base R packages but it is a bit more involved.

Tip: Logarithms and R

The basic `log()` command defaults to the natural log, base *e* i.e. `exp(1)`. The full form of the command is `log(x, base = exp(1))`. If you want log base 2 you can use `log(x, base = 2)`. However, there are two 'convenience' commands: `log2()` and `log10()`, which allow you to determine logs to base 2 or 10 more easily.

8.3   Other diversity indices

The Simpson and Shannon indices are commonly used but they are not the only indices of diversity. Many indices have been devised but here you will see just a few:

- *Berger–Parker dominance* – this is one of the simplest indices and is a measure of dominance or evenness.
- *Fisher's alpha* – this looks at numbers of species with different levels of abundance in a series. The properties of the resulting curve can be used as a diversity index.
- *Rényi entropy* – this is a generalised method of diversity that uses a scale parameter to take rare species into account at different levels.
- *Tsallis entropy* – this is a generalised method of diversity that uses different scales.

You'll explore the various indices in the following sections.

8.3.1 The Berger–Parker dominance index

If you have a completely even community, that is where the species are equally abundant, then no single species is dominant. The Berger–Parker index is a measure of dominance and is very easy to calculate:

$$d = \max(n/N)$$

The dominance d is assessed by looking at the proportion that each species makes towards the total (n being the number of individuals of a species and N being the total number of individuals of all species). The largest proportion is the dominance index. For a completely even community the inverse of the Berger–Parker index will equal the species richness.

It is easy to calculate the Berger–Parker index using your spreadsheet and in the following exercise you can have a go at calculating dominance.

Have a Go: Use Excel to calculate Berger–Parker dominance

For this exercise you'll need the spreadsheet you made earlier. There is a completed version available with the download files, called *Diversity HEven.xls*. You can use this to view the various formulae and can explore the effects of the various indices by altering the data.

1. Open the spreadsheet you created earlier. You should have a column containing counts for various species, a column for the proportions of each species to the total and other columns for square proportions and the logged values. You should also have some rows showing various results such as total abundance, species richness and various measures of evenness. If you did not then you can open the completed file *Diversity HEven.xls* or follow the earlier exercises.

2. Click in cell A28 and type a heading for the dominance result: 'Berger'.

3. Now click in cell B28 and type a formula to calculate the dominance index: =MAX(C2: C11). This formula looks at the range of cells where you stored the proportions and determines the maximum value.

4. In cell A29 type a label for the inverse of the dominance: '1/B'. In cell B29 type a formula to calculate the inverse of the Berger–Parker dominance: =1/B28.

You can now alter the values in the *Count* column and explore the effects of different abundance on the dominance index. Don't forget to save your new spreadsheet.

Tip: Max and Min values in Excel

There are two ways to obtain max and min values in Excel. The MAX and MIN functions are the simplest but LARGE and SMALL are also useful as they allow you to select values other than simply the largest or smallest. The general form of the LARGE function is: =LARGE(*range, which*) where the *which* part is a number. So, setting *which* to 1 returns the 1st largest. The SMALL function works in a similar way.

You can calculate the Berger–Parker dominance index using R quite easily. No special packages are required although the *BiodiversityR* package does allow you to compare the index using grouping variables more easily. In the following exercise you can have a go at using R to determine the Berger–Parker dominance index.

Have a Go: Use R to calculate Berger–Parker dominance

For this exercise you will need the ground beetle data with abundance information that you met earlier. The data are in a CSV file, ready to import to R, called *Ground beetles and abundance.csv*. You will also need the data that contains the site information, *GBsite.csv*. If you already have the data in R you can go directly to step 2.

1. Start by opening R and using the `read.csv()` command to get the data from the *Beetles and habitat.csv* and *Gbsite.csv* files:

```
> gb.biol = read.csv(file.choose(), row.names = 1)
> gb.site = read.csv(file.choose(), row.names = 1)
```

2. Before looking at the ground beetle data, make a simple dataset like so:

```
> a = c(10, 2, 1, 1, 1)    # Small community with dominant
> b = c(3, 3, 3, 3, 3)     # Even community
> divc = rbind(a, b)       # Join community data together
> rm(a, b)                 # Remove original data
> divc                     # Look at community data
   [,1] [,2] [,3] [,4] [,5]
a   10    2    1    1    1
b    3    3    3    3    3
```

3. Use the `prop.table()` command to view the proportions along each row – you will need to set `margin = 1` to get the row proportions like so:

```
> prop.table(divc, margin = 1)
        [,1]       [,2]        [,3]        [,4]        [,5]
a 0.66666667 0.13333333 0.066666667 0.066666667 0.066666667
b 0.20000000 0.20000000 0.200000000 0.200000000 0.200000000
```

4. The `apply()` command can apply a function over the rows of a table (`MARGIN = 1` sets rows) so use this to work out the maximum proportion – this is the Berger–Parker dominance:

```
> apply(prop.table(divc, margin = 1), MARGIN = 1, FUN = max)
        a          b
0.66666667 0.20000000
```

5. The `rbind()` command made a matrix result (a kind of table) but the ground beetle data are in a `data.frame`, which will not work with the `prop.table()` command. You must 'force' the data to be a matrix using the `as.matrix()` command:

```
> apply(prop.table(as.matrix(gb.biol), margin = 1), MARGIN = 1,
    FUN = max)
        E1         E2         E3         E4         E5         E6
0.54265734 0.44277929 0.45665635 0.40369089 0.41915551 0.42286349
        G1         G2         G3         G4         G5         G6
0.48767123 0.39721254 0.39024390 0.29464286 0.37047354 0.31722054
        W1         W2         W3         W4         W5         W6
0.39010989 0.40350877 0.42144638 0.33509934 0.33634312 0.40540541
```

6. Prepare the *BiodiversityR* package:

```
> library(BiodiversityR)
```

7. Use the `diversityresult()` command to calculate the Berger–Parker dominance index for all the sites in the ground beetle dataset:

```
> diversityresult(gb.biol, index = "Berger", method = "s")
        Berger
E1 0.54265734
E2 0.44277929
E3 0.45665635
...
```

8. Now look at the dominance index for just the *Edge* level of the *Habitat*:

```
> diversityresult(gb.biol, index = "Berger", method = "s", y = gb.site,
                factor = "Habitat", level = "Edge")
        Berger
E1 0.54265734
E2 0.44277929
E3 0.45665635
E4 0.40369089
E5 0.41915551
E6 0.42286349
```

9. Combine the sites for the *Edge* habitat together and view the dominance:

```
> diversityresult(gb.biol, index = "Berger", method = "all", y = gb.site,
                factor = "Habitat", level = "Edge")
        Berger
all 0.44393877
```

10. Now compare the dominance across the three habitat types:

```
> diversitycomp(gb.biol, y = gb.site, factor1 = "Habitat",
                index = "Berger")

Habitat n      Berger
  Edge  6 0.44393877
  Grass 6 0.37652682
  Wood  6 0.38262688
```

11. Use the `diversitycomp()` command to determine the standard deviation of the dominance index:

```
> diversitycomp(gb.biol, y = gb.site, factor1 = "Habitat",
                index = "Berger", method = "sd")

Habitat n      Berger
  Edge  6 0.04997868
  Grass 6 0.06815793
  Wood  6 0.03719261
```

12. Use the `diversityresult()` command to calculate the dominance index once more but this time save the result and use it to make a boxplot showing the dominance at the three habitats. Your graph should resemble Figure 8.5:

```
> gb.bp = diversityresult(gb.biol, index = "Berger", method = "s")
> gb.bp$Habitat = gb.site$Habitat # Add habitat as a variable to
                                   the result
```

```
> str(gb.bp)
'data.frame':   18 obs. of  2 variables:
 $ Berger : num   0.543 0.443 0.457 0.404 0.419 ...
 $ Habitat: Factor w/ 3 levels "Edge","Grass",..: 1 1 1 1 1 1 2 2 2 2 ...

> boxplot(Berger ~ Habitat, data = gb.bp)
> title(ylab = "Berger-Parker dominance", xlab = "Habitat (n = 6)")
```

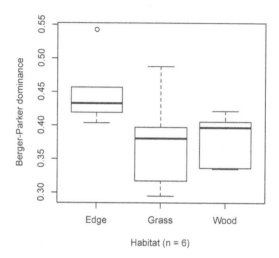

Figure 8.5 Berger–Parker dominance index for ground beetle communities at three different habitat types.

The *BiodiversityR* package is useful because it makes it easy to combine sites by a grouping variable. You can do this using commands from the base R packages but it is a bit more involved.

Tip: Converting objects in R

Sometimes a command requires the target object to be in a particular form. You can coerce the object into the required form using a command of the form as.xxxx() where xxxx is the form you need. For example; as.matrix(), as.table() and as.data.frame().

Tip: Adding columns to data in R

If you have a data.frame you can easily add a new column using the $, e.g. mydata$newcol will add the newcol data as a column to the mydata data frame. However, the new data must be the same length as the data.frame to which it is being added.

The Berger–Parker index is a measure of dominance, so its value gets smaller as the community becomes more even.

8.3.2 Fisher's alpha

Fisher's alpha comes from attempts to understand the patterns of relative abundance of species in a community (Fisher *et al.* 1943, see Chapter 11). Essentially you arrange your community in decreasing order of abundance and then plot a graph of the log abundance against the rank. This is the so-called Fisher log-series. The shape of the curve can be modelled using the formula in Figure 8.6.

$$S = \alpha.\text{Ln}\left(1 + \frac{N}{\alpha}\right)$$

Figure 8.6 Fisher's alpha. *S* is the number of species sampled (species richness); *N* is the total number of individuals and *a* is a constant derived from the data.

The log-series model predicts the number of species at different levels of abundance (Figure 8.7).

$$S_n = \frac{\alpha x^n}{n}$$

Figure 8.7 Fisher's log-series predicts the number of species with various levels of abundance. The variable *x* is a 'nuisance' parameter and is usually estimated from $x = N/(N - a)$.

The variable α is used as a diversity index and is generally calculated using non-linear modelling. The index can really only be used for genuine counts of individuals. The *vegan* package in R will calculate Fisher's alpha and can also determine confidence intervals by using routines in the *MASS* package (which comes as part of R but is not loaded by default). The *BiodiversityR* package can make it easier to use grouping variables in exploring your data. In the following exercise you can have a go at exploring Fisher's alpha for the ground beetle data that you've been using.

Have a Go: Use R to explore Fisher's alpha diversity

For this exercise you will need the ground beetle data with abundance information that you met earlier. The data are in a CSV file, ready to import to R, called *Ground beetles and abundance.csv*. You will also need the data that contains the site information, *GBsite.csv*. If you already have the data in R you can go directly to step 2.

1. Start by opening R and using the `read.csv()` command to get the data from the *Beetles and habitat.csv* and *Gbsite.csv* files:

   ```
   > gb.biol = read.csv(file.choose(), row.names = 1)
   > gb.site = read.csv(file.choose(), row.names = 1)
   ```

2. Prepare the *BiodiversityR* and *vegan* packages by using the `library()` command. The former requires the latter so you only have to load one to get both:

   ```
   > library(BiodiversityR)
   ```

3. Look at Fisher's alpha diversity using the `fisher.alpha()` command:

   ```
   > fisher.alpha(gb.biol)
   ```

```
          E1        E2        E3        E4        E5        E6
   3.1269637 2.4544435 2.7445210 4.8072909 3.7825387 2.9735665
          G1        G2        G3        G4        G5        G6
   7.0632985 5.5483721 4.7528736 7.2616137 6.4372434 5.9442894
          W1        W2        W3        W4        W5        W6
   1.8858721 1.7592441 1.8033635 2.0258620 1.9624758 1.6049145
```

4. The se = TRUE instruction can be used to estimate standard error, however these should not be used for confidence intervals. Look at the standard error for a single sample:

```
> fisher.alpha(gb.biol[1,], se = TRUE)

       alpha         se df.residual code
E1 3.1269637 0.83921324          16    1
```

5. The fisherfit() command carries out the underlying non-linear modelling to compute alpha for single samples. Try this using the first row of the data:

```
> gb.f = fisherfit(gb.biol[1,])
> gb.f

Fisher log series model
No. of species: 17

      Estimate Std. Error
alpha  3.12696    0.83921
```

6. Now determine the confidence intervals using the confint() command. This will load the *MASS* package if it is not already loaded:

```
> confint(gb.f, level = 0.95)
    2.5 %     97.5 %
1.7787646 5.1182650
```

7. The profile() command can be used to investigate the log-likelihood function for the log-series model. It calculates the *tau* statistic, that is, the signed square root of two times log-likelihood profile. The upshot is that you can use this to visualise the fit to the normal distribution. Use the plot() command along with profile() to produce a plot similar to Figure 8.8:

```
> gb.f = fisherfit(gb.biol[1,])
> plot(profile(gb.f))
```

8. Now use the diversityresult() command in the *BiodiversityR* package to look at samples from the *Grass* habitat:

```
>   diversityresult(gb.biol,  y  =  gb.site,  factor  =  "Habitat",
                 level = "Grass", index = "Logalpha", method = "s")
      Logalpha
G1 7.0632985
G2 5.5483721
G3 4.7528736
G4 7.2616137
G5 6.4372434
G6 5.9442894
```

9. The diversitycomp() command can help in using grouping variables. However, some communities can give problems for the non-linear modelling process:

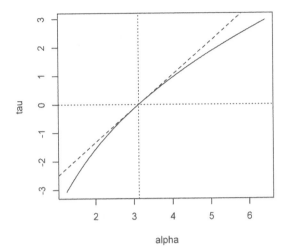

Figure 8.8 Profiling the Fisher log-series for a sample of ground beetles. The dashed line is the assumption of normality. If the profile line is close to the dashed line the standard errors can be used directly.

```
> diversitycomp(gb.biol, y = gb.site, factor1 = "Habitat",
                index = "Logalpha")

Habitat n  Logalpha
  Edge  6 3.7728572
  Grass 6 6.5278714
  Wood  6 1.6060031
Warning messages:
1: In log(p) : NaNs produced
...
```

The `fisherfit()` command should only be used for single samples (but see Section 11.2). The *BiodiversityR* package is useful because it makes it easy to combine sites by a grouping variable. You can do this with commands from the base R packages but it is a bit more involved.

You will see more about patterns of diversity and rank-abundance modelling in Chapter 11.

8.3.3 Rényi entropy

The Rényi entropy is different from the other diversity indices you have met so far in that the result is not a single value. The value of the entropy (think of this as another term for diversity) is assessed over several scales. As the scale becomes larger, rarer species are increasingly downrated. The Rényi entropy is calculated using the formula shown in Figure 8.9.

$$H_\alpha = \frac{Ln \sum p^\alpha}{1 - \alpha}$$

Figure 8.9 The Rényi entropy giving diversity for different scales (a). As a increases so rare species are taken less into account. Common diversity indices are special cases of Rényi entropy.

Some commonly used indices of diversity are related to Rényi: when $\alpha = 1$ you get the Shannon index (H) for example. The relationships are more obvious if you use the so-called Hill numbers (Hill 1973). The relationship is such that:

$$H_{\alpha\text{-Hill}} = \exp(H_\alpha)$$

In other words, to convert a result from 'original' Rényi to the 'Hill' version you take the exponent of the result. In Table 8.3 you can see the 'useful' Hill values and their relationship to Rényi.

Table 8.3 Hill numbers and their relationship to Rényi entropy. To convert a Rényi result into a Hill value $H_{\text{Hill}} = \exp(H)$.

Hill-Rényi	Diversity equivalent
0	Species richness
1	exp(H), exponent of Shannon index
2	1/S, inverse of Simpson's S index
Infinity	1/d, inverse Berger–Parker dominance

As the scale increases so the value of the Rényi entropy decreases, this is because rare species are given less prominence. In Table 8.3 you can see that when the scale is 0 all the rarities are given equal rating and the result is species richness. At the other end of the scale only the most dominant species is accounted for. One conclusion you can draw is that the Shannon index is more sensitive to rare species than Simpson's index.

Using Rényi entropy allows you to plot the diversity in a more meaningful way than a single value would show. The shape of the resulting plot can give insights into the pattern of diversity. A flat shape for example would indicate that the community had high evenness. If the curve fell away steeply with a long 'tail' then the community had a lot of rarities (Figure 8.10).

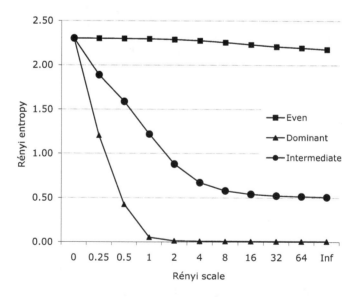

Figure 8.10 Rényi diversity for three different communities. Each has the same number of individuals (1500) and species (10).

Have a Go: Use Excel to calculate Rényi entropy

For this exercise you'll need the spreadsheet you made earlier. There is a completed version available with the download files, called *Diversity BergerParker.xls*. You can use this to view the various formula and can explore the effects of the various indices by altering the data.

1. Open the spreadsheet you created earlier. You should have a column containing counts for various species, a column for the proportions of each species to the total and other columns for square proportions and the logged values. You should also have some rows showing various results such as total abundance, species richness, Simpson's S index and Simpson's D index. If you did not then you can open the completed file *Diversity BergerParker.xls* or follow the earlier exercises.

2. Click in cell G1 and type a heading label for Rényi entropy at scale 0: 'R_0'. In cell H1 add a label: 'R_1'. Repeat the process along the row for all the Rényi scales (0.25, 0.5, 1, 2, 4, 8, 16, 32, 64, Inf).

3. Type a label in cell F14: 'Scale'. Then move to cell G14 and type a zero (0). In cell H14 type the value 0.25. Repeat this process along the row for all the Rényi scales; for infinity you can type a text label: 'Inf'.

4. Now click in cell G2. You need to type a formula that raises the proportion to the power of the Rényi scale (α). You will need to take into account what to do if the proportion is blank so use: =IF($B2=" "," ",$C2^G$14). Notice also that you need to use the $ so that the columns are "fixed". This is because you will copy the formula across columns shortly.

5. Copy the contents of cell G2 down the column to fill G3:G11. In cell G12 type a formula for the sum of the cells above: =SUM(G2:G11) or copy the formula from F12.

6. You can now copy the cells from the G column across to column P. You cannot enter any formulae into column Q because you cannot raise a number to the power of infinity!

7. You should now have proportions raided to appropriate powers and a row of sums. Now click in cell F15 and type a label for the Rényi results: 'Renyi'. In cell G15 type a formula to complete the Rényi equation: =LN(G12)/(1-G14).

8. Copy the cell F15 across the row to fill G15:P15. You cannot fill in cell Q15 as this involves infinity.

9. Cell J15 will contain an error – this is because you are trying to divide by zero. Fix this in one of two ways:

   - Alter the formula in J15 with a simple reference: =B20, which contains the Shannon diversity value.

   - Alter the formulae in cells J2:J11 so that they calculate the proportions (in column C) multiplied by the natural log of the proportions (e.g. =IF($B2="","",$C2*LN(C2)) for the top cell J2). Then use =-J12 as the formula in J15. In other words, calculate Shannon diversity.

10. Click in cell Q15 and type in a formula to determine Rényi entropy for scale infinity: =LN(1/MAX(C2:C11)). This is the natural log of the inverse of the Berger–Parker index.

11. Click in cell F16 and type a label for the Hill numbers: 'Renyi$_{Hill}$'. Now in cell G16 type in a formula to convert the regular Rényi result into a Hill value using =EXP(G15). You can copy this formula along the row to fill in cells H16:Q16.

12. You now have three rows relating to Rényi entropy, the top one gives the scales and the next two give the entropy values. You can use these to create a line graph (not a scatter plot). Highlight the cells G14:Q15 and use the *Insert > Line* button to make a chart with markers. You can edit the appearance as you like (the legend is certainly not required).

You should now save your new spreadsheet. Enter various values into the *Count* column and look at the effects on the Rényi entropy and the shape of the graph.

Using Excel is simple enough but it is not trivial to compare one or more samples and there is a lot of copying and pasting involved. The *vegan* package in R will calculate Rényi entropy readily and the *BiodiversityR* package will allow grouping variables to be used more easily. In the following exercise you can have a go at using R to explore Rényi entropy using the ground beetle data that you are becoming familiar with.

Have a Go: Use R to explore Rényi entropy

For this exercise you will need the ground beetle data with abundance information that you met earlier. The data are in a CSV file, ready to import to R, called *Ground beetles and abundance.csv*. You will also need the data that contains the site information, *GBsite.csv*. If you already have the data in R you can go directly to step 2.

1. Start by opening R and using the `read.csv()` command to get the data from the *Beetles and habitat.csv* and *Gbsite.csv* files:

```
> gb.biol = read.csv(file.choose(), row.names = 1)
> gb.site = read.csv(file.choose(), row.names = 1)
```

2. Prepare the *BiodiversityR* and *vegan* packages by using the `library()` command. The former requires the latter so you only have to load one to get both:

```
> library(BiodiversityR)
```

3. Look at Rényi entropy with Hill numbers by adding the `hill = TRUE` instruction to the `renyi()` command like so:

```
> renyi(gb.biol[1,], hill = TRUE)
        0       0.25        0.5          1          2          4          8
17.000000  9.957486   6.184436   3.550528   2.571619   2.197733   2.008834
       16         32         64        Inf
 1.919425   1.879481   1.860751   1.842784
```

4. Use the `scales` instruction to limit the scale of the results – this time compare the first six sites of the dataset:

```
> renyi(gb.biol[1:6,], scales = c(0,1,2,Inf))
          0         1         2        Inf
E1 2.833213 1.267096 0.9445357 0.6112772
E2 2.639057 1.424505 1.1566577 0.8146838
```

```
E3 2.708050 1.412179 1.1335443 0.7838241
E4 3.218876 1.737146 1.3535217 0.9071058
E5 3.044522 1.636923 1.3042555 0.8695133
E6 2.833213 1.557477 1.2379563 0.8607059
```

5. The `plot()` command for results of the `renyi()` command uses the *lattice* package. Use the following command to customise the plot for the six Edge sites – your graph should resemble Figure 8.11:

```
> plot(renyi(gb.biol[1:6,]),
        strip = function(..., bg) strip.default(..., bg = "gray90"),
        pch = 16, col = 1)
```

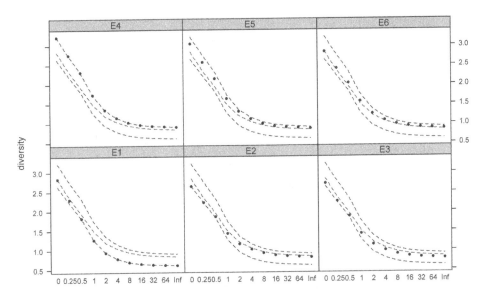

Figure 8.11 Rényi diversity for ground beetles at six Edge habitats. Dashed lines show max, min and median values.

6. Overlaying plots for individual sites is more complicated! Type in the following lines to create a graph resembling Figure 8.12:

```
> gb.r = renyi(gb.biol) # Rényi for all sites
> ## Make a plot without axes!
> plot(as.numeric(gb.r[c("E1"),]), axes = FALSE,
       ylab = "Rényi entropy", xlab = "Scale", pch = 1, col = "black",
       ylim = c(0,3.5))
> ## Add curved line to join points
> lines(spline(1:length(gb.r), gb.r["E1",]), col = "black")
> ## Make custom axes
> axis(2) # Basic y-axis
> axis(1, at = 1:length(gb.r), labels = names(gb.r)) # Custom x-axis
> box() # A bounding box around plot
> ## Add points and then line for site G1
> points(1:length(gb.r), gb.r["G1",], pch = 2, col = "blue")
```

```
> lines(spline(1:length(gb.r), gb.r["G1",]), col = "blue")
> ## Add points/line for site W1
> points(1:length(gb.r), gb.r["W1",], pch = 3, col = "darkgreen")
> lines(spline(1:length(gb.r), gb.r["W1",]), col = "darkgreen")
> ## Make a legend
> legend("topright", bty = "n", col = c("black", "blue", "darkgreen"),
         pch = c(1,2,3), legend = c("E1", "G1", "W1"), lty = 1)
```

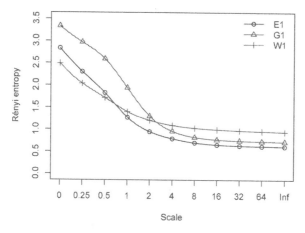

Figure 8.12 Rényi diversity for ground beetles at three habitats. Graph drawn using regular `plot()` command and custom axes.

7. The *BiodiversityR* package allows easier comparison of sites using grouping variables as well as a plotting routine. Start by looking at the *Wood* habitat only:

```
> renyiresult(gb.biol, y = gb.site, factor = "Habitat",
             level = "Wood", method = "s")
         0      0.25      0.5         1         2         4         8       Inf
W1 2.484907 2.034022 1.713855 1.397251 1.200822 1.087585 1.027615 0.9413268
W2 2.397895 2.087555 1.859356 1.582102 1.337719 1.158953 1.035063 0.9075571
W3 2.397895 2.080511 1.839470 1.539511 1.281224 1.103955 0.985282 0.8640627
W4 2.484907 2.141042 1.885299 1.601073 1.394632 1.266443 1.192693 1.0933283
W5 2.484907 2.123128 1.874481 1.616333 1.426570 1.294088 1.208760 1.0896235
W6 2.302585 1.993168 1.785181 1.552470 1.337536 1.156247 1.029842 0.9028677
```

8. The `method = "all"` instruction can combine sites, try it for the *Wood* habitat:

```
> renyiresult(gb.biol, y = gb.site, factor = "Habitat",
             level = "Wood", method = "all")
          0      0.25      0.5         1         2         4         8       Inf
all 2.564949 2.150629 1.866576 1.566727 1.343943 1.194022 1.089308 0.960695
```

9. In the *BiodiversityR* package not all the scales are shown by default. You can use the scales instruction to specify the scales you require. Use the `renyicomp()` command to view all the scales and compare *Habitat*:

```
> renyicomp(gb.biol, y = gb.site, factor = "Habitat",
            scales = c(0, 0.25, 0.5, 1, 2, 4, 8, 16, 32, 64, Inf),
            plotit = FALSE)
          scale
Habitat        0       0.25       0.5          1         2         4         8
  Edge  3.295837 2.683979 2.156552 1.558842 1.209099 1.030054 0.9242941
  Grass 3.610918 3.187267 2.781586 2.162426 1.597457 1.273831 1.1152911
  Wood  2.564949 2.150629 1.866576 1.566727 1.343943 1.194022 1.0893080
          scale
Habitat       16         32         64        Inf
  Edge  0.8661587 0.8382644 0.8249586 0.8120686
  Grass 1.0418804 1.0082746 0.9922702 0.9767660
  Wood  1.0245044 0.9916847 0.9759441 0.9606950
```

10. You can use the `renyiplot()` command to plot results from `renyiresult()` or use the `plotit = TRUE` instruction in the `renyicomp()` command. In either event you must click with the mouse where you want the legend to appear (its top-left corner, use `legend = FALSE` to suppress a legend). Use the `renyiresult()` command to get a result then the `renyiplot()` command to produce a graph resembling Figure 8.13:

```
> gb.rr = renyiresult(gb.biol, method = "s")
> renyiplot(gb.rr[c("E1", "G1", "W1"),])
```

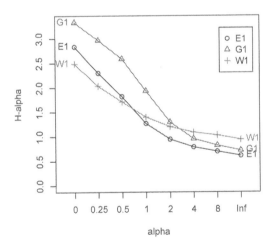

Figure 8.13 Rényi diversity for ground beetles at three habitats. Graph drawn using the `renyiplot()` command in the *BiodiversityR* package.

The *BiodiversityR* package calls on the `renyi()` command from the *vegan* package to carry out the calculations but allows you alternative ways to group and present the results.

There are two other ways to look at Rényi entropy: normalising and accumulating with additional sites.

Rényi entropy and normalising

When you compare Rényi curves for sites you can see not only the shapes of the curves but also their position relative to the y-axis. To be considered completely different in diversity the curves of two sites should not overlap. However, in some cases it would be helpful to be able to compare the shapes of the curves from a 'fixed' starting point.

At scale = 0 the Rényi entropy is related to species richness. If you subtract the Rényi result at scale = 0 from all the other scale results you effectively 'fix' the starting point at zero. This provides a means to standardise your curves and focus on the shapes of the curves.

The *BiodiversityR* package calls this normalising 'evenness' but that is something of a misnomer as the evenness springs from the shape of the curve – what you are doing is standardising one endpoint. The renyiresult() and renyiplot() commands in the *BiodiversityR* package will carry out this standardisation using the evenness = TRUE instruction. The *vegan* package does not have a built-in routine for normalising but it is relatively easy to carry out, as you can see in the following example using the ground beetle data:

```
> renyi(gb.biol[1,])
       0         0.25        0.5          1         2          4          8
2.8332133 2.2983247 1.8220359 1.2670962 0.9445357 0.7874263 0.6975544
      16          32          64        Inf
0.6520254 0.6309958 0.6209800 0.6112772
```

```
> renyi(gb.biol[1,])-renyi(gb.biol[1,])[1]
        0          0.25         0.5           1          2          4
0.0000000 -0.5348887 -1.0111775 -1.5661172 -1.8886777 -2.0457870
        8          16          32          64         Inf
-2.1356590 -2.1811879 -2.2022175 -2.2122333 -2.2219361
```

In this case the result was from a single site but you can carry out a similar procedure using several sites. The difference is that the result object is not a simple vector but a data.frame:

```
> gb.r = renyi(gb.biol)
> gb.r - gb.r[,1]
    0        0.25       0.5         1         2         4         8        16
E1 0 -0.5348887 -1.011177 -1.566117 -1.888678 -2.045787 -2.135659 -2.181188
E2 0 -0.4204514 -0.781962 -1.214552 -1.482400 -1.620277 -1.713793 -1.770173
          32         64        Inf
E1 -2.202218 -2.212233 -2.221936
E2 -1.798093 -1.811442 -1.824373
```

In the preceding example only the first two rows of the result are actually shown. The renyiresult() command in the *BiodiversityR* package will return the 'evenness' directly if you use the evenness = TRUE instruction (only the first two rows of the result are shown):

```
> renyiresult(gb.biol, evenness = TRUE, method = "s")
    0        0.25       0.5         1         2         4         8       Inf
E1 0 -0.5348887 -1.011177 -1.566117 -1.888678 -2.045787 -2.135659 -2.221936
E2 0 -0.4204514 -0.781962 -1.214552 -1.482400 -1.620277 -1.713793 -1.824373
```

The renyiplot() command can plot results in a single window and calculate 'evenness':

```
> renyiplot(gb.r[c(1,2,7,8,13,14),], evenness = TRUE,
  rainbow = FALSE, legend = FALSE)
```

This produces a graph like Figure 8.14a. Compare this with the plot you get if you use the routines in the *vegan* package (Figure 8.14b):

```
> gb.r = renyi(gb.biol[c("W2", "E2", "G2", "G1"),])
> gb.r = gb.r-gb.r[,1]
> class(gb.r) = c("renyi", "data.frame")
> plot(gb.r)
```

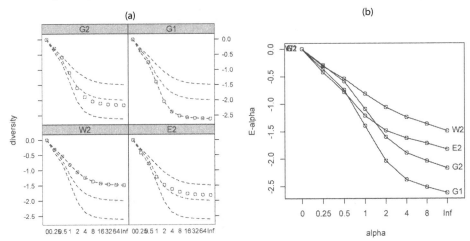

Figure 8.14 Rényi entropy 'normalised' for four samples of ground beetles. Plot (a) shows the result of the `renyiplot()` command in the *BiodiversityR* package. Plot (b) shows the results obtained using the *vegan* package with `renyi()` and `plot()` commands.

It would be possible to create a plot similar to Figure 8.14a using regular commands but you need to define custom axes along the following lines (this produces something like Figure 8.15):

```
> gb.r = renyi(gb.biol)
> gb.r = gb.r-gb.r[,1]

> ## Make a basic plot without axes!
> plot(as.numeric(gb.r["E1",]), axes = FALSE, pch = 1,
        col = "black", ylab = "Standardised Rényi",
    xlab = "Scale", ylim = c(-3,0))

> ## Add the line to join the points
> lines(spline(1:length(gb.r), gb.r["E1",]), col = "black")

> ## Create the axes
> axis(2) # add a basic y-axis
> axis(1, 1:length(gb.r), labels = names(gb.r)) # Custom x-axis
> box() # A bounding box

## Add extra points and lines to join them
> points(1:length(gb.r), gb.r["W1",], col = "darkgreen", pch = 3)
> lines(spline(1:length(gb.r), gb.r["W1",]), col = "darkgreen")
> points(1:length(gb.r), gb.r["G1",], col = "blue", pch = 2)
> lines(spline(1:length(gb.r), gb.r["G1",]), col = "blue")
```

```
> ## Add a legend
> legend("topright", bty = "n",
         col = c("black", "blue", "darkgreen"), pch = c(1,2,3),
         legend = c("E1", "G1", "W1"), lty = 1)
```

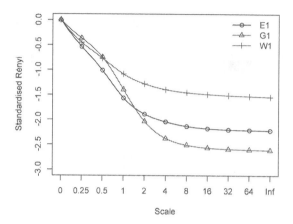

Figure 8.15 Rényi entropy 'normalised' for three samples of ground beetles. Plot was created using 'regular' R commands.

This is quite a performance as you can see!

Rényi entropy and accumulating sites

It is possible to calculate Rényi entropy with accumulating sites, you saw something similar in Section 7.3 when looking at species accumulation curves.

The general method involves several steps:

1. Randomly select some sites (or use them all).
2. Create a permutation to 'shuffle' the sites into a random order.
3. Calculate Rényi entropy for the first site.
4. Add the next site and recalculate.
5. Repeat the process of adding a site and calculating the entropy.
6. Repeat steps 2–5.
7. For each scale of the Rényi entropy calculate the mean value for accumulating sites (i.e. the entropy for one site, for two sites, for *n* sites).
8. Similar to step 7 but for the 95% confidence levels.

You can see that this is quite an involved process and not something you'd want to undertake in Excel. The *vegan* package in R allows the Rényi entropy for accumulating sites to be carried out easily via the `renyiaccum()` command. In the following example you'll see how to do this using some data built-in to the *vegan* package.

1. Start by loading the *vegan* package:

   ```
   > library(vegan)
   ```

2. The data you want are in the *BCI* dataset – this contains 50 samples of tree species

(there are 225 in total). Each sample contains counts of the tree species (> 10 cm diam.) in 1 Ha plots. Load the data using the `data()` command:

```
> data(BCI)
```

3. Create a list of sample sites to use – the `sample()` command will do this (without replacement). The `nrow()` command determines how many rows there are in the dataset. In this case 15 sites will be selected at random:

```
> i = sample(nrow(BCI), 15)
```

4. Now compute the accumulating Rényi entropy for the 15 randomly selected sites. The default for the `renyiaccum()` command uses scales 0, 0.5, 1, 2, 4, and Inf:

```
> BCI.ra = renyiaccum(BCI[i,])
```

5. The `plot()` method for results of `renyiaccum()` uses the lattice package. This is usually installed with R but will be 'loaded' as required:

```
> plot(BCI.ra)
```

You should end up with a plot that looks similar to Figure 8.16. If you try it for yourself you may end up with slightly different results because of the random sampling process. The plot shows the mean entropy and the 95% quantiles.

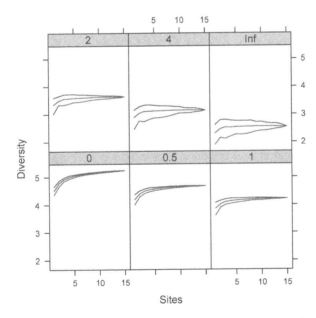

Figure 8.16 Accumulation curves for Rényi entropy for 15 randomly chosen sites (tree species on Barro Colorado Island). Lines show mean entropy at accumulating sites and 95% quantiles.

Tip: Order of panels in *lattice* plots

The panels in a *lattice* plot are labelled in a particular order – starting from the bottom left and working rightwards and up. Use the `as.table = TRUE` instruction to show panels from top to bottom rather than bottom upwards.

There are various options that you can use to help control the appearance of the plot, some of these are shown in Table 8.4.

Table 8.4 Some options for use with *lattice* plotting (e.g. as used by the `renyiaccum()` command in the *vegan* package).

Instruction/example	Result

`col`
`col = c("black", "red",`
`"blue")`
`col = 1:3`
	Sets the colour of the plotted lines (or points). The colours relate to the mean, lower quartile and upper quartile respectively. If you use a single colour it will be used for all lines/points. Colours can be specified as a name or as a number (which relates to the current colour `palette()`)

`type` `type = "l"`	Sets the type of plot produced. The default is for lines but other options include: `type = "p"` for points, `type = "b"` for both and `type = "s"` for 'stair steps' lines.

`pch`
`pch = c("+", ".", ".")`
`pch = 1`
	The plotting character to use if points are to be shown (e.g. if `type = "b"`). Characters can be specified as a number or a character (in quotes). Up to three characters can be specified (mean, lower quartile, upper quartile).

`lty` `lty = c(1, 3, 3)`	The style of line to use. This can be supplied as a number or as a character string: 1 = `"solid"` 2 = `"dashed"`, 3 = `"dotted"`, 4 = `"dotdash"`. Up to three types can be specified: (mean, lower quartile, upper quartile).

Tip: Altering the background colour of the strips in a *lattice* plot

Changing some options for *lattice* plots requires custom functions. To alter the background colour of the strips for example you can use the following custom function as an instruction in a *lattice* command:

```
strip = function(..., bg) strip.default(..., bg = "gray90")
```

8.3.4 Tsallis entropy

The Tsallis entropy is a special case of the general Boltzmann–Gibbs entropy and is used to describe diversity using different scales (Tsallis 1988). The general formula for Tsallis entropy is shown in Figure 8.17.

$$H_q = \frac{1}{1-q}\left(1 - \sum p^q\right)$$

Figure 8.17 Tsallis entropy.

In the formula, q is the scale parameter and at certain values of q the Tsallis entropy results are related to other diversity measures:

- If $q = 0$ the result is species richness – 1.
- If $q = 1$ the result is the Shannon diversity index.
- If $q = 2$ the result is the Simpson's D index.

It is possible to determine Tsallis entropy using Excel but much like for Rényi entropy it is something of a tedious process. The *vegan* package in R allows Tsallis entropy to be computed via the `tsallis()` command:

```
> divc
  [,1] [,2] [,3] [,4] [,5]
a  10    2    1    1    1
b   3    3    3    3    3
> tsallis(divc)
    0      0.2      0.4       0.6       0.8       1       1.2       1.4
a 4 2.919832 2.187421 1.683503 1.331196 1.080574 0.8989462 0.7647237
b 4 3.279873 2.710880 2.259135 1.898648 1.609438 1.3761017 1.1867361
      1.6       1.8        2
a 0.6635124 0.5856189 0.5244444
b 1.0321154 0.9050676 0.8000000
```

The default setting is to use scales 0–2 in 0.2 steps. The `scales` instruction can be used to specify different scales:

```
> tsallis(gb.biol[1:3,], scales = c(0,1,2))
    0       1         2
E1 16 1.267096 0.6111399
E2 13 1.424505 0.6854643
E3 14 1.412179 0.6781096
```

The *vegan* package provides a plotting routine for the results of the `tsallis()` command, which uses the *lattice* package in a similar way to that used by the `renyi()` command you met previously:

```
> gb.t = tsallis(gb.biol[7:12,])
> plot(gb.t, as.table = TRUE)
```

Note that in the preceding example the `as.table` = TRUE instruction was used to present the samples 'top to bottom' (Figure 8.18).

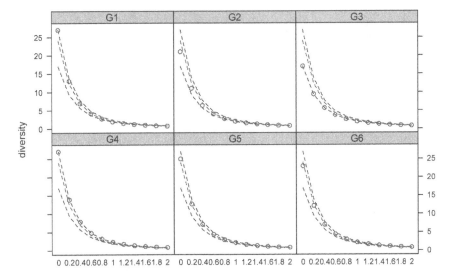

Figure 8.18 Tsallis entropy for six samples of ground beetle communities. Lines show median, max and min values for the dataset.

When you plot the Tsallis entropy the shape of the curve is concave and the value of H decreases asymptotically as q increases, as you can see from Figure 8.18.

In a similar way to the Rényi entropy it is possible to 'convert' the result into a Hill equivalent (Figure 8.19).

$$D_q = \left(1 - \left(q - 1\right)\right) H^{\frac{1}{1-q}}$$

Figure 8.19 Hill numbers equivalent for Tsallis entropy.

Unsurprisingly the Hill version of Tsallis entropy also shows relationships to other diversity measures:

- If $q = 0$ the result is species richness.
- If $q = 1$ the result is exp(H) the exponent of the Shannon diversity.
- If $q = 2$ the result is the inverse Simpson's index, i.e. $1/S$.

You can use the tsallis() command to compute the Hill version of Tsallis entropy using the hill = TRUE instruction. In the following example you can see how Tsallis$_{Hill}$ relates to Shannon and Simpson's indices:

```
> tsallis(gb.biol[13:18,], hill = TRUE, scales = c(0,1,2))
    0       1        2
W1 12 4.044067 3.322849
W2 11 4.865172 3.810341
W3 11 4.662310 3.601044
W4 12 4.958352 4.033491
W5 12 5.034595 4.164391
W6 10 4.723123 3.809643

> ## Species Richness
> rowSums(gb.biol[13:18,] > 0)
W1 W2 W3 W4 W5 W6
12 11 11 12 12 10

> ## Shannon H
> exp(diversity(gb.biol[13:18,], index = "shannon"))
      W1       W2       W3       W4       W5       W6
4.044067 4.865172 4.662310 4.958352 5.034595 4.723123

> ## Inverse Simpson's 1/S
> diversity(gb.biol[13:18,], index = "invsimpson")
      W1       W2       W3       W4       W5       W6
3.322849 3.810341 3.601044 4.033491 4.164391 3.809643
```

Tip: Comments in R

The # character is taken as the beginning of a comment in R. Anything that follows it is ignored until you press Enter and start a new line.

Tsallis entropy and normalising

It is possible to 'normalise' the Tsallis result using the maximum number of species in a sample (Figure 8.20).

$$H_{q(\max)} = \frac{S^{1-q} - 1}{1 - q}$$

Figure 8.20 Calculation for Tsallis entropy 'normalised' by number of species, S.

When using the normalising process results for $q = 0$ end up as 1, whilst for $q = 1$ the results equate to $H_1/\ln(S)$. The `norm = TRUE` instruction in the `tsallis()` command will allow you to conduct the normalisation process:

```
> tsallis(divc, norm = TRUE)
     0       0.2       0.4       0.6       0.8         1       1.2       1.4
a 1 0.890227 0.8069045 0.745198 0.7011285 0.6713983 0.6532556 0.6443923
b 1 1.000000 1.0000000 1.000000 1.0000000 1.0000000 1.0000000 1.0000000
     1.6       1.8         2
a 0.6428665 0.6470443 0.6555556
b 1.0000000 1.0000000 1.0000000
```

The normalisation allows you to visualise evenness as an even community tends to produce a flat line (Figure 8.21):

```
> plot(tsallis(divc, norm = TRUE))
```

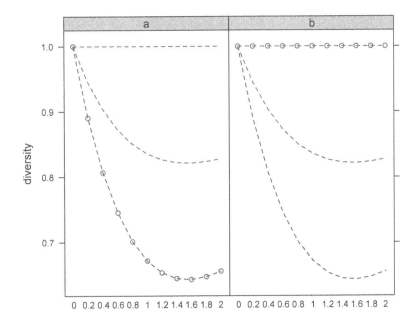

Figure 8.21 Tsallis entropy 'normalised' by number of species for two theoretical communities: (a) has a dominant species whilst (b) is completely even.

Tip: Plotting results of Tsallis entropy

The `plot()` routine for results of the `tsallis()` command uses the *lattice* package in the same way as the `renyi()` command. This means you can use the same options.

Tsallis entropy, evenness and profiling

The calculation of Tsallis entropy 'normalised' by number of species is one way to look at community evenness. If you look at a plot of 'normalised' Tsallis entropy such as Figure 8.21 you can see a convex shape. The 'bottom' of the curve corresponds to the minimum value of q (that is the scale parameter), which is usually called $q*$.

This $q*$ value can be used to create a diversity profile where the Tsallis entropy is given at this scale (that is $q*$). In R, the *vegan* library contains a command `eventstar()`, which calculates the value of $q*$ and three other values:

- The value of evenness based on a normalised Tsallis entropy at $q*$.
- The value of the Tsallis entropy at $q*$.
- The value of the Tsallis entropy at $q*$ converted to Hill numbers.

The basic form of the command is like so:

```
eventstar(x, qmax = 5)
```

In the command `x` is a community data sample (rows as samples and columns as species) or a simple numeric data vector. The `qmax` instruction sets the maximum scale to consider when carrying out the computations – generally the default value of 5 for `qmax` is plenty large enough. The following example shows the `eventstar()` command in operation:

```
> divc
  [,1] [,2] [,3] [,4] [,5]
a   10    2    1    1    1
b    3    3    3    3    3

> eventstar(divc)
       qstar      Estar      Hstar      Dstar
a 1.546469 0.4867863 0.6879814 2.369996
b 4.999922 0.9984195 0.2496048 5.000000
```

The first column, entitled `qstar`, shows the $q*$ values. The other columns show evenness, diversity and Hill number equivalents (effective species) respectively (all calculated for scale = $q*$).

Once you have the $q*$ results you can use them to help you profile the Tsallis diversity for your community data. The $q*$ value is positively associated with rare species in the community, making it is a potentially useful indicator of certain relative abundance distributions of the communities. In the following exercise you can have a go at creating a profile plot showing evenness, diversity (Tsallis entropy) and effective species (Hill number equivalents) using the ground beetle data that should be quite familiar by now.

Have a Go: Make Tsallis profiles

For this exercise you will need the ground beetle data with abundance information that you met earlier. The data are in a CSV file, ready to import to R, called *Ground beetles and abundance.csv*. You will also need the data that contains the site information, *GBsite.csv*. If you already have the data in R you can go directly to step 2.

1. Start by opening R and using the `read.csv()` command to get the data from the *Beetles and habitat.csv* and *Gbsite.csv* files:

```
> gb.biol = read.csv(file.choose(), row.names = 1)
> gb.site = read.csv(file.choose(), row.names = 1)
```

2. Prepare the *vegan* package by using the `library()` command:

```
> library(vegan)
```

3. Start by converting one sample from the community data into a simple numeric variable:

```
> y = as.numeric(gb.biol[6,])
```

4. Now calculate the q^* results for your chosen community:

```
> (z = eventstar(y))
        qstar      Estar     Hstar     Dstar
1 0.7490301 0.1217673 2.190627 5.729814
```

5. Set the scales for use in the `tsallis()` command:

```
> q = seq(0, 2, 0.05)
```

6. Next, determine the various Tsallis profiles:

```
> Eprof = tsallis(y, scales=q, norm=TRUE) # Evenness
> Hprof = tsallis(y, scales=q)            # Basic Tsallis Entropy
> Dprof = tsallis(y, scales=q, hill=TRUE) # Effective species (Hill)
```

7. Make a new graphic window and split it into three parts – one for each profile:

```
> par(mfrow=c(3,1))
```

8. Plot the evenness profile using q as the x values and the evenness results as the y values:

```
> plot(q, Eprof, type="1", main="Evenness")
```

9. Add lines to show the value of q^* (a vertical line) and the value of the minimum evenness (a horizontal line):

```
> abline(v=z$qstar, h=tsallis(y, scales=z$qstar, norm=TRUE), col=2)
```

10. Now plot the diversity profile and add lines for q^* and the Tsallis entropy at a scale of q^*:

```
> plot(q, Hprof, type="1", main="Diversity")
> abline(v=z$qstar, h=tsallis(y, scales=z$qstar), col=2)
```

11. Next, plot the profile for effective species (Hill number equivalents) and add lines to show q^* and the effective species for a scale of q^*:

```
> plot(q, Dprof, type="l", main="Effective number of species")
> abline(v=z$qstar, h=tsallis(y, scales=z$qstar, hill=TRUE), col=2)
```

12. Finally, reset the graphics window back to a single plot:

```
> par(mfrow = c(1,1))
```

The final graph looks like Figure 8.22.

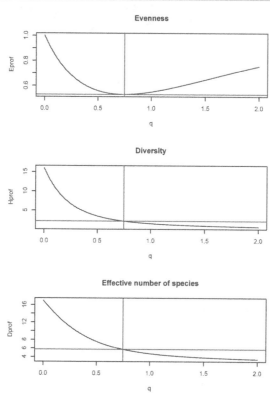

Figure 8.22 Tsallis entropy profiles for a ground beetle community.

The Tsallis q^*, associated evenness, diversity and effective species values make a good summary of a diversity profile.

Tip: Setting graphical parameters

Many graphical parameters can be set via the par() command. You can save current settings to a named object and use this to reset values easily at a later time. For example:

```
> opar = par(mfrow = c(3,1)) # Split graphics window into 3 rows
> # Various commands
> par(opar) # Restore previous settings
```

The opar object holds the settings that were current when you created it.

Tsallis entropy and accumulating sites

It is possible to calculate Tsallis entropy for accumulating sites in much the same way as for the Rényi entropy you saw previously. In the following example you'll see the process for the Barro Colorado island tree data that comes with the *vegan* package:

1. Load the *vegan* package and make sure the *BCI* data is ready:

   ```
   > library(vegan)
   > data(BCI)
   ```

2. There are 50 sites in the data – randomly select 15 rows using the sample() command:

   ```
   > i = sample(nrow(BCI), 15)
   ```

3. Use the randomly selected rows as the data in the accumulation computations:

   ```
   > BCI.ta = tsallisaccum(BCI[i,])
   ```

4. The plot() command uses the *lattice* package for the results of tsallis() and renyi() commands:

   ```
   > plot(BCI.ta, as.table = TRUE, col = c("black", "blue", "blue"),
     lty = c(1,3,3))
   ```

The final result should look like Figure 8.23.

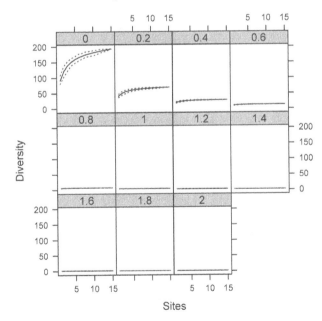

Figure 8.23 Accumulation curves for Tsallis entropy for 15 randomly chosen sites (tree species on Barro Colorado Island). Lines show mean entropy at accumulating sites and 95% quantiles.

There are various additional instructions that you can pass to the plot() command – see Table 8.4 for a list of commonly used options.

8.4   Summary

Topic	Key Points
Diversity indices	A diversity index takes species abundance into account as well as the number of species in a sample. Most diversity measures can be calculated using Excel but it is usually faster and easier to use R.
Simpson's index	Simpson's index is a commonly used measure of diversity. There are several forms of the index with the most popular ranging between 0 and 1 where 1 is maximally diverse. This index is probabilistic and gives the likelihood of a different species being sampled on subsequent events.
	You can calculate the index using the `diversity()` command in the *vegan* package.
Shannon entropy	The Shannon entropy (or Shannon–Wiener or Shannon–Weaver index) derives from information theory and uses the log of species abundance in its calculations. The Shannon entropy has very important properties in measurements of diversity and is linked to many other diversity measures in various ways.
	You can calculate Shannon entropy using the `diversity()` command.
Effective species	The number of effective species is derived from a measure of diversity (e.g. Simpson's or Shannon). If your sample was perfectly even (all species the same abundance) then the value equals the species richness.
	The effective species value is considered to be 'true diversity', especially when calculated from the Shannon entropy.
	For Shannon entropy, H, the effective species are $\exp(H)$.
	For Simpson's D index the effective species are $1/(1 - D)$.
Evenness	Evenness is a measure of dominance. If all species in a sample had the same abundance the evenness would be high. If one species was dominant then the evenness would be lower.
	One way to measure evenness is to take the effective species and divide by species richness.
Berger–Parker dominance index	The Berger–Parker index is a measure of diversity that focusses on the most abundant species in your samples. This means that it is a measure of dominance.
	You can work out the dominance index using basic commands in R but the `diversityresult()` command in the *BiodiversityR* package will also do the job.
Fisher's alpha	Fisher's alpha is a measure of diversity that is derived by using a logarithmic model of species abundance. Essentially you arrange your community in decreasing order of abundance and then plot a graph of the log abundance against the rank.
	You can determine Fisher's alpha using the `fisher.alpha()` and `fisherfit()` commands in the *vegan* package.
Rényi entropy	Rényi entropy is a measure of diversity that is calculated for several scales (not related to geography). As the scale becomes larger, rarer species are increasingly downrated.

You can calculate Rényi entropy using the `renyi()` command.

The Rényi entropy is linked to other diversity measures by so-called Hill numbers, which are derived from the exponent of the Rényi entropy.

You can calculate Rényi entropy for accumulating sites and normalising allows you to compare samples directly.

The `renyiaccum()` command in the *vegan* package will carry out the accumulation (and plot results) but not automatic normalisation. The `renyiresult()` command in the *BiodiversityR* package will also carry out the accumulation but will conduct normalisation too.

Tsallis entropy   Tsallis entropy is a measure of diversity that is calculated for several scales (not geographical). As the scale becomes larger, rarer species are increasingly downrated, so at scale = 0 the entropy is equal to species richness.

The `tsallis()` command will calculate Tsallis entropy.

You can calculate Tsallis entropy for accumulating sites and normalising allows you to compare samples directly; the `tsallisaccum()` command will do both.

The minimum value in a Tsallis profile is known as $q*$; knowing this allows you to calculate diversity statistics at a scale relating to the minimum evenness of a community data sample; the `eventstar()` command will do this.

8.5   Exercises

8.1   The Simpson's and Shannon indices are calculated in a similar manner. What is the main difference between the two?

8.2   The *DeVries* data contains the abundances of tropical forest butterflies for two habitats (canopy and understorey). What is the evenness of the two samples as measured using the Shannon entropy? Note: there are two possible answers.

8.3   Plotting the Rényi (or Tsallis) entropy allows you to see a diversity profile. If the plot 'flatlines' what does this indicate in terms of the diversity?

8.4   The Rényi entropies are related to other diversity measures. When the scale = 0 for example the result = richness – TRUE or FALSE?

8.5   Calculate the values for the Tsallis entropy at minimum evenness for the *DeVries* data.

The answers to these exercises can be found in Appendix 1.

9. Diversity: comparing

Calculating diversity is relatively straightforward (see Chapter 8) but comparing diversity is not so simple. When you have simple species richness things are fine – you can use the richness like any other response variable and use 'classic' hypothesis tests.

The way diversity indices are calculated makes things more challenging. Simply put, the values for diversity do not 'partition' neatly when you combine or split samples. Look at a fairly simple example as illustration of the problem. Here are some simple community data:

```
> m
           A    B   C    D   E   F   G    H    I
Sample1   100   35  0   10   5   2   1    0    0
Sample2    11   33  0   44   0   1   1   22    7
```

You can calculate the diversity for each sample easily enough:

```
> diversity(m, index = "shannon")
     Sample1    Sample2
   0.9950553  1.5027349
```

The mean of these two samples is also easy to determine:

```
> mean(diversity(m, index = "shannon"))
[1] 1.248895
```

Now combine the two samples by adding them together and take the diversity index again:

```
> diversity(colSums(m), index = "shannon")
[1] 1.490202
```

The diversity index of the amalgamated sample is not the same as the mean of the two smaller samples! This can be a problem if you want to use 'regular' statistical hypothesis tests to compare samples. You can use regular hypothesis tests to compare diversity indices but you should consider if your approach is appropriate – for example, is it biologically valid to measure diversity in a quadrat? If you consider that your samples are large enough to be valid then there are two main ways to compare diversity:

- Using a grouping variable.
- Using a continuous variable.

The first is akin to differences tests whilst the second is akin to correlation. If you decide to use regular hypothesis tests then you can check the results for normality and proceed as if you had any response variable. The use of replicated samples for comparisons (and correlations) is covered in Section 9.5.

The two mainstays of diversity, Simpson's index and Shannon entropy, have received special attention and versions of the *t*-test for comparison of two samples have been worked out (see Section 9.2). If you have two simple samples and wish to compare their diversity indices then this is one way you could proceed. You can also use a purely graphical method to visualise differences in diversity indices for multiple sites, foregoing any hypothesis test – you'll see this approach in Section 9.3.

An approach that is becoming increasing popular is use of bootstrapping or randomisation. Randomisation methods work by repeatedly sampling randomly and generating new samples from the original data. If you do this often enough you can begin to estimate the probability of samples being different as you can get confidence intervals for your samples. You can see a range of randomisation and bootstrap methods in Section 9.4. One advantage of bootstrapping is that the distribution of the data is unimportant, so you do not need to worry about having normally distributed data.

In summary, your options depend on what kind of samples you have:

1. One sample for each community:

    - Special version of *t*-test – for comparing diversity using a grouping variable.
    - Correlation – for comparing diversity to a continuous variable.
    - Bootstrapping.

2. Multiple samples for each community:

    - 'Regular' comparisons – for comparisons using grouping variables.
    - Correlations/regression – for comparisons with continuous variables.
    - Bootstrapping – can be used for either.

A graphical representation of a diversity profile using Rényi or Tsallis entropy can show you differences in samples. Although this is not a statistical method it can be useful to visualise your data. This is a good starting point for looking at comparing diversities, which you will see in the next section.

9.1   Graphical comparison of diversity profiles

It is always a good idea to visualise your data and results if you can. Comparisons of diversity are problematic because, as you have seen, of the influence of scale. Any one measure of diversity will take into account rare species to a greater or lesser extent.

Using Rényi or Tsallis diversity helps to overcome the influence of scale (and so rare species) because the entropy is calculated over a range of scales. You saw earlier how to create a graph to compare Rényi entropy (Section 8.3.3) – drawing a plot of Tsallis entropy can be done in a similar manner.

The *BiodiversityR* package contains a command, `renyiplot()`, that allows you to draw a plot slightly more easily but there is no corresponding command for Tsallis entropy. It is fairly easy to produce a few lines in R that will draw a plot. In the following example you can have a go at carrying out the process, which will work for Rényi or Tsallis entropies.

Have a Go: Plot Tsallis entropy curves to compare samples

You'll need the *gb.biol* data for this exercise. These data are in the *CERE.RData* file and show the abundance of ground beetles in three habitats; each habitat has six replicate samples. You'll also need the *vegan* package.

1. Start by ensuring the vegan package is ready for use:

    ```
    > library(vegan)
    ```

2. Now make a result using the `tsallis()` command. Use the first three rows of the ground beetle data:

    ```
    > gb.t = tsallis(gb.biol[c(1,7,13),])
    ```

3. Now you need to rotate the result so that the columns represent the samples and the rows correspond to the scales – the `t()` command will rotate the result and convert to a matrix too (which is useful, as you'll see):

    ```
    > entropy = t(gb.t)
    ```

4. You now need to make a list of labels to use for the x-axis – these are the scales and you can get them from the row names of the result:

    ```
    > q = rownames(entropy)
    ```

5. To make a plot you can use the `matplot()` command, which plots the columns of one matrix against the columns of a second matrix. You need to set the first matrix to be the x-axis (the scales) but tell R to ignore the results! The second matrix will be the entropy calculations:

    ```
    > matplot(q = NULL, entropy, axes = FALSE, type = "l",
              col = 1:3, lty = 1:3,
              xlab = "Scale", ylab = "Diversity")
    ```

6. Now you need to restore the axes. Axis 1 is the bottom and this needs customising to reflect the scale labels:

    ```
    > axis(1, at = 1:length(q), labels = q)
    > axis(2)
    > box()
    ```

7. Finally add a legend to the plot, which should resemble Figure 9.1:

    ```
    > legend("topright", legend = colnames(entropy),
             bty = "n", col = 1:3, lty = 1:3)
    ```

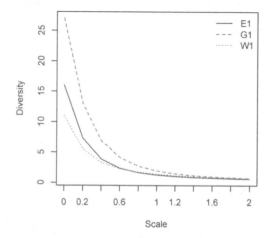

Figure 9.1 Tsallis entropy for three samples of ground beetles.

Although this is not a test of significance the plot does allow you to visualise the situation.

The method outlined in the preceding exercise would work equally well for Rényi entropy, which you can determine via the `renyi()` command in the *vegan* package.

Note: Book examples

You can find examples of spreadsheets and R code in the download file on the companion website. Instructions are in the Introduction and on the website.

Other diversity indices could be plotted to help you visualise the differences between samples. An ideal plot would show you not just the value of the diversity index but give you an impression of the variability in the data. In the following sections you will see how to compare sites using various statistical methods. You will see ways to present the results graphically along with each of these methods.

9.2 A test for differences in diversity based on the *t*-test

A regular *t*-test compares means of two samples – the means being calculated from replicated data. However, you may not have replicated data – you may have a single sample from which you calculate a Shannon diversity index for example. In this case you can compare the Shannon index of two samples using a modified version of the *t*-test (Hutcheson 1970).

The basic form of the *t*-test (Figure 9.2) is like that of a regular *t*-test; the top of the formula being simply the difference between the two Shannon indices.

$$t = \frac{H_a - H_b}{\sqrt{s_{H_a}^2 + s_{H_b}^2}}$$

Figure 9.2 The *t*-test formula used to compare Shannon diversity indices.

The bottom of the formula shows that you require the variance (s^2) of each sample (Figure 9.3). This looks rather complicated but the elements are fairly simple; p is the proportion of each species, N is the total number of individuals and S is the number of different species (the species richness).

$$s_H^2 = \frac{\sum p.(\ln p)^2 - \left(\sum p.\ln p\right)^2}{N} + \frac{S-1}{2N^2}$$

Figure 9.3 Formula to calculate the variance of a Shannon diversity index, where N = total abundance and S = species richness.

Once the formulae have been evaluated you end up with a *t*-value, which you need to compare with a table of critical values. In order to do this you need to know the *degrees of freedom* (this is related to sample size). The formula to calculate degrees of freedom is shown in Figure 9.4.

$$df = \frac{\left(s_{H_a}^2 + s_{H_b}^2\right)^2}{\left(\dfrac{s_{H_a}^2}{N_a} + \dfrac{s_{H_b}^2}{N_b}\right)}$$

Figure 9.4 Formula to determine degrees of freedom in a *t*-test for Shannon indices.

9.2.1 Using Excel to carry out a *t*-test on Shannon indices

It is relatively easy to calculate the formulae for the *t*-test in Excel (although slightly tedious) and shortly you can have a go for yourself. However, before that you should think about how you are going to establish that statistical significance of the resulting *t-value*.

Excel can look up critical values of *t* for you and can also determine the probability of obtaining a *t*-value for any sample size (degrees of freedom). The function that calculates *t*-values is TINV and has the following form:

= TINV(*prob, df*)

You give a numeric value for the probability you want to assess (*prob*), 0.05 is the usual level of significance in ecological work. You also need to indicate the degrees of freedom (*df*). If your value of *t* is greater than (or equal to) the critical value then the result is statistically significant at the probability level you indicated in *prob*.

You can also determine the probability of obtaining a *t*-value for a given number of degrees of freedom (df) using the TDIST function, which has the following general form:

= TDIST(*tval, df, tails*)

You give a value for the *t*-value obtained (*tval*), the degrees of freedom (*df*) and a value for the number of tails (*tails*). Usually you set this to 2 – you are essentially saying that you do not know if one sample is more diverse than the other and you simply wish to know if they are different.

In the following exercise you can have a go at creating a spreadsheet that will carry out a t-test on Shannon indices of two samples.

Have a Go: Make a spreadsheet to carry out a *t*-test between Shannon indices

Start with a blank spreadsheet; usually you will get three worksheets by default. You will use two for calculating the statistics for each site and the third sheet to carry out the final *t*-test and present the results. If you need to make new worksheets you can click on the button at the end of the list of worksheets towards the bottom of the screen, you can also click Shift+F11 or right-click on a worksheet name tab and select *Insert*.

1. Start by typing some headings. In cell A1 type 'Site Name'. In row 2 type headings into columns A–F like so: 'Taxon', 'Count', 'P', 'Ln(P)', 'P*Ln(P)' and 'P*Ln(P)²'. You can make a superscript (or subscript) by clicking in the formula bar after you've entered a cell. Use the mouse to highlight the parts you want superscript and then right-click, choose *Format Cells* and then apply the formatting you want to that selection.

2. Decide how many species you want to leave room for, you can always add more rows so begin by allowing for 20. Type a heading into cell A23, 'Total'.

3. In cell B23 type a formula to work out the total number of individuals: =SUM(B3: B22).

4. Now type a formula in cell C3 to determine the proportion that the species in this row makes to the total: =B3/$B$23. Note that you need to use the dollar signs to 'fix' the B23 cell location into to the formula.

5. Copy the formula in C3 down the entire column to cell C22. Now copy the formula for the sum across from cell B23 into C23. The result will always be 1 unless your spreadsheet is completely blank, when you'll get an error #DIV/0. Fill in values to the first few rows of the *Count* column (B); it helps to have some values in place when creating a spreadsheet.

6. Go to cell D3 and type in a formula to calculate the natural log of the proportion for the first species: = Ln(C3). If the count is zero or blank then the proportion will be 0 and you'll get an error. Modify the formula to take care of blank counts: =IF(B3="","",LN(C3)).

7. The formula in cell D3 will produce a blank ("") if the count is empty but return the natural log of the proportion if not. Copy the formula down the rest of the D column. You can now copy the formula from C23 into cell D23 to give a total.

8. In cell E3 you need to add a formula that multiplies the proportion by the natural log of the proportion (column C * column D). You'll get an error if the original count is blank so take care of it like so: =IF(B3="","",C3*D3). Copy the formula down the rest of the E column. You can now copy the formula from D23 into cell E23 to give a total.

9. In cell F3 you need to add a formula that multiplies the proportion by the square of the natural log of the proportion. Once again you need to take care of possible blank counts: =IF(B3="","",C3*D3^2).

10. The D3^2 part in the formula of step 9 will be evaluated before the * part so you do not need additional parentheses. Copy the formula down the rest of the F column and give the column a total by copying across the formula from cell E23 into F23.

11. In cell A24 type a heading for the number of different species, 'Richness'. This will be the S value from the variance calculation (Figure 9.3). Calculate the species richness by typing in a formula to cell B24: =COUNT(B3:B22).

12. You are now nearly ready to tackle the final calculations, begin by making some headings: in cell A235 type 'SS', in A26 type 'SQ', in A27 type 'H' and in A28 type 'SH'.

13. The SS value will correspond to the sum of the F column so in cell B25 type a formula: =F23.

14. The SQ value will correspond to the sum of column E squared so in cell B26 type a formula: =E23^2.

15. In cell B26 type a formula: =-E23 to represent the Shannon index value; this is the negative sum of column E.

16. The variance formula is complicated but you have simplified it somewhat by working out the SS and SQ values. Type a formula corresponding to Figure 9.3 like so: =((B25-B26)/B23)+((B24-1)/(2*B23^2)). Be especially careful with the parentheses.

17. Now you have the variance calculations for one sample. You can simply copy the entire sheet to a new worksheet; use the mouse or type CTRL+A to highlight all the cells and then use copy. Now paste the cells into Sheet2. You should now have two worksheets each with their own variance calculations.

18. Select the third worksheet – this will contain the final *t*-test calculations and a summary of the results. Begin by typing some headings into column A: 'Site', 'Total', 'Richness', 'H', 'S2H', 't', 'df', 'Crit' and 'p'.

19. In cell B1 type a formula that will collect the name of the first sample from Sheet1: =Sheet1!B1. You can type this directly or use the mouse to select the appropriate cell from the worksheet. Do the same for the second sample name in cell C1: =Sheet2!B1. This will reflect any sample name you type into cell B2 of each sample worksheet.

20. Now use similar formulae to the ones in step 19 to fill in cells B2:B5 and C2:C5. These simply collect the values from the separate sheets into one place as a summary and to help carry out the *t*-test formulae. You will have to type these individually, copy and paste is not a lot of help here.

21. In cell B6 type a formula for the t-test (Figure 9.2) like so: =ABS(B4-C4)/SQRT(B5+C5). Note the ABS part, which makes sure you do not get a negative value (the magnitude of t is important, not its sign).

22. In cell B7 type a formula to determine the *degrees of freedom* (Figure 9.4) like so: =((B5+C5)^2)/((B5/B2)+(C5/C2)).

23. In cell B8 type a formula to determine the critical value of t for the appropriate degrees of freedom at the 5% ($p = 0.05$) level like so: =TINV(0.05,C11).

24. In cell B9 type a formula to determine the probability of obtaining the calculated *t*-value given the degrees of freedom: =TDIST(C10,C11,2).

25. You now have all the elements in place. Note that with small samples the degrees of freedom may be less than one, which will give an error. Save the spreadsheet to a file with an appropriate name.

Later on you can have a go at creating a graph to help visualise the difference in Shannon indices. If you need larger sample sizes then you can simply add new rows but you will need then to copy the formulae for columns A–F into the newly created cells. You might also wish to add borders and other formatting to the spreadsheet in order to make it easier to 'read'.

You might like to carry out a test on some of the ground beetle data as a practice of your new spreadsheet. Open the *Ground beetles.xls* spreadsheet and make a Pivot Table – use a filter to view a single site and copy the abundance data to the clipboard.

Note: Shannon diversity significance spreadsheet

You can find a spreadsheet to calculate the significance of the difference in Shannon indices with the download file on the companion website – the file is called *Shannon stats test.xls*.

Note: Excel and degrees of freedom

The TINV and TDIST functions in Excel are useful to determine critical values of t and statistical probability. However, Excel cannot deal with *degrees of freedom* that are not integer values and rounds up to the nearest integer. This results is small differences to the 'true' values when your *df* are not whole numbers.

9.2.2 Using R to carry out a *t*-test on Shannon indices

It is relatively easy to use R to carry out the computations required to perform the *t*-test on Shannon indices. In the following exercise you can have a go at carrying out the computations for yourself.

Have a Go: Use R to conduct a *t*-test on unreplicated Shannon entropy

You'll need the *DeVries* data for this exercise. The data are in the *CERE.RData* file and show the abundance of tropical forest butterflies for two samples, understory and canopy (Jost 2006).

1. To start with have a look at the data to see what you are dealing with:

```
> DeVries[,1:6]
       Hisache Panprol Neshewi Morachi Taysp.1 Coldirc
canopy    1882    1028      19       5       8     273
under       26     535     984     751     621     250
```

2. Assign the first sample to a variable called *x*, then assign the second sample to a variable called *y*:

```
> x = DeVries[1,]
> y = DeVries[2,]
```

3. Now you need to calculate the total number of individuals in each sample:

```
> Nx <- sum(x, na.rm = TRUE)
> Ny <- sum(y, na.rm = TRUE)
> Nx ; Ny
[1] 5774
[1] 5922
```

4. Now determine the number of species in each sample:

```
> Sx <- sum(x > 0, na.rm = TRUE)
> Sy <- sum(y > 0, na.rm = TRUE)
> Sx ; Sy
[1] 56
[1] 65
```

5. Calculate the proportions that each species contributes to the total abundance for each sample:

```
> Px <- x/Nx
> Py <- y/Ny
```

6. The Shannon indices can be determined without needing the *vegan* package:

```
> Hx <- -sum(log(Px) * Px, na.rm = TRUE)
> Hy <- -sum(log(Py) * Py, na.rm = TRUE)
> Hx ; Hy
[1] 2.640651
[1] 2.982633
```

7. The variance calculations are quite involved so it is easier to split them into separate chunks – start with sample x:

```
> SSx <- (sum(Px*log(Px)^2,na.rm=TRUE))     # intermediate calc
> SQx <- Hx^2# ditto
> Vx <- ((SSx-SQx)/Nx) + ((Sx-1)/(2*Nx^2)) # final variance
> Vx
[1] 0.0003865389
```

8. Repeat the calculations in step 7 for sample y:

```
> SSy <- (sum(Py*log(Py)^2,na.rm=TRUE))
> SQy <- Hy^2
> Vy <- ((SSy-SQy)/Ny) + ((Sy-1)/(2*Ny^2)) # variance
> Vy
[1] 0.0002548043
```

9. The degrees of freedom can be calculated using the variance from steps 7 and 8 along with the total abundances (step 3):

```
> d.f <- ((Vx + Vy)^2)/((Vx/Nx)+(Vy/Ny))
[1] 3.740253
```

10. Now determine the value of the *t*-statistic:

```
> Ht <- abs(Hx - Hy) / sqrt(Vx + Vy) # t-statistic
[1] 13.50384
```

11. You can now calculate the probability using the *t*-value and the degrees of freedom, note that the calculated value is multiplied by two to get the two-tailed probability:

```
> Pval <- pt(Ht, df = d.f, lower.tail = FALSE) * 2
[1] 0.0002608309
```

12. For interest you can determine the critical value for t at $p = 0.05$:

```
> Crit <- qt(0.05/2, df = d.f, lower.tail = FALSE)
[1] 2.854166
```

You can see that the difference in the Shannon entropy is highly significant.

Tip: Reusing R commands

Once you have a series of commands you can save them as plain text and use copy/paste to use them as required. Use a plain text editor like Notepad because smart quotes, which are often used in word processors, are not recognised by R.

By calling your samples by standard names (e.g. *x* and *y*) you are able to use the same lines of R commands over and over again.

Tip: Note: Shannon diversity significance R code

The R code shown here can be found with the download material as part of the *CERE. RData* file. Instructions for using the R functions and data are on the website and also in the Introduction. Once the file is loaded into R the custom function H_sig() allows you to carry out a significance test of the difference between Shannon indices of two samples.

Once you have your result you need to think about the biological relevance of any difference and also how you might represent the result graphically. You'll see graphical methods for displaying *t*-test results in Section 9.3.

9.2.3 Comparing multiple sites using the *t*-test

You saw in the previous section how to use a modified version of the *t*-test to compare the Shannon indices of two sites. If you have several sites then you can still use the *t*-test but you have to compare sites pair by pair. The problem here is that whenever you run multiple tests you are increasing the probability of getting a significant result by random chance. Think of it in terms of dice – your chances of rolling a 6 are 1 in 6 if you have only one die. However, if you have three dice then your chances of rolling a 6 are three times greater.

What you have to do is to make a correction for multiple testing. There are several methods to do this. The simplest is known as the *Bonferroni correction* and to carry it out you multiply each *p*-value by the number of tests carried out.

If you are using Excel then you can simply copy the *p*-values to a new spreadsheet each time you carry out a test. Then you can multiply the list of *p*-values by the number of items in the list. So if you had three comparisons you would multiply each result by 3.

If you are using R you can do something similar – you can take each *p*-value and add it to a numerical vector. Once you are done you can multiply the vector of *p*-values by the number of items in it as the following example shows:

1. Determine the first *p*-value and view the result:

   ```
   > Pval
   [1] 0.003748181
   ```

2. Save the result to a new object:

   ```
   > comp = Pval
   ```

3. Now calculate another *p*-value using a different pair of samples:

   ```
   > Pval
   [1] 0.1226225
   ```

4. Add the new *p*-value to the previous result that was stored in the comp object:

   ```
   > comp = c(comp, Pval)
   > comp
   [1] 0.003748181 0.122622517
   ```

5. Now calculate another *p*-value using a different pair of samples:

   ```
   > Pval
   [1] 0.01103993
   ```

6. Add the third result to the previous *p*-values in the `comp` vector object:

```
> comp = c(comp, Pval)
> comp
[1] 0.003748181 0.122622517 0.011039927
```

7. Now apply a Bonferroni correction by multiplying the vector of results to the number of items in it (that is, the number of comparisons, which is three in this example):

```
> comp * length(comp)
[1] 0.01124454 0.36786755 0.03311978
```

There are other methods of adjusting for multiple comparisons – the Bonferroni correction is very conservative and others are less so. You can apply a range of corrections to a vector of *p*-values by using the `p.adjust()` command:

```
p.adjust(p, method, n)
```

In the command, p is a vector of *p*-values, `method` is the name of the method to use (in quotes – see list below) and n is the number of comparisons. This final value, n, defaults to the number of *p*-values in the vector, p. In most cases you should leave out the n instruction and use the default.

There are several methods that you can apply:

- `"bonferroni"` – the Bonferroni correction.
- `"holm"` – the method of Holm.
- `"hochberg"` – the method of Hochberg.
- `"hommel"` – the method of Hommel.
- `"BH"` – the method of Benjamini & Hochberg.
- `"fdr"` – same as above.
- `"BY"` – the method of Benjamini & Yekutieli.
- `"none"` – do not apply any correction.

For most occasions the `"holm"` method is probably the best to use – the Bonferroni correction is very stringent and dominated by Holm's method.

In the following example you can see the differences between some of the adjustment methods applied to the *p*-values that resulted from the previous example:

```
> comp
[1] 0.003748181 0.122622517 0.011039927

> p.adjust(comp, method = "holm")
[1] 0.01124454 0.12262252 0.02207985

> p.adjust(comp, method = "bonferroni")
[1] 0.01124454 0.36786755 0.03311978

> p.adjust(comp, method = "hoch")
[1] 0.01124454 0.12262252 0.02207985

> p.adjust(comp, method = "hommel")
[1] 0.01124454 0.12262252 0.02207985
```

The *t*-test method of comparison has limitations, not least that you have to adjust the result according to the number of comparisons you made. It is also limited to the Shannon diversity index. There is a formula to calculate confidence intervals for Simpson's index but it is probably better to use more 'direct' methods of estimating differences by using bootstrap-ping methods – two of these are shown in the following section.

9.2.4 Comparing Simpson's index

It is possible to determine the variance in Simpson's index. There are two versions of the formula, the first is an approximation that you can use for large samples (Figure 9.5).

$$s_D^2 = \frac{\sum\left(\frac{n}{N}\right)^3 - \left[\sum\left(\frac{n}{N}\right)^2\right]^2}{0.25N}$$

Figure 9.5 Formula to determine variance in Simpson's diversity index for large samples, where n = number of individuals in a species and N = total number of individuals in sample.

Unfortunately the definition of what is a 'large' sample is somewhat vague. The complete formula is shown in Figure 9.6, and this is probably the 'safest' version to use for any sample size.

$$s_D^2 = \frac{4N(N-1)(N-2)\sum\left(\frac{n}{N}\right)^3 + 2N(N-1)\sum\left(\frac{n}{N}\right)^2 - 2N(N-1)(2N-3)\left[\sum\left(\frac{n}{N}\right)^2\right]^2}{\left[N(N-1)\right]^2}$$

Figure 9.6 Formula to determine variance in Simpson's diversity index, where n = number of individuals in a species and N = total number of individuals in sample.

The formula in Figure 9.6 looks horrible but the computations are fairly simple, just long-winded. Once you have determined the variance of a sample you could calculate a *t-value* for the difference between two samples using the formula shown in Figure 9.2. The problem comes with how to work out the *degrees of freedom*.

The most commonly used method is to take the total number of species and subtract two; more formally this would be written like so:

$$df = (N_1 - 1) + (N_2 - 1)$$

This produces a result that is generally regarded as being 'generous'; however, the version used for the Hutcheson *t*-test (Figure 9.4) is far too conservative.

R and the t-test for Simpson's index

In the following exercise you can have a go at calculating a *t*-value and significance for the difference between two samples of ground beetles.

Have a Go: Use R to carry out a *t*-test between Simpson's Index of two samples

For this exercise you will need the ground beetle data with abundance information that you met earlier. The data are in a CSV file, ready to import to R, called *Ground beetles and abundance.csv*. If you already have the data in R you can go directly to step 2.

1. Start by opening R and using the `read.csv()` command to get the data from the *Beetles and habitat.csv* file:

   ```
   > gb.biol = read.csv(file.choose(), row.names = 1)
   ```

2. Start by selecting two rows of the data to be the samples to compare, rows 1 and 2 will do. Assign one sample to *x* and the other to *y*:

   ```
   > x = gb.biol[1, ]
   > y = gb.biol[2, ]
   ```

3. Determine the number of individuals in each sample:

   ```
   > Nx = sum(x)
   > Ny = sum(y)
   > Nx ; Ny
   [1] 715
   [1] 734
   ```

4. Now calculate the number of species in each sample:

   ```
   > Sx = sum(x > 0)
   > Sy = sum(y > 0)
   > Sx ; Sy
   [1] 17
   [1] 14
   ```

5. Calculate the proportions of each species in its community:

   ```
   > Px = x / Nx
   > Py = y / Ny
   ```

6. Now you can calculate the Simpson index for each sample:

   ```
   > Dx = 1 - sum(Px^2)
   > Dy = 1 - sum(Py^2)
   > Dx ; Dy
   [1] 0.6111399
   [1] 0.6854643
   ```

7. It is best to split the variance calculation into parts; look at sample *x* first:

   ```
   > num1x = 4*Nx*(Nx-1)*(Nx-2)*sum(Px^3)
   > num2x = 2*Nx*(Nx-1)*sum(Px^2)
   > num3x = 2*Nx*(Nx-1)*(2*Nx-3)*sum(Px^2)^2
   > denx = (Nx*(Nx-1))^2
   >    Vx = (num1x + num2x - num3x) / denx
   > Vx
   [1] 0.0001937667
   ```

8. Repeat the variance calculations for the *y* sample:

```
> num1y = 4*Ny*(Ny-1)*(Ny-2)*sum(Py^3)
> num2y = 2*Ny*(Ny-1)*sum(Py^2)
> num3y = 2*Ny*(Ny-1)*(2*Ny-3)*sum(Py^2)^2
>  deny = (Ny*(Ny-1))^2
>    Vy = (num1y + num2y - num3y) / deny
> Vy
[1] 0.0001043976
```

9. Determine the degrees of freedom:

```
> DF = Sx + Sy - 2
> DF
[1] 29
```

10. Compute the *t*-value for the difference between samples:

```
> Dt = abs(Dx - Dy) / sqrt(Vx + Vy)
> Dt
[1] 4.304309
```

11. Calculate the critical value at the 5% (*p* = 0.05) level:

```
> Crit = qt(0.05/2, df = DF, lower.tail = FALSE)
> Crit
[1] 2.045230
```

12. Finally you can look at the significance of the difference:

```
> Pval = pt(Dt, df = DF, lower.tail = FALSE) * 2
> Pval
[1] 0.0001741377
```

You can see that the calculated value of *t* exceeds the critical value and that the result is statistically significant.

Note: R code for Simpson diversity significance test

The R code shown here can be found with the download material as part of the *CERE. RData* file. Instructions for using the R functions and data are on the website and also in the Introduction. Once the file is loaded into R the custom function H_sig() allows you to carry out a significance test of the difference between Simpson indices of two samples – you use the index = "simpson" instruction to tell the command to use Simpson's index (the default is to use Shannon).

Excel and the t-test *for Simpson's index*

It is certainly possible to create a spreadsheet to calculate the statistical difference between Simpson's indices of two samples. The formula for the variance is quite long-winded – it is a good idea to split it into sections to make the computations easier to type into the formula bar. You can also use some special functions akin to the SUM function to help reduce the number of columns of calculated values.

The SUMSQ function for example calculates the squares of each value in a series of values and then adds them together:

= SUMSQ(*range*)

You can specify more than one *range*. If you do, each *range* is squared and summed and the function then returns the sum all the squared ranges.

The SUMPRODUCT function multiplies corresponding elements of series together and returns the sum:

= SUMPRODUCT(*range1, range2, range3*)

If each range is a column for example then the first value in each would be multiplied together, then the second values are multiplied together and so on. The result is the sum of all the values.

In the following example you can have a go at creating a spreadsheet to calculate the statistical difference between the Simpson index of two samples.

Have a Go: Use Excel to carry out a *t*-test between Simpson's indices

Start with a blank spreadsheet. Usually you will get three worksheets by default. You will use two for calculating the statistics for each site and the third sheet to carry out the final *t*-test and present the results. If you need to make new worksheets you can click on the button at the end of the list of worksheets towards the bottom of the screen, you can also click Shift+F11 or right-click on a worksheet name tab and select *Insert*.

1. Start by typing some headings, in cell A1 type 'Site'. Then in cells B2:D2 type three headings: 'Taxon', 'Count' and 'P'.

2. Leave some space for data (20 rows will do to begin with) and type a heading in cell A23, 'Total'. In cell B2 type a formula to determine the sum of the counts: = SUM(B3:B22).

3. Fill in a few values into column B (it is easier to type in formulae when you have real values). Now go to cell C3 and type in a formula that calculates the proportion that this species makes to the total, you'll need to 'fix' the position of the cell reference for the total like so: = B3/$B$23.

4. Copy the formula in cell C3 down the column and then copy across the formula for the sum of the column (or type it afresh). The result should be 1, which is a useful check.

5. In cell A24 type a heading, 'Richness'. In cell B24 type a formula to determine the number of species: = COUNT(B3:B22).

6. In cell A25 type a heading, 'Sum(P^2)'. To get the superscript you can enter the 'plain' text then click the formula bar and highlight the 2. Right-click the highlighted text and select *Format Cells* from the menu.

7. In cell B25 type a formula to calculate the sum of the squares of the proportions: = SUMSQ(C3:C22).

8. In cell A26 type a heading, 'D'. Then, in cell B26 type a formula to calculate the Simpson's index: =1-B25.

9. In cell A27 type a heading, 'Sum(P^3)'. Then, in cell B27 type a formula to calculate

the sum of the cube of the proportions like so: = SUMPRODUCT(C3:C22,C3:C22,C3:C22).

10. To calculate the variance it is easier to split the calculation into parts so begin by making some headings in cells D25:F25, 'Den1', 'Den2' and 'Den3'.

11. In cell D26 type a formula to work out the first part of the denominator for the variance formula (Figure 9.6) like so: =4*B23*(B23-1)*(B23-2)*B27.

12. In cell E26 type a formula for the next section of the denominator of the variance: =2*B23*(B23-1)*B25.

13. In cell F26 type a formula for the third part of the denominator: = 2*B23*(B23-1)*(2*B23-3)*B25^2.

14. You can now assemble the variance formula, start with a heading in cell A28, 'S2D'. Then type a formula into cell B28: =(D26+E26-F26)/(B23*(B23-1))^2.

15. In cell A29 type a heading for the *degrees of freedom*, 'df'. In cell B29 type the formula: =B24-1.

16. Copy everything to the clipboard and then navigate to Sheet2 and copy. You should now have two separate worksheets, one for each of two samples.

17. Now navigate to Sheet3, where you will prepare the final summary and calculations.

18. Start by typing headings into column A starting at cell A1, 'Sites'. Now add more headings like so: 'D', 'S^2_D', 'Richness', 't', 'df', 'Crit' and 'P'.

19. In cell B1 type a formula to get the value entered for the name of the first sample: =Sheet1!B1. It is probably easier to use the mouse to select the 'target' cell, type = to bring up the function editor then navigate to the target and click.

20. Now fill in the cells for 'D', 'S^2_D' and 'Richness' in a similar manner as in step 19.

21. Repeat steps 19 and 20 for the second sample using column B.

22. Now in cell B5 type a formula to calculate the value of *t*: = ABS(B2-C2)/SQRT(B3+C3).

23. In cell B6 type a formula to calculate the degrees of freedom: = (B4-1)+(C4-1).

24. In cell B7 type a formula to calculate the critical value at the 5% (*p* = 0.05) level: =TINV(0.05,B6).

25. Finally you can fill in cell B8 with a formula to calculate the probability of getting a *t*-value of this magnitude at the relevant degrees of freedom: = TDIST(B5,B6,2).

Save your spreadsheet! Later on you can have a go at creating a graph to help visualise the difference in Simpson indices. If you need larger sample sizes then you can simply add new rows but you will need then to copy the formulae for columns A–F into the newly created cells. You might also wish to add borders and other formatting to the spreadsheet in order to make it easier to 'read'.

You might like to carry out a test on some of the ground beetle data as a practice of your new spreadsheet. Open the *Ground beetles.xls* spreadsheet and make a Pivot Table – use a filter to view a single site and copy the abundance data to the clipboard.

Note: Simpson diversity significance spreadsheet

You can find a spreadsheet to calculate the significance of the difference in Shannon indices with the download file on the companion website – the file is called *Simpson stats test.xls*.

9.3 Graphical summary of the *t*-test for Shannon and Simpson indices

You should really show the results of your *t*-test comparing diversity indices in some graphical manner. The simplest way is using a bar chart but you can also use other graphs, as you will see shortly. Part of the *t*-test is calculating the variability/uncertainty in the diversity index. You saw how to calculate variance using a formula for the Shannon index (Figure 9.3) and for Simpson's index (Figure 9.6). Whenever you produce a graph you should try to indicate the variability in the data in some way, error bars are the usual way to do this.

You can use the variance as the measure for your error bars but more commonly the standard deviation or confidence intervals are used. It is easy to determine the standard deviation if you have the variance:

Standard deviation = $\sqrt{\text{variance}}$

The *confidence interval* is perhaps a better way to show the variability in this instance as it allows the reader to visualise the statistical likelihood of the samples being different (or not) more readily than the standard deviation. Whenever you calculate a statistic like the mean or diversity index you are estimating that statistic from a sample. The 'true' value of your calculated value could be different from your measured value. The confidence interval shows the range of values where the 'true' value of your statistic lies (in this case the diversity index). A 95% confidence interval therefore shows the range of values where you can be 95% sure that the real index value lies. If you are comparing samples the confidence intervals that overlap are not likely to be from statistically different samples.

You can calculate a confidence interval using the standard deviation and a critical value (in this case determined from the *t*-statistic):

CI = critical value * std dev.

You saw previously how to calculate a critical value in Excel using the TINV formula and in R using the qt() command. In the comparison you needed the degrees of freedom, which is related to the sample sizes. In order to calculate a confidence interval for each sample you will need to determine the degrees of freedom for each sample separately. You will get the chance to see how to calculate confidence intervals and prepare graphical summaries of the results shortly.

In the following exercise you can have a go at calculating confidence intervals for Shannon indices of two samples of ground beetles using Excel.

Have a Go: Use Excel to calculate confidence intervals for Shannon diversity index

For this exercise you will need the spreadsheet you created earlier. A completed version called *Shannon stats test.xls* is provided with the download material on the companion website. The spreadsheet contains three worksheets, one for each of two samples and a third for the final calculations. You'll also need the *Ground beetles.xls* spreadsheet, which contains data for 18 samples of ground beetles.

1. Start by opening the *Shannon stats test* spreadsheet, either the one you made earlier or the one provided.

2. Open the *Ground beetles.xls* spreadsheet. Click once in the data and then make a Pivot Table (*Insert > Pivot Table*). The data will be highlighted automatically so simply choose to place the competed table in a new worksheet. Make the Pivot Table so that the *Species* form the *Row Labels*, *Sample* forms the *Column Labels* and that *Quantity* is in the *Values* box.

3. Click the triangle next to the Sample label in the completed Pivot Table to bring up the Filter menu. Choose the *E1* sample only and then copy and paste the values into *Sheet1* of your *Shannon Stats test* spreadsheet. Type the name 'E1' in cell B1 to form the sample name.

4. Return to the *Ground beetles* spreadsheet and now alter the filter to display the sample *E2*. Copy the data and paste into the Shannon spreadsheet in *Sheet2*. Type the name of the sample, 'E2', in cell B1.

5. Now you are ready to carry out the calculations and prepare graphs; it is easier to do this with some data. Navigate to *Sheet3* of the *Shannon stats test* spreadsheet.

6. In cell A10 type a label, 'SD', for the standard deviation. In cell B10 type a formula to calculate the standard deviation from the variance: =SQRT(B5). You can copy this formula across into cell C10 so that you now have the standard deviation for both samples.

7. In cell A11 type a label, 'DF', for the degrees of freedom of each sample. In cell B11 type a formula to calculate the degrees of freedom of the single sample: =((B5*2)^2)/ (2*B5/B2)/2. Compare this with the formula in cell B7: you are essentially substituting the value of C5, which is the variance of the other sample, and dividing by 2.

8. Copy the formula in cell B11 into cell C11 so that you calculate the degrees of freedom for the second sample. You should notice that the sum of the two individual degrees of freedom is that same as the overall degrees of freedom (at least to three decimal places).

9. In cell A12 type a label for the critical value, #'CV'. In cell B12 type a formula to calculate the critical value using the degrees of freedom for the single sample: =TINV(0.05,B11). You can also copy the formula across into cell C12 so that you have critical values for both samples. In this case they are the same but this will not always be the case. Note that you used 0.05, which represents the critical value at the 95% level.

10. In cell A13 type a label for the confidence interval, 'CI'. In cell B13 type a formula to calculate the 95% confidence interval: =B12*B10. Copy this formula into cell C13 to make a confidence interval for the second sample.

Save your spreadsheet. You now have the data that you need to create a graphical summary. You can also use different samples but remember that you only have 20 rows so if you need more you'll have to insert rows and make sure any formulae are copied also.

Calculating the degrees of freedom for the Simpson's index t-test is a great deal simpler than for the Shannon index, as it is simply the number of species minus one (for each sample). For practice you could have a go at modifying your earlier spreadsheet. There is a completed version (including graphs) with the download files that accompany this book.

There are two main types of graph that you can use to illustrate differences between the diversity indices, bar charts or 'point' plots. In either case you will need to use the calculated confidence intervals as the values for error bars that help to visualise the variability. In the following exercise you can have a go at making a bar chart and a point plot for the Shannon t-test.

Have a Go: Use Excel to visualise the differences between Shannon indices for two samples

For this exercise you will need to have calculated the confidence intervals for each of the two samples. If you haven't done this yet then look at the previous exercise and work through that first.

1. You should already have some data in your spreadsheet so to begin with you should navigate to *Sheet3*, where the 'final' calculations are found. This should include values for the Shannon diversity indices and the confidence intervals of the two samples.

2. Click once in the worksheet. Ensure that you click a cell that is blank and is also not adjacent to a filled cell; cell B19 will do very well for example.

3. Now you are going to make a vertical bar chart so start the process by using the *Insert > Column* button. You want to select the 2-D Column option and the leftmost option, a simple clustered column chart.

4. A blank chart widow will appear and the toolbar will display the *Chart Tools* menu item. You can now select the data to form the chart. Go to the *Select Data* button (on the *Chart Tools > Design* ribbon menu).

5. You need to click the *Add* button in the *Legend Entries (Series)* section. Click in the *Series Values* box and then use the mouse to select the Shannon index values in the spreadsheet. Then click OK.

6. Now you'll need to select the names for the bars (at present they will simply read 1, 2). Click the *Edit* button in the *Horizontal (Category) Axis Labels* section. Now use the mouse to select the cells that contain the site names and then click OK.

7. Click OK again to return to the spreadsheet and your newly created graph. The legend is not required so simply click it once and use the delete button (you can also right-click it and select *Delete* from the menu).

8. Click once on one of the chart bars and then get ready to make the error bars using the *Chart Tools > Layout > Error Bars* button. You want to choose *More Error Bars Options* from the menu.

9. You want error bars to be both up and down and to have end caps; these options are the usual defaults. At the bottom part of the menu is a section called *Error Amount*; click the *Custom* radio button then the *Specify Value* button to the right of it.

10. You can now choose the values for the positive and negative ranges of the error bars using the *Custom Error Bars* menu. In both cases you want the cells that contain the confidence intervals. You may have to delete anything that appears in the boxes of the menu before selecting the cells with the mouse. Once you are done you can click OK.

11. You can alter the appearance of the error bars using *Line Color* or *Line Style* options but generally black and 0.75pt are fine, so click OK again to return to the spreadsheet and graph.

12. The graph still needs some work to make it acceptable but you have a good starting point. Use the *Chart Tools > Layout* menu to add titles and to edit other graph components. You can also right-click in the chart to bring up a menu allowing you to format individual items (such as the bars themselves).

13. Once you are happy with the bar chart you can try making a point chart. Click in a blank area of a spreadsheet and use the *Insert > Line* button. You want 2-D Line Width Markers, which is generally the first option on the second row. You will remove the line later to leave the markers only!

14. Proceed as before using steps 4–10. You should now have markers joined by a line and with error bars.

15. Now right-click on the line that joins the two markers and select *Format Data Series*. Choose the *Line Color* option in the bar on the left then click the *No line* option. Click *Close* and return to the graph. You should now have two separate markers and error bars.

16. Use the *Chart Tools > Layout > Axes > Primary Vertical Axis* button to format the *y* axis. You want to select *More Primary Vertical Axis Options* (or right-click the axis and select *Format Axis*).

17. Use the Axis Options menu to alter the appearance of the axis; most importantly you can alter the minimum and maximum values displayed. Choose 0.6 as a min and 2.0 as a max to focus on the data in the chart for example. Click the *Close* button to return to the graph.

You now have a bar chart (what Excel calls a column chart) and a 'point' chart. The point chart is useful because it focuses on the differences between the samples, when the bars of a bar chart are tall and the error bars are small this is more helpful than the standard bar chart.

Repeat the exercise using the Simpson's index – you will see that the point chart highlights the differences between the samples better than the bar chart (Figure 9.7).

Note: Diversity spreadsheets

Versions that include the calculations for confidence intervals and graphs are included with the download material. The spreadsheets are called *Shannon stats test.xls* and *Simpson stats test.xls*.

The R program has powerful and flexible graphical capabilities and you can easily calculate the confidence intervals and produce a graphical output incorporating them. You can

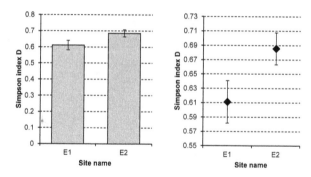

Figure 9.7 Comparison of Simpson's diversity between two samples of ground beetles. Error bars show 95% CI. The point chart highlights differences between diversity more clearly than the bar chart.

produce bar charts using the `barplot()` command and point plots using `plot()`. The error bars can be added by using either the `segments()` or `arrows()` commands.

Note: Excel and degrees of freedom

Excel can only handle degrees of freedom in whole numbers. This is not a problem with the Simpson's index but with Shannon entropy you usually get values for degrees of freedom that are not integers. Excel will give you a result but it rounds values down so you may get slightly different values for calculations in Excel compared to R.

Both `segments()` and `arrows()` commands add sections of line to existing plots. In both cases you supply the *x*, *y* co-ordinates of the starting and ending points. The main difference between the two commands is that you can specify 'end caps' by using the `arrows()` command. In the following exercise you can have a go at calculating confidence intervals and adding error bars to a bar chart for yourself.

Have a Go: Use R to make a bar chart of differences in Shannon diversity index

For this exercise you will need the ground beetle data. This is part of the *CERE.RData* file, which you can get via the companion website.

1. You will need to calculate the Shannon index and confidence interval for each of the two samples in the comparison. Start by assigning one sample to a simple variable:

```
> x = gb.biol[1, ]
```

2. Now calculate the Shannon index for the sample:

```
> Nx = sum(x)                               # Total individuals
> Px = x/Nx                                 # Proportions
> Hx = -sum(log(Px)*Px, na.rm = TRUE)       # Shannon Index
> Sx = sum(x>0)                             # No Species
```

3. Calculate the variance of the sample:

```
> SSx = sum(Px*log(Px)^2, na.rm = TRUE)
> SQx = Hx^2
>  Vx = ((SSx-SQx)/Nx) + ((Sx-1)/(2*Nx^2)) # Variance
```

4. You need the degrees of freedom for the single sample, so use a modified version of Figure 9.4:

```
> DFx = ((Vx*2)^2)/((Vx/Nx)+(Vx/Nx))/2 # degrees of freedom
```

5. Use the degrees of freedom of the single sample to work out a critical value for t at the 95% level:

```
> CVx = qt(0.05/2, df = DFx, lower.tail = FALSE) # Critical value
```

6. Now determine the standard deviation and the 95% confidence interval for this sample:

```
> SDx = sqrt(Vx) # Std Dev
> CIx = CVx*SDx # 95% Conf Int
```

7. Save the Shannon index value to a new variable:

```
> H = Hx
> H
[1] 1.267096
```

8. Save the confidence interval to a new variable:

```
> CI = CIx
> CI
[1] 0.3087971
```

9. Now select a second sample and assign this to the variable x:

```
> x = gb.biol[2, ]
```

10. Now repeat the steps 2–6 to work out the Shannon index and confidence interval for this new sample. Check the new values:

```
> Hx
[1] 1.424505
> CIx
[1] 0.4038871
```

11. Append the second Shannon index value to the first one:

```
> H = c(H, Hx)
> H
[1] 1.267096 1.424505
```

12. Append the second confidence interval to the first one:

```
> CI = c(CI, CIx)
> CI
[1] 0.3087971 0.4038871
```

13. You now have two variables, H and CI, which contain the statistics for the two samples that you need to plot. However, you will need to ensure that the y-axis is

tall enough to accommodate the height of the bars and the error bar. Work out the maximum value the y-axis needs to be:

```
> TV = max(H + CI)
> TV
[1] 1.828393
```

14. Now you can make a bar chart and set the limits of the y-axis appropriately:

```
> bp = barplot(H, ylim = c(0, TV), names = c("E1", "E2"))
```

15. Add titles to the axes:

```
> title(ylab = "Shannon index, H", xlab = "Sample name")
```

16. Finally you can add the error bars using the arrows() command:

```
> arrows(bp, H+CI, bp, H-CI, length = 0.1, angle= 90, code = 3)
```

The final graph should resemble Figure 9.8.

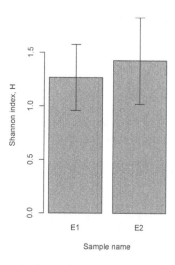

Figure 9.8 Shannon diversity indices of two samples of ground beetles as a bar chart. Error bars show 95% confidence intervals.

This basic bar chart can be altered in appearance in a variety of ways, as can the error bars themselves. The arrows() command can accept various graphical parameters, such as lty, lwd and col, which alter line type, width and colour.

Note: R code to calculate confidence intervals for Shannon diversity index

The commands used to calculate the confidence intervals for the Shannon diversity index have been bundled into a custom function called *H_CI*, which comes as part of the *CERE. RData* file.

The bar chart is useful and the commands that you used in the preceding exercise can be called to create a chart containing more than two samples; you can simply keep appending to the *H* and *CI* variables. However, as you saw earlier with the Simpson's index, sometimes a bar chart is not the most effective way to visualise differences in samples. In the following exercise you can have a go at making a point chart as an alternative. You will have to make customised graph axes here using the `axis()` command.

Have a Go: Use R to make a point chart of differences in Shannon diversity index

For this exercise you will need the ground beetle data. This is part of the *CERE.RData* file, which you can get via the companion website. You will build on the preceding exercise and so it is useful to have the R commands that you used easy to hand.

1. Start by making variables to hold the Shannon index values for your sites. If you completed the preceding exercise you will have a variable, *H*, already. If not then complete the preceding exercise:

```
> H
[1] 1.267096 1.424505
```

2. Now make sure that you have a variable holding the confidence intervals for your sites, from the preceding exercise the variable *CI* will be available:

```
> CI
[1] 0.3087971 0.4038871
```

3. Calculate the top and bottom ranges for the *y*-axis so that you can maximise the area occupied by the data:

```
> TV = max(H + CI)
> BV = min(H - CI)
> TV ; BV
[1] 1.828393
[1] 0.958299
```

4. Use the `pretty()` command to make the *y*-axis into 'pretty' intervals (that is, sensible ones):

```
> AL = pretty(c(BV, TV))
> AL
[1] 0.8 1.0 1.2 1.4 1.6 1.8 2.0
```

5. Now make a point plot. You will need to turn off the axes (you will add them back shortly). You also need to set the limits of both *x* and *y*-axes to get a 'nice' fit:

```
> plot(1:2, H,
        ylim = c(min(AL), max(AL)),
        xlim = c(0.5, 2.5),
        axes = FALSE,
        xlab = "Site name",
        ylab = "Shannon index")
```

6. Put in the *y*-axis; there is no need to modify this one:

```
> axis(2)
```

7. Put in an *x*-axis; this time you need to customise it to 'match' the data. The `at` part describes the number of items to appear on the axis and the `labels` part describes the axis tick labels:

```
> axis(1, at = 1:2, labels = c("E1", "E2"))
```

8. Add a bounding box around the plot to make is neat:

```
> box()
```

9. Finally, add the error bars:

```
> arrows(1:2, H+CI, 1:2, H-CI,
         length = 0.1,
         angle = 90,
         code = 3)
```

The final plot should resemble Figure 9.9.

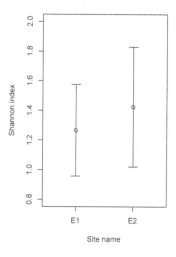

Figure 9.9 Shannon diversity indices of two samples of ground beetles as a point plot. Error bars show 95% confidence intervals.

If you had more than two samples (by appending extra values to the *H* and *CI* variables) you can easily create a point plot with minor modification. You simply alter the `1:2` values in the commands to reflect the number of samples that you actually have.

Note: R code to calculate confidence intervals for Simpson's diversity index

The commands used to calculate the confidence intervals for Simpson's diversity index have been bundled into a custom function called *H_CI*, which comes as part of the *CERE. RData* file. You need to use the `index = "simpson"` instruction to tell the command to use Simpson's index (the default is to use Shannon).

Although it is possible to keep appending diversity index values to a variable and therefore create a larger plot, it is better to use the flexibility of R to help you automate the process – this is illustrated in the following section.

9.3.1 Comparing multiple sites using graphs

One of the strengths of R is that you can take 'snippets' of commands and make reusable custom commands from them. Once you have a custom command it can be used in a variety of situations – one way is to repeatedly apply your custom function to several samples, such as the rows of a dataset.

You can make a custom function using the `function()` command – the basic way of using it is like so:

```
function(arg1, arg2, ...) {commands to be carried out}
```

Usually you assign a name to your new function. The main body of the command follows the list of arguments – generally the body of the command is enclosed in curly brackets as these allow the commands to 'spill over' several lines. The following example shows the lines of commands that make up the custom function `H_CI()`, which calculates confidence intervals for the Shannon diversity index.

```
H_CI = function(x) {

Nx = sum(x)                              # Total individuals
Px = x/Nx                                # Proportions
Hx = -sum(log(Px)*Px, na.rm = TRUE)      # Shannon Index
Sx = sum(x>0)                            # No Species

SSx = sum(Px*log(Px)^2, na.rm = TRUE)
SQx = Hx^2
Vx = ((SSx-SQx)/Nx) + ((Sx-1)/(2*Nx^2))       # Variance
DF = ((Vx*2)^2)/((Vx/Nx)+(Vx/Nx))/2           # degrees of
freedom
CV = qt(0.05/2, df = DF, lower.tail = FALSE)  # Critical value
SD = sqrt(Vx)                                 # Std Dev
CI = CV*SD                                    # 95% Conf Int

result = CI    # Make a result object
return(result) # Sets the value that the function returns

} # end
```

Notice that most of these commands are calculations and that the main body of the function starts with a curly bracket { and continues over several lines until the closing }. The final calculated value is the one that is 'returned' unless you explicitly give the name of an object to be returned. So, in this example `CI` would be the returned value but it is assigned a name (`result`) and returned explicitly anyway.

You can make your custom command by typing directly to the R console or you can use a text editor and then save the file with a .R extension. The text file can be called into action by using the `source()` command. If you use `file.choose()` as well you can simply select the text file from disk:

```
> source(file.choose())
```

Now that you have a custom function you can use it on any sample of data. In this example there is only a single argument to the function, the name of the community data sample. The command can be used simply:

```
> H_CI(gb.biol[1,])
[1] 0.3087971
```

If you want to use the command on several samples at once you have to take a slightly different approach.

Note: R functions for confidence intervals of Shannon and Simpson indices

A custom function for calculating confidence intervals is provided with the download file *CERE.RData*. The general idea is illustrated in the text – the command is called H_CI() and calculates confidence intervals for both Shannon and Simpson's indices. The default uses Shannon but you can alter this using the index = "simpson" instruction.

The apply() command

The apply() command allows you to use a function repeatedly over the rows (or columns) of a data object. The basic form of the command is:

```
apply(X, MARGIN, FUN)
```

You supply the name of the data object (X) and the name of the function (FUN) that you want to use. The MARGIN part relates to either rows (1) or columns (2). The command works for data frames or matrix data objects that are rectangular, that is, each row is the same length and so are the columns. To use the ground beetle community data, for example, you would use the following:

```
> apply(gb.biol, MARGIN = 1, FUN = H_CI)
        E1        E2        E3        E4        E5        E6        G1
 0.3087971 0.4038871 0.4120979 0.2465022 0.2744394 0.3084946 0.3095574
        G2        G3        G4        G5        G6        W1        W2
 0.3977260 0.5296750 0.3978153 0.3547690 0.3996990 0.7502927 0.7051761
        W3        W4        W5        W6
 0.6160593 0.9373282 1.2364349 1.3108941
```

Some functions can accept additional instructions. The diversity() command, for example (in the vegan package), can calculate Shannon or Simpson's indices. You can specify additional instructions that are relevant to the function you are applying like so:

```
> apply(gb.biol, MARGIN = 1, FUN = diversity, index = "shannon")
       E1       E2       E3       E4       E5       E6       G1       G2
 1.267096 1.424505 1.412179 1.737146 1.636923 1.557477 1.935163 2.000272
       G3       G4       G5       G6       W1       W2       W3       W4
 1.929426 2.360250 2.139567 2.205175 1.397251 1.582102 1.539511 1.601073
       W5       W6
 1.616333 1.552470
```

If you assign the results of your apply() commands to named objects you can use these to make a graph comparing multiple sites quite easily. In the following exercise you can have a go at making a bar chart to compare Shannon diversity for all 18 samples in the ground beetle dataset.

Have a Go: Use R to produce a bar chart of Shannon diversity for multiple sites

For this exercise you will need the ground beetle data *gb.biol* and the custom function *H_CI*. Both are included in the *CERE.RData* file, which you can get from the companion website.

1. Prepare the *vegan* package:

   ```
   > library(vegan)
   ```

2. Make a new variable to hold the Shannon index results for the data:

   ```
   > H = apply(gb.biol, MARGIN = 1, FUN = diversity, index = "shannon")
   ```

3. Now use the H_CI() command to determine the confidence intervals – save the result to a new variable:

   ```
   > CI = apply(gb.biol, MARGIN = 1, FUN = H_CI)
   ```

4. Create a basic bar chart of the Shannon indices using the barplot() command. Use the cex.names instruction to make the bar labels a bit smaller than standard:

   ```
   > bp = barplot(H, ylim = c(0, max(H+CI)), cex.names = 0.8)
   ```

5. Add the error bars (95% confidence intervals) using the arrows() command:

   ```
   > arrows(bp, H+CI, bp, H-CI, length = 0.1, angle = 90, code = 3)
   ```

6. Finish the plot by adding some axis titles:

   ```
   > title(xlab = "Site name", ylab = "Shannon index, H")
   ```

The final bar chart should resemble Figure 9.10.

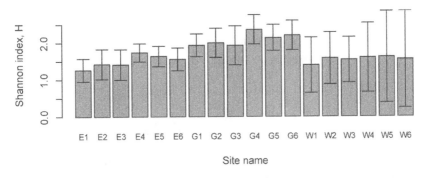

Figure 9.10 Shannon diversity indices for 18 samples of ground beetles. Error bars are 95% confidence intervals.

Notice that in this case you did not need to specify the names for the bars because they were taken from the original data (each row has a site name).

Tip: Using subsets of data

If you have a data object representing many samples you can use square brackets to create a subset. A `data.frame` is two-dimensional so you need to specify rows and columns (in that order), for example:

```
gb.biol[1:6, ] # Use rows 1-6 and all columns.
gb.biol[c(1, 7, 13), ] # Use rows 1, 7, and 13 (and all columns).
gb.biol[-1:-6, ] # Use all rows except 1-6.
gb.biol[c("E1", "G1", "W1"), ] # Use rows with names E1, G1 and W1
```

You can use the `apply()` command to help you to produce a point chart using similar commands to those you tried previously. In the following exercise you can have a go at making a point chart to show Shannon indices and 95% confidence intervals for all 18 samples in the ground beetle data.

Have a Go: Use R to produce a point plot of Shannon diversity for multiple sites

For this exercise you'll need the ground beetle data and the custom function H_CI(), both of which are contained in the *CERE.RData* file on the companion website.

1. Start by preparing the *vegan* package:

```
> library(vegan)
```

2. Now use the `apply()` command to make a new result containing the Shannon index values for all the samples in the *gb.biol* data:

```
> H = apply(gb.biol, MARGIN = 1, FUN = diversity, index = "shannon")
```

3. Use the `apply()` command again to get the confidence intervals, H_CI will be the function used to perform the calculations:

```
> CI = apply(gb.biol, MARGIN = 1, FUN = H_CI)
```

4. Work out the extent of the *y*-axis and then use the `pretty()` command to split the values into "nice" intervals for the axis:

```
> TV = max(H + CI) # Top value for y-axis
> BV = min(H - CI) # Bottom value for y-axis
> AL = pretty(c(BV, TV)) # Create axis intervals
```

5. You now need to determine how many samples there are to go onto the plot:

```
> NS = length(CI) # Number of samples
> NS
[1] 18
```

6. Create the basic plot and suppress the original axes – alter the plotting symbol character and size and add suitable titles:

```
> plot(1:NS, H, ylim = c(min(AL), max(AL)),
        xlim = c(0.5, NS+0.5), axes = FALSE,
        xlab = "Site names", ylab = "Shannon index, H",
        pch = 18, cex = 2)
```

7. Put in an x-axis – set the number of intervals to display and the labels:

```
> axis(1, at = 1:NS, labels = names(H))
```

8. Put in the y-axis and add a bounding box to the entire plot:

```
> axis(2)
> box()
```

9. Use the `arrows()` command to add error bars using the confidence intervals:

```
> arrows(1:NS, H+CI, 1:NS, H-CI, length = 0.1, angle = 90, code = 3)
```

10. The `abline()` command can add horizontal lines, which you can use for gridlines. Use the `seq()` command to create a sequence for the placement of the gridlines:

```
> abline(h = seq(0.5, 2.5, 0.5), lty = 3, col = "gray50")
```

The final graph should resemble Figure 9.11.

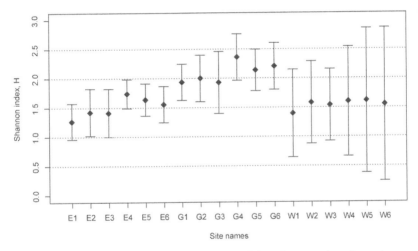

Figure 9.11 Shannon index values for 18 ground beetle samples. Error bars are 95% confidence intervals.

In the point plot you used the pch instruction to alter the plotting symbol and cex to make these symbols larger. Try using different values and experiment with the look of the plot. You can also alter the error bars – try using lwd = 2 (or other values) to alter the width of the bars for example.

The confidence intervals are highly influenced by the degrees of freedom and therefore by species richness. Because of the way the degrees of freedom are calculated, the confidence intervals for the Shannon calculations tend to be a lot larger than those for the Simpson index. It is probably better to use randomisation or bootstrapping methods to determine confidence intervals as these methods do not rely on degrees of freedom and so are not sensitive to species richness (see Section 9.4).

Working with 'disparate' samples

Sometimes your samples are 'separate', that is they are not part of a single dataset. In most cases your samples are in a single dataset and, assuming rows are samples, the rows are all the same length (with 'missing' species represented as zero). If you have to compare separate samples using R you will need an approach that can handle the differences in samples; this may well mean making a `list` object.

The `sapply()` command allows you to apply a function repeatedly over elements of a `list` object. The general form of the command is:

```
sapply(X, FUN)
```

A `list` can be a useful object because the elements that go to make up a `list` can be quite different from one another. This means that you can 'stitch together' several samples of differing lengths for example, allowing you to use the `sapply()` command.

However, it can also be tricky to get out the elements you require from a `list` because of this variety. One particular problem is one of names – `matrix` and data frames can have row names, usually relating to the site names, but simple vectors cannot. The simplest way around this is to create names as you make your list:

```
> sapply(list(E1 = gb.biol[1,], Y.96=bf.biol[1,]), FUN = H_CI)
      E1       Y.96
0.3087971 0.7220790
```

In the following example it is not necessary to make new names because the elements of the `list` are from the same `data.frame` and therefore have row names:

```
> sapply(list(gb.biol[1,], gb.biol[7,]), FUN = H_CI)
[1] 0.3087971 0.3095574

> sapply(list(gb.biol[1,], gb.biol[7,]), FUN = diversity, index
= "shannon")
[1] 1.267096 1.935163
```

However, you will then need to use the `sapply()` command to get the rownames separately:

```
> sapply(list(gb.biol[1,], gb.biol[7,]), FUN = rownames)
[1] "E1" "G1"
```

You can still provide names as part of the `list()` command:

```
> sapply(list(E1 = gb.biol[1,], G1 = gb.biol[7,]), FUN = diversity)
      E1       G1
1.267096 1.935163
```

Now if you use this as the basis for a `barplot()` you will not have to specify the names explicitly and do not need to run the `sapply()` command for the rownames.

Note: List objects

When you make a list object the items you specify form the elements of the `list`. If these elements themselves consist of smaller parts then the smaller parts remain within their original element. For example the `list()` command that follows produces a `list` of two items:

```
> dat = list(bf.biol[1:4,], gb.biol[1:6,])
```

Each element of the `list` consists of several rows, the first contains four rows and the second contains six rows.

The `sapply()` command is most useful for when you only have a few 'odd' samples to compare that are of unequal size. If you have more samples it would be easier to return to your original data and combine the new samples with the old. You can then recreate your Pivot Tables and community data to make one large 'set'.

It is possible to do this in R. You need several steps:

1. Convert the community data to biological recording format, where you have three columns representing, *Sample, Species* and *Abundance*:

   ```
   brf = as.data.frame(as.table(as.matrix(comm)))
   ```

2. Repeat for as many datasets as you require.
3. Combine the data frames using the `rbind()` command. You'll have to make sure that the sample names (site names) are different for each `data.frame`, otherwise you'll end up mixing data from multiple samples.
4. Use cross-tabulation to reform the biological records into a community dataset that contains the combination of samples:

   ```
   xtabs(Freq ~ Samples + Species, data)
   ```

Although you can combine data using R it is probably better to return to Excel to do this. Excel makes a better database or data management system and it is easier to add new variables and so on (see Chapter 6).

9.4 Bootstrap comparisons for unreplicated samples

When you have unreplicated samples, that is, you only have one sample for each site, you have to rely on a formula to calculate the variance of your Shannon or Simpson's diversity indices. These formulae are estimates of the variance but are fraught with issues. If you have worked through the preceding exercises you may have noticed that the confidence intervals for some samples are very large. The confidence intervals are based on the calculations for variance and degrees of freedom and are especially sensitive to species richness. A better approach is to use a method of resampling. The idea is to use the inherent properties of the data to create a whole series of new samples that allow you to determine the variability more effectively.

The process is called *bootstrapping* and works more or less like this:

1. Take your original community sample.
2. Work out how many individuals there are in the sample.
3. Randomly select individuals (with replacement) from the original sample so that you have a new sample with the same number of individuals.
4. Repeat the sampling process many times (usually 1000 or more).
5. Calculate the statistic of interest from each of the new samples (e.g. Shannon diversity index).
6. Assess variability using the variability of the new (bootstrapped) samples.

It would be very tedious to attempt this in Excel but R can carry out the process quite easily. In fact there is a dedicated command, `boot()`, which is designed to help you undertake the process. The command is found in the package *boot*, which is installed quite simply from within R:

```
> install.packages("boot")
```

This assumes that you are connected to the Internet. The command will access, download and install the package for you. Once you have it you can load the commands using the `library()` command:

```
> library(boot)
```

In the following sections you'll see how to use the `boot()` command to determine confidence intervals for Shannon and Simpson indices as well as how to carry out the process 'from scratch'.

9.4.1 Calculating the bootstrap statistics for diversity indices

Bootstrapping is a way of generating lots of samples using a single sample as a starting point for a random process of resampling. This method does not rely on any underlying pattern of distribution and it is becoming a more commonly used method because it gives 'reliable' results, especially when you use many re-samples.

Here is a simple community sample:

```
> x = c(9, 5, 4, 2, 1, 1) # a sample
> names(x) = LETTERS[1:6] # basic names
> x
A B C D E F
9 5 4 2 1 1
```

In this instance there are only six 'species' and 22 individuals – not very realistic but sufficient to see the processes of bootstrapping in action. You want to make a new sample based on random sampling from the original. The new sample should contain the same number of individuals as the original sample. To start with you need to convert the data into an individual-based sample rather than the species-based one you have at present:

```
> cx = rep(1:length(x), x) # convert to individuals
> cx
 [1] 1 1 1 1 1 1 1 1 1 2 2 2 2 2 3 3 3 3 4 4 5 6
```

You can see that there are nine 1s, five 2s and so on. The species have been converted into simple numbers. It not matter not having species names because you only need their abundances to determine the diversity indices. You can convert the individual-based sample back to a species-based one using the `table()` command:

```
> table(cx)
cx
1 2 3 4 5 6
9 5 4 2 1 1
```

The species names have been lost but this is unimportant for the calculations. The new samples need to have the same number of individuals as the original so you need to determine this:

```
> Nx = sum(x)
> Nx
[1] 22
```

To generate a new random sample you use the `sample()` command:

```
> Rx = sample(cx, size = Nx, replace = TRUE) # new sample
> Rx
 [1] 4 3 2 1 2 4 6 1 1 1 2 6 2 1 1 3 1 3 2 4 4 2
```

Notice that the instruction `replace = TRUE` was used. This allows a previously placed species to be placed again (i.e. replaced). If this was not used (the default is `replace = FALSE`) then you'd simply end up with your original sample. Now you can work out the Shannon index from the new sample:

```
> Tx = table(Rx)         # convert to species-based sample
> Px = prop.table(Tx)    # proportions
> Hx = -sum(Px*log(Px))  # Shannon Index
> Hx
[1] 1.51835
```

Each time you run the `sample()` command you get a new sample and therefore a slightly different Shannon index. You need to rerun the resampling process lots of times and create many values for Hx, the Shannon index. Once you have many, say 1000, results you can work out the 95% confidence intervals from the 0.025 and 0.975 *quantiles*.

Bootstrapping using loops

To run the commands 1000 times or more manually would be very tedious so you can use the `for()` command to make a loop that will run the commands for you and 'build' a result. In the following exercise you can have a go at making a simple copy 'n' paste script that will build a bootstrap sample for the Shannon index.

Have a Go: Use a loop to make a bootstrap sample of Shannon diversity

For this exercise you will use basic commands to create a bootstrap sample of Shannon diversity. You do not need any additional R packages or data for this exercise. Once you have completed it you will have a series of commands that can be used as a copy 'n' paste script, which will work for any data (as long as it is called *x*).

1. Start by making some data. Use something simple to start with (species names are not required), as you can assign *x* to any community sample later:

    ```
    > x = c(9, 5, 4, 2, 1, 1)
    ```

2. Now calculate the number of individuals in the sample – this value will be used to set the size of the re-samples:

    ```
    > Nx = sum(x)            # number of individuals
    > Nx
    [1] 22
    ```

3. Calculate the Shannon index of the original sample. Note that *x* is already in species-based form:

    ```
    > Px = prop.table(x)     # proportions
    > Ho = -sum(Px*log(Px))  # Shannon Index
    > Ho
    [1] 1.511330
    ```

4. Now you need to make a blank vector object to hold the result from the bootstrapping process. The elements will start off as 0 and will be filled in as you go along:

```
> Hb = vector(mode = "numeric", length = 1000)
```

5. Create a loop that will run 1000 times using the `for()` command. Within the loop you need to add the commands that will calculate the Shannon index and add it to the *Hb* result object. Don't forget to type the curly brackets:

```
> for(i in 1:1000) {                          # bootstrap loop
    Rx = sample(cx, size = Nx, replace = TRUE) # new sample
    Tx = table(Rx)                             # convert to sample
    Px = prop.table(Tx)                        # proportions
    Hx = -sum(Px*log(Px))                      # Shannon Index
Hb[i] = Hx                                     # Add to result object
}
```

6. Use the `quantile()` command to work out the 95% confidence intervals:

```
> quantile(Hb, c(0.025, 0.975)) # CI
    2.5%      97.5%
1.031618 1.657994
```

7. The standard deviation of the bootstrapped sample is the standard error of your estimate of *H*, the Shannon index:

```
> sd(Hb)        # Standard Error
[1] 0.1608566
```

8. The mean of the bootstrap sample is the estimate of the Shannon index:

```
> mean(Hb)
[1] 1.385407
```

9. The difference between the mean of the bootstrap estimate and the original Shannon index is called the bias (or offset):

```
> mean(Hb) - Ho  # Bias (or offset)
[1] -0.1259230
```

If you take all these commands and put them into a text editor you can use them to create a bootstrap sample of Shannon index for any sample. You simply need to make a copy of the data you require and call it *x*.

Tip: Making copy 'n' paste scripts

You can use the script editor in Windows or Mac to help you make scripts. In Windows you can open this using *File > New Script*. In Mac use *File > New Document*. You can type commands into the script window or use copy and paste to move commands from the console (or any other program) into the script window. Commands can be 'run' by using Ctrl+R in Windows and Cmd+Enter for Mac.

It is easy to alter the script in the preceding exercise to calculate Simpson's index:

```
> Dx = 1-sum(Px^2)
```

Compare this to the corresponding command (step 3) in the preceding exercise:

```
> Hx = -sum(Px*log(Px))
```

Bootstrapping using the `boot ()` *command*

The *boot* package provides a range of commands that help streamline the bootstrapping process – the command you'll use most often is `boot()`. The most basic form of the command requires three elements:

```
boot(data, statistic, R)
```

The first element is the `data` that you require the bootstrapped results from. The `statistic` part is a function that calculates the statistic you require – for example, the Shannon diversity index. The final part, `R`, is a number that will be how many times the `statistic` will be generated – the number of bootstrap replicates.

The `statistic` function that you make must operate on individual-based data samples otherwise the resampling process cannot operate correctly. The `statistic` function itself should only have two arguments, the data and an index value. For calculating the Shannon index the custom function would look something like the following:

```
shannon1 <- function(data,i) {
  pi <- table(data[i])/length(data[i])
  -sum(log(pi)*pi)
  }
```

You can see that the function contains only two instructions, the `data` and `i`, an index. The first line of the function body calculates the proportions and assigns the result to a variable `pi`. The next line uses the proportions contained in the `pi` variable to calculate the Shannon diversity index.

It is easier to see what is happening by using an example. Start with some simple data:

```
> x = c(9, 5, 4, 2, 1, 1)
> x
[1] 9 5 4 2 1 1
```

This sample is a species-based sample (there are nine individuals of the first species, five of the second species and so on) and you need an individual-based sample:

```
> cx = rep(1:length(x), x)
> cx
[1] 1 1 1 1 1 1 1 1 1 2 2 2 2 2 3 3 3 3 4 4 5 6
```

The species-based sample can be recreated using the `table()` command:

```
> table(cx)
cx
1 2 3 4 5 6
9 5 4 2 1 1
```

The proportions of each species can therefore be calculated by dividing the table by the number of species in the sample – which can be gathered using the `length()` command:

```
> pi = table(cx)/length(cx)
> pi
cx
     1          2          3          4          5          6
0.40909091 0.22727273 0.18181818 0.09090909 0.04545455 0.04545455
```

You could also use the sum of the table itself:

```
> table(cx)/sum(table(cx))
```

Once you have the proportions it is easy to calculate the Shannon index:

```
-sum(log(pi)*pi)
[1] 1.511330
```

The custom function returns the value of the Shannon diversity index – if you use the function by itself you simply ignore the `i` part like so:

```
> cx
 [1] 1 1 1 1 1 1 1 1 1 2 2 2 2 2 3 3 3 3 4 4 5 6
> shannon1(cx)
[1] 1.511330
```

Now when you get to use the custom function as part of the `boot()` command the `i` part is used as an index value and you get your Shannon result many times (`i` times in fact, where `i` is the number of replicates you decided to use).

The `boot()` command gives you a simple result that contains three main elements:

- The original value of the statistic – in this case the Shannon index of the original sample.
- The bias, that is the difference between the original statistic and the mean of the bootstrap values.
- The standard error of the statistic – that is the standard error of *H*.

You can easily determine the confidence intervals from the bootstrap values and in the following exercise you can have a go at doing this for some ground beetle data.

Have a Go: Use the `boot()` command to determine bootstrap statistics for Shannon diversity

You will need the ground beetle data for this exercise. The *gb.biol* data are contained in the *CERE.RData* file, which you can get from the companion website. You will also need the *boot* package. If you already have this installed then you can go to step 2.

1. If you haven't yet got the *boot* package then do so now using the `install.packages()` command:

```
> install.packages("boot")
```

2. Make sure the *boot* package is loaded and ready for use:

```
> library(boot)
```

3. Prepare the data, which are in species-based layout, by converting to an individual-based layout:

```
> x = gb.biol[1,]
> cx = rep(1:length(x), x)
```

4. Set-up the custom function to calculate the Shannon diversity index from an individual-based sample:

```
> shannon1 <- function(data,i) {
  pi <- table(data[i])/length(data[i])
 -sum(log(pi)*pi)
 }
```

5. Now use the `boot()` command to make a bootstrap sample of 1000 replicates using the `shannon1()` custom function:

```
> bx = boot(cx, shannon1, R = 1000)
```

6. View the general output of your result by typing its name:

```
> bx
ORDINARY NONPARAMETRIC BOOTSTRAP

Call:
boot(data = cx, statistic = shannon1, R = 1000)

Bootstrap Statistics :
    original       bias   std. error
t1* 1.267096 -0.01434183 0.04241785
```

7. You can view the original statistic (the Shannon diversity) separately because the result object contains several parts (note that you type a zero, 0, in the following command not the letter O:

```
> bx$t0
[1] 1.267096
```

8. Now review the custom function you used:

```
> bx$statistic
function(data,i) {
    pi <- table(data[i])/length(data[i])
 -sum(log(pi)*pi)
   }
```

9. Get the 95% confidence intervals by using the `quantile()` command on the 1000 replicates of the Shannon diversity index you created:

```
> quantile(bx$t, c(0.025, 0.975))
    2.5%     97.5%
1.168600 1.334862
```

10. The mean of the bootstrap sample is not quite the same as the original Shannon diversity index:

```
> mean(bx$t)
[1] 1.252754
```

11. Calculate the bias by subtracting the mean of the bootstrap sample from the original Shannon diversity:

```
> mean(bx$t)-bx$t0
[1] -0.01434183
```

Now that you have a series of commands to produce a bootstrap sample you can easily save these to make a copy 'n' paste script that you can use on any sample.

Note: R script for bootstrapping Shannon diversity

The R commands that calculate a bootstrap sample for Shannon diversity are available as a custom function `H_boot()`, which is part of the *CERE.RData* file.

Now that you have made a script to calculate a bootstrap sample for the Shannon diversity index it is a relatively easy process to alter it for use with Simpson's diversity index. You only have to alter one line in the custom function:

```
simp1 = function(data, i) {
  pi <- table(data[i])/length(data[i])
1-sum(pi^2)
  }
```

The third line is the only difference apart from the name of the function itself. In the following exercise you can have a go at making a copy 'n' paste script that you can save and use to make a bootstrap sample for Simpson's diversity index.

Have a Go: Make a copy 'n' paste script for Simpson's diversity index bootstrapping

You don't need any data for this exercise although the ground beetle data (*gb.biol*) would be useful to practise with. The beetle data are part of the *CERE.RData* file, which you can get from the companion website. You will need the *boot* package to conduct the bootstrapping. If you already have this installed then you can go to step 2.

1. If you haven't yet got the *boot* package then do so now using the `install.packages()` command:

   ```
   > install.packages("boot")
   ```

2. Make sure the *boot* package is loaded and ready for use:

   ```
   > library(boot)
   ```

3. Open a script window so that you can type your commands into it. In Windows use *File > New Script*. In Mac use *File > New Document*. In Linux you will have to use a separate editor.

4. You are going to assume that the 'target' data are called *x* – any sample that you have can be copied to a variable called x before you run the script. Now type the following commands into the script editor as they appear below:

   ```
   cx = rep(1:length(x), x)
   simp1 = function(data, i){
    pi <- table(data[i])/length(data[i])
    1-sum(pi^2)
    }
   bx = boot(cx, simp1, R = 1000)
   qx = quantile(bx$t, c(0.025, 0.975))
   ```

5. Now save the script to a file with a sensible name – give the file a .R extension, which will identify it as an R script.

Your commands will now be saved and can be used over and over again. You can open the script file using R or a text editor. If you use a text editor then you can use copy and paste to run your commands. In the script editor of Windows you can use Ctrl+R to run highlighted commands (use Cmd+Enter in a Mac).

Note: R script for bootstrapping Simpson's diversity

The R commands that calculate a bootstrap sample for Simpson's diversity are available as a custom function `H_boot()`, which is part of the *CERE.RData* file. Note that this is not a copy 'n' paste script but carries out the same commands. The command can work out bootstrap statistics for Shannon or Simpson's indices; the default is `index = "shannon"`, you use `index = "simpson"` for the alternative.

Now you are able to take individual samples and use the bootstrapping procedures to create new samples that allow you to estimate variability in the Shannon or Simpson's diversity indices. The confidence intervals allow you to compare samples because non-overlapping confidence intervals will be statistically significant.

In the following exercise you can use the bootstrap confidence intervals to assess differences in diversity indices using some ground beetle data.

Have a Go: Use bootstrap confidence intervals to assess differences in diversity indices

For this exercise you will need the *CERE.RData* file, which is available on the companion website. Follow the instructions on the website and in the Introduction to load the file into your copy of R. You'll also need the *boot* package. If you already have this package then you can go to step 2.

1. If you haven't yet got the *boot* package then do so now using the `install.packages()` command:

    ```
    > install.packages("boot")
    ```

2. Make sure the *boot* package is loaded and ready for use:

    ```
    > library(boot)
    ```

3. Use the `H_boot()` command to view the Simpson's diversity and 95% confience intervals for the first row of the ground beetle data – the command operates on species-based data:

    ```
    > H_boot(gb.biol[1,], index = "simpson")

    H =  0.6111399
         2.5%      97.5%
    0.5830903 0.6353307
    ```

4. Note that the command shows you the original value of the diversity index and the confidence intervals. Look now at another sample:

    ```
    > D_boot(gb.biol[2,])

    D =  0.6854643
         2.5%      97.5%
    0.6641219 0.7039575
    ```

5. You can see that the confidence intervals do not overlap so we can assume that

there is a probable difference in the Simpson's index between these two samples. Look at another sample:

```
> H_boot(gb.biol[7,])

H =  0.7259298
      2.5%       97.5%
0.6760203 0.7613267
```

6. You can see from steps 3–5 that there is no overlap of the confidence intervals between samples 1 and 2, 1 and 7, or 2 and 7 but that the intervals overlap between samples 1 and 7.

7. Repeat the steps 3–5 using the H_boot() command for the Shannon diversity:

```
> H_boot(gb.biol[1,])

H = 1.267096
      2.5%       97.5%
1.170963 1.338600

> H_boot(gb.biol[2,])

H =  1.424505
      2.5%       97.5%
1.350877 1.490447

> H_boot(gb.biol[7,])

H =  1.935163
      2.5%       97.5%
1.747147 2.024886
```

This time you can see that none of the confidence intervals overlap – so all the samples are probably different from one another. This highlights the differences between diversity indices and the importance of scale.

Note: Randomness

Your results will probably look slightly different from those shown here. Whenever you use a random process R will generate random numbers, which will inevitably be different each time!

You should be careful about comparing many sites and making statistical inference. As you saw previously, the more comparisons you make the greater the chance of getting a significant difference. The comparisons of confidence intervals are helpful but not a statistical hypothesis test in quite the same way as the *t*-test is. When reporting the results of comparing confidence intervals you should be careful not to use the term 'significantly different' because you haven't actually run a significance test.

What you have done is to provide robust estimates of the diversity indices but it would be helpful to visualise these differences graphically.

9.4.2 Graphing the results of bootstrap diversity indices

In Section 9.2 you saw how to make bar and point charts of Shannon and Simpson's diversity with confidence intervals calculated using formulae to determine variance (Figures 9.3 and 9.6). The confidence intervals were very dependent on species richness and compared to the bootstrap confidence intervals, quite large.

In the following section you'll see how to make bar charts and point plots of bootstrap confidence intervals.

Calculating bootstrap confidence intervals for multiple samples

In the preceding section you saw how to calculate bootstrap confidence intervals, and you did this for several samples individually. Ideally you need to be able to calculate confidence intervals for several samples at once so that you can use the results in a plot. The key is the apply() command, which you met earlier. With a little bit of coercion you can make the commands to calculate confidence intervals into a custom function that can be used with apply() – allowing you to produce results for several samples at once.

In order to use the apply() command you will need to first make a custom function to calculate the confidence intervals for a single sample. You have already used all the appropriate commands so it should be just a matter of assembly.

Here is a custom function to determine bootstrap confidence intervals for the Shannon diversity index:

```
Happ = function(x) {
 cx = rep(1:length(x), x)
 shannon1 = function(data, i){
  pi <- table(data[i])/length(data[i])
  -sum(log(pi)*pi)
  }
 bx = boot(cx, shannon1, R = 1000)
 qx = quantile(bx$t, c(0.025, 0.975))
 }
```

These command lines will produce a command called Happ(), which will carry out a bootstrap sampling process and then calculate the 95% confidence intervals. The function consists of several lines:

- The first line sets the name of the function (Happ) and provides the required arguments (the name of the data, x). The main body of the function lies between the curly brackets but notice that there is another function embedded with the main one and that has its own curly brackets.
- The second line of the command turns the original data from a species-based sample into an individual-based sample.
- Lines 3–6 are another function and calculate the Shannon index from the individual-based sample.
- Line 7 carries out the bootstrap calculations and produces a sample with 1000 replicates.
- The penultimate line computes the confidence intervals.
- The last line is the closing curly bracket and denotes the end of the custom function.

If you run this function nothing appears to happen!

```
> Happ(gb.biol[1,])
```

You need to assign the result to a named object:

```
> CI = Happ(gb.biol[1,])
> CI
     2.5%      97.5%
1.160204 1.344330
```

You really want the command to produce a result for several samples and the way to do this is using Happ() as the named function in the apply() command:

```
> apply(gb.biol[1:3,], MARGIN = 1, FUN = Happ)
           E1       E2       E3
2.5%  1.169233 1.339387 1.313603
97.5% 1.339371 1.492625 1.478916
```

In the preceding example you can see the command used to create confidence intervals for three rows of the ground beetle data. Note that there are two lines in the result, the first row for the lower confidence interval and the second row for the upper.

It is a simple process to alter the custom function Happ() to one that will produce bootstrap confidence intervals for the Simpson's index. In your script editor or text editor type the following:

```
Dapp = function(x) {
  require(boot)
cx = rep(1:length(x), x)
simp1 = function(data, i){
  pi <- table(data[i])/length(data[i])
1-sum(pi^2)
}
bx = boot(cx, simp1, R = 1000)
qx = quantile(bx$t, c(0.025, 0.975))
}
```

You can then copy and paste this into R and the new custom command, Dapp() will be created. The command works just the same as the Happ() command you saw earlier, and you can use it with the apply() command to produce confidence intervals for Simpson's diversity index for several samples:

```
> apply(gb.biol[7:12,], MARGIN = 1, FUN = Dapp)
            G1        G2        G3        G4        G5        G6
2.5%  0.6788617 0.7347358 0.7274587 0.8181822 0.7674832 0.7904784
97.5% 0.7663847 0.8070330 0.8080904 0.8652229 0.8269594 0.8430746
```

As before there are two rows in the result, one for the lower confidence interval and one for the upper.

Note: Functions to calculate bootstrap confidence intervals for multiple sites

The custom functions shown in the text to work out bootstrap confidence intervals are reproduced in the *CERE.RData* file. They appear as commands Happ() and Dapp().

Bar charts and bootstrap confidence intervals

Bar charts are familiar to everyone and you can use similar commands to those you used in Section 9.2 to create a bar chart showing the diversity index values and confidence intervals. In the following exercise you can have a go at making a bar chart of Shannon diversity and bootstrap confidence intervals, which you can compare to the previous bar chart (Figure 9.10).

Have a Go: Make a bar chart of Shannon diversity and bootstrap confidence intervals

For this exercise you'll need the *vegan* and *boot* packages. You will also need the ground beetle data *gb.biol*, which are part of the *CERE.RData* file. The file also contains the Happ() and Dapp() custom functions, which you saw how to make for yourself earlier.

1. Prepare the *vegan* and *boot* packages:

    ```
    > library(vegan)
    > library(boot)
    ```

2. Use the diversity() command in the *vegan* package to calculate the Shannon diversity index for the ground beetle data:

    ```
    > gb.H = diversity(gb.biol)
    ```

3. Now use the Happ() command to carry out a bootstrap sampling and calculate the confidence intervals. The apply() command will allow the function to operate on all the rows of the ground beetle data. The command will take a few moments to complete as you are carrying out 18,000 resampling operations:

    ```
    > gb.Hci = apply(gb.biol,1,Happ)
    ```

4. Look at the confidence intervals and note that there are two rows – the first for the lower confidence interval and the second for the upper:

    ```
    > gb.Hci
               E1        E2        E3        E4        E5        E6        G1
    2.5%  1.161547 1.335187 1.324862 1.644317 1.552209 1.469612 1.737714
    97.5% 1.336963 1.492346 1.482941 1.801947 1.697502 1.621607 2.043163
               G2        G3        G4        G5        G6        W1        W2
    2.5%  1.810226 1.718905 2.173347 1.952421 2.028589 1.342450 1.520598
    97.5% 2.108435 2.050405 2.449763 2.232835 2.291781 1.441722 1.638344
               W3        W4        W5        W6
    2.5%  1.469022 1.530115 1.559085 1.488845
    97.5% 1.593480 1.658248 1.661129 1.601376
    ```

5. Now make the basic bar chart using the barplot() command. You will need to alter the *y*-axis to accommodate the tallest bar and its confidence interval:

    ```
    > bp = barplot(gb.H,
                   ylim = c(0, max(gb.Hci)),
                   xlab = "Site",
                   ylab = "Shannon index, H")
    ```

6. Finally you can add the error bars to indicate the confidence intervals using the
    `arrows()` command – the final graph should resemble Figure 9.12:

```
> arrows(bp, gb.Hci[1,], bp, gb.Hci[2,],
         length = 0.1,
         angle = 90,
         code = 3)
```

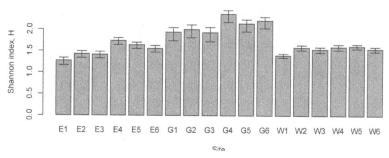

Figure 9.12 Shannon index values for 18 ground beetle samples. Error bars are 95% confidence intervals from bootstrap sampling.

Notice that you did not add the value of the confidence interval to the diversity index
result because this time you calculated the exact value of the upper and lower bounds
rather than a magnitude for the 'range' about the index.

   If you compare to the bar chart that used the formula (Figure 9.3) to determine variance and thus confidence intervals (Figure 9.10) you will notice two things; the bootstrap confidence intervals are much smaller, and they are asymmetric.

If you want to produce a bar chart for Simpson's index you can simply use the same commands but substitute the `Dapp()` command for calculation of the confidence intervals. The
`diversity()` command can work out the Simpson's index:

```
> gb.H = diversity(gb.biol, index = "simpson")
> gb.Hci = apply(gb.biol,1,Dapp)
```

The only other change needed is to the *y*-axis title.

Point plots and bootstrap confidence intervals

The bar chart is familiar but not necessarily the best method to highlight differences
between samples. You can see from Figure 9.12, for example, that the error bars appear
quite small. A point chart would be a better way to visualise differences between samples as the plot focuses on the error bars. You saw previously how to make a point
plot (Section 9.2) and you can use the similar commands to make one using bootstrap
confidence intervals. In the following exercise you can have a go at making a point plot
using the ground beetle data once again so that you can compare this with the earlier
plot (Figure 9.11).

Have a Go: Make a point plot of Shannon diversity and bootstrap confidence intervals

For this exercise you'll need the *vegan* and *boot* packages. You will also need the ground beetle data *gb.biol*, which are part of the *CERE.RData* file. The file also contains the `Happ()` and `Dapp()` custom functions, which you saw how to make for yourself earlier.

1. Prepare the *vegan* and *boot* packages:

```
> library(vegan)
> library(boot)
```

2. Use the `diversity()` command in the *vegan* package to calculate the Shannon diversity index for the ground beetle data:

```
> H = diversity(gb.biol)
```

3. Now use the `Happ()` command to carry out a bootstrap sampling and calculate the confidence intervals. The `apply()` command will allow the function to operate on all the rows of the ground beetle data. The command will take a few moments to complete as you are carrying out 18,000 resampling operations:

```
> CI = apply(gb.biol, 1, Happ)
```

4. Look at the confidence intervals and note that there are two rows – the first for the lower CI and the second for the upper:

```
> gb.Hci
          E1       E2       E3       E4       E5       E6       G1
2.5%  1.161547 1.335187 1.324862 1.644317 1.552209 1.469612 1.737714
97.5% 1.336963 1.492346 1.482941 1.801947 1.697502 1.621607 2.043163
          G2       G3       G4       G5       G6       W1       W2
2.5%  1.810226 1.718905 2.173347 1.952421 2.028589 1.342450 1.520598
97.5% 2.108435 2.050405 2.449763 2.232835 2.291781 1.441722 1.638344
          W3       W4       W5       W6
2.5%  1.469022 1.530115 1.559085 1.488845
97.5% 1.593480 1.658248 1.661129 1.601376
```

5. Now calculate the upper and lower limits required for the *y*-axis:

```
> TV = max(CI[2,]) # Top value for y-axis
> BV = min(CI[1,]) # Bottom value for y-axis
```

6. Use the `pretty()` command to split the *y*-axis into 'neat' chunks:

```
> AL = pretty(c(BV, TV)) # Create axis intervals
```

7. Determine how many samples are required to plot by looking at the number of Shannon index values in the H, result object:

```
> NS = length(H) # Number of samples
```

8. You now have all the data required for the plot – use the `plot()` command to start the plotting process:

```
> plot(1:NS, H,
        ylim = c(min(AL), max(AL)),
        xlim = c(0.5, NS+0.5),
        axes = FALSE,
        xlab = "Site names",
        ylab = "Shannon index, H")
```

9. You now have a chart with no axes so add the y-axis using the intervals calculated in steps 5 and 6:

```
> axis(2, at = AL)
```

10. Add an x-axis using the appropriate number of tick marks and using labels taken from the Shannon index result, H:

```
> axis(1, at = 1:NS, labels = names(H))
```

11. Now add in the error bars representing the bootstrap confidence intervals by using the arrows() command:

```
> arrows(1:NS, CI[2,], 1:NS, CI[1,],
         length = 0.1, angle = 90, code = 3)
```

12. Finish off the plot with some gridlines and a surrounding box:

```
> abline(h = AL, lty = 3, col = "gray50")
> box()
```

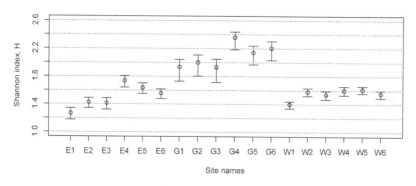

Figure 9.13 Shannon index values for 18 ground beetle samples. Error bars are 95% confidence intervals from bootstrap sampling.

Your figure should resemble Figure 9.13. You can see that compared to the point plot using variance-derived confidence intervals (Figure 9.11) the error bars are smaller and also asymmetric.

If you want to produce a bar chart for Simpson's index you can simply use the same commands but substitute the Dapp() command for calculation of the confidence intervals. The diversity() command can work out the Simpson's index:

```
> H = diversity(gb.biol, index = "simpson")
> CI = apply(gb.biol,1,Dapp)
```

The only other change needed is to the *y*-axis title. You can also experiment with other graphical instructions to alter the plotting character or thickness of the error bars for example (using `pch` and `lwd` instructions).

Note: R script for plotting bootstrap diversity and confidence intervals

The commands for calculating bootstrap diversity and confidence intervals and for producing point plots have been made into a custom function `plot_H()`. This is available as part of the *CERE.RData* file. The command uses the Shannon index by default but you can use Simpson's index via the `index = "simpson"` instruction. The `plot_H()` command actually calls another custom function called `H_ci()`, which carries out the bootstrapping part.

So far you have seen how bootstrapping can be used to resample some community data to produce confidence intervals. This is a good way to get an idea of variability as you are not relying on any particular formula. You have already seen how the variance formulae for Simpson's and Shannon indices are sensitive to species richness. However, the bootstrapping methods you've seen so far do not produce a *p*-value for differences between samples. It is possible to do this, as shown in the next section.

9.4.3 Statistical tests using bootstrap statistics

If you can resample a community data sample to produce confidence intervals then you can apply the same approach to the comparison of two samples and get an idea of statistical significance.

Comparing bootstrap samples

You have already seen how to create bootstrap samples for Shannon and Simpson's diversity indices, so you will use these samples as the starting point for investigating differences between samples. You can use the custom function `H_boot()` to carry out the bootstrapping. This is available as part of the *CERE.RData* file and utilises the `boot()` command from the *boot* package.

One approach would be to compare the bootstrap samples directly by subtracting one from the other. This would give you a new sample that represented differences in diversity index for the many bootstrapped resamples you carried out. You could then compare how many of these differences were larger than the original difference – this would be a kind of *p*-value.

The `H_boot()` command will carry out a bootstrapping routine and produce the results you need (it works for Simpson's or Shannon indices). Running the command produces an immediate result:

```
> Hbx = H_boot(gb.biol[1,])

H =   1.267096
     2.5%      97.5%
 1.170726 1.339600
```

You can see the original Shannon diversity index value and the confidence intervals. However, the result object contains other statistics that were not presented – you can see what is available using the `names()` command:

```
> names (Hbx)
[1] "boot.stats" "CI"
```

The result is split into two parts "boot.stats" and "CI". You can access these parts using the $ like so:

```
> Hbx$boot.stats
```

```
ORDINARY NONPARAMETRIC BOOTSTRAP


Call:
boot (data = cx, statistic = shannon1, R = R)


Bootstrap Statistics :
     original      bias    std. error
t1* 1.267096 -0.0135911  0.0444591
```

You can see that the first part contains the results from the boot () command. This is itself subdivided:

```
> names (Hbx$boot.stats)
 [1] "t0"         "t"         "R"         "data"      "seed"
 [6] "statistic" "sim"       "call"      "stype"     "strata"
[11] "weights"
```

The parts you are most interested in are t0 and t, which are the original Shannon index and the bootstrapped Shannon indices respectively. The R part contains the number of replicates in the bootstrap sample and the statistic is the function that was used to work out the bootstrap samples.

So, once you have saved the result of the H_boot () command to a named object you can easily get the original Shannon diversity index value and extract the sample that contains the bootstrapped values:

```
> Hbx$boot.stats$t0
[1] 1.267096
> mean (Hbx$boot.stats$t)
[1] 1.253505
> Hbx$boot.stats$R
[1] 1000
```

A good start is to make two result objects containing the two community samples of interest. You already have one sample so you need to run the H_boot () command again for the second sample:

```
> Hby = H_boot (gb.biol[2,])

H =   1.424505
     2.5%      97.5%
1.340480 1.486502
```

It would be helpful to make a new sample that represented the difference between the two bootstrapped samples since you are interested in differences between the Shannon diversity indices:

```
> diff = Hbx$boot.stats$t - Hby$boot.stats$t
```

You also need the difference between the original samples diversity indices – the abs () command will ensure that the difference is positive (you are only interested in the magnitude):

```
> x0 = Hbx$boot.stats$t0
> y0 = Hby$boot.stats$t0
> diff0 = abs(x0 - y0)
> diff0
[1] 0.1574092
```

If you compare the difference between the original index values and the sample representing the bootstrapped differences you can get an idea of how likely it is that the difference was down to chance:

```
> mean(abs(diff) > diff0)
[1] 0.531
```

This is not quite correct! Look at the confidence intervals:

```
> Hbx$CI
    2.5%     97.5%
1.170726 1.339600
> Hby$CI
    2.5%     97.5%
1.340480 1.486502
```

There is not really any overlap (if you try this your values may be slightly different) so the p-value of 0.531 cannot be correct. What has happened is that we are comparing the wrong values. The bootstrap statistics represent multiple resampling of the original communities – some of the Shannon indices will be lower than the original and some greater (about half the time greater and half lower in fact). Each comparison between the two bootstrap samples is therefore going to vary about the mean of the original difference and result in a p-value of about 0.5.

All is not lost – you have a new sample that represents differences between re-samples of the original communities. What you need to do is to 'adjust' the sample to represent the differences under the null hypothesis situation:

```
> H0 = diff - mean(diff)
```

You can now get an approximate p-value by comparing how many times your new 'null' sample has a value greater than the original difference:

```
> mean(abs(H0) > diff0)
[1] 0.006
```

You can carry out the same kind of analysis on any two bootstrapped samples – as long as you adjust the differences to represent a null hypothesis. The p-value can be obtained by inspection of the samples – comparing the bootstrapped differences to the original – but sometimes you get 0 as a result. This would happen when none of the bootstrapped differences were larger than the original difference. You can think of this as being a very small p-value but it is likely that you'd get something >0 if you carried out more replicates, and zero as a probability is not very realistic or satisfactory. In such cases you can calculate a z-score.

Calculating z-scores

A z-score is a way of standardising a deviation. In other words it is a way to determine how many standard deviations an observation is away from the mean. Once you have a z-score you can determine a p-value using the normal distribution and an appropriate degrees of freedom. The general formula for working out a z-score is shown in Figure 9.14.

$$z = \frac{x - \mu}{\sigma}$$

Figure 9.14 The formula for calculating a z-score, x is an observation, μ is the mean and σ the standard deviation of a sample.

For the comparison of two bootstrapped samples you can calculate a z-score by substituting into the equation (Figure 9.14):

- Replace x with your original difference in Shannon diversity.
- Replace μ with the mean of your bootstrapped null sample.
- Replace σ with the standard deviation of the bootstrapped null sample.

Once you have a z-score you will need the degrees of freedom. This is problematic as the value will influence the result heavily. The number of individuals is too high a value and results in unrealistically small p-values. The number of species is better and produces slightly more conservative p-values. You take the total number of species in both samples and subtract 2 to get the degrees of freedom. You can then determine the p-value using the pt () command:

```
pt(z, df, lower.tail = FALSE) * 2
```

You need to multiply the result by 2 because you have determined the one-tailed p-value. It is best to use the p-value you obtain from the bootstrapping directly but if you do get a 0 then at least you can report an alternative.

In the following exercise you can have a go at making a copy 'n'paste script that you can use to determine the z-score and significance of the difference between the Shannon diversity index of two samples.

Have a Go: Make a copy 'n' paste script to calculate the z-score for differences between Shannon diversity of two samples

For this exercise you will need the CERE.RData file, which contains the H_boot () command. You'll also need the *boot* and *vegan* packages. If these packages are installed but not loaded the H_boot () command will do so for you. If you haven't got the packages installed you will need to use the install.packages () command to get them.

```
> install.packages(c("vegan", "boot"))
```

1. Start by assigning simple variables to 'hold' two community samples to compare:

```
> x = gb.biol[1, ]
> y = gb.biol[2, ]
```

2. Now work out the degrees of freedom from the number of species in each sample:

```
> DF = sum(x > 0) + sum(y > 0) - 2
> DF
[1] 29
```

3. Use the H_boot() command to carry out a bootstrap sampling for each of the two community samples – the command will give you some results right away:

```
> xH = H_boot(x) # Bootstrap sample1

H =  1.267096
     2.5%      97.5%
1.176235 1.338760
> yH = H_boot(y) # Bootstrap sample2

H =  1.424505
     2.5%      97.5%
1.344516 1.488202
```

4. Use the results to work out the original diversity index values and the difference between the samples:

```
> x0 = xH$boot.stats$t0
> y0 = yH$boot.stats$t0
> diff0 = abs(x0 - y0)
> diff0
[1] 0.1574092
```

5. Now make a new sample from the difference between the two bootstrapped samples:

```
> diff = xH$boot.stats$t - yH$boot.stats$t
```

6. Create a null model by subtracting the mean of the new sample from every replicate:

```
> H0 = diff - mean(diff)
```

7. Compare the null model to the original Shannon index difference to work out a simulated p-value:

```
> Psim = mean(abs(H0) > diff0)
> Psim
[1] 0.003
```

8. Use the null model to work out a z-score:

```
> Z = (diff0-mean(H0))/sd(H0)
> Z
[1] 2.800452
```

9. Finally, calculate a p-value from the z-score using the degrees of freedom derived from the number of species:

```
> Pz = pt(Z, df = DF, lower.tail = FALSE)*2
> Pz
[1] 0.008988498
```

You can see that the two *p*-values are different. You should report the `Psim` value unless this turns out to be zero, in which case you can choose to report the *z*-score and *p*-value derived from that.

You can easily modify the commands in the preceding exercise so that you calculate bootstrap Simpson's indices – you simply add the instruction `index = "simpson"` to the `H_boot()` command.

Using Monte Carlo simulation to compare diversity indices using species-based samples

The term *Monte Carlo* is used to indicate simulation procedures, such as the bootstrap analyses you've already seen, as well as other process that involve randomisation. Earlier you saw how to compare bootstrap samples for diversity indices (Shannon and Simpson) and determine statistical significance as well as calculating a *z*-score.

The bootstrap methods you have seen so far convert the species-based samples into individual-based samples before carrying out the randomisation processes (the bootstrapping). However, this is not the only way you can proceed, and in this section you'll see how to take species-based samples and compare their diversity without having to convert the samples.

You can think of your community data as a matrix with columns representing the species and rows for the sites (i.e. samples). The individual values in the matrix are the various abundances. Instead of converting each row (sample) to an individual-based sample you can randomise the entire matrix. What you need to do is randomise the values so that the row and column totals remain the same (the values within the individual cells can alter). R provides a command `r2dtable()` that can carry this out for you. It is easier to see how this works with a simple example.

Since the object of the exercise is to compare two samples you can start by making a data matrix that contains two communities:

```
> x = c(100, 35, 0, 10, 5, 2, 1, 0, 0)
> y = c(11, 33, 0, 44, 0, 1, 1, 22, 7)
> m = rbind(x, y)
> rownames(m) = c("Sample1", "Sample2")
> colnames(m) = paste("Sp.", LETTERS[1:9], sep = "")
```

The first two commands simply create two simple numeric vectors of data – notice that they are the same length. The next command joins the two samples together to form a matrix. The penultimate commands add some simple labels to the rows. The last command makes labels for the columns by joining together the `Sp.` label with an upper case letter (from A to I). The final result looks like this:

```
> m
        Sp.A Sp.B Sp.C Sp.D Sp.E Sp.F Sp.G Sp.H Sp.I
Sample1  100   35    0   10    5    2    1    0    0
Sample2   11   33    0   44    0    1    1   22    7
```

The `rd2table()` command needs three values:

```
r2dtable(n, r, c)
```

- n, the number of times you want to randomise.
- r, the row totals (for a two-row matrix this would be two values).
- c, the column totals (one for each column of the matrix).

The row and column totals are easily determined:

```
> rt = rowSums(m)
> ct = colSums(m)
> rt
Sample1 Sample2
    153     119
> ct
Sp.A Sp.B Sp.C Sp.D Sp.E Sp.F Sp.G Sp.H Sp.I
 111   68    0   54    5    3    2   22    7
```

You can also check that the row and column totals sum to the same value:

```
> sum(rt) ; sum(ct)
[1] 272
[1] 272
```

Now you have everything you need to create a new random 2-D table:

```
> r2dtable(1, rt, ct)
[[1]]
      [,1] [,2] [,3] [,4] [,5] [,6] [,7] [,8] [,9]
[1,]   55   42    0   33    4    3    0   12    4
[2,]   56   26    0   21    1    0    2   10    3
```

Notice that you lose the row and column names – this is not really a problem because when you come to calculate the diversity indices from the matrix (one for each row of course) you are only interested in the final result.

There are two main ways you can use this random table process to look at differences in communities:

- Make 1000+ random tables and determine the diversity differences from those.
- Make one random table and determine the diversity then repeat 1000+ times.

Which choice you make is up to you but the second method is more space efficient – you can use a loop and 'build' your result. In the following exercise you can have a go at using this random table approach to compare the Shannon diversity of two samples.

Have a Go: Use random tables in a Monte Carlo process to compare Shannon diversity

For this exercise you will need the *vegan* package. If you haven't got the package installed you will need to use the install.packages() command to get it. You will also need the ground beetle data to work with; these are part of the *CERE.RData* file, which you can get from the companion website.

```
> install.packages("vegan")
```

1. Prepare the *vegan* package so its commands are available:

```
> library(vegan)
```

2. You need a matrix with two rows for the community data. Make one from the two rows of the *gb.biol* data:

```
> M = rbind(gb.biol[17,], gb.biol[18,])
```

3. Work out the row totals for the community matrix you've just made:

```
> RT = rowSums(M)
```

4. Now work out the column totals for the matrix:

```
> CT = colSums(M)
```

5. Calculate the degrees of freedom from the total number of species:

```
> DF = sum(M > 0) - 2
```

6. Use the `diversity()` command to determine the Shannon diversity index of both samples and thus the difference between them. You will need to use the `as.numeric()` command to 'strip off' the headings and ensure you get a plain number as the result:

```
> Ho = as.numeric(abs(diversity(M)[1] - diversity(M)[2]))
> Ho
[1] 0.06386276
```

7. Now make a blank vector object as a container to hold the results, which you will build shortly. The length instruction can be set to any value but use 1000 for now, which will result in 1000 replicates:

```
> Hb = vector(mode = "numeric", length = 1000)
```

8. Start a loop that runs for 1000 repeats (to match the length of the result vector from step 7) – start with the `for()` command and then a single curly bracket (delete any closing bracket that appears:

```
> for(i in 1:1000) {
```

9. R should now be displaying a + cursor, which indicates that extra input is required before the current command line can be run. Use the `rd2table()` command to make a random table from the community matrix:

```
+ Mb = r2dtable(1, RT, CT)
```

10. The result of the `rd2table()` command is a list so now convert the result from step 9 into a matrix using the `as.matrix()` command:

```
+ Mb = as.matrix(Mb[[1]])
```

11. Now use the `diversity()` command to work out the difference between the Shannon diversity of the two samples in the current random table:

```
+ Hd = as.numeric(abs(diversity(Mb)[1] - diversity(Mb)[2]))
```

12. Add the diversity index difference you just calculated in step 11 to the result object. The square brackets will ensure that the value is added at the appropriate position:

```
+ Hb[i] = Hd
```

13. The loop is still 'operational' so finish it by entering the closing curly bracket. Once you do this the loop itself will run for another 999 times until it has gone around 1000 times:

```
+ }
```

14. Now you can use the Hb result to explore the differences between the two samples and compute the final statistics. Start with the confidence intervals:

```
> Q = quantile(Hb, c(0.025, 0.975))
> Q
        2.5%          97.5%
0.001338166 0.089981066
```

15. Now determine how many times the simulated result was greater than the original difference – dividing by the number of replicates give a simulated p-value:

```
> Ps = sum(Hb > Ho) / 1000 # You can also use mean(Hb > Ho)
> Ps
[1] 0.112
```

16. Calculate a z-score for the simulated sample:

```
> Z = (Ho - mean(Hb)) / sd(Hb)
> Z
[1] 1.302702
```

17. Finally you can calculate a p-value based on the z-score using the degrees of freedom based on the number of species:

```
> Pz = pt(Z, df = DF, lower.tail = FALSE)*2
> Pz
[1] 0.2074775
```

You can see that the two p-values are different but that the result is not – the Shannon diversity index is not significantly different between the two samples. Since you got a simulated p-value here you should not bother reporting the p-value from the z-score. If you get a simulated p-value of 0 then you could try increasing the number of replicates or simply use the p-value from the z-score.

You can see from the preceding exercise that it is easy to modify the commands to compare Simpson's diversity index or inverse Simpson's, as alternatives to Shannon, since these are used in the diversity() command.

Note: R code for Monte Carlo simulation

The commands needed to create the Monte Carlo simulation are packaged into several custom commands as part of the CERE.RData file. The main function is a command H_bss(), which also has special summary() and print() routines. The command can handle Shannon, Simpson and inverse Simpson indices.

If you use this Monte Carlo approach and wish to visualise the relationship between the two samples then you can use one of the previous bootstrapping processes to determine confidence intervals for each sample.

9.5 Comparisons using replicated samples

You have seen so far that comparing diversity indices requires some idea of the variability so that you can estimate confidence intervals. You know how to estimate variability for Shannon and Simpson's indices using formulae (Figures 9.3 and 9.6), and also how to use randomisation methods to create new samples and therefore estimate variability from those (Section 9.4).

When you have replicated data things can become a bit easier as you can now use a more 'usual' array of statistical tests, for example:

- For comparing two samples – *t*-test or *U*-test.
- For comparing multiple samples – ANOVA, Kruskal–Wallis or multiple pairwise tests with *p*-value corrections.
- For comparing diversity with some variable – correlation or regression.

The first step is to decide if your samples are appropriate for the question at hand. Is it really appropriate to determine diversity of a grassland using a 1 m² quadrat for example? If the answer is 'no' then you might be able to amalgamate samples into larger and more appropriate sampling units.

Once you have appropriate data samples you need to arrange your data so that you can carry out the analysis that you require. This will involve Pivot Tables in Excel (Section 4.2.7) or cross-tabulation of some kind in R (Section 6.3.3).

When you finally have your data arranged you can carry out your analysis. An important part of any analysis is the graphical overview, which you can do from your Pivot Tables (Section 6.3.2) or using R (Section 6.3.3).

In the following sections you will some examples of the different kinds of analysis and get the chance to have a go for yourself.

9.5.1 Comparing two or more samples

If you are comparing samples and looking for differences in diversity indices then you can use 'regular' statistical methods for your analysis. The main decision you have is whether your data are normally distributed or not.

Two samples

If you only have two samples to compare (that is, two sets of replicated observations) then you would use the *t*-test or the *U*-test to compare the average diversity index. You could conduct the *t*-test using Excel but it is not so easy to carry out a *U*-test. In the following exercise you can have a go at preparing some data for analysis and then carrying out an analysis using R. You'll have to create a subset of the data because there are more than two levels of the grouping variable.

Have a Go: Carry out a comparison of diversity between two samples of replicated data

For this exercise you'll need the ground beetle data that are part of the *CERE.RData* file. You will start by arranging the data from the original biological records. You could of course do this using Excel and then import into R. You'll also need the *vegan* package.

1. Start by looking at the *gb* data, which contains the 'raw' biological records of the ground beetles:

```
> head(gb)
                Species Quantity Sample     Abbr Max.Ht Habitat
1 Pterostichus melanarius         4     E5  Pte.mel    2.5    Edge
2 Leistus spinibarbis             2     E6  Lei.spi    2.0    Edge
3 Carabus violaceus               1     E6  Car.vio    2.0    Edge
4 Platynus assimilis              4     E5  Pla.ass    2.5    Edge
5 Agonum muelleri                 1     E5  Ago.mue    2.5    Edge
6 Calathus rotundicollis         94     E5  Cal.rot    2.5    Edge
```

2. First of all you should make the community data so that you can assess the diversity index – you will need to use the `xtabs()` command for this:

```
> comm = xtabs(Quantity ~ Sample + Abbr, data = gb)
```

3. Now convert the data from a `table` to a `data.frame`. You'll have to make the data into a matrix first otherwise you'll end up converting back to the original biological recording format:

```
> class(comm) = "matrix"
> comm = as.data.frame(comm)
```

4. You've got the biological data ready but now you need to get the grouping variable. Make a new data object to hold the *Habitat* variable:

```
> envi = aggregate(Quantity ~ Habitat + Sample, data = gb, FUN = sum)
> head(envi)
  Habitat Sample Quantity
1    Edge     E1      715
2    Edge     E2      734
3    Edge     E3      646
4    Edge     E4      867
5    Edge     E5      971
6    Edge     E6      901
```

5. Although it is not strictly necessary you can now make the row names reflect the *Sample* column. Do this by assigning the row names to the *Sample* variable, which you can then remove:

```
> rownames(envi) = envi$Sample
> envi = envi[,-2]
> head(envi)
   Habitat Quantity
E1    Edge      715
E2    Edge      734
E3    Edge      646
E4    Edge      867
E5    Edge      971
E6    Edge      901
```

6. It is helpful to put all your data into one place so first compute the Shannon diversity and then assemble the results into a new `data.frame` with the grouping variable (*Habitat*). You will need the *vegan* library loaded for this:

```
> library(vegan)
> H = diversity(comm, index = "shannon")
> dat = data.frame(H = H, site = envi$Habitat)
> dat
          H    site
E1 1.267096   Edge
E2 1.424505   Edge
E3 1.412179   Edge
E4 1.737146   Edge
E5 1.636923   Edge
E6 1.557477   Edge
G1 1.935163  Grass
G2 2.000272  Grass
G3 1.929426  Grass
G4 2.360250  Grass
G5 2.139567  Grass
G6 2.205175  Grass
W1 1.397251   Wood
W2 1.582102   Wood
W3 1.539511   Wood
W4 1.601073   Wood
W5 1.616333   Wood
W6 1.552470   Wood
```

7. You can use the square brackets to 'extract' a sample for a particular site – first look at a vector and then a data.frame:

```
> dat$H[dat$site == "Edge"]
[1] 1.267096 1.424505 1.412179 1.737146 1.636923 1.557477

> dat[dat$site == "Edge",]
          H site
E1 1.267096 Edge
E2 1.424505 Edge
E3 1.412179 Edge
E4 1.737146 Edge
E5 1.636923 Edge
E6 1.557477 Edge
```

8. Use the tapply() command to look at the normality of all the samples using the Shapiro–Wilk test:

```
> tapply(dat$H, INDEX = dat$site, FUN = shapiro.test)

$Edge
    Shapiro-Wilk normality test
data: X[[1L]]
W = 0.9755, p-value = 0.9272

$Grass
    Shapiro-Wilk normality test
data: X[[2L]]
W = 0.9104, p-value = 0.4389
```

```
$Wood
   Shapiro-Wilk normality test
data: X[[3L]]
W = 0.8213, p-value = 0.09059
```

9. None of the samples are significantly different from a normal distribution – use a quantile–quantile plot to visualise each sample:

```
> qqnorm(dat$H[dat$site == "Edge"]) # QQ plot for "Edge"
> qqline(dat$H[dat$site == "Edge"]) # Add line
```

10. There are three *sites* here so you'll need to use a `subset` in order to carry out a *t*-test on a pair of samples:

```
> t.test(H ~ site, data = dat, subset = site%in% c("Edge", "Grass"))

   Welch Two Sample t-test

data:  H by site
t = -5.9694, df = 10, p-value = 0.0001376
alternative hypothesis: true difference in means is not equal to 0
95 percent confidence interval:
 -0.8089698 -0.3692063
sample estimates:
 mean in group Edge mean in group Grass
           1.505888            2.094976
```

11. Try to visualise the result with a `boxplot()` command results in a problem as you display the unused *site* variable as well as the `subset`:

```
> boxplot(H ~ site, data = dat, subset = site%in% c("Edge", "Grass"))
```

12. You need to make a `subset` of the data as a separate object:

```
pd = subset(dat, subset = site %in% c("Edge", "Grass"))
```

13. The unused levels of the *site* variable are still in the result but now you can use the `droplevels()` command to exclude them. Use the `boxplot()` command and add some titles. Your final graph should resemble Figure 9.15:

```
> boxplot(H ~ droplevels(site), data = pd)
> title(xlab = "Habitat", ylab = "Shannon diversity index, H")
```

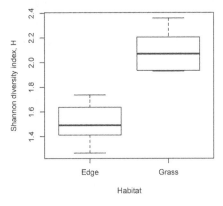

Figure 9.15 Shannon index for ground beetle communities at two habitats. Strip = median Shannon index, box = IQR, whiskers = extremes.

The box-whisker plot (Figure 9.15) shows the raw data expressed as the median, inter-quartiles and extreme values.

You can easily carry out a *U*-test instead by using the `wilcox.test()` command. Other indices can be compared by altering the `diversity()` command in step 6.

Multiple comparisons

If you have more than two samples to compare then you have two choices:

- Use analysis of variance (or the non-parametric equivalent).
- Use several pairwise tests and then adjust the *p*-values for multiple comparisons.

As with the two-sample case you need to determine if your data are normally distributed and select the most appropriate method of analysis. Generally ANOVA would be preferred to the multiple pairwise approach because the post-hoc methods for conducting the pairwise comparisons are less conservative. You can use Excel for some processes – especially the Pivot Table preparation of data and graphical overview – but when it comes to the analyses it is generally easier to use R. In the following exercise you can have a go at preparing and analysing multiple replicated samples using R.

Have a Go: Carry out a comparison of diversity between multiple samples of replicated data

For this exercise you'll need the ground beetle data that are part of the *CERE.RData* file. You will start by arranging the data from the original biological records. You could of course do this using Excel and then import into R. You'll also need the *vegan* package.

1. Start by extracting the community data from the original biological records:

```
> comm = xtabs(Quantity ~ Sample + Abbr, data = gb)
```

2. Now compute the Shannon diversity index for each of the samples – you'll need the *vegan* library loaded:

```
> library(vegan)
> H = diversity(comm)
> H
        E1       E2       E3       E4       E5       E6       G1       G2
1.267096 1.424505 1.412179 1.737146 1.636923 1.557477 1.935163 2.000272
      G3       G4       G5       G6       W1       W2       W3       W4
1.929426 2.360250 2.139567 2.205175 1.397251 1.582102 1.539511 1.601073
      W5       W6
1.616333 1.552470
```

3. Go back to the original biological records and make a new object that contains the grouping variable for the *Habitat*:

```
> envi = aggregate(Quantity ~ Habitat + Sample, data = gb, FUN = sum)
```

4. Make a new object for the *Habitat* variable:

```
> site = envi$Habitat
> site
```

```
 [1] Edge Edge Edge Edge Edge Edge Grass Grass Grass
[10] Grass Grass Grass Wood Wood Wood Wood Wood Wood
Levels: Edge Grass Wood
```

5. Use the `tapply()` command to check the normality of the samples with a series of Shapiro–Wilk tests. This is similar to step 8 of the preceding exercise:

```
> tapply(H, INDEX = site, FUN = shapiro.test)
```

6. Since the data appear normally distributed you can use analysis of variance using the `aov()` command:

```
> Haov = aov(H ~ site)
> summary(Haov)
            Df Sum Sq Mean Sq F value Pr(>F)
site         2 1.29571 0.64786  30.028 5.692e-06 ***
Residuals   15 0.32362 0.02157
---
Signif. Codes:  0 `***' 0.001 `**' 0.01 `*' 0.05 `.' 0.1 ` ' 1
```

7. Carry out a post-hoc test using the `TukeyHSD()` command:

```
> TukeyHSD(Haov)
 Tukey multiple comparisons of means
  95% family-wise confidence level

Fit: aov(formula = H ~ site)

$site
                  diff       lwr        upr       p adj
Grass-Edge  0.58908809  0.3688140  0.8093622 0.0000132
Wood-Edge   0.04223582 -0.1780383  0.2625099 0.8733261
Wood-Grass -0.54685227 -0.7671264 -0.3265782 0.0000309
```

8. Visualise the situation using the `boxplot()` command. You should add appropriate titles to produce a graph similar to Figure 9.16:

```
> boxplot(H ~ site)
> title(ylab = "Shannon diversity index, H", xlab = "Habitat")
```

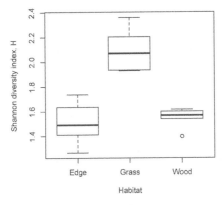

Figure 9.16 Shannon index for ground beetle communities at three habitats. Strip = median Shannon index, box = IQR, whiskers = extremes < 3/2 * IQR.

9. Try an alternative approach and use multiple *t*-tests with a correction for multiple comparisons:

```
> pairwise.t.test(H, g = site, p.adjust.method = "bonferroni")

    Pairwise comparisons using t tests with pooled SD

data:  H and site

        Edge    Grass
Grass 1.4e-05 -
Wood 1       3.3e-05

P value adjustment method: bonferroni
```

Notice that in Figure 9.16 there is an outlier in the *Wood* sample. You can use the `range = 0` instruction to force the whiskers to extend to the max/min (rather than 1.5 times the IQR).

Bootstrap analysis

You saw earlier how bootstrapping could be used to get an idea of variability for a single sample (Section 9.4.1), and also how to use it to carry out statistical testing (Section 9.4.3). It is possible to use bootstrapping methods when you have replicated data and therefore to compare several sites (each with multiple replicates). You can also amalgamate replicates from each site to produce one large sample, which you then use in a 'regular' bootstrapping process. The latter approach might be especially useful of you have observations based on quite small sampling units.

In the following exercises you can have a go at the two approaches. First you can use a convenient R package to help bootstrap analyse multiple sites at once. Later you can try amalgamating samples and carrying out some bootstrapping.

The following exercise uses a convenient R package called *simboot*. This is designed expressly to carry out confidence interval and *p*-value assessment for Shannon and Simpson diversity indices when using replicated data and a grouping variable.

Have a Go: Use a grouping variable with replicated data to compare diversity indices using bootstrapping

You'll need the *simboot* package for this exercise and the ground beetle data that are part of the *CERE.RData* file. The *simboot* package requires packages *boot* and *mvtnorm* but these will usually be installed when you use `install.packages()` to get *simboot*.

1. Start by preparing the *simboot* package – the packages *boot* and *mvtnorm* will be loaded alongside *simboot* when you run the command:

```
> library(simboot)
Loading required package: boot
Loading required package: mvtnorm
```

2. Now prepare the community data from the original biological records:

```
> comm = xtabs(Quantity ~ Sample + Abbr, data = gb)
```

3. The sbdiv() command that you will use shortly needs the data to be in the form of a data.frame so you need to alter the current xtable result:

```
> class(comm) = "matrix"
> comm = as.data.frame(comm)
```

4. Make a new object to hold the *Habitat* information, which will be used as the grouping variable:

```
> envi = aggregate(Quantity ~ Habitat + Sample, data = gb, FUN = sum)
```

5. The bootstrapping procedure does not need you to determine the diversity index because the sbdiv() command calculates the Shannon (or Simpson) index for you. You need to specify the community data, the grouping variable and the index to use as well as the type of post-hoc test to run:

```
> sbdiv(comm, f = envi$Habitat, theta = "Shannon", type = "Tukey")
$conf.int
               estimate   lower   upper
Grass - Edge      0.620   0.405   0.835
Wood  - Edge      0.037  -0.179   0.252
Wood  - Grass    -0.583  -0.799  -0.368

$p.value
               adj. p raw p
Grass - Edge   0.000 0.000
Wood  - Edge   0.908 0.674
Wood  - Grass  0.000 0.000

$conf.level
[1] 0.95

$alternative
[1] "two.sided"
```

In order to visualise the situation the boxplot() command is the simplest, giving you the median diversity index, inter-quartiles and range.

Compare the results obtained using the analysis of variance, pairwise *t*-tests and bootstrap analysis via sbdiv() – they are all in broad accordance.

You might decide that it is not really appropriate to assess biodiversity using small samples and that a better approach would be to amalgamate samples from the various sites so that you have one large sample for each site. In the following exercise you can have a go at amalgamating some samples and then using some bootstrapping methods to assess differences in the diversity.

Have a Go: Use bootstrapping on amalgamated samples

For this exercise you'll use some custom functions in the *CERE.RData* file (`H_bss()` and `H_boot()` commands). You will use the ground beetle data so that you can compare results to the other approaches that you've seen. You will also need the *vegan* package, this will be loaded for you once you run `H_bss()` or `H_boot()` but you can use the `library()` command before then.

1. Start by preparing the data from the original biological records. You need to use the `xtabs()` command for this:

   ```
   > amal = xtabs(Quantity ~ Habitat + Abbr, data = gb)
   ```

2. The object you created is a table object – you can see this by using the `class()` command, which reveals that the object actually holds two classes:

   ```
   > class(amal)
   [1] "xtabs" "table"
   ```

3. You do not need to convert the data into a `data.frame` or a matrix because the `H_bss()` and `H_boot()` commands can 'handle' the table. Use the `H_boot()` command to work out the 95% confidence intervals for one of the sites:

   ```
   > H_boot(amal[1,])

   H =  1.558842
      2.5%     97.5%
   1.522005 1.590767
   ```

4. Now look at the confidence intervals for another sample – you can use the name like so:

   ```
   > H_boot(amal["Wood",])

   H =  1.566727
      2.5%     97.5%
   1.543017 1.589500
   ```

5. Complete the set and work out CI for the remaining sample:

   ```
   > H_boot(amal[2,])

   H =  2.162426
      2.5%     97.5%
   2.090923 2.220895
   ```

6. Use the `H_bss()` command to carry out a bootstrap significance test between two of the samples:

   ```
   > H_bss(amal[c("Edge", "Grass"), ])
           H.orig.1 H.orig.2 Statistic  Z.score       2.5%    97.5% P(sim)
   shannon 1.558842 2.162426 0.6035832 24.12843 0.001511143 0.08893133    0
                  P(z) df
   shannon 3.236972e-33 62
   ```

7. Repeat step 6, but this time assign the result to a named object:

```
> E.G = H_bss(amal[c("Edge", "Grass"), ])
```

8. Run the H_bss() command twice more to produce results for the other two comparisons:

```
> E.W = H_bss(amal[c("Edge", "Wood"), ])
> G.W = H_bss(amal[c("Grass", "Wood"), ])
```

9. The results of the H_bss() command contain several items, use the names() command to see what they are:

```
> names(E.G)
[1] "H.orig"    "statistic" "z.score"   "CI"        "p.sim"       "p.z"
[7] "df"        "Index"
```

10. Look at the p-values based on the z-score for the three comparisons:

```
> E.G$p.z ; E.W$p.z ; G.W$p.z
[1] 4.505781e-33
[1] 0.4649685
[1] 6.576236e-30
```

11. Combine the three p-values into one object – give the result some names to keep track of the comparisons:

```
> Hp = c(E.G$p.z, E.W$p.z, G.W$p.z)
> names(Hp) = c("EG", "EW", "GW")
> Hp
            EG             EW             GW
4.505781e-33   4.649685e-01   6.576236e-30
```

12. You need to adjust the p-values for multiple comparisons – try a couple of different adjustment methods:

```
> p.adjust(Hp, method = "bonferroni")
            EG             EW             GW
1.351734e-32   1.000000e+00   1.972871e-29
```

```
> p.adjust(Hp, method = "holm")
            EG             EW             GW
1.351734e-32   4.649685e-01   1.315247e-29
```

If you wish to visualise the results then you'll need to make a point plot that shows the Shannon index values and the 95% confidence intervals – look back to Section 9.4.2 for the appropriate commands. If you wanted to explore Simpson's index instead then you can simply run the commands in the exercise using the H_bss() and H_boot() commands with the index = "simpson" instruction.

Note: Different test options

Of course you should not run all these tests on your data at once and 'pick the best' – the options are merely presented to show you that there are options.

9.5.2 Correlating diversity

In the examples you've seen so far, diversity has been compared between samples or groups of samples. In statistical terms you can think of the diversity (measured as a diversity index) as being the response variable and the groups as being the predictor variable. These terms are often called the *dependent variable* and the *independent variable*. In this book the terms *response* and *predictor* are preferred. An independent variable is so-called because it is independent of the dependent variable but this can lead to confusion. When you have several response variables they may not actually be independent of one another.

In the previous examples the predictor variables were categorical – in fact the same data were used (the ground beetles) – and you compared diversity with habitat or site. There are often occasions where you have a variable that is a continuous variable, such as time, temperature, soil pH or concentration of some chemical. In such cases the analytical approach is one of *correlation* or *regression*. There are some subtle differences between them:

- *Correlation* – you only have two variables to compare, diversity and something else. The correlation does not imply cause and effect. The relationship does not have to be linear.
- *Regression* – you can use more than two variables, diversity and more than one predictor variable. You are implying some mathematical relationship between the response variable and the predictors; this does not have to be linear.

The first step is to determine if the correlation or regression you want to undertake is 'sensible' – are your samples comparable in respects other than the predictors you are looking at? For example, if you wanted to compare diversity and soil depth it would not be sensible to use communities from mixed habitats, e.g. desert, grassland, woodland.

The next step is to arrange your data appropriately – a task suited to the Pivot Table in Excel or to cross-tabulation commands in R. You can carry out parametric correlation and regression using Excel but you cannot conduct non-parametric analysis easily.

In most respects you are treating the diversity index as a 'regular' response variable so your approaches to the analysis aren't any different from any other branch of analytics. The general approaches to analysis are:

- *Correlation* – when you only have one predictor variable.
  - Normally distributed data – Pearson's product moment.
  - Non-parametric data – Spearman rho or Kendall tau rank correlations.
- *Regression* – when you have one or more predictor variables.
  - Normally distributed data – regular linear regression (can be curvilinear).
  - Non-parametric data – non-linear regression or generalised linear modelling (usually with a Poisson distribution).
- Binomial data – Logistic regression (generalised linear modelling).

These methods of analysis are not strictly only community analysis so you won't see all these methods covered in detail, but you can have a go at some in the following exercises.

In the exercise that follows you can have a go at carrying out a simple correlation using the butterfly data you met earlier.

Have a Go: Correlating diversity index to a variable

For this exercise you will need the butterfly data that are part of the *CERE.RData* file. You will also need the *vegan* package to work out the diversity index.

1. Start by preparing the *vegan* package:

```
> library(vegan)
```

2. Look at the butterfly biological records – use the head() command to see the first few lines:

```
> head(bf)
      Spp   Yr Qty
1 Lg.skip 1996   3
2 Lg.wht  1996   4
3 Sm.wht  1996   2
4 GV.wht  1996 119
5 Or.tip  1996  90
6 Sm.cop  1996   9
```

3. Now convert the biological records into a community data object using the xtabs() command:

```
> comm = xtabs(Qty ~ Yr + Spp, data = bf)
```

4. The row names are years and so can be used as a continuous variable. Look at the row names:

```
> rownames(comm)
 [1] "1996" "1997" "1998" "1999" "2000" "2001" "2002" "2003v "2004" "2005"
```

5. Convert the row names into a numerical variable for the correlation:

```
> pv = as.numeric(rownames(comm))
> pv
 [1] 1996 1997 1998 1999 2000 2001 2002 2003 2004 2005
```

6. Calculate the Shannon diversity index for the community data:

```
> H = diversity(comm, index = "shannon") # Shannon diversity as resp var
> H
    1996     1997     1998     1999     2000     2001     2002     2003
2.081733 2.099876 1.770715 1.619179 1.705982 1.636914 1.449775 1.155682
    2004     2005
1.560148 1.714095
```

7. Visualise the relationship between Shannon diversity and year with a scatter plot (Figure 9.17):

```
> plot(H ~ pv, ylab = "Shannon diversity, H", xlab = "Year")
```

8. Use a Spearman rho rank correlation test to see if there is a statistically significant correlation:

```
> cor.test(~ H + pv, method = "spearman")

    Spearman's rank correlation rho

data: H and pv
```

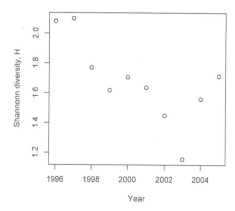

Figure 9.17 Shannon index for butterfly communities and year of observation for a site in SE England.

```
S = 274, p-value = 0.04403
alternative hypothesis: true rho is not equal to 0
sample estimates:
       rho
-0.660606
```

9. You should not really add a straight line to the correlation because you used a non-parametric analysis. However, as an exploratory method use the `abline()` command to add a straight line based on the linear fit:

```
> abline(lm(H ~ pv), lty = "dashed")
```

10. The fit does not look very good so try a locally weighted polynomial regression as a scatter plot smoother – your final plot should resemble (Figure 9.18):

```
> lines(lowess(pv, H), lty = 1)
```

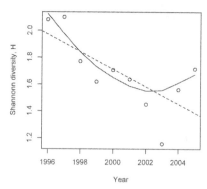

Figure 9.18 Shannon index for butterfly communities and year of observation for a site in SE England. Dashed line = linear best-fit. Solid line = locally-weighted polynomial regression.

You can see that the relationship between Shannon diversity and year of observation is not really straightforward.

Tip: Smooth lines

Adding lines to a graph can often result in a 'choppy' appearance because the line is formed from segments of straight line. You can use the `spline()` command to interpolate and smooth your lines into curves e.g.:

```
> lines(spline(lowess(x, y)), lty = 1)
```

The command above takes the `lowess()` line and smoothes it out.

In the following exercise you can have a go at performing a regression analysis of some lichen data to various soil chemical parameters. This will involve a model building process, which is useful when you have a lot of parameters to assess.

Have a Go: Build a regression model of diversity and soil chemistry

In this exercise you'll use some data that is built into the *vegan* package. There are two parts to the data: *varespec* shows pasture lichen communities and *varechem* shows some soil characteristics for the samples in the *varespec* data.

1. Start by loading the *vegan* library and ensuring that the data are ready to use:

```
> library(vegan)
> data(varespec)
> data(varechem)
```

2. Look at the variables in the *varespec* data and also the row names, which correspond to site/sample names:

```
> names(varechem)
 [1] "N"        "P"        "K"        "Ca"       "Mg"       "S"
 [7] "Al"       "Fe"       "Mn"       "Zn"       "Mo"       "Baresoil"
[13] "Humdepth" "pH"

> rownames(varechem)
 [1] "18" "15" "24" "27" "23" "19" "22" "16" "28" "13" "14" "20" "25" "7"
[15] "5"  "6"  "3"  "4"  "2"  "9"  "12" "10" "11" "21"
```

3. The columns of the *varespec* data are species names and the rows are the samples (which correspond to the row names of the *varechem* data). Calculate the Shannon diversity of the lichen samples:

```
> H = diversity(varespec, index = "shannon")
> H
      18       15       24       27       23       19       22       16
2.017763 1.837101 1.834652 1.871294 2.146992 1.980967 2.014880 2.041299
      28       13       14       20       25        7        5        6
1.238434 1.966392 2.084649 2.111599 2.121102 1.527634 1.210139 1.634510
       3        4        2        9       12       10       11       21
1.387198 1.813530 1.096279 1.189561 1.564995 1.192426 1.921902 2.283356
```

4. Try visualising all the relationships between the variables with a pairs plot:

```
> pairs( ~ H + ., data = varechem)
```

5. The graph you produced in step 4 is far too congested so narrow down your options by producing a correlation matrix:

```
> cor(H, varechem, method = "pearson")
                N            P          K         Ca        Mg            S
[1,] -0.07076834 -0.04095678 0.1861217 0.3234471 0.2741854 0.01771047
             Al           Fe         Mn         Zn            Mo  Baresoil
[1,] -0.342186 -0.3996103 0.2410095 0.04453363 0.001068949 0.8016919
      Humdepth          pH
[1,] 0.4979519 -0.3607835
```

6. Look to see which of the correlations has a value larger than 0.3:

```
> cor(H, varechem, method = "pearson")> 0.3
         N     P     K    Ca    Mg     S    Al    Fe    Mn    Zn    Mo
[1,] FALSE FALSE FALSE  TRUE FALSE FALSE FALSE FALSE FALSE FALSE FALSE
      Baresoil Humdepth    pH
[1,]      TRUE     TRUE FALSE
```

7. Identify the variables that have correlation coefficients > 0.3 by determining their column numbers:

```
> i = which(cor(H, varechem, method = "pearson") > 0.3)
> i
[1] 4 12 13
```

8. Turn the column numbers into a single text item containing the column names:

```
> cn = paste(colnames(varechem)[i], collapse = "+")
> cn
[1] "Ca+Baresoil+Humdepth"
```

9. Now join the columns you just identified in steps 7 and 8 to the name of the response variable. This will be used as a `formula` for the `pairs()` command:

```
> pf = paste("~H+", cn, collapse = "")
> pf
[1] "~H+ Ca+Baresoil+Humdepth"
```

10. Now use the `pairs()` command to display the response variable and the selected predictors using the formula you made in steps 7–9. You can also set the lower part of the plot to display a scatter plot smoother – the final plot should resemble Figure 9.19:

```
> pairs(as.formula(pf), data = varechem, lower.panel = panel.smooth)
```

11. Now make a blank regression model that contains an intercept only:

```
> vlm = lm(H ~ 1, data = varechem)
```

12. Use the `add1()` command to see which of the variables will influence the model the most:

```
> add1(vlm, scope = varechem, test = "F")
Single term additions
```

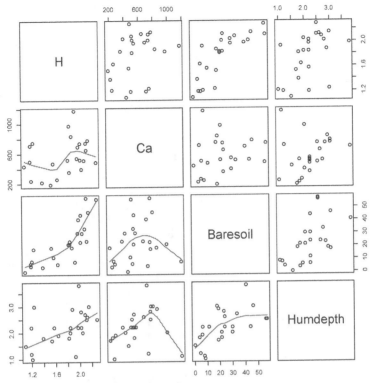

Figure 9.19 Shannon index and various soil characteristics for lichen pasture communities. Lines show locally weighted polynomial regression as a scatter plot smoother.

```
Model:
H ~ 1
           Df Sum of Sq     RSS     AIC F value     Pr(F)
<none>                   3.0463 -47.539
P           1   0.00511 3.0412 -45.580  0.0370   0.84930
K           1   0.10553 2.9407 -46.386  0.7895   0.38388
Ca          1   0.31869 2.7276 -48.191  2.5705   0.12313
Mg          1   0.22901 2.8173 -47.415  1.7884   0.19479
S           1   0.00096 3.0453 -45.547  0.0069   0.93454
Al          1   0.35669 2.6896 -48.528  2.9176   0.10169
Fe          1   0.48645 2.5598 -49.715  4.1808   0.05303 .
Mn          1   0.17694 2.8693 -46.976  1.3567   0.25659
Zn          1   0.00604 3.0402 -45.587  0.0437   0.83630
Mo          1   0.00000 3.0463 -45.539  0.0000   0.99604
Baresoil    1   1.95786 1.0884 -70.240 39.5746 2.486e-06 ***
Humdepth    1   0.75534 2.2909 -52.378  7.2536   0.01328 *
pH          1   0.39652 2.6497 -48.886  3.2921   0.08327 .
```

13. The *Baresoil* variable has the smallest *AIC* value, largest *F* and smallest *p*-value so add that to the model you made in step 11:

```
> vlm = lm(H ~ Baresoil, data = varechem)
> summary(vlm)
```

```
Call:
lm(formula = H ~ Baresoil, data = varechem)

Residuals:
     Min       1Q    Median        3Q       Max
-0.42033 -0.16967   0.00201   0.15194   0.44062

Coefficients:
             Estimate Std. Error t value Pr(>|t|)
(Intercept) 1.346401   0.079077  17.027 3.74e-14 ***
Baresoil    0.017748   0.002821   6.291 2.49e-06 ***
```

14. Use the `addl()` command again to see which of the remaining variables will influence the model the most:

```
> addl(vlm, scope = varechem, test = "F")
Single term additions

Model:
H ~ Baresoil
            Df Sum of Sq     RSS     AIC F value  Pr(F)
<none>                   1.08840 -70.240
P            1  0.008268 1.08013 -68.423  0.1608 0.6925
K            1  0.008152 1.08025 -68.421  0.1585 0.6946
Ca           1  0.102959 0.98544 -70.625  2.1941 0.1534
Mg           1  0.022195 1.06620 -68.735  0.4372 0.5157
S            1  0.006798 1.08160 -68.391  0.1320 0.7200
Al           1  0.001734 1.08666 -68.279  0.0335 0.8565
Fe           1  0.004161 1.08424 -68.332  0.0806 0.7793
Mn           1  0.006179 1.08222 -68.377  0.1199 0.7326
Zn           1  0.000239 1.08816 -68.246  0.0046 0.9465
Mo           1  0.001753 1.08664 -68.279  0.0339 0.8557
Humdepth     1  0.002482 1.08592 -68.295  0.0480 0.8287
pH           1  0.018261 1.07014 -68.646  0.3584 0.5558
```

15. You can call a halt to the model-building process here because none of the remaining variables add significantly to the model. However, you can now visualise the relationship with a plot – add a scatter plot smoother instead of a straight line. The results should resemble Figure 9.20:

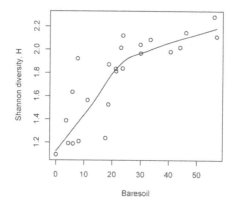

Figure 9.20 Shannon index and % Bare soil for lichen pasture communities. Line = locally weighted polynomial smoother.

```
> plot(H ~ Baresoil, data = varechem, ylab = "Shannon diversity, H")
> lines(spline(lowess(varechem$Baresoil, H)))
```

You will see that the regression model has a reasonable R^2 value of 0.627 and is highly significant. However, the lowess() line shows a definite kink.

9.6 Summary

Topic	Key Points
Mean diversity	Diversity indices do not partition evenly; the mean of two separate samples is not the same as the diversity of the two samples combined. This had led to various approaches to define the variability in diversity index for single samples. There are versions of the *t*-test but bootstrapping (randomisation) is seen as most reliable.
Graphical summary of diversity profiles	Rényi and Tsallis entropy profiles are useful graphical summaries of diversity, which you can use to compare two or more samples. Using Rényi or Tsallis diversity helps to overcome the influence of scale (and so rare species) because the entropy is calculated over a range of scales. The tsallis() and renyi() commands in the *vegan* package can create the entropy profiles but you'll need to make the sample comparison graphs for yourself. The matplot() command is useful to help plot the results.
A *t*-test for differences in Shannon index	A special version of the *t*-test can be used for comparison of the Shannon index between two samples. The important element is the estimation of the variance of the index. The degrees of freedom are also estimated from the number of species in the samples.
	If you need to compare several samples you must conduct multiple pairwise tests and modify the resulting *p*-values; a common approach is the Bonferroni correction.
	You can use the p.adjust() command to correct *p*-values for multiple tests.
A *t*-test for differences in Simpson's index	There is a variant of the *t*-test designed to test the difference between Simpson's index. This is similar in approach to that used for comparison of Shannon indices.
	The H_sig() command (part of the download file for this book) can carry out *t*-tests using Simpson's or Shannon indices.
Graphical summary of the *t*-test	In order to summarise graphically the difference between two samples (Simpson's or Shannon) you need to determine confidence intervals: CI = SD * critical value. You can then use the CI as the basis for error bars on a bar chart (or some other plot).
Graphical summaries for multiple sites	You can use the apply() command to get the confidence intervals for multiple samples. The custom functions H_CI() and D_CI() carry out the CI calculations (they are part of the download file for this book).
	CIs are sensitive to species richness when derived from the *t*-test formulae. They tend to be large for Shannon and small for Simpson's index. Bootstrapped CIs are probably the more reliable CIs.

Topic	Key Points
Bootstrapping for unreplicated samples	Bootstrapping is a form of randomisation where samples are 'shuffled about' many times in order to simulate and determine variability. The underlying distribution of the data is unimportant.
	You can use `for()` loops to carry out randomisation (with the `sample()` command).
	The `boot()` command in the *boot* package carries out bootstrapping.
Bootstrap statistics and multiple samples	You can use the `apply()` command to carry out CI calculations for multiple sites by using the appropriate function. The `plot_H()` command (part of the download file for this book) will do this and plot the results for Simpson's or Shannon indices.
Statistical inference and bootstrapping	You can use a Monte Carlo approach to resampling your data to produce statistical inference. Differences between bootstrapped samples can be used as the basis for a null hypothesis. The proportion of times that the actual difference is larger than your null is a simulated *p*-value.
	A *z*-score is one way to obtain statistical significance by 'standardising deviation'. This is useful for times when you get a 0 result as a *p*-value from randomisation.
	The custom command `H_bss()` will carry out a bootstrap analysis and compare two samples for Simpson's or Shannon indices. The command is part of the download file for this book.
Hypothesis testing with replicated samples	If your samples contain replicated data you can use 'regular' hypothesis tests. For differences you might use the *t*-test, *U*-test or ANOVA for example. You can also use correlation or regression. However, it is important that your data are 'biologically sensible', with appropriate samples that express the diversity.
Bootstrapping with replicated data	Bootstrapping can be used with replicated data and a grouping variable. The *simboot* package contains the `sbdiv()` command that can carry out a bootstrapping analysis of Simpson's or Shannon indices using replicated data and a grouping variable.
Correlating diversity	You can use a model-building approach to explore the link between diversity indices and other variables.
	The `add1()` command can help you determine which variables will significantly affect a regression model.
	You can use the `lowess()` command to add a locally weighted polynomial regression line instead of a traditional straight line of best fit.

9.7 Exercises

9.1 Tsallis or Rényi entropy plots are useful because the graphs help highlight evenness in the samples being compared – TRUE or FALSE?

9.2 You can use a regular *t*-test to compare diversity indices from two sites – TRUE or FALSE?

9.3 You have four samples, one each from four different habitats. You wish to compare their Shannon entropy. What are your options?

9.4 Which of the following statements about bootstrapping are true?

A) The underlying distribution in not important.
B) You can compare two single samples.
C) You can compare replicated data samples.
D) You can compare Rényi entropy.
E) None of the above.

5. When you display a graph of diversity against another variable, using a locally weighted polynomial regression line is useful because it 'follows' the points rather than being a straight line of best fit – TRUE or FALSE?

The answers to these exercises can be found in Appendix 1.

10. Diversity: sampling scale

Diversity is some sort of measure of how many species are present in a sample (e.g. species richness, Chapter 7). Some measures of diversity also take into account the relative abundance of species to form an index of diversity (Chapter 8). You can measure diversity by using samples of varying sizes and you might generally suppose that the larger the sample the better your estimate of diversity will be. You can split the assessment of diversity into different levels of measurement, according to the sampling scale over which you are operating. So, measurement of diversity in a single quadrat is one scale (a small one), whilst measurement of diversity in an entire habitat is another (large scale).

The sampling scale you use gives rise to different measures of diversity:

- *Alpha diversity* – the smallest unit of diversity sampling scale. This might be a single quadrat or the average of several quadrats in a single habitat.
- *Gamma diversity* – the largest unit of diversity sampling scale. This is the total diversity of the samples you have, these samples might be all from one habitat or, more likely, from several habitats in a landscape.
- *Beta diversity* – an intermediate sampling scale that links the *alpha* and *gamma* scales of diversity. Unlike *alpha* and *gamma* diversity, *beta* diversity is calculated indirectly from your samples and represents changes in diversity between samples.

So, *alpha* and *gamma* diversity are the basic partitions (sampling scales) of diversity and *beta* diversity describes the link or change between the two scales of measurement. This chapter is largely about *beta* diversity. There are various ways to determine *beta* diversity, as you will see shortly.

10.1 Calculating *beta* diversity

There are various ways to calculate measures of *beta* diversity – the exact relationship between *alpha*, *gamma* and *beta* diversity varies according to your measure of diversity. In general your approach is affected by the sort of data you have got:

- Presence-absence data – when you only have lists of species your options are somewhat limited. The general approach is to look at samples and determine how many species are unique to each sample and how many species are shared. Essentially you are looking at species richness.

- Abundance or frequency data – when you have abundance data you can determine an index of diversity, which can be used in calculations of *beta* diversity.

It is also possible to examine *beta* diversity without actually measuring diversity (see Section 10.7)! Later you will see how to use abundance data to work out differences in species composition using various indices of dissimilarity. However, before that you need to
explore *beta* diversity by more 'conventional' methods.

10.1.1 Presence-absence data

If you only have a list of species for a site then obviously you only have presence-absence
data. Even if you have abundance information you can reduce your data to simple binary
form – this approach was widely used until fairly recently. In general you should not
reduce the 'sensitivity' of your data unless you have a really good reason (e.g. to compare
results with another study).

In order to undertake the analyses you need to have appropriate data and you should
think about this before collecting any data if possible:

- Scale of samples – you will inevitably have to take samples, so you should ensure
  that the size of your samples is appropriate for your study. If you already have
  data then you could combine samples to make larger units of measurement if you
  deem the original units to be too small.
- Arranging your data – generally some form of *biological recording* format is preferable for original data. You will then need to use Pivot Tables (in Excel) or cross-
  tabulation in R to rearrange your data into a community structure.

It is certainly possible to examine *beta* diversity using Excel but it is often quite a tedious
process, especially when you have many samples. For the most part you will see how to
use R to analyse *beta* diversity but Excel has its uses, not least as a data management tool.

There are a couple ways to think about *beta* diversity:

- True *beta* diversity – here the overall diversity of a landscape is subdivided into
  smaller sampling units and *beta* diversity represents how many subunits you
  would need if no species were shared between units.
- Species turnover – here the *beta* diversity is represented as the increment in species diversity from sample to sample.

In the next section you'll see how to determine true *beta* diversity.

True beta *diversity*

So-called true *beta* diversity is a multiplicative measure. It can be represented like so:

$$\beta = \gamma/\alpha$$

Here, the *gamma* diversity is the total species diversity of all your measurements – usually
these would be from different habitats in a landscape. The *alpha* diversity is the mean of
the species diversity for each group of samples – usually the mean of samples from each

habitat. Now if the habitats did not have any shared species the *beta* diversity would equal the number of groups (habitats). The smaller the value of the *beta* diversity (compared to the number of groups), the more shared species there are between the groups (habitats) in the landscape.

Calculating beta diversity between habitats or sample groups

In the following exercise you can have a go at making a spreadsheet to calculate true *beta* diversity using presence-absence data for the case when you have only a single sample per group (habitat).

Have a Go: Make a spreadsheet to calculate simple true *beta* diversity from presence-absence data

For this exercise you will not need anything other than your spreadsheet. You'll be making a template that will determine *beta* diversity – you can paste in community data as presence-absence (i.e. 1s for presence and 0s or blank for absence) as you like and can modify the number of rows and columns to suit.

1. In cell B2 type a label for the first sample, 'S1' will do nicely.

2. In cells B3:B4 type labels for two more samples, and call them 'S2' and 'S3'.

3. You need to leave some room for actual data. To begin with just 12 rows will suffice; in cell A14 type a label, 'Total', which will represent the total number of species in each sample.

4. In the cells B2:D13 type some data; these will be presence-absence so simply use 1 for presence. You could use 0 (zero) or leave cells blank for absence; it is easier to see the data if you leave blanks but the choice is yours.

5. In cell B14 type a formula to determine the species richness to the first sample: =COUNTIF(B2:B13,">0"). Copy this formula into cells C14:D14. You could use the COUNT formula but this would also count 0 values as an item; using SUM is 'safer' and would work whether you have blank or 0 as your absence data but it works only if you are using presence-absence data.

6. In cell A15 type a label for the *alpha* diversity result, 'Alpha'.

7. In cell B15 type a formula to work out the *alpha* diversity, which is the average of the species richness for the three samples: =AVERAGE(B14:D14).

8. In cell A16 type a label for the *gamma* diversity, 'Gamma'.

9. In cell B16 you want a formula that represents *gamma* diversity, the total species richness of the three samples combined. It is possible to make a single formula to do this but it is long and involved. The formula would be: =SUM(B14:D14)-SUMPRODUCT(B2:B13,C2:C13)-SUMPRODUCT(C2:C13,D2:D13). The problem is that each time you add a new sample you have to add a new element to the formula. It is better to use an intermediate step, which allows more flexibility in the long run.

10. In cell F1 type a label for the intermediate richness calculations, 'Rich'.

11. In cell F2 type a formula to 'flag' the row as having data for a species present or not: =IF(SUM(B2:D2)>0,1,"").

12. Copy the formula in cell F2 down the column to fill cells F3:F13. Each cell in the 'Rich' column will contain a 1 if the row (columns B, C and D) is 'occupied' and be blank if not.

13. In cell F14 type a formula to work out the total number of species in the data: =SUM(F2:F13).

14. You can now use the value of cell F14 as your *gamma* diversity result in cell B16: =F14.

15. In cell A17 type a label for the *beta* diversity result, 'Beta'.

16. True *beta* diversity is *gamma* ÷ *alpha* so in cell B17 type a formula to calculate this: =B16/B15.

You can now explore the *beta* diversity by altering the data in the cells B2:D13. Note that if there are no species in common your value for *beta* diversity is always equal to the number of samples (in this case 3).

It would be easy enough to use the approach in this exercise by using a Pivot Table to arrange your data by habitat (or some other grouping). The *alpha* diversity comes from the COUNTIF formula on the columns (see step 5) and the *gamma* diversity from the total number of species found via the IF formula (see step 11).

If you are using R then calculating species richness, and so determining *beta* diversity from presence-absence data is easy if you use the `specnumber()` command in the *vegan* package. In the following exercise you can have a go at calculating *beta* diversity for yourself.

Have a Go: Use R to calculate true *beta* diversity from presence-absence data

You will need the *vegan* package for this exercise; if you haven't already got this then type the following command:

```
> install.packages("vegan")
```

1. Start by making the commands in the *vegan* package accessible:

```
> library(vegan)
```

2. Make some simple data to work with; start by making three samples:

```
> a <- c(rep(1,5), rep(0,7))
> b <- c(rep(0,4), rep(1,4), rep(0,4))
> c <- c(rep(0,7), rep(1,5))
```

3. Join the three samples together to make one community dataset. Note that the samples are all the same length:

```
> comm <- rbind(a,b,c)
> comm
```

	[,1]	[,2]	[,3]	[,4]	[,5]	[,6]	[,7]	[,8]	[,9]	[,10]	[,11]	[,12]
a	1	1	1	1	1	0	0	0	0	0	0	0
b	0	0	0	0	1	1	1	1	0	0	0	0
c	0	0	0	0	0	0	0	1	1	1	1	1

4. Use the `specnumber()` command to work out the richness for each sample, you can determine the mean value directly – this is the *alpha* diversity:

```
> alpha <- mean(specnumber(comm))
> alpha
[1] 4.666667
```

5. Now work out the *gamma* diversity, this is the total number of species. You have 12 columns on your data so you know that the number of columns equals the number of species. However, a column could conceivably contain zeroes so use the `specnumber()` command again to work out species frequencies (use MARGIN = 2). You then count how many species have a frequency of greater than zero:

```
> gamma <- length(specnumber(comm, MARGIN = 2) > 0)
> gamma
[1] 12
```

6. You can now work out the true *beta* diversity:

```
> beta <- gamma/alpha
> beta
[1] 2.571429
```

In this exercise you used some 'made-up' data but you can use the same approach on any community dataset that has rows as samples and columns representing species presence-absence.

In the preceding exercises you looked at three simple samples and the assumption was that they were from different habitats. This does not have to be the case and the calculated *beta* diversity could represent the variability in diversity (species richness) within one habitat type (that is, between samples). Often you will have several samples from the various habitats and will need to combine them in some way. If your samples are very small then it might be better to combine them so that you end up with one sample from each habitat. If your samples are large enough then you can use mean values.

It starts to become rather more difficult to use Excel to work out *beta* diversity when you have multiple samples from different habitats. To demonstrate this you can have a go at working out the true *beta* diversity using the ground beetle data, which contains multiple samples from three habitat types.

Have a Go: Use Excel to calculate *beta* diversity for species richness using replicated habitat data

You will need the data *Ground beetles and habitat.xls* spreadsheet for this exercise.

1. Open the *Ground beetles and habitat.xls* spreadsheet. Click once somewhere in the block of data then start a Pivot Table by clicking on *Insert > Pivot Table*. The data should be selected so click *OK* to place the Pivot Table in a new worksheet.

2. Make a Pivot Table by dragging the field labels to the appropriate boxes in the *Pivot Table Field List* menu. Drag the *Abbr* field to the *Row Labels* box, drag the *Quantity* field to the *Values* box. Drag the *Habitat* field to the *Column Labels* box and then also drag the *Sample* field so that it is underneath. Your field list box should resemble Figure 10.1.

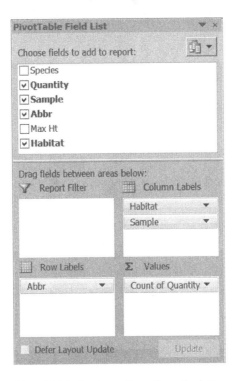

Figure 10.1 Setting up a Pivot Table for *beta* diversity calculation.

3. You need to alter the data to show presence-absence so click on the field in the *Values* box that says 'Sum of Quantity'. Choose *Value Field Settings* and alter the *Summarize value field by* part to *Count*.

4. The data should now show 1s and blanks to represent presence-absence. The columns sums are the species richness for each sample. The subtotals for the habitats give the frequency of each species. Go to cell Y5, which is after the grand total for the rows. Type a label 'Edge' to represent the presence or absence of the species in the *Edge* habitat.

5. Type labels for the other two habitats, 'Grass' and 'Wood' in cells Z5:AA5.

6. In cell Y6 you need to type a formula to determine the presence or absence of a species (in row 6) for the *Edge* habitat: =IF(SUM(B6:G6)>0,1,0). There will be a 1 if any of the samples for the Edge habitat contain a 1 but blank if they are all empty (or zero).

7. In cell Z6 type a formula similar to that in step 6 for the *Grass* habitat: =IF(SUM(I6:N6)>0,1,0).

8. In cell AA6 type a formula for the *Wood* habitat: =IF(SUM(P6:U6)>0,1,0).

9. Use the mouse to highlight/select cells Y6:AA6 and then copy to the clipboard. Paste the formulae down the rest of the three columns, down to row 53.

10. In cell X54 type a label, 'Gamma' for the total species richness for each habitat. Underneath (cells X55:X56) type labels, 'Alpha' and 'Gamma' to hold the other results.

11. In cell Y54 type a formula to calculate the *gamma* diversity. This is the sum of the column: =SUM(Y6:Y53). Now copy the formula into the cells Z54:AA54.

12. The *alpha* diversity is the mean species richness for the samples in each habitat. So, in cell Y55 type in a formula: =AVERAGE(B54:G54). This takes the column totals for the *Edge* habitat samples.

13. In cell Z55 type a similar formula to step 12 for the Grass habitat: =AVERAGE(I54: N54). Notice that you cannot simply copy the previous formula.

14. In cell AA55 type a formula for the Wood habitat: =AVERAGE(P54:U54).

15. The *beta* diversity is *gamma* ÷ *alpha* so in cell Y56 type a formula to calculate this: =Y54/Y55. You can copy this formula into cells Z56:AA56.

You should see the results: 1.486, 1.521 and 1.147. This shows that the extent of species overlap is fairly small – remember that with no overlap you would see the number of samples (6).

So, using Excel to determine *beta* diversity is fairly straightforward but, of course, the more samples you have the more 'individual' your formulae will become. Using R gives a bit more flexibility because you do not have to create formulae for individual groups of samples. In the following exercise you can have a go at calculating the *beta* diversity using the same ground beetle data.

Have a Go: Use R to calculate *beta* diversity for species richness using replicated habitat data

You will need the *CERE.RData* file for this exercise; you'll be looking at the ground beetle data. There are two data items *gb.biol* contains the community data and *gb.site* contains the habitat information. You'll also need the *vegan* package.

1. Start by making sure that the *vegan* package is loaded and ready:

   ```
   > library(vegan)
   ```

2. The *gamma* diversity can be found for the habitats by with the specnumber() command, which allows a grouping variable to be used:

   ```
   > gamma <- specnumber(gb.biol, gb.site$Habitat)
   > gamma
    Edge  Grass  Wood
     27    37    13
   ```

3. The *alpha* diversity needs to be calculated from the mean of the samples for each habitat. This time the `tapply()` command can be used to work out the mean, using the `specnumber()` command to get the individual sample richness:

```
> alpha <- tapply(specnumber(gb.biol), gb.site$Habitat, FUN=mean)
> alpha
     Edge       Grass        Wood
18.166667   24.333333   11.333333
```

4. The *beta* diversity is simply *gamma ÷ alpha*:

```
> gamma/alpha
     Edge       Grass        Wood
1.4862385   1.5205479   1.1470588
```

Of course these steps use the data already in community format – to get the data from the original biological records (*gb*) you would use the following steps:

```
> gb.biol <- xtabs(Quantity ~ Sample + Abbr, data = gb)
> gb.site <- aggregate(Quantity ~ Sample + Habitat, data=gb, FUN=sum)
```

So far you have seen how to use a simple comparison between samples or how to use groups of samples. However, there are times when your samples are too small to be 'sensible' as a measure of diversity. At these times you want to amalgamate samples into larger groups.

Amalgamating samples

If you need to amalgamate samples then a Pivot Table in Excel is the ideal way to go. Whatever grouping variable you use, your samples will be summed for each species. You can then save your Pivot Table and create a new data file. If you need to amalgamate samples using R then your approach will depend on what data you have.

   If you imported the original biological records then you can simply use the `xtabs()` command to make a new community data object using the grouping variable you want. For example, here are some data in biological recording format:

```
> head(gb)
                   Species Quantity Sample   Abbr Max.Ht Habitat
1 Pterostichus melanarius         4     E5 Pte.mel    2.5    Edge
2    Leistus spinibarbis           2     E6 Lei.spi    2.0    Edge
3       Carabus violaceus          1     E6 Car.vio    2.0    Edge
4    Platynus assimilis            4     E5 Pla.ass    2.5    Edge
5       Agonum muelleri            1     E5 Ago.mue    2.5    Edge
6 Calathus rotundicollis          94     E5 Cal.rot    2.5    Edge
```

To make a grouped dataset you can use the `xtabs()` command like so:

```
> gb.grp = xtabs(Quantity ~ Habitat + Abbr, data = gb)
```

You end up with grouped rows for each of the species (columns), here's what the first four columns look like:

```
> gb.grp[,1:4]
        Abbr
Habitat    Aba.par    Acu.dub    Ago.afr    Ago.ful
   Edge       2146          0          0          0
   Grass       709          3         20          3
   Wood       2013          0          0          0
```

Your result is a `table` and you might prefer to convert it to a `data.frame` object:

```
> gb.grp = as.matrix(gb.grp)        # Convert to matrix
> class(gb.grp) = "matrix"          # Make correct class
> gb.grp = as.data.frame(gb.grp)    # Convert to data frame
```

If your data were imported as a community dataset right from the outset then you'll need a different approach. The simplest way is to use the `rowsum()` command. This allows you to sum the rows of a `data.frame` (or `matrix`) using a grouping variable. The `rowSum()` command needs two instructions, the original data and the grouping variable:

```
rowSum(x, group)
```

In the following exercise you can have a go at using a grouping variable to amalgamate the ground beetle community data according to habitat:

Have a Go: Use a grouping variable to amalgamate samples from community data

You'll need the ground beetle data from the *CERE.RData* file for this exercise. The *gb.biol* object is the main community dataset, whilst the *gb.site* object contains the *Habitat* grouping variable.

1. Start by looking at the *gb.site* data, which contain the grouping variable you will use:

```
> head(gb.site)
      Max.Ht    Habitat    index
E1      5.5        Edge        1
E2     15.0        Edge        2
E3     15.0        Edge        3
E4      3.0        Edge        4
E5      2.5        Edge        5
E6      2.0        Edge        6
```

2. The *gb.biol* data contain the abundances set out sample by sample:

```
> gb.biol[1:8,1:6]
      Aba.par    Acu.dub    Ago.afr    Ago.ful    Ago.mue    Ago.vid
E1       388          0          0          0          0          0
E2       325          0          0          0          0          0
E3       295          0          0          0          0          0
E4       350          0          0          0          2          0
E5       407          0          0          0          1          0
```

E6	381	0	0	0	0	0
G1	178	1	1	1	0	1
G2	114	0	1	0	0	0

3. Use the `rowSum()` command to amalgamate the samples according to the *Habitat* variable from the *gb.site* data:

```
> gbd = rowSum(gb.biol, group = gb.site$Habitat)
```

4. You can see the result by typing the new name (here you show only the first four columns):

```
> gbd[,1:4]
          Aba.par    Acu.dub    Ago.afr    Ago.ful
Edge        2146          0          0          0
Grass        709          3         20          3
Wood        2013          0          0          0
```

Now your new data object can be used for further analysis.

There is a potential problem with this approach. As the sample size increases the total number of species also increases. Recall Section 7.3.2 where you looked at rarefaction as a way to 'standardise' sample size for species richness estimation. This effect of sample size can affect your *beta* diversity calculations because *beta* diversity will tend to increase with sample size. For this reason it is usual to calculate *beta* diversity for pairs of samples, as you will see next.

Calculating beta diversity between pairs of samples

Because species richness tends to increase with sample size *beta* diversity will also increase. It is common to compare *beta* diversity between pairs of samples to overcome this issue.

When you compare *beta* diversity using species presence-absence you are looking to compare the number of shared species to the number of unique species. There are many algorithms that are used to compare shared species. In these algorithms you have three values:

- *a*, the number of species shared between the two samples.
- *b*, the number of species unique to the first sample.
- *c*, the number of species unique to the second sample.

The most commonly used algorithm is:

$$\beta = (b + c) \div (2 * a + b + c)$$

This algorithm equates to the true *beta* diversity -1 (in other words you need to +1 to get true β) and is sometimes called *species turnover*, which you will meet shortly.

It is possible to undertake the required calculation using Excel but it is quite tedious and the more samples you have to compare the more tedious it becomes. In the following exercise you can have a go at comparing three samples, amalgamated from three habitats, for the ground beetle data you've been using. This is a useful exercise, as you will see, because you have to use a *Calculated Field* in your Pivot Table, which gives great flexibility.

Have a Go: Use Excel to compare *beta* diversity between pairs of samples

You will need the data *Ground beetles and habitat.xls* spreadsheet for this exercise.

1. Start by opening the data in your spreadsheet and making a Pivot Table in a new worksheet. You want to end up with the species as the rows (you can use the abbreviated names) and the *Habitat* field as the columns. Drag the *Quantity* field to the *Values* box.

2. The data will show the total abundances for each habitat. You really need these data to show the presence-absence. Try altering the *Sum of Quantity* field item in the *Values* box to show *Count* rather than *Sum* – click the field and select *Value Field Settings*.

3. Now your data table shows the frequency for each species in each habitat – the maximum value is 6 because there are six samples for each of the three habitats. Click once in the main data section of the Pivot Table. Now click the *Pivot Table Tools > Options > Formulas > Calculated Field* button and a new dialogue window should appear.

4. Now that you have the *Insert Calculated Field* window open you can create a custom formula to use in the Pivot Table. In the *Name* box type a simple name: 'PresAbs' will do nicely.

5. The *Formula* section will probably show = 0, click in the box and backspace to delete the 0 but leave the = in place. Type IF (and then click in the *Fields* window and select the *Quantity* item. Click the *Insert Field* button and *Quantity* should appear in your formula. Click in the Formula section and complete the formula so that it reads: =IF(Quantity > 0, 1, 0).

6. Now that your formula is ready you can click the *OK* button to place it into the Pivot Table. You should see 1s and 0s appear in the table and the label *Sum of Pres-Abs*. This will also appear in the *Values* box of the *Pivot Table Field List* window.

7. Remove the original *Quantity* field from the Pivot Table – either drag the *Sum of Quantity* field out of the *Values* box or un-tick the *Quantity* item in the *Choose fields...* box.

8. Now you have ones and zeroes representing the presence or absence of all the species for each of the three habitats. Notice that the row and column totals are all set to 1. This is not helpful so remove all totals using *Pivot Table Tools > Design > Grand Totals*.

9. In cell A53 type a label for the *alpha* diversity (species richness), 'Alpha'. Now in cell B53 type a formula to calculate the *alpha* diversity: =SUM(B5:B52). Copy this formula into cells C53:D53.

10. Now in cell F4 type a label 'Rich'. You are going to determine the number of species in total, the *gamma* diversity. In cell F5 type a formula to determine if a species is present in that row: =IF(SUM(B5:D5)>0,1,0). Copy the formula down the rest of the row to cell F52.

11. In cell F53 type a formula to work out total gamma diversity: =SUM(F5:F52). In cell E53 you can type a label for it, 'Gamma'.

12. You need to determine the number of shared species for each pair of comparisons (there will be three). Start with some labels to help keep track of the comparisons; in cell B55 type 'Edge'. Then make labels in the adjacent cells C55:D55 for 'Grass' and 'Wood'. Also make labels In cells A56:A58, 'Edge', 'Grass' and 'Wood'. You now have a matrix that you can fill in.

13. In cell C56 type a formula to work out the number of shared species between the *Edge* and *Grass* habitats: =SUMPRODUCT(B5:B52,C5:C52). The formula multiplies one column by the other, one row at a time. If a row has 1 in both columns the result is 1 (shared), if one value is 0 then the result is 0 (not shared). The formula then adds all the results together.

14. Repeat step 13 for the other two comparisons. In cell D56 type: =SUMPRODUCT(B5: B52,D5:D52). In cell D57 type: =SUMPRODUCT(C5:C52,D5:D52). You now have three values corresponding to *a*, the number of shared species for each of the pairwise comparisons.

15. Now go to cell B60 and type more labels for the matrix that will hold the calculated values for the unique species, this is similar to step 12.

16. Go to cell C61; this will hold the number of species unique to the *Edge* habitat (the row label), in the comparison between the *Edge* and *Grass* habitats. Type a formula to calculate it, you need to take the number of species in the *Edge* habitat and subtract the number of species shared with the *Grass* habitat like so: =B53-C56.

17. In cell D61 type a similar formula to step 16 to work out the unique species for the *Edge* habitat in the *Edge-Wood* comparison: =B53-D56.

18. In cell D62 type a formula to work out the unique species to the *Grass* habitat in the *Grass-Wood* comparison: =C53-D57.

19. In cell B62 type the formula: =C53-C56. In cell B63 type: =D53-D56. In cell C63 type: =D53-D57. You now have six values for *b* and *c*, the number of unique species in each of the pairwise comparisons.

20. You can now work out the *beta* diversity for the three pairwise comparisons. Go to cell B57 and type a formula: =(C61+B62)/(2*C56+C61+B62)+1. This give beta for the *Edge-Grass* comparison.

21. In cell B58 type the formula: =(D61+B63)/(2*D56+D61+B63)+1.

22. In cell C58 type the formula: =(D62+C63)/(2*D57+D62+C63)+1.

You now have the *beta* diversity for the three comparisons but if you divide *gamma* by *alpha* you will not get the same result. This is because you have determined pairwise values but your original *gamma* and *alpha* results are for all samples together. What you need to do is to work out the total number of species in each pairwise comparison to get a *gamma* value for each. Then you need to determine *alpha* by taking the mean of the species richness for each comparison. If you do this then you find that you get the same *beta* results for the comparisons.

Using Excel is slightly awkward because it is not easy to copy formulae when you are using a matrix. You cannot get the $ to 'fix' rows and columns easily because you need to copy formulae across columns and down rows. In the preceding exercise you only had three comparisons but the original data contain 18 samples – calculating pairwise *beta* diversity values would be a thankless task. In this case it is a great deal easier to use R because there is a command that will perform the calculations for you in the *vegan* package.

In the following exercise you can have a go at using R to work out *beta* diversity using the `betadiver()` command in the *vegan* package.

Have a Go: Use R to compare *beta* diversity for pairs of samples

You will need the *vegan* package for this exercise and the ground beetle data in the *CERE.RData* file.

1. Start by making sure the *vegan* package is loaded and ready to use:

    ```
    > library(vegan)
    ```

2. Make a community dataset from the ground beetle data by amalgamating samples by habitat. You can do this in two ways, the first uses the original biological records and the second takes the sample community data and amalgamates it using the habitat information in the *gb.site* data:

    ```
    > gbt <- xtabs(Quantity ~ Habitat + Abbr, data = gb)
    > gbt <- comm_grp(gb.biol, groups = Habitat, data = gb.site)
    ```

3. Now use the `betadiver()` command to calculate the *beta* diversity:

    ```
    > betadiver(gbt, method = "w") +1
               Edge    Grass
    Grass  1.46875
    Wood   1.45000 1.64000
    ```

4. If you need to work out the *gamma* diversity for a single pair then you can use something like the following:

    ```
    > sum(specnumber(gbt[c(1,3),], MARGIN = 2) > 0)
    [1] 29
    ```

5. You can work out the *alpha* diversity for a pair using something like the following:

    ```
    > mean(specnumber(gbt[c(1,3),]))
    [1] 20
    ```

In most cases you are not really interested in the pairwise *gamma* or *alpha* results, the *beta* diversity is what you really want as this shows the variability in species composition between samples.

The `betadiver()` command will calculate the *a*, *b* and *c* components of species co-occurrence for you and use them to calculate the *beta* diversity using several algorithms. You can use the command to give you the components or the beta result and can also use the result to create a plot, helping you to visualise the data.

Beta diversity algorithms

There are a number of ways of representing *beta* diversity in terms of shared numbers of species. The various algorithms all have one thing in common, they use the *a*, *b* and *c* values that represent shared and unique species.

The method you saw in the preceding exercise used the formula $\beta = (b + c) \div (2 * a + b + c)$, this was developed by Whittaker (1960), hence the method = "w" part of the betadiver() command. The betadiver() command can accept various instructions:

```
betadiver(x, method = NA, order = FALSE, help = FALSE)
```

The instructions are:

- x – the community data with rows as samples and columns as species.
- method = NA – the algorithm to use for calculation. If this is omitted or set to NA the command produces a result with three parts corresponding to the *a*, *b* and *c* components.
- order = FALSE – if set to TRUE the samples are ordered in increasing number of species.
- help = FALSE – if this is TRUE the command shows a list of algorithms.

Omitting the method give the *a*, *b* and *c* components:

```
> betadiver(gbt)
$a
        Edge    Grass
Grass    17
Wood     11       9

$b
        Edge    Grass
Grass    20
Wood      2       4

$c
        Edge    Grass
Grass    10
Wood     16      28
```

If help = TRUE is specified you can see the list of possible algorithms:

```
> betadiver(help = TRUE)
1 "w" = (b+c)/(2*a+b+c)
2 "-1" = (b+c)/(2*a+b+c)
3 "c" = (b+c)/2
4 "wb" = b+c
5 "r" = 2*b*c/((a+b+c)^2-2*b*c)
6 "I" = log(2*a+b+c)-2*a*log(2)/(2*a+b+c)-
((a+b)*log(a+b)+(a+c)*log(a+c))/(2*a+b+c)
7 "e" = exp(log(2*a+b+c)-2*a*log(2)/(2*a+b+c)-
((a+b)*log(a+b)+(a+c)*log(a+c))/(2*a+b+c))-1
8 "t" = (b+c)/(2*a+b+c)
9 "me" = (b+c)/(2*a+b+c)
10 "j" = a/(a+b+c)
11 "sor" = 2*a/(2*a+b+c)
12 "m" = (2*a+b+c)*(b+c)/(a+b+c)
```

```
13 "-2" = pmin(b,c)/(pmax(b,c)+a)
14 "co" = (a*c+a*b+2*b*c)/(2*(a+b)*(a+c))
15 "cc" = (b+c)/(a+b+c)
16 "g" = (b+c)/(a+b+c)
17 "-3" = pmin(b,c)/(a+b+c)
18 "l" = (b+c)/2
19 "19" = 2*(b*c+1)/((a+b+c)^2+(a+b+c))
20 "hk" = (b+c)/(2*a+b+c)
21 "rlb" = a/(a+c)
22 "sim" = pmin(b,c)/(pmin(b,c)+a)
23 "gl" = 2*abs(b-c)/(2*a+b+c)
24 "z" = (log(2)-log(2*a+b+c)+log(a+b+c))/log(2)
```

These algorithms correspond to those reviewed in Koleff *et al.* (2003), some are the same (e.g. 1 and 2, 8 and 9) and all give subtly different results. The most commonly used is that of Whittaker (1960), which corresponds to the method = "w".

These algorithms produce a matrix of values, the *beta* diversities of the pairwise comparisons, which are essentially showing the difference in diversity between pairs. Such a matrix is known as a *dissimilarity matrix* (you can also get similarity matrices). You will meet these again in Section 10.6 where you'll look at overlap and similarity in *beta* diversity. You will also see dissimilarities in Chapter 12, which is concerned with the analysis of similarity (and dissimilarity) in a more general sense.

Species turnover

Some measures of *beta* diversity are also known as species turnover values. The Whittaker measure for example that you met earlier is such a measure. In general you can consider three main measures of species turnover:

- Absolute turnover.
- Whittaker turnover.
- Proportional turnover.

These measures of turnover have one thing in common; they all use measures of *alpha* and *gamma* diversity, calculated pairwise. In order to calculate them for yourself you need an easy way to calculate these values. The *vegan* package contains a useful command designdist(), that allows you to perform the necessary calculations for *alpha* and *beta*.

The basic form of the command you will use is like so:

```
designdist(x, method = "(A+B-2*J)/(A+B)", terms = "binary")
```

In the command the instructions are:

- x – the community data, with samples as rows and species as columns.
- method – the method to use, where J = shared quantities, A = total for the first sample, B = total for the second sample.
- terms – by default the terms are specified as presence-absence. It is possible to use other forms (see Chapter 12).

Using the designdist() command you can calculate *alpha* diversity using "(A+B)/2" as the method. The *gamma* diversity can be calculated using method = "J".

In the following exercise you can have a go at calculating *alpha* and *gamma* diversity using the designdist() command.

Have a Go: Use R to calculate *alpha* and *gamma* diversity for pairs of samples in a community dataset

You will need the *vegan* package for this exercise and the *CERE.RData* file, containing the ground beetle data.

1. Start by making sure that the *vegan* package of commands is ready for use:

   ```
   > library(vegan)
   ```

2. Use the rowSum() command to make a new community dataset with samples amalgamated by habitat:

   ```
   > gbt = rowSum(gb.biol, group = gb.site$Habitat)
   ```

3. Now calculate the *alpha* diversity for pairs of samples:

   ```
   > alpha <- designdist(gbt, method = "(A+B)/2", terms = "binary")
   > alpha
           Edge    Grass
   Grass    32
   Wood     20       25
   ```

4. Now calculate the *gamma* diversity for pairs of samples:

   ```
   > gamma <- designdist(gbt, method = "A+B-J", terms = "binary")
   > gamma
           Edge Grass
   Grass    47
   Wood     29      41
   ```

5. Calculate true *beta* diversity using the values you just computed:

   ```
   > gamma/alpha
              Edge        Grass
   Grass    1.46875
   Wood     1.45000    1.64000
   ```

If you keep these values you can use them to determine various measures of species turnover.

You will now see how to use values for *alpha* and *gamma* diversity to determine the various sorts of turnover.

Absolute species turnover

Absolute species turnover can be expressed in terms of *alpha* and *beta* diversity like so:

$$\beta_A = \gamma - \alpha$$

You can see readily that this is an additive measure rather than a multiplicative one. The absolute species turnover tells you how much more species diversity there is in the entire dataset compared with the individual samples.

In the preceding exercise you calculated *gamma* and *alpha* diversities so working out the absolute species turnover is easy:

```
> gamma - alpha
         Edge    Grass
Grass     15
Wood       9       16
```

Whittaker species turnover

Whittaker species turnover can be expressed in terms of *alpha* and *beta* diversity like so:

$$\beta_W = (\gamma - \alpha)/\alpha = \gamma/\alpha - 1$$

This is very similar to true *beta* diversity. You can think of this as a measure of how many times the species composition changes completely between the individual samples of the community dataset.

Using the values you calculated in the preceding exercise you can determine Whittaker species turnover:

```
> gamma/alpha -1
         Edge       Grass
Grass    0.46875
Wood     0.45000    0.64000
```

Proportional species turnover

Proportional species turnover can be expressed in terms of *alpha* and *beta* diversity like so:

$$\beta_P = (\gamma - \alpha)/\gamma = 1 - \alpha/\gamma$$

This quantifies what proportion of the species diversity in the dataset is not contained in an average sample. For the data used in the preceding exercise the proportional turnover is easily determined:

```
> 1 - alpha/gamma
         Edge         Grass
Grass    0.3191489
Wood     0.3103448    0.3902439
```

Visualising beta diversity

Having a matrix of *beta* diversity results is very nice but when you have a lot of samples the matrix is not easily read and interpreted. You need a way to visualise your results. There are two main ways you can do this:

- *Cluster dendrogram* – this shows the samples like a 'family tree'.
- *Ternary plot* – this shows the a, b and c components from the *beta* diversity calculations in a triangular fashion.

These methods produce radically different results and give different insights into the *beta* diversity you are observing.

Cluster dendrograms to visualise beta diversity

A dendrogram is like a 'family tree' and shows the relationship between samples. If you use *beta* diversity as the index of relationship then you can visualise how the species composition alters between samples. You will see more about creating cluster dendrograms in Chapter 12.

To make a cluster dendrogram you will need to start with a matrix of dissimilarity (or similarity). Then you need to 'convert' this matrix into a dendrogram, which you can then plot. The `betadiver()` command can make the initial matrix of *beta* values (there are other ways; see Chapter 12).

The command to 'convert' the matrix to a dendrogram is `hclust()`, which is part of the basic *stats* package of R. The `hclust()` command can use various clustering algorithms (see Chapter 12) but you'll stick to the default, `"complete"` in the following exercise, where you can have a go at making a cluster dendrogram to visualise *beta* diversity for the ground beetle community data.

Have a Go: Use R to make a cluster dendrogram to visualise *beta* diversity

You'll need the ground beetle data in the *CERE.RData* file for this exercise. You will also need the *vegan* package to make the *beta* diversity matrix.

1. Start by preparing the *vegan* package:

   ```
   > library(vegan)
   ```

2. Create a matrix of *beta* diversity values using the `betadiver()` command:

   ```
   > gb.beta <- betadiver(gb.biol, method = "w")
   ```

3. Now make a dendrogram object using the `hclust()` command:

   ```
   > gb.clus <- hclust(gb.beta)
   ```

4. The cluster object you just made has a special `class`, `"hclust"`, which has a special plotting routine. To visualise the dendrogram you use the `plot()` command:

   ```
   > plot(gb.clus, hang = -1)
   ```

Your dendrogram should resemble Figure 10.2.

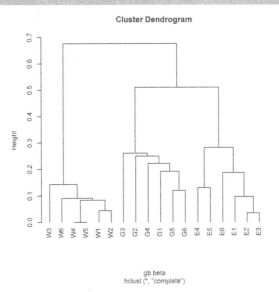

Figure 10.2 Beta diversity for species richness in ground beetle communities (18 samples from three habitat types). Algorithm for beta diversity: Whittaker.

Notice that the dendrogram has automatic titles, you can alter them by adding title instructions in the `plot()` command:

```
> plot(gb.clus, hang = -1, main = "Beta diversity",
        ylab = "Beta values", xlab = "Beetle community sample",
        sub = "Complete joining cluster")
```

Note that this plot uses a subtitle. To suppress a title completely use an empty pair of quotes.

Note: Typing multiple lines

If you wish to type multiple lines then you simply omit part of the command, usually the closing parenthesis), and carry on a new line. R displays a + as the cursor to tell you that it is expecting more to be entered before the command is evaluated.

You can also use curly brackets to enclose blocks of commands, such as when writing a custom function. If you omit the closing } then R will not evaluate your commands until the final } is typed.

In the dendrogram you produce the value on the y-axis is the distance between the pairs of samples, in the case of the preceding exercise this was Whittaker's *beta* diversity. The actual value is not the length of the 'icicles' but the position of the horizontal cross-bar that joins a pair of samples.

Ternary plots to visualise beta diversity

A ternary plot is the general name given to a triangular plot, where values are plotted in three axes. In the case of the *beta* diversity you have a, b and c values relating to the shared species and the unique species in pairs of samples.

To make a ternary plot you'll need three values, which you can get from the `betadiver()` command in the *vegan* package. To get these values you do not specify any `method` (or set `method = NA`), for example:

```
> gb.dis <- betadiver(gb.biol, method = NA)
```

The result has a special class, `"betadiver"`, which has its own plotting routine. You can also extract the a, b and c values using the `scores()` command, which will allow you to get the co-ordinates in x, y format:

```
> scores(gb.dis, triangular = TRUE)
            x           y
[1,]  0.5882353  0.7131974
[2,]  0.5588235  0.7641401
[3,]  0.3461538  0.5329387
[4,]  0.4047619  0.7010682
[5,]  0.5000000  0.6062178
[6,]  0.3281250  0.3518228
```

The default is `triangular = TRUE` but if you set it to `FALSE` you get the co-ordinates in a, b, c format:

```
> scores(gb.dis, triangular = FALSE)
```

```
        a   b   c
[1,]   14   0   3
[2,]   15   0   2
[3,]   16   9   1
[4,]   17   4   0
[5,]   14   3   3
[6,]   13   15  4
```

In the preceding examples only the first few rows of the results are shown.

You can use the scores with other commands, the *vcd* package for example has a more flexible command for drawing ternary plots, `ternaryplot()`. In the following exercise you can have a go at drawing a couple of ternary plots using the ground beetle data.

Have a Go: Use R to make a ternary plot to visualise beta diversity

You'll need the ground beetle data in the *CERE.RData* file for this exercise. You will need the *vegan* package to make the *beta* diversity matrix. You will also use the *vcd* package as one way to make a ternary plot.

1. Start by getting the *vcd* package. The package will require others (*MASS*, *grid* and *colorspace*), if you do not already have them the installation command should download and install them for you:

   ```
   > install.packages("vcd")
   ```

2. Now prepare the packages *vegan* and *vcd*:

   ```
   > library(vegan)
   > library(vcd)
   ```

3. Create a *beta* diversity result that contains the *a*, *b* and *c* values representing shared and unique species:

   ```
   > gb.dis <- betadiver(gb.biol)
   ```

4. Use the `plot()` command to produce a basic ternary plot from the `betadiver()` result, your plot should resemble Figure 10.3:

   ```
   > plot(gb.dis)
   ```

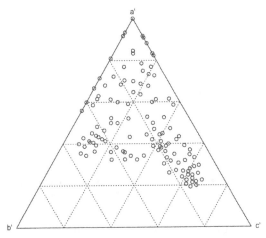

Figure 10.3 Beta diversity for species richness in ground beetle communities (18 samples from three habitat types), a' = continuity (shared species), b' = species gains, c' = species losses.

5. The plot produced by `betadiver()` is fairly basic (Figure 10.3). You can use the `ternaryplot()` command in the *vcd* package to get better control over the figure. Start by getting the *a, b, c* scores from the *beta* diversity result:

```
> dscore <- scores(gb.dis, triangular = FALSE)
```

6. Use the `expression()` command to make a label for the plot:

```
> lab <- expression("Ground beetle "*beta*" diversity")
```

7. Now use the `ternaryplot()` command to draw a ternary diagram of the *beta* diversity, your figure should resemble Figure 10.4:

```
> lab <- expression("Ground beetle "*beta*" diversity")
> ternaryplot(dscore[,c(2,3,1)], col = 1, pch = 1,
              labels = "outside", cex = 0.6, main = lab,
              dimnames = c("b'", "c'", "a'"))
```

Ground beetle β diversity

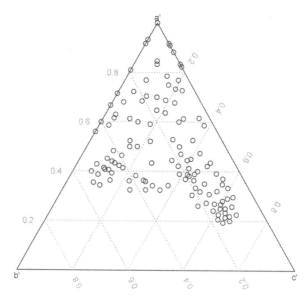

Figure 10.4 Beta diversity for species richness in ground beetle communities (18 samples from three habitat types), a′ = continuity (shared species), b′ = species gains, c′ = species losses.

The `ternaryplot()` command takes the columns in a particular order, the first forms the bottom-left corner, the second column forms the bottom-right corner and the third column forms the top corner of the triangle. So, you need to specify the order in which you want them to appear; it is usual to have a′ at the top.

The `ternaryplot()` command displays values for the gridlines, this makes it easier for readers to estimate the values along any one axis.

The a′ component of the ternary plot represents the shared species between pairs of samples, think of this as being 'continuity'. The b′ component represents the species in the first sample of a pair. The c′ component represents the species in the second sample of a pair. So, points which lie close to a vertical line through the middle of the plot show 'stability'. Points that lie to the left represent species gains and points to the right, species losses. The graph overall therefore represents changes in *alpha* (sample) diversity.

10.1.2 Abundance data

When you have abundance data rather than simple presence-absence, you can make more 'sensitive' judgements about differences in communities. You saw earlier how to assess diversity using various indices or entropy (Chapter 8). The relationship between the components of diversity, *alpha*, *beta* and *gamma*, is slightly different when you use an entropy or index of diversity.

Relationship between components of diversity when using diversity indices

When you are dealing with presence-absence data, the relationship between the components of diversity is multiplicative. In other words:

$\alpha \times \beta = \gamma$

When you are dealing with entropies or diversity indices the relationship between the components changes, Table 10.1 shows the relationships for some commonly used measures of diversity.

Table 10.1 The relationship between components of diversity for various measures of diversity.

Type of diversity	Components of diversity
Species richness	$\beta = \gamma/\alpha$
Shannon entropy	$\beta = \gamma - \alpha$
Simpsons index (Gini-Simpson)	$\beta = (\gamma - \alpha) \div (1 - \alpha)$
Rényi entropy	$\beta = \gamma - \alpha$
Tsallis entropy	$\beta = (\gamma - \beta) \div [1 - (q - 1)(\alpha)]$

Now *beta* diversity is a measure of how species composition alters between sampling units, the smaller the value the more similar samples are. The *gamma* diversity is a measure of the overall diversity of the samples, pooled together. The *alpha* diversity is the mean of the diversity of the subunits across your samples. Once you have the *alpha* and *gamma* components of diversity, using your chosen measure, you can determine *beta* diversity using the appropriate formula from Table 10.1.

Calculating beta diversity using the Shannon entropy

To determine the *beta* diversity using the Shannon diversity index (or entropy) you first need to determine over what scale of measurement you wish to calculate. If, for example, you have several samples from each of several habitats you have two main choices: you can average the diversity in each habitat or you can combine samples from within each habitat. You can also look at all the samples regardless of habitat or any kind of grouping.

There is one additional issue you have to take into account. If you are amalgamating samples that may have differences in abundance then the final calculated (*gamma*) diversity will be more heavily influenced by the samples with the larger abundances. You need some way to 'even out' any differences in abundance. You can do this by *normalising* your samples. You can do this by dividing each abundance by the total abundance for the sample it comes from. In this way each sample sums to 1.

The simplest way to normalise your data is to use the `prop.table()` command:

```
prop.table(x, margin = 1)
```

The following example shows how this works and the result when applied to the Shannon entropy. First let's make some data that are an order of magnitude different in species abundance:

```
> a = c(10, 9, 8, 2, 1, 1, 1)
> b = c(100, 90, 75, 25, 9, 8, 6)
```

Now make a community dataset by combining the samples into a matrix:

```
> cd = rbind(a, b)
> cd
   [,1] [,2] [,3] [,4] [,5] [,6] [,7]
a    10    9    8    2    1    1    1
b   100   90   75   25    9    8    6
```

Normalise these data by dividing each item by its row total:

```
> cd.n = prop.table(cd, margin = 1)
> cd.n
     [,1]    [,2]    [,3]     [,4]     [,5]     [,6]     [,7]
a  0.3125  0.2812  0.2500  0.06250  0.03125  0.03125  0.03125
b  0.3195  0.2875  0.2396  0.07987  0.02875  0.02556  0.01917
```

The *gamma* diversity can be obtained by using the column sums. Let's compare the original data and the normalised data:

```
> diversity(colSums(cd))
[1] 1.5416797
```

```
> diversity(colSums(cd.n))
[1] 1.5533921
```

You can see that there is a difference. The difference appears small but it can be very important. In this case the second sample, *b*, is not exactly 10 times the abundance of sample *a*. If it were so then the normalisation would not be necessary – real biological data are rarely obliging.

The `decostand()` command in the *vegan* package also allows you to carry out normalisation:

```
decostand(comm, method = "total")
```

The command allows you to carry out other normalisation processes as well but we are interested only in the row sums method at present. The advantage of the `decostand()`

command is that it will operate on a `data.frame` object whereas the `prop.table()` command requires your data to be in the form of a `matrix`.

In the following exercise you can have a go at determining *beta* diversity for the ground beetle community data; here you have three habitats, each of which contains six samples. You will use community normalisation to 'even up' the differences in abundance.

Have a Go: Determine *beta* diversity using Shannon entropy

For this exercise you will need the *vegan* and *BiodiversityR* packages. You'll also need the ground beetle data that are contained in the *CERE.RData* file.

1. Start by preparing the *vegan* and *BiodiversityR* packages, since the former is required by the latter you only have to load the latter:

   ```
   > library(BiodiversityR)
   ```

2. Normalise the samples and carry out the analyses on the normalised data:

   ```
   > gb.n = decostand(gb.biol, method = "total")
   ```

3. Use the `diversity()` command to calculate the mean Shannon entropy for samples within each habitat. This is the *alpha* diversity:

   ```
   > alpha <- tapply(diversity(gb.n), gb.site$Habitat, FUN = mean)
   > alpha
         Edge      Grass       Wood
   1.5058876  2.0949757  1.5481234
   ```

4. Now use the `diversitycomp()` command to work out the Shannon entropy for each habitat. The result is the *gamma* diversity for each habitat, that is, the diversity of the pooled samples for each habitat. Note that the command produces a result with several columns, you want the second:

   ```
   > gamma <- diversitycomp(gb.n, y = gb.site, factor1 = "Habitat")[,2]
   > gamma
         Edge      Grass       Wood
   1.5429305  2.1515716  1.5730569
   ```

5. The *beta* diversity is *gamma* minus *alpha* so calculate that now:

   ```
   > gamma - alpha
          Edge        Grass         Wood
   0.037042917  0.056595896  0.024933493
   ```

6. Now work out the mean Shannon entropy for all the samples without using a grouping variable. This is the *alpha* diversity:

   ```
   > alpha <- mean(diversity(gb.biol))
   > alpha
   [1] 1.7163289
   ```

7. You can amalgamate all the samples together to work out a single diversity index. This will be the *gamma* diversity:

   ```
   > gamma <- diversity(colSums(gb.n))
   > gamma
   [1] 1.8741544
   ```

8. Now calculate the *beta* diversity amongst the samples:

```
> gamma - alpha
[1] 0.15782553
```

9. In step 4 you worked out the Shannon entropy for each of the three habitats. This pooled the samples and then carried out the Shannon computations. Repeat this step but this time work out the *alpha* diversity between habitats by taking a mean:

```
> alpha <- mean(diversitycomp(gb.n, y = gb.site,
              factor1 = "Habitat", index = "Shannon",
              method = "all")[,2])
> alpha
[1] 1.755853
```

10. You already have the *gamma* diversity of the entire pooled dataset from step 7:

```
> gamma <- diversity(colSums(gb.n))
> gamma
[1] 1.8741544
```

11. Work out the *beta* diversity for the pooled data:

```
> gamma - alpha
[1] 0.11830143
```

Try the same calculations without first normalising the data. You'll find the results slightly different, but plausible, until you get to the final step. Without normalised data the final result would be negative – this cannot be correct.

In the preceding exercise you partitioned your data in several ways to work out components of *beta* diversity in slightly different ways. The components of diversity were related additively, i.e. *alpha* + *beta* = *gamma*.

It would be useful if the components of diversity were related in a multiplicative manner, as they were for species richness. The problem lies in the measure of diversity.

You are using an index of diversity and this does not change in a regular fashion with increased numbers of species. Here is an illustration that shows this in action. Use R to make a simple community dataset, the first sample contains eight species, all equally abundant. The second sample contains 16 species, equally abundant:

```
> a <- c(rep(5,8), rep(0,8))
> b <- rep(5,16)
> a;b
[1] 5 5 5 5 5 5 5 5 0 0 0 0 0 0 0 0
[1] 5 5 5 5 5 5 5 5 5 5 5 5 5 5 5 5
```

It ought to be intuitively obvious that the second sample is twice as diverse as the first one because it has twice as many species (all with identical abundance). However, calculating the Shannon entropy gives a different result:

```
> library(vegan)
> diversity(rbind(a, b), index = "shannon")
       a         b
2.0794415  2.7725887
```

The Simpson's index fares no better:

```
> diversity(rbind(a, b), index = "shannon")
        a           b
2.0794415   2.7725887
```

You can check the actual species richness, which definitely gives the correct result:

```
> specnumber(rbind(a, b))
 a    b
 8   16
```

What is needed is a slightly different approach that uses 'species equivalents' or 'effective species'.

Using effective species to determine beta diversity

True diversity is a diversity measure that records the number of species in such a way that a sample with twice as many equally abundant species as another sample will have a diversity twice as high as that other sample. Fortunately it is quite easy to 'convert' the regular diversity index measures into a measure of true diversity (Table 10.2).

Table 10.2 Converting diversity indices into effective species – a measure of true diversity.

Diversity index (H)	Effective species
Species richness	H
Shannon entropy	$\exp(H)$
Simpson's index (Gini-Simpson)	$1/(1 - H)$
Rényi entropy	$\exp(H)$
Tsallis entropy	$[(1 - (q - 1) H)]1/(1 - q)$

If you look at the same simple communities as earlier you can see how this works:

```
> a <- c(rep(5,8), rep(0,8))
> b <- rep(5,16)
> a;b
 [1] 5 5 5 5 5 5 5 5 0 0 0 0 0 0 0 0
 [1] 5 5 5 5 5 5 5 5 5 5 5 5 5 5 5 5
```

The Shannon entropy can be converted by taking the exponent:

```
> exp(diversity(rbind(a, b), index = "shannon"))
 a    b
 8   16
```

Now you get the 'correct' result, the second sample is twice as diverse as the first. The Simpson index works as follows:

```
> 1/(1-diversity(rbind(a, b), index = "simpson"))
 a    b
 8   16
```

The `diversity()` command can determine the effective species directly:

```
> diversity(rbind(a, b), index = "invsimpson")
 a  b
 8 16
```

The Rényi entropy is similar to Shannon and the exponent works for all scales

```
> exp(renyi(rbind(a,b)))
    0  0.25 0.5   1   2   4   8  16  32  64 Inf
a   8    8   8   8   8   8   8   8   8   8   8
b  16   16  16  16  16  16  16  16  16  16  16
```

The Tsallis entropy needs a bit more work, since you have to set a single scale to get the result:

```
> q <- 2 # set scale
> (1-(q-1)*tsallis(rbind(a,b), scale = q))^(1/(1-q))
21 22
 8 16
```

Once you have the diversity in terms of *effective species* you can partition the diversity in the same way as you did for species richness:

$$\beta = \gamma/\alpha$$

In the following exercise you can have a go at using effective species to calculate the *beta* diversity of the ground beetle communities.

Have a Go: Use effective species to calculate *beta* diversity

You will need the *vegan* and *BiodiversityR* libraries for this exercise as well as the ground beetle data in the *CERE.RData* file.

1. Start by preparing the packages. The *BiodiversityR* package requires *vegan* so you only need to load the former to get the latter too:

    ```
    > library(BiodiversityR)
    ```

2. Normalise the samples to even out differences in abundance:

    ```
    > gb.n = decostand(gb.biol, method = "total")
    ```

3. Use the `diversity()` command to calculate the mean Shannon entropy for samples within each habitat. This is the *alpha* diversity:

    ```
    > alpha <- tapply(diversity(gb.n), gb.site$Habitat, FUN = mean)
    > alpha
         Edge      Grass       Wood
    1.5058876  2.0949757  1.5481234
    ```

4. Now use the `diversitycomp()` command to work out the Shannon entropy for each habitat. The result is the *gamma* diversity for each habitat, that is, the diversity of the pooled samples for each habitat. Note that the command produces a result with several columns; you want the second:

```
> gamma <- diversitycomp(gb.n, y=gb.site, factor1="Habitat")[,2]
> gamma
     Edge      Grass       Wood
1.5429305  2.1515716  1.5730569
```

5. Calculate the effective species values by taking the exponent of the results:

```
> exp(alpha) ; exp(gamma)
     Edge      Grass       Wood
4.5081533  8.1252434  4.7026370
     Edge      Grass       Wood
4.6782799  8.5983608  4.8213642
```

6. Using effective species *beta* diversity is *gamma* ÷ *alpha* so determine this now:

```
> exp(gamma) / exp(alpha)
     Edge      Grass       Wood
1.0377376  1.0582281  1.0252469
```

Once you have the diversity index (or entropy) you can convert it to effective species and then the *beta* diversity can be determined from *gamma* ÷ *alpha*.

It should not be possible for the mean *alpha* diversity to exceed the overall diversity (the *gamma* diversity). If you see this in your calculations it is a sign that you should normalise your community data using `prop.table()` or `decostand()` commands.

10.2 Additive diversity partitioning

You saw earlier how the components of diversity can be additive (Section 10.1.2). The *vegan* package contains a command called `adipart()`, which can calculate the *alpha, beta* and *gamma* components across a range of scales in an hierarchical sampling design. By using a randomisation process similar to the bootstrapping you met previously (Section 9.4) it is possible to get some idea of the significance of differences in the components of the diversity across the hierarchy of samples that are compared to the overall *gamma* diversity.

At the highest level of sampling the components of diversity are:

$$\beta_m = \gamma - \alpha_m$$

At every lower level in the hierarchy the components are:

$$\beta_i = \alpha_i + 1 - \alpha_i$$

This means that the additive partition of diversity is:

$$\gamma = \alpha_1 - \Sigma\beta_i$$

The procedure allows you to compare diversity components at different scales of measurement. For example, the ground beetle data contains a simple hierarchical structure, there are three habitats (the highest level) and within each habitat there are six samples (the lowest level).

The command can work in several ways and the general form of the command is like so:

```
adipart(formula, data, index = "shannon",
        weights = "unif", relative = FALSE,
        nsimul = 99)
```

You can see that the command can accept various instructions:

- formula – in the form y ~ x, the left hand side of the formula, y, is the community dataset, with rows as samples and columns as species. The right side of the formula, x, represents the hierarchy so you can specify several variables, separated by + signs.
- data – the data.frame where the variables for the hierarchy are to be found.
- index – the diversity index to use, can be one of "shannon" (the default), "simpson" or "richness".
- weights – the relative weights of the samples, "unif" (the default) keeps weights equal whereas "prop" alters weights according to relative abundances.
- relative – by default this is FALSE but if you set it to TRUE the *alpha* and *beta* components are given relative to the value for *gamma*.
- nsimul – this sets the number of permutations to use, the default is 99.

In the following example you can have a go at calculating the diversity components using the ground beetle data again.

Have a Go: Use additive diversity partitioning to assess significance of diversity components

You will need the *vegan* package for this exercise as well as the ground beetle data in the *CERE.RData* file.

1. Start by preparing the *vegan* package:

    ```
    > library(vegan)
    ```

2. Make a factor object that represents all the samples, this is the lowest level in the sampling hierarchy. You'll need a factor with one level and the same size as the number of samples:

    ```
    > length(rownames(gb.biol))
    [1] 18

    > set <- factor(rep(1,18))
    > set
     [1] 1 1 1 1 1 1 1 1 1 1 1 1 1 1 1 1 1 1
    Levels: 1
    ```

3. The set object will form the lowest level of hierarchy. The highest level is the habitat, which you already have in the *gb.sites* object:

    ```
    > gb.sites$Habitat
     [1] Edge Edge Edge Edge Edge Edge Grass Grass Grass Grass Grass
    ```

```
[12] Grass Wood   Wood   Wood   Wood   Wood   Wood
Levels: Edge Grass Wood
```

4. Look at the diversity components using species richness:

```
> adipart(gb.biol ~ Habitat + set, data = gb.site,
         index = "richness", nsimul = 999, weights = "unif")

adipart with 999 simulations
with index richness, weights unif

          statistic         z      2.5%        50%     97.5% Pr(sim.)
alpha.1  25.66667  -15.45370  37.33333  39.00000  40.66670.001 ***
gamma    48.00000    0.00000  48.00000  48.00000  48.0000    1.000
beta.1   22.33333   15.45370   7.33333   9.00000  10.66670.001 ***
---
Signif. codes:  0 '***' 0.001 '**' 0.01 '*' 0.05 '.' 0.1 ' ' 1
```

5. Now look at the Shannon entropy:

```
> adipart(gb.biol ~ Habitat + set, data = gb.site,
         index = "shannon", nsimul = 99, weights = "unif")

adipart with 99 simulations
with index shannon, weights unif

           statistic          z        2.5%        50%    97.5% Pr(sim.)
alpha.1  1.76266512   2.35923547  1.73551383  1.74784150  1.76116 0.03 *
gamma    1.75378497   0.00000000  1.75378497  1.75378497  1.75378 1.00
beta.1  -0.00888016  -2.35923547 -0.00737457  0.00594347  0.01827 0.03 *
---
Signif. codes:  0 '***' 0.001 '**' 0.01 '*' 0.05 '.' 0.1 ' ' 1
```

6. Notice that in step 5 you got a negative *beta* diversity – this is because you did not normalise the data:

```
> gb.n = decostand(gb.biol, method = "total")
```

7. Now run step 5 again using the normalised data:

```
> adipart(gb.n ~ Habitat + set, data = gb.site,
         index = "shannon", nsimul = 99, weights = "unif")

adipart with 99 simulations
with index shannon, weights unif

          statistic         z      2.5%        50%     97.5% Pr(sim.)
alpha.1  1.755853  6.502572  0.858340  1.098612  1.22358  0.01 **
gamma    1.874154  0.000000  1.874154  1.874154  1.87415  1.00
beta.1   0.118301 -6.502572  0.650579  0.775542  1.01581  0.01 **
---
Signif. codes:  0 '***' 0.001 '**' 0.01 '*' 0.05 '.' 0.1 ' ' 1
```

Using more replicates in your simulations will generally result in 'more robust' *p*-values but using more simulations will take more processing power and take longer. You won't necessarily get significant values simply by using more replicates of course!

The results of the `adipart()` command show you the significance of the diversity components *alpha* and *beta* relative to the overall diversity *gamma*.

Note: Normalising community data

It is often useful (and indeed necessary) to normalise or standardise your data before undertaking an analysis. This can help with problems caused by unequal sample sizes and large variations in species abundance. The `decostand()` command in the *vegan* package can carry out a range of standardisation processes.

It is sometimes easier to use a separate data object to show the hierarchical nature of your sampling, as a kind of 'map' – in the following example you can have a go at this and use it with the `adipart()` command.

Have a Go: Calculate additive diversity components using a hierarchy 'map'

You will need the *vegan* package for this exercise as well as the ground beetle data in the *CERE.RData* file.

1. Start by preparing the *vegan* package:

   ```
   > library(vegan)
   ```

2. Normalise the data by row sums:

   ```
   > gb.n = decostand(gb.biol, method = "total")
   ```

3. The top level of the hierarchy needs to be the individual samples:

   ```
   > L1 = as.factor(rownames(gb.n))
   > L1
    [1] E1 E2 E3 E4 E5 E6 G1 G2 G3 G4 G5 G6 W1 W2 W3 W4 W5 W6
   Levels: E1 E2 E3 E4 E5 E6 G1 G2 G3 G4 G5 G6 W1 W2 W3 W4 W5 W6
   ```

4. The second level of the hierarchy relates to the habitats:

   ```
   > L2 = gb.site$Habitat
   > L2
    [1] Edge  Edge Edge Edge Edge Edge Grass Grass Grass Grass
   Grass Grass
   [12] Grass Wood Wood Wood Wood Wood Wood
   Levels: Edge Grass Wood
   ```

5. The third (and final) level of the hierarchy relates to the samples taken together:

   ```
   > L3 = as.factor(rep(1,length(rownames(gb.n))))
   > L3
    [1] 1 1 1 1 1 1 1 1 1 1 1 1 1 1 1 1 1 1
   Levels: 1
   ```

6. Now combine the hierarchy into a single `data.frame` object:

   ```
   > levs = data.frame(L1, L2, L3)
   > names(levs) <- c("sample", "habitat", "overall")
   ```

7. Look at the structure of the hierarchy:

```
> str(levs)
'data.frame':  18 obs. of  3 variables:
 $ sample : Factor w/18 levels "E1","E2","E3",..: 1 2 3 4 5 6 7 8 9 10 ...
 $ habitat: Factor w/3 levels "Edge","Grass",..: 1 1 1 1 1 2 2 2 2 ...
 $ overall: Factor w/1 level "1": 1 1 1 1 1 1 1 1 1 ...
```

8. Finally you can examine the hierarchical diversity partitioning:

```
> adipart(gb.n ~ ., data = levs, index = "shannon")

adipart with 99 simulations
with index shannon, weights unif

         statistic            z    2.5%        50%    97.5% Pr(sim.)
alpha.1 1.7163289  0.0000000 0.0000000 0.0000000 0.00000 0.01 **
alpha.2 1.7558530  6.5151111 0.8843258 1.0695429 1.22358 0.01 **
gamma   1.8741544  0.0000000 1.8741544 1.8741544 1.87415 1.00
beta.1  0.0395241 -9.8460649 0.8843258 1.0695429 1.22358 0.01 **
beta.2  0.1183014 -6.5151111 0.6505788 0.8046115 0.98983 0.01 **
---
 Signif. codes:  0 '***' 0.001 '**' 0.01 '*' 0.05 '.' 0.1 ' ' 1
```

You can see that the *alpha* diversity amongst the first level of the hierarchy is significant as is the *alpha* diversity amongst the second level (habitats). For the *beta* diversity you can see that it is also significant for both levels of the hierarchy. The third level of the hierarchy is all the samples pooled together (this forms the *gamma* component) so you do not get a value for *alpha* or *beta* for this.

Try the preceding exercise without normalising the data and you will see that you get quite different results.

10.3 Hierarchical partitioning

The hiersimu() command works in a similar fashion to the adipart() command in Section 10.2 but doesn't compare the diversities to the overall *gamma* diversity. Thus you can use it to compare diversity between different levels in a sampling hierarchy. In effect you are looking at comparing *alpha* diversity across the levels of the sampling hierarchy.

The hiersimu() command uses similar instructions to the adipart() command (Section 10.2) but the differences are notable:

```
hiersimu(formula, data, FUN, location = "mean",
         relative = FALSE, nsimul = 99)
```

The various instructions affect how the command operates:

- formula – in the form y ~ x, the left hand side of the formula, y, is the community dataset, with rows as samples and columns as species. The right side of the formula, x, represents the hierarchy so you can specify several variables, separated by + signs.
- data – the data.frame where the variables for the hierarchy are to be found.

- FUN – the function to use to calculate the diversity. You cannot pass additional parameters to FUN so if you need to compare Simpson's index for example you will need to specify the function yourself.
- location – the default is "mean", but you can also specify "median" to determine the scale of the samples.
- relative – by default this is FALSE but if you set it to TRUE the *alpha* components are given relative to the value for *gamma*.
- nsimul – this sets the number of permutations to use, the default is 99.

The hiersimu() command can do more than compare diversity across hierarchies, which is why the FUN instruction allows other functions to be specified. The drawback is that you cannot specify to use the diversity() command with anything other than the default settings and must create your own custom functions.

In the following example you can have a go at using the hiersimu() command to compare *alpha* diversity for the ground beetle communities.

Have a Go: Compare *alpha* diversity using hierarchical sampling

You will need the *vegan* package and the *CERE.RData* file for this exercise.

1. Prepare the *vegan* package:

```
> library(vegan)
```

2. The top level of the hierarchy is the individual samples. Make a factor from the sample names:

```
> sample <- as.factor(rownames(gb.biol))
> sample
 [1] E1 E2 E3 E4 E5 E6 G1 G2 G3 G4 G5 G6 W1 W2 W3 W4 W5 W6
Levels: E1 E2 E3 E4 E5 E6 G1 G2 G3 G4 G5 G6 W1 W2 W3 W4 W5 W6
```

3. The next level of the hierarchy will be the habitats – you already have these in the *gb.site* data object, where they are a factor:

```
> habitat <- gb.site$Habitat
> habitat
 [1] Edge Edge Edge Edge Edge Edge Grass Grass Grass Grass Grass Grass
[13] Wood Wood Wood Wood Wood Wood
Levels: Edge Grass Wood
```

4. The last level is all the samples combined together. You want to make a factor with one single level, which is the same length as the number of samples. Start by getting the number of samples in the dataset:

```
> rnam <- length(rownames(gb.biol))
> rnam
[1] 18
```

5. Now make an object that contains the correct number of items from step 4:

```
> rlen <- rep(1, rnam)
> rlen
 [1] 1 1 1 1 1 1 1 1 1 1 1 1 1 1 1 1 1 1
```

6. Make a `factor` object for the lowest and final part of the sampling hierarchy:

```
> overall <- as.factor(rlen)
> overall
 [1] 1 1 1 1 1 1 1 1 1 1 1 1 1 1 1 1 1 1 1
Levels: 1
```

7. You could have constructed the overall factor object in one go rather than use steps 4–6:

```
> overall <- as.factor(rep(1,length(rownames(gb.biol))))
```

8. Put the sampling hierarchy into a single `data.frame` object to make a hierarchy 'map':

```
> levs <- data.frame(sample, habitat, overall)
```

9. Normalise the data before running the hierarchical analysis:

```
> gb.n = decostand(gb.biol, method = "total")
```

10. Now use the `hiersimu()` command to look at the *alpha* diversity:

```
> hiersimu(gb.n ~ ., data = levs, FUN = diversity)

hiersimu with 99 simulations
          statistic        z      2.5%       50%    97.5% Pr(sim.)
sample    1.716329 0.000000 0.000000  0.000000  0.00000  0.01 **
habitat   1.755853 6.123213 0.820618  1.088421  1.22358  0.01 **
overall   1.874154 0.000000 1.874154  1.874154  1.87415  1.00
---
Signif. codes:  0 '***' 0.001 '**' 0.01 '*' 0.05 '.' 0.1 ' ' 1
```

11. You could also achieve the same result as step 9 by specifying the variables directly since they are separate objects:

```
> hiersimu(gb.n ~ sample + habitat + overall, FUN = diversity)
```

12. The `FUN` part of the command uses any function – try using the `specnumber()` command to look at *alpha* diversity and species richness:

```
> hiersimu(gb.n ~ ., data = levs, FUN = specnumber)

hiersimu with 99 simulations
          statistic        z      2.5%       50%     97.5% Pr(sim.)
sample     17.944  -16.177   22.278    22.833    23.444   0.01 **
habitat    25.667  -15.566   37.150    39.000    40.517   0.01 **
overall    48.000    0.000   48.000    48.000    48.000   1.00
---
Signif. codes:  0 '***' 0.001 '**' 0.01 '*' 0.05 '.' 0.1 ' ' 1
```

13. The `FUN` part of the command uses any function but you cannot specify additional instructions so must 'accept' the defaults. However, you can specify your own custom command easily enough. Use the `diversity()` command as the basis for a custom command to look at Simpson's index:

```
> simp <- function(x) {
  diversity(x, index = "simpson")
  }
```

14. Now use your custom `simp()` function/command in the `hiersimu()` command to look at Simpson's index:

```
> hiersimu(gb.n ~ ., data = levs, FUN = simp)

hiersimu with 99 simulations

           statistic         z       2.5%        50%     97.5%  Pr(sim.)
sample      0.739236  0.000000   0.000000   0.000000   0.00000  0.01 **
habitat     0.744858  2.324962   0.500000   0.629630   0.68519  0.01 **
overall     0.751509  0.000000   0.751509   0.751509   0.75151  1.00
---
Signif. codes:  0 `***' 0.001 `**' 0.01 `*' 0.05 `.' 0.1 ` ' 1
```

The `hiersimu()` command can use any command to get a result but the command must produce a single value (e.g. diversity index) as its result.

Tip: Use all variables without typing names

The full stop can be used as a wildcard character and take the place of all the variables in a data object if you are using a `formula`.

This approach is similar to the bootstrapping approach taken in Section 9.4 – there you explored differences in (*alpha*) diversity between sample groups. The `hiersimu()` command allows you to explore a more complex hierarchical sampling structure, most often one based on geographical scale.

10.4   Group dispersion

Another way to look at *beta* diversity is to examine the homogeneity of variance of the various sampling groups. Because you have many species (you are looking at communities) this is a multivariate approach – you are looking at the dispersion of the various sampling groups. This method is somewhat analogous to Levene's test for homogeneity of variances.

The method essentially plots dissimilarity between sampling groups using principal co-ordinates and reduces the complicated multi-dimensional space to a two-dimensional space. Difference between the groups are then assessed.

The `betadisper()` command in the *vegan* package carries out analysis of group dispersion. The command produces results that can be used in further analyses and there are also plotting methods that allow easy visualisation of the results. The basic command looks like this:

```
betadisper(x, groups, type = "median")
```

The command has three basic instructions:

- x – the command operates on a dissimilarity matrix, not a community dataset. You must first create a dissimilarity, the `betadiver()` command is one way to make a dissimilarity matrix.

- `groups` – the grouping variable to use in comparing *beta* diversity.
- `type` – the default `"median"` compares group medians, the alternative is `"centroid"`.

The important thing to note is that the command operates on a dissimilarity matrix rather than a community dataset directly. You saw the `betadiver()` command in Section 10.1.1, and found it was able to produce a matrix using various measures of *beta* diversity by using presence-absence data. The `betadisper()` command is able to use the result of `betadiver()` as its dissimilarity measure.

There are other commands that can make dissimilarity matrices, the `dist()` command and `vegdist()` for example. The former is part of the *stats* package and comes as part of the basic distribution of R. The latter is part of the *vegan* package. You will see dissimilarity (and similarity) measures more fully in Chapter 12.

The `betadisper()` command produces a result that you can explore more fully using other commands, including plotting. In the following exercise you can have a go at exploring *beta* diversity for yourself.

Have a Go: Use multivariate homogeneity of group dispersion to explore *beta* diversity

For this exercise you will need the *vegan* package and the ground beetle data from the *CERE.RData* file.

1. Prepare the *vegan* package:

   ```
   > library(vegan)
   ```

2. You need to make a dissimilarity matrix from the community data – use the `betadiver()` command to compute Whittaker's *beta* diversity for pairs of samples:

   ```
   > gb.b <- betadiver(gb.biol, method = "w")
   ```

3. Now make a multivariate dispersion model using the `betadisper()` command; use the *Habitat* variable for the grouping:

   ```
   > mod <- betadisper(gb.b, group = gb.site$Habitat)
   ```

4. Look at the model you just created, typing its name gives some basic information:

   ```
   > mod

     Homogeneity of multivariate dispersions

   Call: betadisper(d = gb.b, group = gb.site$Habitat)

   No. of Positive Eigenvalues: 10
   No. of Negative Eigenvalues: 6

   Average distance to centroid:
      Edge    Grass    Wood
   0.10558  0.13343  0.05274

   Eigenvalues for PCoA axes:
    PCoA1     PCoA2    PCoA3  PCoA4  PCoA5   PCoA6  PCoA7  PCoA8 PCoA9 PCoA10
   0.9906    0.3088   0.0969 0.0546 0.0383 0.0325 0.0188 0.0075 0.0024 0.0001
   ```

```
PCoA11     PCoA12  PCoA13  PCoA14   PCoA15   PCoA16
-0.0027  -0.0047 -0.0091 -0.0159  -0.0205 -0.0375
```

5. Use the `plot()` command to create a visualisation of the groups and their disper-
    sions, use the `hull = TRUE` instruction to add a 'boundary' around each group.
    Your graph should resemble Figure 10.5:

    ```
    > plot(mod, hull = TRUE)
    ```

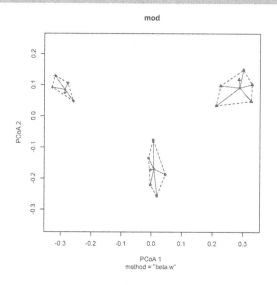

Figure 10.5 Multivariate homogeneity of group variance for ground beetle communities
in three habitats using Whittaker's *beta* diversity.

6. Now use the `boxplot()` command to visualise the distances to the centroid (the
    variances) for each group. Your figure should resemble Figure 10.6:

    ```
    > boxplot(mod)
    ```

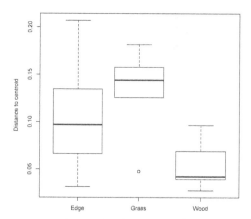

Figure 10.6 Distance to centroid in multivariate homogeneity of group variance analysis for
ground beetle communities in three habitats using Whittaker's *beta* diversity.

7. Carry out an analysis of the variance using the `anova()` command:

```
> anova(mod)
Analysis of Variance Table

Response: Distances
           Df    Sum Sq    Mean Sq  F value  Pr(>F)
Groups      2  0.020155  0.0100777   4.6695  0.02651 *
Residuals  15  0.032373  0.0021582

---
Signif. codes:  0 '***' 0.001 '**' 0.01 '*' 0.05 '.' 0.1 ' ' 1
```

8. The `anova()` command compares groups overall so now use a Tukey post-hoc test to look at the groups pairwise:

```
> TukeyHSD(mod)
  Tukey multiple comparisons of means
    95% family-wise confidence level

Fit: aov(formula = distances ~ group, data = df)

$group
                 diff          lwr          upr       p adj

Grass-Edge   0.02784992  -0.04181858   0.09751842  0.5651107
Wood-Edge   -0.05283670  -0.12250520   0.01683180  0.1540077
Wood-Grass  -0.08068662  -0.15035512  -0.01101812  0.0226201
```

Here you used the Whittaker measure of *beta* diversity but any measure of dissimilarity could be used to explore *beta* diversity, see Section 10.7. Note that this time you did not need to normalise the data because you were not calculating a diversity index directly.

Note: Calculating Whittaker's beta diversity

The `betadiver()` command in the *vegan* package can calculate Whittaker's *beta* diversity and make a dissimilarity matrix:

```
> betadiver(gb.biol, method = "w")
```

You can also make the same dissimilarity matrix using the `vegdist()` command:

```
> vegdist(gb.biol, method = "bray", binary = TRUE)
```

You'll see more about alternative dissimilarity measures and *beta* diversity in Section 10.7. Chapter 12 puts similarity, dissimilarity and cluster analysis centre stage.

10.5 Permutation methods

You've already seen permutation used to look at *beta* diversity; in Section 10.2 you looked at the `adipart()` command as a way of splitting up diversity into *alpha*, *beta* and *gamma* components.

There are several ways you can set about using permutation tests to look at *beta* diversity using a dissimilarity matrix as a starting point.

10.5.1 Permutation and homogeneity of group dispersion

In Section 10.4 you saw how to use the `betadisper()` command to look at multivariate homogeneity of group dispersion. You used the `anova()` command to look at the analysis of variance – the command actually performs a permutation test rather than a 'traditional' ANOVA.

You can use the `permutest()` command to carry out a permutation test on the result of your `betadisper()` command. You simply execute the `betadisper()` command instead of the `anova()` command on your dispersion model result. You can alter the number of permutations and also carry out a pairwise analysis. In the following example you can have a go at using the permutation test for yourself.

Have a Go: Carry out a permutation test on a multivariate group dispersion

For this exercise you will need the *vegan* package and the ground beetle data from the *CERE.RData* file.

1. Prepare the *vegan* package:

    ```
    > library(vegan)
    ```

2. You need to make a dissimilarity matrix from the community data – use the `betadiver()` command to compute Whittaker's *beta* diversity for pairs of samples:

    ```
    > gb.b <- betadiver(gb.biol, method = "w")
    ```

3. Now make a multivariate dispersion model using the `betadisper()` command; use the *Habitat* variable for the grouping:

    ```
    > mod <- betadisper(gb.b, group = gb.site$Habitat)
    ```

4. Carry out a permutation test on the dispersion model result:

    ```
    > permutest(mod)
    Permutation test for homogeneity of multivariate dispersions
    No. of permutations: 999
    **** STRATA ****
    Permutations are unstratified
    **** SAMPLES ****
    Permutation type: free
    Mirrored permutations for Samples?: No
    Response: Distances
               Df  Sum Sq     Mean Sq    F   N.Perm  Pr(>F)
    Groups      2  0.020155  0.0100777  4.6695  999 0.016 *
    Residuals  15  0.032373  0.0021582
    ---
    Signif. codes:  0 '***' 0.001 '**' 0.01 '*' 0.05 '.' 0.1 ' ' 1
    ```

5. The default number of permutations is 999 but you can alter it. Examine the samples pairwise and set the permutations to a different value:

```
> permutest(mod, pairwise = TRUE, control = permControl(nperm = 99))
Permutation test for homogeneity of multivariate dispersions
No. of permutations: 99
**** STRATA ****
Permutations are unstratified
**** SAMPLES ****
Permutation type: free
Mirrored permutations for Samples?: No

Response: Distances
           Df    Sum Sq    Mean Sq F N.Perm Pr(>F)
Groups      2 0.020155 0.0100777 4.6695  99 0.03 *
Residuals  15 0.032373 0.0021582
---
Signif. codes:  0 '***' 0.001 '**' 0.01 '*' 0.05 '.' 0.1 ' ' 1

Pairwise comparisons:
(Observed p-value below diagonal, permuted p-value above
diagonal)
          Edge       Grass Wood
Edge             0.4500000 0.07
Grass 0.3926020             0.02
Wood  0.0778734 0.0037874
```

6. Run the command in step 5 again but this time save the result to a named object:

```
> modr <- permutest(mod, pairwise = TRUE)
```

7. The result contains several components:

```
> names(modr)
[1] "tab"     "pairwise"   "groups"    "control"
```

8. Look at the pairwise part of the result, which contains the observed and permuted *p*-values:

```
> modr$pairwise
$observed
Edge-Grass   Edge-Wood Grass-Wood
0.39260196  0.07787345 0.00378736

$permuted
Edge-Grass Edge-Wood Grass-Wood
     0.400     0.096      0.010
```

9. Look to see which of the permuted *p*-values is <0.05 and therefore statistically significant:

```
> modr$pairwise$permuted < 0.05
Edge-Grass Edge-Wood Grass-Wood
     FALSE     FALSE       TRUE
```

Since you were looking at a `betadisper()` result you can of course use the `anova()`, `TukeyHSD()`, `plot()` and `boxplot()` commands as you did in the preceding exercise.

Tip: Viewing results objects

The results of many commands produce objects that contain multiple components. Simply typing the name of the result object may not show you all that the result contains. You can usually view the components by using the names() command. The str() command will also help but can produce a lengthy output.

Once you know the names of the result components you can use the $ to extract individual items. These may themselves also be split into additional components.

The permutest() command simply provides a different way of looking at the result and putting a value to the significance of the differences in the *beta* diversities.

10.5.2 Permutational multivariate analysis of variance (MANOVA)

In analysis of variance you examine the sums of squares as a way to explore variability according to some experimental design. When you have many species, such as in a community dataset, you need to run some form of multivariate analysis of variance (MANOVA). The adonis() command in the *vegan* package can carry out such an analysis using a dissimilarity matrix as its starting point using a permutation process.

The dissimilarity matrix allows the command to work out 'co-ordinates' for the various experimental groups – the group centroids. The command first calculates the centroids for the various experimental groups and then determines the sums of squares of deviations from these points. The permutation part comes next – significance tests are carried out using permutations of the raw data and compared to the sums of squares that were calculated originally.

The basic form of the command is like so:

```
adonis(formula, data, permutations = 999,
       method = "bray", strata = NULL, ...)
```

The various instructions allow you to specify the experimental design you wish to test:

- formula – you specify a formula of the form Y ~ A + B * C where Y is a dissimilarity matrix (the response variable) and A, B and C are factors or continuous variables (predictor variables).
- data – the data where the predictor variables are to be found.
- permutations – the number of permutations to use in the significance tests, the default is 999.
- method – you can specify the response variable as a community dataset instead of a dissimilarity matrix, in which case you must specify the algorithm to use to create the matrix. This must be one of the methods available to the vegdist() command. The default is "bray", the Bray–Curtis method.
- strata – you can specify one or more predictor variables that are 'controlled', that is the permutations (randomisations) will not occur. This is analogous to a random treatment effect. Essentially you treat this variable as random and look at variability within each group when calculating the significance.
- ... – other instructions can be given, usually these will relate to the method of creating the dissimilarity matrix.

In the following exercise you can have a go at using the MANOVA process to look at Whittaker's *beta* diversity in ground beetle communities.

Have a Go: Use permutational multivariate analysis of variance to examine *beta* diversity

You'll need the *vegan* package and the ground beetle data in the *CERE.RData* file for this exercise.

1. Start by preparing the *vegan* package:

   ```
   > library(vegan)
   ```

2. Use the `betadiver()` command to make a dissimilarity matrix using Whittaker's *beta* diversity:

   ```
   gb.b <- betadiver(gb.biol, method = "w")
   ```

3. Now carry out a permutational MANOVA using the `adonis()` command:

   ```
   > adonis(gb.b ~ gb.site$Habitat)
   Call:
   adonis(formula = gb.b ~ gb.site$Habitat)
                   Df SumsOfSqs MeanSqs F.Model R2      Pr(>F)
   gb.site$Habitat  2  1.24443  0.62221 43.298  0.85236 0.001 ***
   Residuals       15  0.21556  0.01437         0.14764
   Total           17  1.45999                  1.00000
   ---
   Signif. codes:  0 `***' 0.001 `**' 0.01 `*' 0.05 `.' 0.1 ` ' 1
   ```

   This is a simple experimental design, you only have a single predictor variable, *Habitat*.

You can run the `adonis()` command to explore *beta* diversity using other dissimilarity matrices other than the Whittaker's. In Section 10.7 you will see how to use a variety of measures of dissimilarity with the `adonis()` command and others.

10.5.3 Multi-response permutation procedure

The *multi-response permutation procedure* (MRPP) is allied to analysis of variance in that it compares dissimilarity between and within groups. If two groups of sampling units are really different (e.g. in their species composition), then the average of the within-group compositional dissimilarities ought to be less than the average of the dissimilarities between two random collection of sampling units drawn from the entire population.

The procedure can be carried out using the `mrpp()` command in the *vegan* package. First the pairwise distance between all the samples is calculated. Then the overall weighted mean of the within-groups means of these distances is determined. Then the permutation kicks in and rearranges samples. The means of the permutations are compared to the overall weighted mean to determine significance.

The `mrpp()` command uses a dissimilarity matrix to conduct the analysis. If you supply a community dataset, `mrpp()` will convert it to a dissimilarity matrix using the Euclidean metric (unless you specify another). The command has a plotting method, which will produce a dendrogram. You can also compute the dissimilarities within and between groups using the `meandist()` command.

In the following exercise you can have a go at using MRPP to look at Whittaker's *beta* diversity in the ground beetle dataset.

Have a Go: Use multi-response permutation procedure to explore *beta* diversity

You'll need the *vegan* package and the ground beetle data in the *CERE.RData* file for this exercise.

1. Start by preparing the *vegan* package:

    ```
    > library(vegan)
    ```

2. Use the betadiver() command to make a dissimilarity matrix using Whittaker's *beta* diversity measure:

    ```
    > gb.b <- betadiver(gb.biol, method = "w")
    ```

3. Now carry out an MRPP using the *Habitat* variable as the grouping:

    ```
    > mrpp(gb.b, gb.site$Habitat)
    Call:
    mrpp(dat = gb.b, grouping = gb.site$Habitat)

    Dissimilarity index: beta.w
    Weights for groups:  n

    Class means and counts:

              Edge    Grass    Wood
    delta   0.1658   0.2121   0.07704
    n       6        6        6

    Chance corrected within-group agreement A: 0.5973
    Based on observed delta 0.1517 and expected delta 0.3766

    Significance of delta: 0.001
    Based on   999  permutations
    ```

4. Use the meandist() command to calculate the mean dissimilarities within and between groups:

    ```
    > gb.dist <- meandist(gb.b, gb.site$Habitat)
    > gb.dist
                 Edge        Grass         Wood
    Edge    0.1658346   0.4162264   0.40856934
    Grass   0.4162264   0.2120832   0.58607936
    Wood    0.4085693   0.5860794   0.07703746
    attr(,"class")
    [1] "meandist" "matrix"
    attr(,"n")
    grouping
     Edge Grass  Wood
        6     6     6
    ```

5. The result of the meandist() command has a class "meandist", that has its own plotting routine. Visualise the relationship between the sample groups: your dendrogram should resemble Figure 10.7:

    ```
    > plot(gb.dist)
    > title(ylab = "Whittaker beta", xlab = "Sample grouping")
    ```

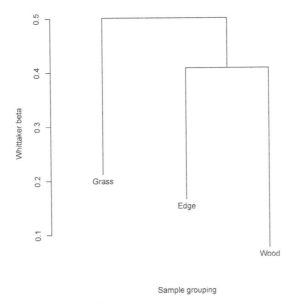

Figure 10.7 Mean distances based on Whittaker's *beta* diversity for MRPP of ground beetle communities at three habitats. Terminal branches show within-group dissimilarities.

The terminal branches of the dendrogram show the within-group dissimilarities.

So far you've focused on differences in *beta* diversity; in the next section you will look at overlap and similarity between samples.

10.6 Overlap and similarity

So far you have looked at diversity as a way to compare samples, largely focusing on differences between samples. For example *beta* diversity is a measure of changes in diversity between different sampling scales, which can be between habitats or simply between sampling units. Low *beta* diversity means that one community dominates (all the sampling units are similar) and high *beta* means communities have little overlap and the sampling units are more different from one another.

So, *beta* diversity is a measure of change in diversity between samples. If you take the inverse of this ($1/\beta$) you would have a measure of the stability of communities – a measure of similarity rather than of difference. If you think in terms of species richness or effective species than the measure of similarity is:

$$1/\beta = \alpha/\gamma$$

You first find out the average diversity of the communities you are interested in, the *alpha* diversity. Then you determine the total diversity of the pooled samples from all the communities, the *gamma* diversity. If there is little similarity, few shared species for example, the total diversity will be a lot larger than the average. The ratio of *alpha* divided by *gamma* will be 1 when all the communities are identical in species composition.

In the following sections you'll see how to look at measuring community overlap and also a method for analysing the significance of it.

10.6.1 Overlap

The traditional measure of overlap is to use α/γ as you saw earlier. However, this produces misleading results, especially when overlap is small. An example should make this clearer. Consider two communities that have little overlap:

```
> a = c(1:11, rep(0,10))
> b = c(rep(0,10), 1:11)
> comm = rbind(a,b)
> colnames(comm) = LETTERS[1:21]
> comm
  A B C D E F G H I  J  K L M N O P Q R S  T  U
a 1 2 3 4 5 6 7 8 9 10 11 0 0 0 0 0 0 0 0  0  0
b 0 0 0 0 0 0 0 0 0  0  1 2 3 4 5 6 7 8 9 10 11
```

The two communities have exactly the same number of species and in the same relative proportions but there is only one species in common. Intuitively we expect the overlap to be very small. Look at the calculations:

1. Work out *alpha* diversity using species richness, this is the mean number of species between the two samples:

```
> alpha = mean(specnumber(comm))
> alpha
[1] 11
```

2. Now work out *gamma* diversity, which is the total number of species in the two samples pooled:

```
> gamma = specnumber(colSums(comm))
> gamma
[1] 21
```

3. The overlap should be *alpha ÷ gamma*:

```
> alpha / gamma
[1] 0.5238095
```

The problem is that the result is close to 50% but there is only one common species out of 21 in total. What you really want is a value for overlap that is one when the communities are identical and 0 when they are completely different. This can be achieved with some tinkering to the formula to produce the following:

$$(\alpha/\gamma - 1/N)/(1 - 1/N)$$

In this case N is the number of communities. The *alpha* and *beta* diversities must be measured in terms of species rather than an entropy or diversity index. This means that you must 'convert' your *alpha* and *beta* values to effective species values before evaluating the overlap.

In the preceding example you used species richness so the overlap can be calculated without any further conversion:

```
> o = (alpha/gamma-1/2)/(1-(1/2))
> o
[1] 0.04761905
```

Now the overlap is very small, which is much more consistent with the data.

The calculations for determining overlap are generally simple enough and you can carry them out using Excel. However, it is tedious to do when you have more than a couple of samples and so most of the time it is sensible to use R. In the following exercise you can have a go at using R to determine overlap of some communities.

Have a Go: Use R to work out overlap between two communities

You will need the *vegan* package for this exercise as well as the *CERE.RData* file.

1. Start by preparing the *vegan* package:

   ```
   > library(vegan)
   ```

2. Look at the *DeVries* data, which show two forest butterfly communities (Jost 2006). One sample is for species in the canopy and the other is for species in the understorey:

   ```
   > str(DeVries)
    num [1:2, 1:74] 1882 26 1028 535 19 ...
    - attr(*, "dimnames")=List of 2
    ..$ : chr [1:2] "canopy" "under"
    ..$ : chr [1:74] "Hisache" "Panprol" "Neshewi" "Morachi" ...
   ```

3. Examine the Simpson's index for the two communities:

   ```
   > diversity(DeVries, index = "simpson")
        canopy        under
   0.84915281   0.91989925
   ```

4. Now determine the *alpha* diversity, the mean of the two Simpson's indices:

   ```
   > mH = mean(diversity(DeVries, index = "simpson"))
   > mH
   [1] 0.88452603
   ```

5. The *gamma* diversity is the Simpson's index of the two samples pooled together:

   ```
   > tH = diversity(colSums(DeVries), index = "simpson")
   > tH
   [1] 0.93182598
   ```

6. Using *alpha ÷ gamma* gives an unrealistically high value for overlap:

   ```
   > mH/tH
   [1] 0.9492395
   ```

7. Work out the effective species value for the *alpha* diversity result:

   ```
   > m = 1/(1-mH)
   > m
   [1] 8.6599603
   ```

8. Now work out the effective species value for the *gamma* diversity result:

```
> t = 1/(1-tH)
> t
[1] 14.668344
```

9. Finally you can determine the overlap from the effective species values:

```
> o = (m/t-1/2)/(1-(1/2))
> o
[1] 0.18076866
```

You can see that there is a big difference in the apparent overlap between the two samples using the two methods. The basic *alpha* ÷ *gamma* result gives an overlap of 95% but the 'modified' value is 18%. As an additional exercise you can have a go at calculating overlap using normalised samples.

In the preceding exercise you examined the overlap between two sites but it is certainly possible to extend the analysis to cover more. In the following exercise you can have a go at determining overlap between ten samples of plant data.

Have a Go: Examine overlap between multiple communities

For this exercise you'll need the *vegan* package and the *CERE.RData* file.

1. Start by preparing the *vegan* package:

```
> library(vegan)
```

2. Look at the *psa* data, which shows the abundances of various plant species at ten sampling sites:

```
> str(psa)
'data.frame':   183 obs. of  3 variables:
 $ Site   : Factor w/ 10 levels "ML1","ML2","MU1",..
 $ Species: Factor w/ 76 levels "Achillea millefolium",..
 $ Qty    : num ...
```

3. Note that the species names are full scientific names. This is not a problem but it will make it hard to view the columns, which will be very wide. Use the abbreviate() command to make a new vector of plant names that are shorter:

```
Abbr = abbreviate(psa$Species, minlength = 6, method = "both.sides")
```

4. Now make a community dataset using the abbreviated names:

```
> psa.biol = xtabs(Qty ~ Site + Abbr, data = psa)
> psa.biol[1:4,1:8]
     Abbr
Site Achllm Agpdmp Agrstc Agrsts Anthrs Arctmm Arrhne Bdnscr
 ML1      6      0    0.0      0      0      0      0      0
 ML2      3      0    8.0      0      0      0      0      0
 MU1      5      0    8.0      0      0      0      0      0
 MU2      4      0    5.6      0      0      0      0      0
```

5. Start with species richness and calculate the *alpha* diversity from the mean richness for each sample:

```
> alpha = mean(specnumber(psa.biol))
> alpha
[1] 18.3
```

6. Now determine the *gamma* diversity from the species richness of all samples pooled together:

```
> gamma = specnumber(colSums(psa.biol))
> gamma
[1] 76
```

7. Work out how many samples there are:

```
> length(rownames(psa.biol))
[1] 10
```

8. Now determine the overlap:

```
> o = (alpha/gamma-1/10)/(1-(1/10))
> o
[1] 0.15643275
```

9. Look at the *alpha* diversity again but use the Shannon entropy:

```
> alpha = mean(diversity(psa.biol))
> alpha
[1] 2.4908694
```

10. Now determine the *gamma* diversity:

```
> gamma = diversity(colSums(psa.biol))
> gamma
[1] 3.5731368
```

11. Convert the *alpha* and *gamma* diversities into effective species:

```
> m = exp(alpha)
> t = exp(gamma)
> m ; t
[1] 12.071766
[1] 35.628175
```

12. You know that there are ten samples from step 7 so now you can work out the overlap:

```
> o = (m/t-1/10)/(1-(1/10))
> o
[1] 0.26536267
```

The final overlap works out to be around 26%, but if you had simply used *alpha ÷ gamma* the result would have been nearer 70%. Try out this exercise using normalised data and see if there is any appreciable difference in the result.

In the examples you've seen so far the communities being investigated for overlap have all been treated as independent. However, you may well have some grouping factor. The approach to analysis of overlap when you have groups of samples is simply to amalgamate the sample by group and then to treat these pooled samples as the data in your calculations.

In the following exercise you can have a go at exploring overlap using some plant data that have a grouping variable.

Have a Go: Examine overlap in communities using grouping variables

You will need the *vegan* package for this exercise. The package contains the commands you'll need as well as the data. You will also need the *BiodiversityR* package to help compute the diversity indices for groups of samples.

1. Start by preparing the *vegan* and *BiodiversityR* packages. The former is required by the latter so you only have to load the latter:

```
> library(BiodiversityR)
```

2. You will be using the Dutch dune meadow data for this exercise. The biological data are in the *dune* data object and the habitat variables are in the *dune.env* object. Make sure that the data are available:

```
> data(dune)
> data(dune.env)
```

3. Look at the two data objects to familiarise yourself with their contents. The *dune* data consists of 20 samples of plant abundance data (there are 30 species). The *dune. env* data has the same 20 samples but five variables:

```
> rownames(dune)
 [1] "2"  "13" "4"  "16" "6"  "1" "8" "5" "17" "15" "10" "11" 9" "18" "3"
[16] "20" "14" "19" "12"  "7"
> names(dune)
 [1] "Belper" "Empnig" "Junbuf" "Junart" "Airpra" "Elepal" "Rumace" "Viclat"
 [9] "Brarut" "Ranfla" "Cirarv" "Hyprad" "Leoaut" "Potpal" "Poapra" "Calcus"
[17] "Tripra" "Trirep" "Antodo" "Salrep" "Achmil" "Poatri" "Chealb" "Elyrep"
[25] "Sagpro" "Plalan" "Agrsto" "Lolper" "Alogen" "Brohor"
> head(dune.env)
        A1    Moisture       Management    Use Manure
    2  3.5           1       BF Haypastu          2
   13  6.0           5       SF Haypastu          3
    4  4.2           2       SF Haypastu          4
   16  5.7           5       SF  Pasture          3
    6  4.3           1       HF Haypastu          2
    1  2.8           1       SF Haypastu          4
```

4. Work out how many samples there are. You are going to be grouping the samples by *Management* so you need to know how many levels there are in the *Management* factor variable:

```
> nlevels(dune.env$Management)
[1] 4
```

5. Now use the `diversitycomp()` command in the *BiodiversityR* package to get the *alpha* diversity for samples using the *Management* grouping variable – use the default Shannon index:

```
> alpha = diversitycomp(dune, y = dune.env, factor1 = "Management",
                        method = "all")[,2]
> alpha
      BF        HF        NM        SF
2.5667565  2.8704344  2.8072648  2.6845198
```

6. Note that in step 5 you used only the second column of the results (the first contains the number of samples in each group). You want the mean Shannon index across the groups so work this out now:

```
> alpha = mean(alpha)
> alpha
[1] 2.7322439
```

7. Use all the samples pooled together to determine the overall *gamma* diversity:

```
> gamma = diversity(colSums(dune))
> gamma
[1] 3.0819694
```

8. Convert the Shannon diversity results to effective species using the `exp()` command:

```
> m = exp(alpha)
> t = exp(gamma)
> m ; t
[1] 15.367330
[1] 21.801295
```

9. Finally you can determine the overlap (remember that $N = 4$ because there are four levels of *Management*):

```
> o = (m/t-1/4)/(1-(1/4))
> o
[1] 0.60650872
```

Your final result shows that there is about 61% overlap (in terms of Shannon diversity index) between the samples when grouped by *Management*.

Note: Example data in R

There are many examples of data in R. To access a dataset you can 'prepare' it using the `data()` command:

```
> data(dune)
```

Once prepared like this a data object is visible using the `ls()` command. A complete list of available data can be obtained by leaving the parentheses empty:

```
> data()
```

If you type the name of a specific package you can obtain a listing of data for that package (even if it is not loaded), you need to put the package name in quotes:

```
> data(package = "vegan")
```

The data usually have a help entry so typing `help(name)` will bring up the appropriate information. Once you are finished with a data object use the `rm()` to remove it. The data objects that are visible when you use `ls()` can be treated like any other object and can be saved and altered (but you cannot change the original, you are working on a copy).

Overlap can be a useful measure that is complimentary to *beta* diversity.

10.6.2 Analysis of similarity

You can think of *beta* diversity as a measure of similarity or, if you take the inverse, of overlap. In the previous section you looked at overlap between samples in terms of species richness and effective species. You also saw how to use grouping variables to examine overlap.

It is possible to get an idea of the statistical significance of similarity or overlap when you are using grouping variables. One way is to use a dissimilarity matrix based on the original data and then to examine the ranks of the dissimilarities both between groups and within groups. Permutation can be used to assess the significance of the difference between the between and within group dissimilarity. If the species composition is really different between sample groups then the dissimilarities between the groups will be larger than the dissimilarities within the groups. In the *vegan* package the `anosim()` command can carry out the computations required.

The starting point for the analysis is a matrix of dissimilarity. You can use various methods to create a dissimilarity matrix, such as the `dist()` or `vegdist()` commands. You can also specify the input data as a regular community dataset and then provide a method from the `vegdist()` command to apply, then `anosim()` will create the dissimilarity matrix for you.

The general form of the command is like so:

```
anosim(dat, grouping, permutations=999, distance="bray", strata)
```

There are several instructions that you can give to the command:

- `dat` – the data to use in the analysis, which should be a dissimilarity matrix or a community dataset with rows as samples and columns as species.
- `grouping` – a grouping variable.
- `permutations` – the number of permutations to use, the default is 999.
- `distance` – if the input data are a community dataset then you can specify the algorithm to use to create a dissimilarity matrix. This should be one of the options used for the `vegdist()` command.
- `strata` – you can specify which observations are permuted by specifying a variable here.

In the following exercise you can have a go at using `anosim()` to examine similarity between sample groups using the ground beetle data.

Have a Go: Examine statistical significance of differences in species composition between sampling groups

You will need the *vegan* package and the *CERE.RData* file for this exercise.

1. Start by preparing the *vegan* package:

```
> library(vegan)
```

2. Make a dissimilarity matrix using species richness and Whittaker's measure of *beta* diversity:

```
> gb.b <- betadiver(gb.biol, method = "w")
```

3. Now use the `anosim()` command to look at the significance of the grouping variable on the species composition:

```
> gb.ano <- anosim(gb.b, gb.site$Habitat)
> gb.ano
Call:
anosim(dat = gb.b, grouping = gb.site$Habitat)
Dissimilarity: beta.w

ANOSIM statistic R: 0.9988
      Significance: 0.001

Based on   999  permutations
```

4. The `anosim()` command produces a result that has its own class "anosim", which has a dedicated plotting command – use this to visualise the within and between group dissimilarities. Your plot should resemble Figure 10.8:

```
> plot(gb.ano)
```

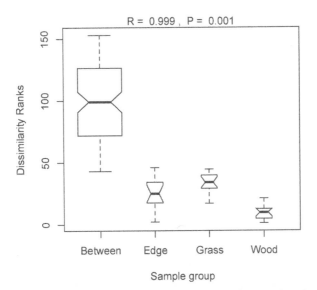

Figure 10.8 Analysis of similarity. Dissimilarity ranks between and within sampling groups for ground beetle communities using Whittaker's *beta* diversity.

5. Now use the Bray–Curtis algorithm to carry out an analysis of similarity:

```
>  gb.ano  <-  anosim(gb.biol,  grouping  =  gb.site$Habitat,
                   distance = "bray")
```

6. The `anosim()` command produces a result that has its own `summary()` command:

```
> summary(gb.ano)
Call:
anosim(dat = gb.biol, grouping = gb.site$Habitat, distance = "bray")
Dissimilarity: bray

ANOSIM statistic R: 0.7885
       Significance: 0.001

Based on  999  permutations

Empirical upper confidence limits of R:
  90%    95% 97.5%    99%
0.119 0.167 0.201 0.276

Dissimilarity ranks between and within classes:
            0%    25%   50%     75% 100%   N
Between 13  63.75  99.5  126.25  153 108
Edge     2   8.50  23.0   30.50   59  15
Grass    8  21.50  60.0   72.50   81  15
Wood     1   9.50  22.0   55.00   70  15
```

7. Now try using the `vegdist()` command to make a dissimilarity matrix using the Morisita algorithm:

```
> gb.mor = vegdist(gb.biol, method = "morisita")
```

8. Use the dissimilarly matrix you made in step 7 as the basis for an analysis of similarity:

```
> gb.ano = anosim(gb.mor, grouping = gb.site$Habitat)
> gb.ano

Call:
anosim(dat = gb.mor, grouping = gb.site$Habitat)
Dissimilarity: morisita

ANOSIM statistic R: 0.6827
       Significance: 0.001

Based on  999  permutations
```

You can see that the `anosim()` command is quite flexible and allows you to specify the dissimilarities in several ways.

The `anosim()` command allows you to examine differences in species composition (essentially *beta* diversity) between sampling groups. You can think of this as an alternative approach to the permutational multivariate analysis of variance that you met in Section 10.5.2, where you used the `adonis()` command.

In Section 10.5.2 you used the `adonis()` command with Whittaker's measure of *beta* diversity. The command can use dissimilarities much like the `anosim()` command and you can specify either a dissimilarity matrix or a community dataset along with a `method` to create a dissimilarity. In the following example you can see how to specify the Bray–Curtis method to a community dataset (ground beetle data):

```
> adonis(gb.biol ~ Habitat, data = gb.site, method = "bray")

Call:
adonis(formula = gb.biol ~ Habitat, data = gb.site, method = "bray")

          Df  SumsOfSqs  MeanSqs  F.Model   R2     Pr(>F)
Habitat   2   1.30897    0.65448  45.503   0.8585  0.001 ***
Residuals 15  0.21575    0.01438           0.1415
Total     17  1.52472                      1.0000

---
Signif. codes:  0 '***' 0.001 '**' 0.01 '*' 0.05 '.' 0.1 ' ' 1
```

Since *beta* diversity is essentially a measure of differences in species composition, any kind of dissimilarity measure can be used. In the next section you'll see some examples of this. Differences between samples (in terms of species composition) are used to create dissimilarity matrices, which can be used in analysis of similarity. In this section similarity was used to get an idea of overlap in species composition and beta diversity but the analysis of similarity can be extended to allow you to determine clusters of similar samples. Earlier you used presence-absence data to create a dendrogram of *beta* diversity (Section 10.1.1), and in Chapter 12 you will extend the analysis of similarity and explore other methods of creating clusters of samples based on similarities of one sort or another.

10.7 *Beta* diversity using alternative dissimilarity measures

In this chapter you've seen how to look at *beta* diversity, that is differences in diversity (species composition) between samples using species richness and also using diversity indices, such as Shannon entropy or Simpson's index. In Section 10.6 you looked at overlap and similarity and began to use other methods of assessing differences in species composition.

Any dissimilarity measure will also be a measure of differences in species composition between samples and therefore of *beta* diversity. The upshot is that you can use many of the approaches you've met in this chapter along with various dissimilarity measures. In the following exercise you can have a go at exploring differences in species composition (*beta* diversity) using some of the methods you've seen earlier and using various algorithms to create matrices of dissimilarity.

Have a Go: Explore differences in species composition (*beta* diversity) using various dissimilarity measures

You'll need the *vegan* package and the *CERE.RData* file for this exercise.

1. Start by preparing the *vegan* package:

   ```
   > library(vegan)
   ```

2. Make a dissimilarity matrix using the Bray–Curtis method via the vegdist() command:

   ```
   > gb.b <- vegdist(gb.biol, method = "bray")
   ```

3. Use the dissimilarity matrix you made in step 2 in an analysis of multivariate homogeneity of group variance via the `betadisper()` command:

```
> mod <- betadisper(gb.b, group = gb.site$Habitat)
```

4. Now view the ANOVA table for the model from step 3:

```
> anova(mod)
Analysis of Variance Table

Response: Distances
          Df    Sum Sq    Mean Sq  F value  Pr(>F)
Groups     2  0.004652  0.0023260   1.0567  0.3721
Residuals 15  0.033016  0.0022011
```

5. Make a new dissimilarity matrix using the Gower algorithm:

```
> gb.go <- vegdist(gb.biol, method = "gower")
```

6. Use the Gower matrix from step 5 in an analysis of multivariate homogeneity of group variance via the `betadisper()` command, then perform a permutation test to assess the significance of the model:

```
> mod <- betadisper(gb.go, group = gb.site$Habitat)
> permutest(mod)

Permutation test for homogeneity of multivariate dispersions
No. of permutations: 999

Response: Distances
          Df    Sum Sq    Mean Sq       F  N.Perm      Pr(>F)
Groups     2  0.037382  0.0186912  12.254     999  0.001 ***
Residuals 15  0.022880  0.0015254
---
Signif. codes:  0 '***' 0.001 '**' 0.01 '*' 0.05 '.' 0.1 ' ' 1
```

7. Now use the `mrpp()` using a Euclidean distance metric:

```
> mrpp(gb.biol, grouping = gb.site$Habitat, distance = "euclidean")
Call:
mrpp(dat = gb.biol, grouping = gb.site$Habitat, distance =
"euclidean")

Dissimilarity index: euclidean
Weights for groups:  n
Class means and counts:

       Edge  Grass  Wood
delta 97.22  49.46   136
n         6      6     6

Chance corrected within-group agreement A: 0.5479
Based on observed delta 94.23 and expected delta 208.4
Significance of delta: 0.001
Based on  999 permutations
```

8. Use a Raup–Crick metric as the dissimilarity in an analysis of similarity using the `anosim()` command:

```
> anosim(gb.biol, grouping = gb.site$Habitat, distance = "raup")
Call:
anosim(dat = gb.biol, grouping = gb.site$Habitat, distance = "raup")
Dissimilarity: raup

ANOSIM statistic R: 0.9782
      Significance: 0.001

Based on 999 permutations
```

9. Finally, use a Chao metric as the dissimilarity with the `adonis()` command:

```
> adonis(gb.biol ~ gb.site$Habitat, method = "chao")
Call:
adonis(formula = gb.biol ~ gb.site$Habitat, method = "chao")
                 Df SumsOfSqs MeanSqs F.Model     R2 Pr(>F)
gb.site$Habitat   2  0.126429 0.063215 1527.8  0.99511 0.001 ***
Residuals        15  0.000621 0.000041 0.00489
Total            17  0.127050                 1.00000
---
Signif. codes:  0 '***' 0.001 '**' 0.01 '*' 0.05 '.' 0.1 ' ' 1
```

Notice that the choice of metric that you use to make your dissimilarities can affect the result!

There are many choices for the algorithm you use to create a dissimilarity matrix. As you have seen in the preceding exercise, the results can be affected by your choice of algorithm. You'll meet the various methods of making dissimilarities in Chapter 12.

10.8 *Beta* diversity compared to other variables

Using *beta* diversity is one way to compare species composition. You can also use various dissimilarity measures to produce a matrix that is essentially a pairwise measure of differences in species composition and so *beta* diversity (Section 10.7). If you wish to compare *beta* diversity to some other variable you need a way to correlate a dissimilarity matrix with a variable.

This is where the *Mantel test* comes in. The Mantel test allows you to compare two dissimilarity matrices. The `mantel()` command in the *vegan* package will carry out Mantel tests:

```
mantel(xdis, ydis, method="pearson", permutations=999, strata)
```

The command can accept several instructions:

- `xdis` – a dissimilarity matrix.
- `ydis` – a dissimilarity matrix. It does not matter which dissimilarity is from the community data and which from the environmental data.
- `method` – the method of correlation, the default is to use the Pearson product moment but you can also specify `"spearman"` or `"kendall"` to carry out rank-based correlation.
- `permutations` – the number of permutations to use, defaults to 999.

- `strata` – you can specify which observations are permuted by specifying a variable here.

The *ade4* package can also carry out Mantel tests. There are two different commands:

```
mantel.rtest(m1, m2, nrepet = 99)
mantel.randtest(m1, m2, nrepet = 999)
```

In both commands you specify two dissimilarity matrices and the number of permutations required.

In most cases you will have several environmental factors that you wish to link to beta diversity. In the following exercise you can have a go at carrying out Mantel tests on some lichen communities and associated soil characteristics.

Have a Go: Carry out Mantel tests to explore the relationship between habitat factors and species composition

You'll need the *vegan* package for this exercise, the package contains the `mantel()` command as well as the data for the lichen communities and the associated soil characteristics. You will also use the *ade4* package.

1. Start by preparing the *vegan* package and the data as well as the *ade4* package:

   ```
   > library(ade4)
   > library(vegan)
   > data(varespec, varechem)
   ```

2. Make a dissimilarity matrix using the Bray–Curtis metric for the lichen species data:

   ```
   > biol <- vegdist(varespec, method = "bray")
   ```

3. Now make a dissimilarity matrix for the environmental data using the Euclidean distance metric, use the `scale()` command to rescale the variables:

   ```
   > env <- vegdist(scale(varechem), method = "euclidean")
   ```

4. Use the `mantel()` command to carry out a Mantel test:

   ```
   > m1 <- mantel(biol, env)
   > m1

   Mantel statistic based on Pearson's product-moment correlation

   Call:
   mantel(xdis = biol, ydis = env)

   Mantel statistic r: 0.3047
         Significance: 0.001

   Empirical upper confidence limits of r:
      90%    95%  97.5%    99%
   0.122  0.146  0.177  0.202

   Based on 999 permutations
   ```

5. Try the `mantel.rtest()` command to carry out another Mantel test:

```
> m2 <- mantel.rtest(biol, env, nrepet = 999)
> m2
Monte-Carlo test
Observation: 0.3047454
Call: mantelnoneuclid(m1 = m1, m2 = m2, nrepet = nrepet)
Based on 999 replicates
Simulated p-value: 0.001
```

6. Now try the mantel.randtest() command to carry out yet another variant of the Mantel test:

```
> m3 <- mantel.randtest(biol, env)
> m3
Monte-Carlo test
Call: mantel.randtest(m1 = biol, m2 = env)

Observation: 0.3047454

Based on 999 replicates
Simulated p-value: 0.002
Alternative hypothesis: greater

     Std.Obs    Expectation       Variance
3.453280829   0.006327043    0.007467708
```

7. The commands in the *ade4* package have a plotting method, try this to visualise the results. Your graph should resemble Figure 10.9:

```
> plot(m3)
```

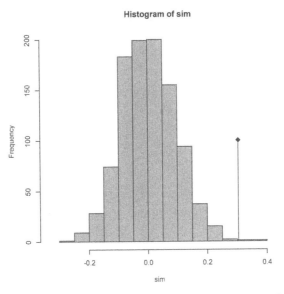

Figure 10.9 Results of Mantel test for lichen pasture communities and soil characteristics. Histogram shows the simulated correlation values and the vertical line shows the Mantel statistic ($r = 0.305$, $p = 0.003$).

The choice of metric to use in creating the dissimilarities will affect the result.

In the preceding exercise you used a matrix of several soil characteristics. This can make the results hard to interpret, since all you know is that the soil characteristics affect the species composition. It is possible to use a single variable in a Mantel test: you simply need to create the dissimilarity matrix using a single variable. In the following exercise you can have a go at using a single variable in some Mantel tests.

Have a Go: Carry out a Mantel test using a single habitat variable

You'll need the *vegan* and *ade4*packages for this exercise.

1. Start by preparing the *vegan* and *ade4*packages:

```
> library(ade4)
> library(vegan)
```

2. Make a Bray–Curtis dissimilarity matrix from the ground beetle community data:

```
> biol <- vegdist(gb.biol, method = "bray")
```

3. Now make a dissimilarity matrix using only a single variable, *Habitat*:

```
> env <- vegdist(gb.site["Max.Ht"], method = "euclidean")
```

4. Use the mantel() command to carry out a Mantel test:

```
> m1 <- mantel(biol, env)
> m1

Mantel statistic based on Pearson's product-moment correlation

Call:
mantel(xdis = biol, ydis = env)

Mantel statistic r: 0.3616
      Significance: 0.003

Empirical upper confidence limits of r:
    90%     95%    97.5%      99%
 0.0846  0.1591  0.2456   0.2716

Based on 999 permutations
```

5. Now use the mantel.rtest() command to carry out another Mantel test:

```
> m2 <- mantel.rtest(biol, env, nrepet = 999)
> m2
Monte-Carlo test
Observation: 0.3615859
Call: mantelnoneuclid(m1 = m1, m2 = m2, nrepet = nrepet)
Based on 999 replicates
Simulated p-value: 0.001
```

6. Try the mantel.randtest() command to conduct yet another Mantel test:

```
> m3 <- mantel.randtest(biol, env)
> m3
```

```
Monte-Carlo test
Call: mantel.randtest(m1 = biol, m2 = env)

Observation: 0.3615859

Based on 999 replicates
Simulated p-value: 0.002
Alternative hypothesis: greater

      Std.Obs    Expectation      Variance
  5.109932017  -0.005102198  0.005149481
```

You can of course use the `plot()` command to visualise the two results from the *ade4* commands for your *m2* and *m3* Mantel tests.

Using different dissimilarity metrics will affect your results of course. Choosing an appropriate measure of dissimilarity is something of a 'black art'; everyone seems to have their own favourite.

Note: Commands in packages

One you have enabled a command package using the `library()` command, the package contents are available for use. This means that you can access the commands and any datasets simply by using the appropriate name. However, it can be useful to remind yourself where the particular commands are. You can do this by preceding a command name with the package and two colons like so:

```
vegan::diversity()
```

If you are keeping notes this can be a useful way to keep track of where the commands you use originate.

10.9 Summary

Topic	Key Points
Scales of diversity	Diversity can be measured at different scales. *Alpha* is the diversity of individual sampling units. *Gamma* is the diversity of all sampling units taken together (amalgamated). *Beta* diversity is the change from one sampling unit to another.
	The general relationship is $\gamma = \alpha \times \beta$.
Types of data	You can use presence-absence or abundance data to calculate *beta* diversity. You can also work out *beta* diversity directly using dissimilarity (i.e. you do not need to determine *alpha* or *gamma* diversity).
True beta diversity	True *beta* diversity is taken as the number of samples you'd need such that there were no shared species between samples.
Amalgamating samples	You can use Pivot tables in Excel to amalgamate samples. In R you can use `xtabs()` and `rowSums()` commands.
Species turnover	The increment in species from one sample to another is called species turnover, a measure of *beta* diversity.

Topic	Key Points
	There are three main kinds of turnover: absolute, Whittaker and proportional.
	Because *beta* diversity generally increases with sample size species turnover is often carried out pairwise between samples.
	The `betadiver()` command in the vegan package can calculate pairwise *beta* diversity using various metrics.
	The `designdist()` command can calculate *alpha* and *gamma* diversity.
Visualising turnover	You can visualise turnover in two main ways.
	Dendrogram: samples that are close together in the dendrogram show little turnover. The `hclust()` command can produce dendrograms.
	Ternary: samples near the centre show little turnover. The graph overall represents changes in *alpha* (sample) diversity. The `ternaryplot()` command in the *vcd* package will draw ternary plots. Obtain the data using `betadiver()` and `scores()` commands (from *vegan*).
Abundance data	With abundance data your measures of diversity will generally be in the form of diversity indices. This alters the scale relationship such that $\gamma = \alpha + \beta$ for Shannon entropy.
Normalisation	*Gamma* diversity is affected by absolute abundance so data must be normalised for comparing samples. Use the `prop.table()` or `decostand()` commands (the latter is in *vegan*) to normalise community data.
Effective spp	Converting diversity indices into *effective species* values returns the scale relationship to multiplicative so that $\gamma = \alpha \times \beta$. For the Shannon index $\exp(H)$ converts the entropy to effective species.
Additive partitioning	It is possible to use a sampling hierarchy to partition diversity additively (using a randomisation process). The adipart() command in the *vegan* package allows you to compare diversity components at different scales of measurement.
Hierarchical partitioning	The `hiersimu()` command in the *vegan* package can compare (*alpha*) diversity across hierarchies. It is similar to the `adipart()` command but does not compare results to the highest level of diversity (*gamma*).
Group dispersion	You can examine the homogeneity of variance across sampling groups using the `betadisper()` command in the *vegan* package. This gives you a way to explore *beta* diversity using a grouping variable.
	You start with a dissimilarity matrix and can use the result of the `betadisper()` command as the starting point.
	The `anova()` command can carry out a test of significance on the result of `betadisper()` and there are also plotting methods for visualising results.
Permutation	There are several ways to use permutations to explore *beta* diversity:
	The `permutest()` command takes the result of the `betadisper()` command (group dispersion) and provides a test of significance of *beta* diversity using a grouping variable.
	The `adonis()` command carries out permutational MANOVA using a dissimilarity matrix as a starting point.

The `mrpp()` command carries out a multi-response permutational procedure to compare dissimilarity within and between sampling groups. The `meandist()` command is a 'helper' for `mrpp()` and finds sampling group means from the dissimilarity matrix.

Overlap and similarity

Low *beta* diversity indicates that one community is dominant (i.e. all sampling units are similar). High *beta* diversity means communities have little overlap and sampling units are different.

Overlap can be calculated from: $(\alpha/\gamma - 1/N)/(1 - 1/N)$, where N is the number of communities under comparison.

The inverse of overlap is similarity. You can use the `anosim()` and `adonis()` commands in the *vegan* package to examine the dissimilarities between and within sampling groups. Both commands take a dissimilarity matrix as a starting point.

Dissimilarity metrics

Any dissimilarity measure will also be a measure of differences in species composition between samples and therefore of *beta* diversity.

Commands such as `anosim()`, `adonis()`, `betadisper()` and `mrpp()` use dissimilarity matrices as a starting point for analysis.

Mantel tests

You can compare *beta* diversity to other variables using a Mantel test via the `mantel()` command in the *vegan* package. The test compares the correlation between two (dissimilarity) matrices.

The *ade4* package contains two commands for Mantel tests; `mantel.rtest()` and `mantel.randtest()`.

The choice of dissimilarity metric will influence your results.

10.10 Exercises

10.1 You need to determine three quantities in order to calculate species turnover using presence-absence data. What are these three quantities?

10.2 How can you partition the scales of diversity multiplicatively rather than additively for indices of diversity?

10.3 Which of the following statements is **not** true of normalising community data?

A) You do it so that *gamma* diversity does not exceed total *alpha* diversity.
B) You can use the `decostand()` command to normalise community data.
C) It allows diversity components to partition correctly.
D) Normalising data uses relative species abundances.

10.4 In analyses of beta diversity you can use any metric to calculate a matrix of dissimilarity – TRUE or FALSE?

10.5 Look at the link between vegetation height and ground beetle *beta* diversity. The species abundance data are in the *gb.biol* data object and the vegetation height is in the *gb.site* data. Is there a significant link?

The answers to these exercises can be found in Appendix 1.

11. Rank abundance or dominance models

One way of looking at the diversity of a community is to arrange the species in order of abundance and then plot the result on a graph. If the community is very diverse then the plot will appear 'flat'. You met this kind of approach in Chapter 8 when looking at evenness and drew an evenness plot in Section 8.3.4 using a Tsallis entropy profile. In *dominance* plots the species abundance is generally represented as the log of the abundance.

Various models have been proposed to help explain the observed patterns of dominance plots. In this chapter you'll see how to create these models and to visualise them using commands in the *vegan* command package. Later in the chapter you will see how to examine Fisher's log-series (Section 11.2) and Preston's lognormal model (Section 11.3) but first you will look at some dominance models.

11.1 Dominance models

Rank–abundance dominance (RAD) models, or dominance/diversity plots, show logarithmic species abundances against species rank order. They are often used as a way to analyse types of community distribution, particularly in plant communities.

The *vegan* package contains several commands that allow you to create and visualise RAD models.

11.1.1 Types of RAD model

There are several models in common use; each takes the same input data (logarithmic abundance and rank of abundance) and uses various parameters to fit a model that describes the observed pattern.

There are five basic models available via the *vegan* package:

- Lognormal.
- Preemption.
- Broken stick.
- Mandelbrot.
- Zipf.

The `radfit()` command carries out the necessary computations to fit all the models to a community dataset. The result is a complicated object containing all the models applied to each sample in your dataset. You can then determine the 'best' model for each sample that you have.

The *vegan* package also has separate commands that allow you to interrogate the models and visualise them. You can also construct a specific model for a sample or entire dataset. The various models are:

- *Lognormal* – plants are affected by environment and each other. The model will tend to normal, growth tends to be logarithmic so Lognormal model is likely.
- *Preemption* (Motomura model or geometric series) – resource partitioning model. The most competitive species grabs resources, which leaves less for other species.
- *Broken stick* – assumes abundance reflects partitioning along a gradient. This is often used as a *null* model.
- *Mandelbrot* – cost of information. Abundance depends on previous species and physical conditions (the costs). Pioneers therefore have low costs.
- *Zipf* – cost of information. The forerunner of Mandelbrot (a subset of it with fewer parameters).

The models each have a variety of parameters, in each case the abundance of species at rank r (a_r) is the calculated value.

Broken stick model

The broken stick model has no actual parameters: the abundance of species at rank r is calculated like so:

$$a_r = J/S \ \Sigma(1/x)$$

In this model J is the number of individuals and S is the number of species in the community. This gives a null model where the individuals are randomly distributed among observed species, and there are no fitted parameters.

Preemption model

The (niche) preemption model (also called Motomura model or geometric series) has a single fitted parameter. The abundance of species at rank r is calculated like so:

$$a_r = J\alpha(1 - \alpha)^{(r-1)}$$

In this model J is the number of individuals and the parameter α is a decay rate of abundance with rank. In a regular RAD plot (see Section 11.1.3) the model is a straight line.

Lognormal model

The lognormal model has two fitted parameters, the abundance of species at rank r is calculated like so:

$$a_r = \exp(\log(\mu) + \log(\sigma) \times N)$$

This model assumes that the logarithmic abundances are distributed normally. In the model, N is a normal deviate and μ and Σ are the mean and standard deviation of the distribution.

Zipf model

In the Zipf model there are two fitted parameters, the abundance of species at rank r is calculated like so:

$$a_r = J \times P_1 \times r^\gamma$$

In the Zipf model J is the number of individuals, P_1 is the proportion of the most abundant species and γ is a decay coefficient.

Mandelbrot model

The Mandelbrot model adds one parameter to the Zipf model, the abundance of species at rank r is calculated like so:

$$a_r = Jc \, (r + \beta)^\gamma$$

The addition of the β parameter leads to the P_1 part of the Zipf model becoming a simple scaling constant c.

Summary of models

Much has been written about the ecological and evolutionary significance of the various models. If your data happen to fit a particular model it does not mean that the underlying ecological theory behind that model must exist for your community. Modelling is a way to try to understand the real world in a simpler and predictable fashion. The models fall into two basic camps:

- Models based on resource partitioning.
- Models based on statistical theory.

The resource-partitioning models can be further split into two, operating over ecological time or evolutionary time.

The broken stick model is an ecological resource-partitioning model. It is often used as a null model because it assumes that there are environmental gradients, which species partition in a simple way.

The preemption model is an evolutionary resource-partitioning model. It assumes that the most competitive species will get a larger share of resources regardless of when it arrived in the community.

The lognormal model is a statistical model. The lognormal relationship appears often in communities. One theory is that species are affected by many factors, environmental and competitive – this leads to a normal distribution. Plant growth is logarithmic so the lognormal model 'fits'. Note that the normal distribution refers to the abundance-class histogram.

The Zipf and Mandelbrot models are statistical models related to the cost of information. The presence of a species depends on previous conditions; environmental and previous species presence – these are the costs. Pioneer species have low costs – they do not need the presence of other species or prior conditions. Competitor species and late-successional species have higher costs, in terms of energy, time or ecosystem organisation.

You can think of the difference between lognormal and Zipf/Mandelbrot models as being how the factors that affect the species operates:

- Lognormal: factors apply simultaneously.
- Zipf/Mandelbrot: factors apply sequentially.

Most of the models assume you have genuine counts of individuals. This is fine for animal communities but not so sensible for plants, which have more plastic growth. In an ideal situation you would use some kind of proxy for biomass to assess plant communities. Cover scales are not generally viewed as being altogether suitable but of course if these are all the data you've got, then you'll probably go with them! In the next section you will see how to create the various models and examine their properties.

11.1.2 Creating RAD models

There are two main ways you could proceed when it comes to making RAD models:

- Make all RAD models and compare them.
- Make a single RAD model.

In the first case you are most likely to prepare all the possible models so that you can see which is the 'best' for each sample. In the second case you are most likely to wish to compare a single model between samples.

The `radfit()` command in the *vegan* package will prepare all five RAD models for a community dataset or single sample. You can also prepare a single RAD model using commands of the form `rad.xxxx()`, where xxxx is the name of the model you want (Table 11.1).

Table 11.1 RAD models and their corresponding R commands (from the vegan package).

RAD model	Command
Lognormal	`rad.lognormal()`
Pre-emption	`rad.preempt()`
Broken stick	`rad.null()`
Mandelbrot	`rad.zipfbrot()`
Zipf	`rad.zipf()`

You'll see how to prepare individual models later but first you will see how to prepare all RAD models for a sample.

Preparing all RAD models

The `radfit()` command allows you to create all five common RAD models for a community dataset containing multiple samples. You can also use it to obtain models for a single sample.

RAD model overview

To make a model you simply use the `radfit()` command on a community dataset or sample. If you are looking at a dataset with several samples then the data must be in the form of a `data.frame`. If you have a single sample then the data can be a simple `vector` or a `matrix`.

The result you see will depend on whether you used a multi-sample dataset or a single sample. For a dataset with several samples you see a row for each of the five models – split into columns for each sample:

```
> gb.rad = radfit(gbt)
> gb.rad

Deviance for RAD models:

                 Edge       Grass      Wood
Null         6410.633   1697.424   2527.32
Preemption    571.854    422.638    155.43
Lognormal     740.107     72.456    856.94
Zipf          931.124    132.885   1427.66
Mandelbrot    229.538     45.899    155.43
```

If you only used a single sample then the result shows a row for each model with columns showing various results:

```
> gb.rad.E1 <- radfit(gb.biol[1,])
> gb.rad.E1

RAD models, family poisson
No. of species 17, total abundance 715

                 par1        par2     par3   Deviance AIC      BIC
Null           828.463                        888.755  888.755
Preemption       0.5215                         86.117  148.409 149.242
Lognormal        1.5238     2.4142              96.605  160.897 162.563
Zipf             0.63709  -2.0258              105.544  169.836 171.502
Mandelbrot    3399.6       -5.3947  3.9929     39.999  106.291 108.791
```

In any event you end up with a result object that contains information about each of the RAD models. You can explore the result in more detail using a variety of 'helper' commands and by using the $ syntax to view the various result components.

RAD model components

Once you have your RAD model result you can examine the various components. The result of the radfit() command is a type of list, which contains several layers of components. The top 'layer' is a result for each sample:

```
> gbt.rad

Deviance for RAD models:

                 Edge       Grass      Wood
Null         6410.633   1697.424   2527.32
Preemption    571.854    422.638    155.43
Lognormal     740.107     72.456    856.94
Zipf          931.124    132.885   1427.66
Mandelbrot    229.538     45.899    155.43

> names(gbt.rad)
[1] "Edge"  "Grass"  "Wood"
```

For each named sample there are further layers:

```
> names(gbt.rad$Edge)
[1] "y"     "family"    "models"
```

The $models layer contains the five RAD models:

```
> names(gbt.rad$Edge$models)
[1] "Null"    "Preemption"    "Lognormal"    "Zipf"    "Mandelbrot"
```

Each of the models contains several components:

```
> names(gbt.rad$Edge$models$Mandelbrot)
[1] "model"          "family"      "y"               "coefficients"
[5] "fitted.values"  "aic"         "rank"            "df.residual"
[9] "deviance"       "residuals"   "prior.weights"
```

So, by using the $ syntax you can drill down into the result and view the separate components. AIC values, for example, are used to determine the 'best' model from a range of options. The AIC values are an estimate of information 'lost' when a model is used to represent a situation. In the following exercise you can have a go at creating a series of RAD models for the 18-sample ground beetle community data. You can then examine the details and compare models.

Have a Go: Create multiple RAD models for a community dataset

For this exercise you will need the ground beetle data in the *CERE.RData* file. You will also need the *vegan* package.

1. Start by preparing the *vegan* package:

```
> library(vegan)
```

2. Make a series of RAD models for the ground beetle data – you may get warnings, which relate to the fitting of some of the generalised linear models – do not worry overly about these:

```
> gb.rad <- radfit(gb.biol)
```

3. View the result by typing its name – you will see the deviance for each model/sample combination:

```
> gb.rad

Deviance for RAD models:

                 E1        E2        E3       E4        E5       E6       G1
Null        828.4633  546.8046  532.3123  874.3219  893.7626  701.7052  331.3146
Preemption   86.1171   74.6780   49.5901  155.1716  101.2498   75.4722  131.4807
Lognormal    96.6051  144.1771  110.9866  138.4723  148.3373  137.1522   27.4805
Zipf        105.5441  184.9127  145.8725  162.6555  197.7341  186.7730   14.7817
Mandelbrot   39.9992   74.6780   49.5901   63.0773  106.0696   75.4718   14.5470
                 G2        G3        G4       G5        G6       W1       W2
Null        155.8671   85.2082  132.7137  199.5453  135.7377  684.9151  272.1441
Preemption   51.0619   23.3040   42.5072   62.1768   45.4607   99.0215   25.4618
```

```
Lognormal   13.7845    9.7973   14.3199   15.8622  18.7468 286.6316  92.0102
Zipf        11.6686   10.9600   23.3319   19.1257  25.5595 398.0188 179.2625
Mandelbrot   4.2693    3.0943    4.7256    7.0041   8.9452  99.0212  25.4566
                 W3        W4        W5        W6
Null        270.9451 296.5025 330.6709  204.388
Preemption   20.5931  42.4143  47.5003   32.311
Lognormal    75.1871 153.1896 174.7615   78.464
Zipf        143.4578 238.3210 282.7760  168.513
Mandelbrot   20.5906  42.4127  47.4973   32.299
```

4. Use the `summary()` command to give details about each sample and the model details – the list is quite extensive:

```
> summary(gb.rad)

*** E1 ***

RAD models, family poisson
No. of species 17, total abundance 715

                 par1     par2     par3 Deviance      AIC      BIC
Null                                     828.463  888.755  888.755
Preemption     0.5215                     86.117  148.409  149.242
Lognormal      1.5238   2.4142            96.605  160.897  162.563
Zipf          0.63709  -2.0258           105.544  169.836  171.502
Mandelbrot     3399.6  -5.3947   3.9929   39.999  106.291  108.791
```

5. Look at the samples available for inspection:

```
> names(gb.rad)
 [1] "E1" "E2" "E3" "E4" "E5" "E6" "G1" "G2" "G3" "G4" "G5" "G6" "W1" "W2"
[15] "W3" "W4" "W5" "W6"
```

6. Use the $ syntax to view the RAD models for the *W1* sample:

```
> gb.rad$W1

RAD models, family poisson
No. of species 12, total abundance 1092

                par1          par2         par3 Deviance      AIC  BIC
Null                                            684.915 737.908  737.908
Preemption  0.47868                             99.022 154.015  154.499
Lognormal    3.2639        1.7828              286.632 343.625  344.594
Zipf        0.53961       -1.6591              398.019 455.012  455.982
Mandelbrot      Inf -1.3041e+08 2.0021e+08     99.021 158.014  159.469
```

7. From step 6 you can see that the preemption model has the lowest AIC value. View the AIC values for all the models and samples:

```
> sapply(gb.rad, function(x) unlist(lapply(x$models, AIC)))
                 E1       E2       E3       E4       E5       E6       G1
Null        888.7550 604.7068 589.1388 965.4858 973.9772 771.2113 418.5341
Preemption  148.4088 134.5803 108.4166 248.3355 183.4645 146.9783 220.7002
Lognormal   160.8968 206.0793 171.8131 233.6363 232.5519 210.6583 118.7000
Zipf        169.8358 246.8149 206.6991 257.8194 281.9487 260.2791 106.0012
Mandelbrot  106.2909 138.5802 112.4166 160.2413 192.2843 150.9779 107.7665
```

```
                G2        G3       G4        G5        G6        W1        W2
Null         229.47852 144.66473 226.4738  286.5904  19.38683 737.9082 325.96427
Preemption   126.67329  84.76055 138.2674  151.2219 131.10975 154.0146  81.28201
Lognormal     91.39596  73.25383 112.0800  106.9072 106.39589 343.6247 149.83040
Zipf          89.27999  74.41651 121.0921  110.1708 113.20861 455.0118 237.08266
Mandelbrot    83.88072  68.55087 104.4857  100.0491  98.59426 158.0142  85.27674
                W3        W4       W5        W6
Null         323.27808 350.63206385.2040  252.71046
Preemption    74.92607  98.54388104.0333   82.63346
Lognormal    131.52006 211.31915233.2946  130.78639
Zipf         199.79075 296.45061341.3091  220.83585
Mandelbrot    78.92356 102.54224108.0304   86.62174
```

8. Look at what models are available for the *G1* sample:

```
> names(gb.rad$G1$models)
[1] "Null"    "Preemption"    "Lognormal"    "Zipf"    "Mandelbrot"
```

9. View the lognormal model for the *G1* sample:

```
> gb.rad$G1$models$Lognormal

RAD model: Log-Normal
Family: poisson
No. of species:  28
Total abundance: 365

    log.mu  log.sigma    Deviance         AIC         BIC
 0.8149253  2.0370676  27.4805074  118.6999799  121.3643889
```

The $ syntax allows you to explore the models in detail and, as you saw in step 7, you can also get a summary of the 'important' elements of the models.

The structure of the result for a single sample is the same as for the multi-sample data but you have one less level of data – you do not have the sample names.

By comparing the AIC values for the various models you can determine the 'best' model for each sample, as you saw in step 7 of the preceding exercise.

The `radfit()` command assumes that your data are genuine count data and therefore are integers. If you have values that are some other measure of abundance then you'll have to modify the model fitting process by using a different distribution `family`, such as Gamma. This is easily carried out by using the `family = Gamma` instruction in the `radfit()` command. In the following exercise you can have a go at making RAD models for some non-integer data that require a *Gamma* fit.

Have a Go: Create RAD models for abundance data using a *Gamma* distribution

You will need the *vegan* package for this exercise and the *bf.biol* data, which are found in the *CERE.RData* file.

1. Start by preparing the *vegan* package:

```
> library(vegan)
```

2. The *bf.biol* data were prepared from the *bf* data. The `xtabs()` command was used and the result is a table object that has two classes – look at the data class:

```
> class(bf.biol)
[1] "xtabs" "table"
```

3. You need to get these data into a `data.frame` format so that the `radfit()` command can prepare a series of models for each sample:

```
> bfs =as.matrix(bf.biol)
> class(bfs) = "matrix"
> bfs = as.data.frame(bfs)
```

4. Now make a `radfit()` result using a Gamma distribution, since the data are not integers. You will get warnings that the generalised linear model did not converge:

```
> bfs.rad = radfit(bfs, family = Gamma)
```

5. Look at the RAD models you prepared:

```
> bfs.rad

Deviance for RAD models:

                  1996     1997     1998     1999     2000     2001
Null           6.75808  5.90335  9.76398 16.10717 14.07862 12.34689
Preemption     0.82914  1.24224  1.16693  2.23559  2.61460  0.52066
Lognormal      2.32981  1.95389  3.34543  3.85454  2.67501  1.27117
Zipf           3.80899  5.85292  5.03517  2.57252  2.50295  4.76591
Mandelbrot     0.65420  1.19315  1.13141  0.80418  1.08860  0.51195
                  2002     2003     2004     2005
Null          18.68292 22.86051 18.99316 14.3380
Preemption     1.84098  4.44566  2.68697  1.9573
Lognormal      2.71852  3.41538  4.32521  3.0185
Zipf           3.26510  3.39652  3.14484  2.2478
Mandelbrot     0.95159  2.62833  1.11120  0.6803
```

The RAD models prepared using the Gamma distribution can be handled like the models you made using the Poisson distribution (the default).

It is useful to visualise the models that you make, and you'll see how to do this shortly (Section 11.1.3) but before that you will see how to prepare single RAD models.

Preparing single RAD models

Rather than prepare all five RAD models you might prefer to examine a single model. You can use the $ syntax to get the single models from a `radfit()` result but this can be a bit tedious.

The *vegan* package contains several commands that allow you to create individual RAD models (Table 11.1). These commands are designed to operate on single samples rather than data frames containing multiple samples. However, with some coercion you can make results objects containing a single RAD model for several samples. An advantage of single models is that you can use various 'helper' commands to extract model components. In the following exercise you can have a go at making some single RAD model results.

Have a Go: Create single RAD models for community data

For this exercise you will need the ground beetle data in the *CERE.RData* file. You will also need the *vegan* package.

1. Start by preparing the *vegan* package:

    ```
    > library(vegan)
    ```

2. Make a preemption model for the *W1* sample:

    ```
    > gbW1.pe = rad.preempt(gb.biol["W1",])
    > gbW1.pe

    RAD model: Preemption
    Family: poisson
    No. of species:  12
    Total abundance: 1092

            alpha    Deviance           AIC           BIC
        0.4786784  99.0215048  154.0145519  154.4994586
    ```

3. Get the fitted values from the model:

    ```
    > fitted(gbW1.pe)
     [1] 522.7168308 272.5035660 142.0619905 74.0599819 38.6090670
     [6]  20.1277400  10.4930253   5.4702405  2.8517545  1.4866812
    [11]   0.7750390   0.4040445
    ```

4. Now make a Mandelbrot model for the *G1* sample:

    ```
    > gbG1.zb = rad.zipfbrot(gb.biol["G1",])
    ```

5. Use some helper commands to extract the AIC and coefficients from the model:

    ```
    > AIC(gbG1.zb)
    [1] 107.7665

    > coef(gbG1.zb)
            c         gamma         beta
    0.6001759  -1.6886844  0.1444777
    ```

6. Now make a lognormal model for the entire dataset – use the `apply()` command like so:

    ```
    > gb.ln = apply(gb.biol, MARGIN = 1, rad.lognormal)
    ```

7. Use the `names()` command to see that the result contains a model for each sample:

    ```
    > names(gb.ln)
     [1] "E1" "E2" "E3" "E4" "E5" "E6" "G1" "G2" "G3" "G4" "G5" "G6" "W1"
    [14] "W2" "W3" "W4" "W5" "W6"

    > names(gb.ln$E1)
     [1] "model"          "family"        "y"               "coefficients"
     [5] "fitted.values"  "aic"           "rank"            "df.residual"
     [9] "deviance"       "residuals"     "prior.weights"
    ```

8. View the model for the *E1* sample:

    ```
    > gb.ln$E1
    ```

```
RAD model: Log-Normal
Family: poisson
No. of species:  17
Total abundance: 715

     log.mu  log.sigma   Deviance        AIC        BIC
    1.523787   2.414170  96.605129 160.896843 162.563270
```

9. Now view the coefficients for all the samples:

```
> sapply(gb.ln, FUN = coef)
                  E1        E2        E3        E4        E5        E6
log.mu      1.523787 2.461695 2.107499 1.686040 2.052081 2.371504
log.sigma2.414170 1.952583 2.056276 2.120549 2.092042 1.995912
                  G1        G2        G3        G4        G5        G6
log.mu      0.8149253 1.249582 1.284923 1.369442 1.286567 1.500785
log.sigma 2.0370676 1.761516 1.651098 1.588253 1.760267 1.605790
                  W1        W2        W3        W4        W5        W6
log.mu      3.263932 3.470459 3.269124 3.178878 3.357883 3.525677
log.sigma   1.782800 1.540651 1.605284 1.543145 1.526280 1.568422
```

The result you obtained in step 6 is a `list` object – so you used the `sapply()` command to get the coefficients in step 9.

Tip: The `lapply()` and `sapply()` commands

The `lapply()` command operates on `list` objects and allows you to use a function over the components of the `list`. The result is itself a `list`, with `names` the same as the original components. The `sapply()` command is very similar but the result is a `matrix` object, which may be more convenient.

In the preceding exercise you used the `AIC()`, `coef()` and `fitted()` commands to extract the AIC values, coefficients and the fitted values from the models. Other 'helper' commands are `deviance()` and `resid()`, which produce the deviance and residuals respectively.

Comparing different RAD models

You've already seen how to look at the different models and to compare them, by looking at AIC values for example. The `sapply()` command is particularly useful as it allows you to execute a command over the various elements of the result (which is a kind of `list`).

It would be useful to determine if different models were significantly different from one another. The model with the lowest AIC value is considered to be the 'best' and it is easy to see which this is by inspection:

```
> sapply(mod, function(x) unlist(lapply(x$models, AIC)))
                Edge      Grass      Wood
Null        6535.1052 1850.7079 2604.8112
Preemption   698.3265  577.9220  234.9224
Lognormal    868.5789  229.7401  938.4294
Zipf        1059.5962  290.1691 1509.1537
Mandelbrot   360.0098  205.1833  238.9203
```

It is a little more difficult to get a result that shows the 'best' model for each sample. In the following exercise you can have a go at making a result object that contains the lowest AIC value and the matching model name for every sample.

Have a Go: Create a result object that shows the lowest AIC and matching model name for a `radfit()` result

For this exercise you will need the ground beetle data in the *CERE.RData* file. You will also need the *vegan* package.

1. Start by preparing the *vegan* package:

```
> library(vegan)
```

2. Make a RAD result using the `radfit()` command:

```
> gb.rad = radfit(gb.biol)
```

3. Typing the result name gives the deviance but you need the AIC values as a result object so use the `sapply()` command like so:

```
> gb.aic = sapply(gb.rad, function(x) unlist(lapply(x$models, AIC)))
```

4. View the result: you get five rows, one for each model, and a column for each sample:

```
> gb.aic[, 1:6]
                 E1       E2       E3       E4       E5       E6
Null        888.7550 604.7068 589.1388 965.4858 973.9772 771.2113
Preemption  148.4088 134.5803 108.4166 248.3355 183.4645 146.9783
Lognormal   160.8968 206.0793 171.8131 233.6363 232.5519 210.6583
Zipf        169.8358 246.8149 206.6991 257.8194 281.9487 260.2791
Mandelbrot  106.2909 138.5802 112.4166 160.2413 192.2843 150.9779
```

5. The result is a matrix so convert it to a data.frame:

```
> gb.aic = as.data.frame(gb.aic)
```

6. Now make an index that shows which AIC value is the lowest for every sample:

```
> index <- apply(gb.aic, MARGIN = 2, function(x) which(x == min(x)))
> index
E1 E2 E3 E4 E5 E6 G1 G2 G3 G4 G5 G6 W1 W2 W3 W4 W5 W6
 5  2  2  5  2  2  4  5  5  5  5  5  2  2  2  2  2  2
```

7. You want to make a new vector that contains the lowest AIC values. You need a simple loop:

```
> aicval <- numeric(0)
> for(i in 1:ncol(gb.aic)) {
    aicval[i] <- gb.aic[which(gb.aic[,i] == min(gb.aic[,i])), i]
    }
```

8. In step 7 you set-up a 'dummy' vector to receive the results. Then the loop runs for as many columns as there are in the AIC results. Each time around the loop gets the minimum AIC value for that column and adds it to the dummy vector. View the result:

```
> aicval
 [1] 106.29094 134.58026 108.41663 160.24125 183.46446 146.97826
 [7] 106.00117  83.88072  68.55087 104.48570 100.04912  98.59426
[13] 154.01455  81.28201  74.92607  98.54388 104.03335  82.63346
```

9. Assign names to the minimum AIC values by using the original sample names:

```
> names(aicval) <- colnames(gb.aic)
> aicval
       E1        E2        E3        E4        E5        E6        G1
106.29094 134.58026 108.41663 160.24125 183.46446 146.97826 106.00117
       G2        G3        G4        G5        G6        W1        W2
 83.88072  68.55087 104.48570 100.04912  98.59426 154.01455  81.28201
       W3        W4        W5        W6
 74.92607  98.54388 104.03335  82.63346
```

10. Now use the index value you made in step 6 to get the names of the RAD models that correspond to the lowest AIC values:

```
> monval <- rownames(gb.aic)[index]
```

11. Assign the sample names to the models so you can keep track of which model belongs to which sample:

```
> names(monval) <- colnames(gb.aic)
> monval
           E1            E2            E3            E4            E5
 "Mandelbrot"  "Preemption"  "Preemption"  "Mandelbrot"  "Preemption"
           E6            G1            G2            G3            G4
 "Preemption"       "Zipf"  "Mandelbrot"  "Mandelbrot"  "Mandelbrot"
           G5            G6            W1            W2            W3
 "Mandelbrot"  "Mandelbrot"  "Preemption"  "Preemption"  "Preemption"
           W4            W5            W6
 "Preemption"  "Preemption"  "Preemption"
```

12. Now assemble the results into a new data.frame:

```
> gb.models <- data.frame(AIC = aicval, Model = monval)
```

Your final result has two columns, one containing the lowest AIC and a column containing the name of the corresponding RAD model. The row names of the data.frame contain the sample names so there is no need to make an additional column.

Note: Minimum AIC values and RAD model

The steps in the exercise that created the data.frame containing the minimum AIC values and the corresponding RAD model name are packaged into a custom function rad_aic(), which is part of the *CERE.RData* file.

The minimum AIC values allow you to select the 'best' RAD model but how do you know if there is any significant difference between models? It may be that several models are equally 'good'. There is no practical way of determining the statistical difference between RAD models because they are based on different data (the models use different parameters). This means you cannot use the `anova()` command for example, like you might with `lm()` or `glm()` models.

If you have replicate samples, however, you can use analysis of variance to explore differences between models. The process involves looking at the variability in the model deviance between replicates. In the following exercise you can have a go at exploring variability in RAD models for a subset of the ground beetle data with six replicates.

Have a Go: Perform ANOVA on RAD model deviance

For this exercise you will need the ground beetle data in the *CERE.RData* file. You will also need the *vegan* package.

1. Start by preparing the *vegan* package:

```
> library(vegan)
```

2. Make a RAD model result from the first six rows of the ground beetle data (relating to the *Edge* habitat) using the `radfit()` command:

```
> Edge.rad = radfit(gb.biol[1:6, ])
```

3. Extract the deviance from the result:

```
> Edge.dev = sapply(Edge.rad, function(x)
                unlist(lapply(x$models, deviance)))
> Edge.dev
               E1         E2         E3         E4        E5        E6
Null       828.46326 546.80456 532.31232 874.32187 893.7626 701.70519
Preemption  86.11705  74.67805  49.59012 155.17162 101.2498  75.47216
Lognormal   96.60513 144.17705 110.98661 138.47233 148.3373 137.15216
Zipf       105.54412 184.91272 145.87254 162.65549 197.7341 186.77304
Mandelbrot  39.99923  74.67802  49.59008  63.07732 106.0696  75.47178
```

4. Now rotate the result so that the models form the columns and the replicates (samples) are the rows:

```
> Edge.dev = t(Edge.dev)
> Edge.dev
        Null  Preemption  Lognormal      Zipf  Mandelbrot
E1  828.4633    86.11705   96.60513  105.5441    39.99923
E2  546.8046    74.67805  144.17705  184.9127    74.67802
E3  532.3123    49.59012  110.98661  145.8725    49.59008
E4  874.3219   155.17162  138.47233  162.6555    63.07732
E5  893.7626   101.24981  148.33726  197.7341   106.06963
E6  701.7052    75.47216  137.15216  186.7730    75.47178
```

5. You will need to use the `stack()` command to rearrange the data into two columns, one for the deviance and one for the model name. However, first you need to convert the result into a `data.frame` object:

```
> Edge.dev = as.data.frame(Edge.dev)
> class(Edge.dev)
[1] "data.frame"
```

6. Now you can use the `stack()` command and alter the column names:

```
> Edge.dev = stack(Edge.dev)
> names(Edge.dev) = c("deviance", "model")
```

7. You now have a `data.frame` that can be used for analysis. Before that though you should alter the names of the RAD models as they are quite long – make them shorter:

```
> levels(Edge.dev$model) = c("Log", "Mb", "BS", "Pre", "Zip")
```

8. Now carry out an ANOVA on the deviance to see if there are significant differences between the RAD models – use the logarithm of the deviance to help normalise the data:

```
> Edge.aov = aov(log(deviance) ~ model, data = Edge.dev)
> summary(Edge.aov)
            Df  Sum Sq   Mean Sq  F value   Pr(>F)
model        4  20.9193   5.2298   65.545  6.937e-13 ***
Residuals   25   1.9948   0.0798
---
Signif. codes:  0 '***' 0.001 '**' 0.01 '*' 0.05 '.' 0.1 ' ' 1
```

9. Use the `TukeyHSD()` command to explore differences between the individual models:

```
> TukeyHSD(Edge.aov, ordered = TRUE)
  Tukey multiple comparisons of means
    95% family-wise confidence level
    factor levels have been ordered

Fit: aov(formula = log(deviance) ~ model, data = Edge.dev)

$model
              diff         lwr        upr       p adj
Pre-Mb   0.2700865  -0.20887308  0.7490461  0.4776096
Log-Mb   0.6773801   0.19842055  1.1563397  0.0028170
Zip-Mb   0.9053575   0.42639796  1.3843171  0.0000820
BS-Mb    2.3975406   1.91858106  2.8765002  0.0000000
Log-Pre  0.4072936  -0.07166594  0.8862532  0.1233203
Zip-Pre  0.6352710   0.15631147  1.1142306  0.0053397
BS-Pre   2.1274541   1.64849457  2.6064137  0.0000000
Zip-Log  0.2279774  -0.25098216  0.7069370  0.6345799
BS-Log   1.7201605   1.24120094  2.1991201  0.0000000
BS-Zip   1.4921831   1.01322353  1.9711427  0.0000000
```

10. Visualise the differences between the models by using a box-whisker plot. Your plot should resemble Figure 11.1:

```
> boxplot(log(deviance) ~ model, data = Edge.dev, las = 1)
> title(xlab = "RAD model", ylab = "Log(model deviance)")
```

11. You can see from the results that there are differences between the RAD models. The Mandelbrot model has the lowest overall deviance but it is not significantly different from the preemption model. You can see this more clearly if you plot the Tukey result directly; the following command produces a plot that looks like Figure 11.2:

```
> plot(TukeyHSD(Edge.aov, ordered = TRUE), las = 1)
```

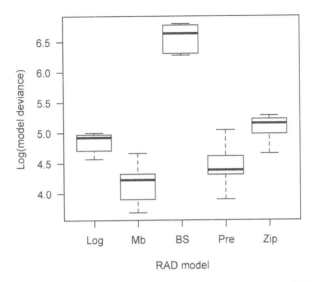

Figure 11.1 Model deviance (log deviance) for various RAD models. Log = lognormal, Mb = Mandelbrot, BS = broken stick, Pre = preemption, Zip = Zipf–Mandelbrot.

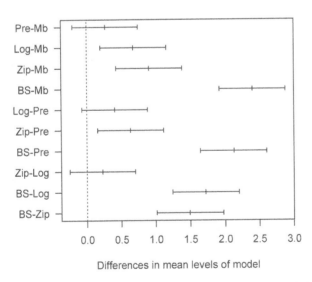

Figure 11.2 Tukey HSD result of pairwise differences in RAD model for ground beetle communities.

In this case the deviance of the original RAD models was log transformed to help with normalising the data – notice that you do not have to make a transformed variable in advance, you can do it from the `aov()` and `boxplot()` commands directly.

Tip: Reordering variables

In many cases the `factor` variables that you use in analyses are unordered. R takes them alphabetically and this can be 'inconvenient' for some plots. Use the `reorder()` command to alter the order of the various `factor` levels. The command works like so:

```
reorder(factor, response, FUN = mean)
```

So, if you want to reorder the `factor` to show a boxplot in order of mean value you would use the command to make a new 'version' of the original `factor`, which you then use in your `boxplot()` command.

Note: ANOVA for RAD models

The commands required for conducting ANOVA on RAD models are bundled into a custom command, `rad_test()`, which is part of the *CERE.RData* file. The command includes `print()`, `summary()` and `plot()` methods, the latter allowing you to produce a boxplot of the deviance or a plot of the post-hoc results.

11.1.3 Visualising RAD models using dominance/diversity plots

You need to be able to visualise the RAD models as dominance/diversity plots. The `radfit()` and `rad.xxxx()` commands have their own `plot()` routines, some of which use the *lattice* package. This package comes as part of the basic R installation but is not loaded until required.

Most often you will have the result of a `radfit()` command and will have a result that gives you all five RAD models for the samples in your community dataset. In the following exercise you can have a go at comparing the DD plots for all the samples in a community dataset.

Have a Go: Visualise RAD models from multiple samples

For this exercise you will need the ground beetle data in the *CERE.RData* file. You will also need the *vegan* package. The *lattice* package will be used but this should already be installed as part of the normal installation of R.

1. Start by preparing the *vegan* package:

    ```
    > library(vegan)
    ```

2. Make a RAD model from the ground beetle data using the `radfit()` command:

    ```
    > gb.rad = radfit(gb.biol)
    ```

3. Make a plot that shows all the samples and selects the best RAD model for each one. The `plot()` command will utilise the *lattice* package, which will be readied if necessary. The final graph should resemble Figure 11.3:

    ```
    > plot(gb.rad)
    ```

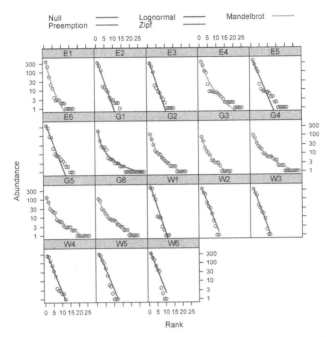

Figure 11.3 RAD models for 18 samples of ground beetles. Each panel shows the RAD model with the lowest AIC.

4. Now make a plot that shows all samples but for a single RAD model (*Preemption*), your graph should resemble Figure 11.4:

```
> plot(gb.rad, model = "Preemption")
```

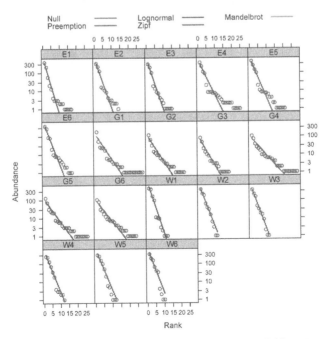

Figure 11.4 Dominance/diversity for the preemption RAD model for 18 samples of ground beetles.

It is not easy to alter the colours on the plots – these are set by the `plot()` command internally.

If you have a single sample and a `radfit()` result, containing the five RAD models, you can use the `plot()` command to produce a slightly different looking plot. In this instance you get a single plot window with the 'fit lines' superimposed onto a single plot window. If you want a plot with separate panels you can use the `radlattice()` command to produce one. In the following exercise you can have a go at visualising RAD models for a single sample.

Have a Go: Visualise RAD models for a single sample

For this exercise you will need the ground beetle data in the *CERE.RData* file. You will also need the *vegan* package. The *lattice* package will be used but this should already be installed as part of the normal installation of R.

1. Start by preparing the *vegan* package:

   ```
   > library(vegan)
   ```

2. Make a RAD model from the ground beetle data using the `radfit()` command:

   ```
   > gb.rad = radfit(gb.biol)
   ```

3. Now use the `plot()` command to view all the RAD models for the *E1* sample; your plot should resemble Figure 11.5:

   ```
   > plot(gb.rad$E1)
   ```

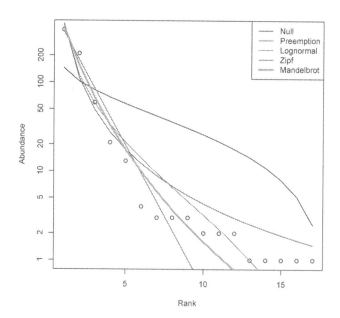

Figure 11.5 Dominance/diversity for RAD models in a sample of ground beetles.

4. You can compare the five models in separate panels by using the `radlattice()` command; your graph should resemble Figure 11.6:

```
> radlattice(gb.rad$E1)
```

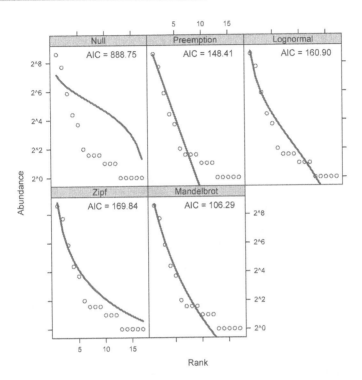

Figure 11.6 Dominance/diversity for RAD models in a sample of ground beetles.

In this exercise you selected a single sample from a `radfit()` result that contained multiple samples. You can make a result for a single sample easily by simply specifying the appropriate sample in the `radfit()` command itself, e.g.

```
> radfit(gb.biol["E1", ]
```

However, it is easy enough to prepare a result for all samples and then you are able to select any one you wish.

Tip: Plot axes in log scale

To plot both axes in a log scale you can simply use the `log = "xy"` instruction as part of your `plot()` command. Note though that this only works for plots that operate in a single window and not the *lattice* type plots.

In the preceding exercise you used the default colours and line styles. It is not trivial to alter them because they are built-in to the commands and not 'available' as separate user-controlled instructions. However, you can produce a more customised graph by producing single RAD model results. These single-model results have `plot()`, `lines()` and `points()` methods, which allow you fine control over the graphs you produce.

Customising DD plots

The regular `plot()` and `radlattice()` commands allow you to compare RAD models for one or more samples. However, you may wish to visualise particular model-sample combinations and produce a more 'targeted' plot. In the following exercise you can have a go at making a more selective plot.

Have a Go: Make a selective dominance/diversity plot

For this exercise you will need the ground beetle data in the *CERE.RData* file. You will also need the *vegan* package. The *lattice* package will be used but this should already be installed as part of the normal installation of R.

1. Start by preparing the *vegan* package:

    ```
    > library(vegan)
    ```

2. Make a lognormal RAD model for the *E1* sample of the ground beetle community data:

    ```
    > m1 = rad.lognormal(gb.biol["E1",])
    ```

3. Make another lognormal model but for the *E2* sample:

    ```
    > m2 = rad.lognormal(gb.biol["E2",])
    ```

4. Now make a broken stick model for the *E3* sample:

    ```
    > m3 = rad.null(gb.biol["E3",])
    ```

5. Start the plot by looking at the *m1* model you made in step 2:

    ```
    > plot(m1, pch = 1, lty = 1, col = 1)
    ```

6. Add points from the *m2* model (step 3) and then add a line for the RAD model fit. Use different colour, plotting symbols and line type from the plot in step 5:

    ```
    > points(m2, pch = 2, col = 2)
    > lines(m2, lty = 2, col = 2)
    ```

7. Now add points and lines for the m3 model (step 4) and use different colours and so on:

    ```
    > points(m3, pch = 3, col = 3)
    > lines(m3, lty = 3, col = 3)
    ```

8. Finally add a legend; make sure that you match up the colours, line types and plotting characters. Your final graph should resemble Figure 11.7:

    ```
    > legend(x = "topright",
            legend = c("Lognormal E1", "Lognormal E2", "Broken Stick E3"),
            pch = 1:3, lty = 1:3, col = 1:3, bty = "n")
    ```

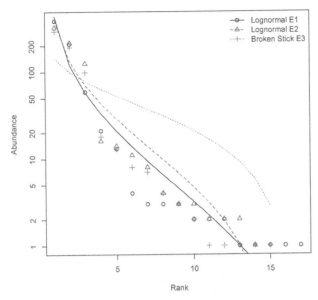

Figure 11.7 Dominance/diversity for different RAD models and samples of ground beetles.

The graph you made is perhaps not a very sensible one but it does illustrate how you can build a customised plot of your RAD models.

Identifying species on plots

Your basic dominance/diversity plot shows the log of the species abundances against the rank of that abundance. You see the points relating to each species but it might be helpful to be able to see which point relates to which species. The plot() commands that produce single-window plots (i.e. not ones that use the *lattice* package) of RAD models allow you to identify the points, so you can see which species are which. The identify() command allows you to use the mouse to essentially add labels to a plot. The plot needs to be of a specific class, "ordiplot", which is produced when you make a plot of an RAD model. This class of plot is also produced when you plot the results of ordination (see Chapter 14).

In the following exercise you can have a go at making a plot of an RAD model and customising it by identifying the points with the species names.

Have a Go: Identify the species from points of a dominance/diversity plot

For this exercise you will need the ground beetle data in the *CERE.RData* file. You will also need the *vegan* package.

1. Start by preparing the *vegan* package:

   ```
   > library(vegan)
   ```

2. Make a Zipf model of the *E1* sample:

   ```
   > gb.zipf = rad.zipf(gb.biol["E1",])
   ```

3. Now make a plot of the RAD model but assign the result to a named object:

    ```
    > op = plot(gb.zipf)
    ```

4. The plot shows basic points and a line for the fitted model. You will redraw the plot and customise it shortly but first look at the *op* object you just created:

    ```
    > op
    $species
              rnk   poi
    Aba.par    1    388
    Pte.mad    2    210
    Neb.bre    3     59
    Pte.str    4     21
    Cal.rot    5     13
    Pte.mel    6      4
    Pte.nige   7      3
    Ocy.har    8      3
    Car.vio    9      3
    Poe.cup   10      2
    Pla.ass   11      2
    Bem.man   12      2
    Sto.pum   13      1
    Pte.obl   14      1
    Pte.nigr  15      1
    Lei.ful   16      1
    Bem.lam   17      1

    attr(,"class")
    [1] "ordiplot"
    ```

5. The *op* object contains all the data you need for a plot. The `identify()` command will use the row names as the default labels but first, redraw the plot and display the points only (the default is `type = "b"`). Also, suppress the axes and make a little more room to fit the labels into the plot region:

    ```
    > op = plot(gb.zipf, type = "p", pch = 43, xlim = c(0, 20), axes = FALSE)
    ```

6. You may get a warning message but the plot is created anyhow. The `type = "p"` part turned off the line to leave the points only (`type = "l"` would show the line only). Now add in the *y*-axis and set the axis tick positions explicitly:

    ```
    > axis(2, las = 1, at = c(1,2,5,10,20, 50, 100, 200, 400))
    ```

7. Now add in the *x*-axis but shift its position in the margin so it is one line outwards. You can also specify the axis tick positions using the `pretty()` command, which works out neat intervals:

    ```
    > axis(1, line = 1, at = pretty(0:20))
    ```

8. Finally you get to label the points. Start by typing the command:

    ```
    > identify(op, cex = 0.9)
    ```

9. Now the command will be waiting for you to click with the mouse in the plot region. Select the plot window by clicking in an outer margin or the header bar – this 'activates' the plot. Position your mouse cursor just below the top-most point and click once. The label appears just below the point. Now move to the next point and click just to the right of it – the label appears to the right. The position of the label relative

to the point will depend on the position of the mouse. In this way you can position the labels so they do not overlap. Click on the other points to label them. Your final graph should resemble Figure 11.8.

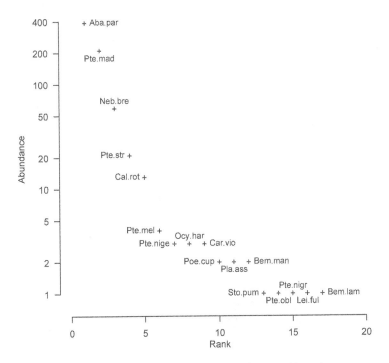

Figure 11.8 Dominance/diversity plot for a sample of ground beetles.

If you only wish to label some points then you can stop identification at any time by pressing the Esc key.

Tip: Labels and the `identify()` command

By default the `identify()` command takes its labels from the data you are identifying, usually the row names. You can specify other labels using the `labels` instruction. You can also alter the label appearance using basic graphical parameters, e.g. `cex` (character expansion/size) and `col` (colour).

The `as.rad()` command allows you to reassemble your data into a form suitable for plotting in a DD plot. The command takes community abundance data and reorders the species in rank order, with the most abundant being first. The result holds a class, `"rad"`, that can be used in a `plot()` command:

```
> as.rad(gb.biol[1,])
Aba.par Pte.mad Neb.bre Pte.str Cal.rot Pte.mel Pte.nige
    388     210      59      21      13       4       3
Ocy.har  ar.vio Poe.cup Pla.ass Bem.man Sto.pum Pte.obl
      3       3       2       2       2       1       1
```

```
Pte.nigr  Lei.ful  Bem.lam
       1        1        1
attr(,"class")
[1] "rad"
```

The resulting `plot()` would contain only the points, plotted as the log of the abundance against the rank.

One of the RAD models you've seen is the lognormal model. This was one of the first RAD models to be developed and in the following sections you will learn more about lognormal data series.

11.2 Fisher's log-series

The lognormal model you've seen so far stems from the original work of Fisher – you met this earlier (in Section 8.3.2) in the context of an index of diversity. The *vegan* package uses non-linear modelling to calculate Fisher's log-series (look back at Figures 8.6 and 8.7).

The `fisherfit()` command in the *vegan* package carries out the main model fitting processes. You looked at this in Section 8.3.2 but in the following exercise you can have a go at making Fisher's log-series and exploring the results with a different emphasis.

Have a Go: Explore Fisher's log-series

You will need the *vegan* and *MASS* packages for this exercise. The *MASS* package comes as part of the basic distribution of R but is not loaded by default. You will also use the ground beetle data in the *CERE.RData* file.

1. Start by preparing the *vegan* and *MASS* packages:

    ```
    > library(vegan)
    > library(MASS)
    ```

2. Make a log-series result for all samples in the ground beetle community dataset:

    ```
    > gb.fls <- apply(gb.biol, MARGIN = 1, fisherfit)
    ```

3. You made Fisher's log-series models for all samples – see the names of the components of the result:

    ```
    > names(gb.fls)
     [1] "E1" "E2" "E3" "E4" "E5" "E6" "G1" "G2" "G3" "G4" "G5" "G6"
    [13] "W1" "W2" "W3" "W4" "W5" "W6"
    ```

4. Look at the result for the *G1* sample:

    ```
    > gb.fls$G1

    Fisher log series model
    No. of species: 28

             Estimate  Std. Error
    alpha     7.0633       1.5389
    ```

5. Look at the components of the result for the *G1* sample:

```
> names(gb.fls$G1)
[1] "minimum"    "estimate"     "gradient"    "hessian"
[5] "code"       "iterations"   "df.residual" "nuisance"
[9] "fisher"
```

6. Look at the $fisher component:

```
> gb.fls$G1$fisher
 1 2  3  4  5  6 17  23  25  52 178
12 3  3  1  2  1  1   2   1   1   1

attr(,"class")
[1] "fisher"
```

7. You can get the frequency and number of species components using the as.fisher() command:

```
> as.fisher(gb.biol["G1",])
 1 2  3  4  5  6 17  23  25  52 178
12 3  3  1  2  1  1   2   1   1   1
attr(,"class")
[1] "fisher"
```

8. Visualise the log-series with a plot and also look at the profile to ascertain the normality, split the plot window in two and produce a plot that resembles Figure 11.9:

```
> opt = par(mfrow = c(2,1))
> plot(gb.fls$G1)
> plot(profile(gb.fls$G1))
> par(opt)
```

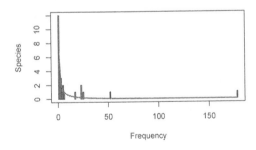

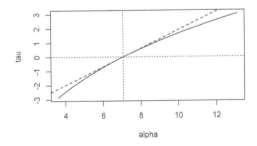

Figure 11.9 Fisher's log-series (top) and profile plot (bottom) for a sample of ground beetles.

9. Use the `confint()` command to get the confidence intervals of all the log-series — you can use the `sapply()` command to help:

```
> sapply(gb.ff, confint)
            E1        E2        E3        E4        E5E6
2.5%   1.778765  1.315205  1.501288  3.030514  2.287627  1.695922
97.5%  5.118265  4.196685  4.626269  7.266617  5.902574  4.852549
            G1        G2        G3        G4        G5        G6
2.5%   4.510002  3.333892  2.687344  4.628881  4.042329  3.658648
97.5% 10.618303  8.765033  7.892262 10.937909  9.806917  9.205348
            W1        W2        W3        W4        W5        W6
2.5%   0.9622919 0.8672381 0.8876847  1.029564 0.9991796 0.7614008
97.5%  3.3316033 3.1847201 3.2699293  3.595222 3.4755296 2.9843201
```

The `plot()` command produces a kind of bar chart when used with the result of a `fisherfit()` command. You can alter various elements of the plot such as the axis labels. Try also the `bar.col` and `line.col` instructions, which alter the colours of the bars and fitted line.

Fisher's log-series can only be used for counts of individuals and not for other forms of abundance data. You must have integer values for the `fisherfit()` command to operate.

Tip: Convert abundance data to log-series data

The `as.fisher()` command in the *vegan* package allows you to 'convert' abundance data into Fisher's log-series data.

Fisher's model seems to imply infinite species richness and so 'improvements' have been made to the model. In the following section you'll see how Preston's lognormal model can be used.

11.3 Preston's lognormal model

Preston's lognormal model (Preston 1948) is a subtle variation on Fisher's log-series. The frequency classes of the x-axis are collapsed and merged into wider bands, creating octaves of doubling size, 1, 2, 3–4, 5–8, 9–16 and so on. Furthermore, for each frequency half the species are transferred to the next highest octave. This makes the data appear more lognormal by reducing the lowest octaves (which are usually high).

In the *vegan* package the `prestonfit()` command carries out the model fitting process. By default the frequencies are split, with half the species being transferred to the next highest octave. However, you can turn this feature off by using the `tiesplit = FALSE` instruction.

The expected frequency *f* at an abundance octave *o* is determined by the formula shown in Figure 11.10.

$$f = S0 \times \exp\frac{\left(-\left(\log_2(o) - \mu\right)^2\right)}{2 \times \sigma^2}$$

Figure 11.10 Preston's lognormal model. Expected frequency for octaves (o), where μ is the location of the mode, δ is the mode width (both in log2 scale) and S0 is expected number of species at mode.

The lognormal model is usually truncated at the lowest end, with the result that some rare species may not be recorded – this truncation is called the *veil line*.

The `prestonfit()` command fits the truncated lognormal model as a second degree log-polynomial to the octave pooled data using Poisson (when `tiesplit = FALSE`) or quasi-Poisson (when `tiesplit = TRUE`) error distribution.

The `prestondistr()` command uses an alternative method: fitting a left-truncated normal distribution to log2 transformed non-pooled observations with direct maximisation of log-likelihood.

Both commands have plotting routines, which produce a bar chart and fitted line. You can also add extra lines to the plots. In the following exercise you can have a go at exploring Preston's log-series.

Have a Go: Explore Preston's lognormal models

You will need the *vegan* package for this exercise. You will also use the ground beetle data in the *CERE.RData* file.

1. Start by preparing the *vegan* package:

    ```
    > library(vegan)
    ```

2. Make a Preston lognormal model using the combined data from the ground beetle samples:

    ```
    > gb.oct = prestonfit(colSums(gb.biol))
    > gb.oct

    Preston lognormal model
    Method: Quasi-Poisson fit to octaves
    No. of species: 48

         mode      width        S0
     3.037810   4.391709   5.847843

    Frequencies by Octave
                      0          1          2          3          4          5
    Observed   2.000000   4.500000  10.000000   9.500000   2.000000   6.000000
    Fitted     4.603598   5.251002   5.686821   5.847627   5.709162   5.292339
                      6          7          8         10         11         12
    Observed   4.000000   2.000000   4.000000   1.000000   1.000000   1.000000
    Fitted     4.658067   3.892659   3.088657   1.664425   1.130409   0.728936
                     13
    Observed   1.0000000
    Fitted     0.4462991
    ```

3. For comparison use the `prestondistr()` command to make an alternative model:

    ```
    > gb.ll = prestondistr(colSums(gb.biol))
    > gb.ll

    Preston lognormal model
    Method: maximized likelihood to log2 abundances
    No. of species: 48

         mode      width        S0
     2.534594   3.969009   5.931404
    ```

```
Frequencies by Octave
                        0          1          2          3          4          5
Observed  2.000000  4.500000  10.000000  9.500000   2.000000  6.000000
Fitted    4.837308  5.504215   5.877843  5.890765   5.540596  4.890714
                        6          7          8         10         11         12
Observed  4.000000  2.000000   4.000000  1.000000   1.0000000  1.000000
Fitted    4.051531  3.149902   2.298297  1.011387   0.6099853  0.345265
                       13
Observed  1.0000000
Fitted    0.1834074
```

4. Both versions of the model have the same components. Look at these using the names() command:

```
> names(gb.oct)
[1] "freq"        "fitted"      "coefficients"      "method"

> names(gb.ll)
[1] "freq"        "fitted"      "coefficients"      "method"
```

5. View the Quasi-Poisson model in a plot – use the yaxs = "i" instruction to 'ground' the bars:

```
> plot(gb.oct, bar.col = "gray90", yaxs = "i")
```

6. The plot shows a histogram of the frequencies of the species in the different octaves. The line shows the fitted distribution. The vertical line shows the mode and the horizontal line the standard deviation of the response. Add the details for the alternative model using the lines() command:

```
> lines(gb.ll, line.col = "blue", lty = 2)
```

7. Now examine the density distribution of the histogram and add the density line to the plot; your final graph should resemble Figure 11.11:

```
> den = density(log2(colSums(gb.biol)))
> lines(den$x, ncol(gb.biol)*den$y, lwd = 2, col = "darkgreen", lty = 3)
```

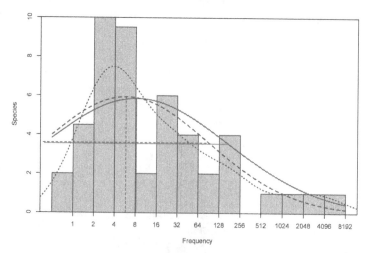

Figure 11.11 Preston's log-series for ground beetle communities. Solid line = quasi-Poisson model, dashed line = log likelihood model, dotted line = density.

8. You can see that neither model is a really great fit! Have a look at the potential number of 'unseen' species:

```
> veiledspec(gb.oct)
Extrapolated    Observed      Veiled
    64.37529    48.00000    16.37529

> veiledspec(gb.ll)
Extrapolated    Observed      Veiled
    59.01053    48.00000    11.01053
```

9. The Preston model is truncated (usually at both ends) and so other estimates of 'unseen' species may be preferable, try the `specpool()` command:

```
> specpool(gb.biol)
       Species    chao chao.se    jackl jackl.se    jack2    boot
All        48 53.33333 4.929127 55.55556 3.832931 57.64706 51.81422
       boot.se       n
All 3.021546      18
```

If you want a sample-by-sample estimate of unseen species then try the `estimateR()` command wrapped in `apply()` e.g. `apply(gb.biol, 1, estimateR)`. Look back to Section 7.3.4 for more details.

Tip: Axis extensions

By default R usually adds a bit of extra space to the ends of the x and y axes (about 4% is added to the ends). This is controlled by the `xaxs` and `yaxs` graphical parameters. The default is `xaxs = "r"` and the same for `yaxs`. You can 'shrink' the axes by setting the value to `"i"`. for the axis you require.

You can convert species count data to Preston octave data by using the `as.preston()` command:

```
> as.preston(colSums(gb.biol))
   0    1    2    3    4    5    6    7    8   10   11   12   13
 2.0  4.5 10.0  9.5  2.0  6.0  4.0  2.0  4.0  1.0  1.0  1.0  1.0
attr(,"class")
[1] "preston"
```

This makes a result object that has a special class, `"preston"`, which currently does not have any specific commands associated with it. However, you could potentially use such a result to make your own custom functions.

11.4 Summary

Topic	Key Points
RAD or dominance-diversity models	Rank-abundance dominance models arrange species in order of abundance. and show log(abundance) against the rank of abundance. The flatter the curve the more diverse the community is. Various models have been proposed to help explain the observed patterns of dominance plots.
	The `radfit()` command in the *vegan* package calculates a range of RAD models.

Broken stick model	Broken stick: This gives a null model where the individuals are randomly distributed among observed species, and there are no fitted parameters. The model is seen as an ecological resource-partitioning model. The `rad.null()` command calculates broken stick models.
Preemption model	The preemption model (also called Motomura model or geometric series) has a single fitted parameter. The model tends to produce a straight line. The model is seen as an evolutionary resource-partitioning model. The `rad.preempt()` command calculates preemption models.
Lognormal model	The lognormal model has two fitted parameters. It assumes that the log of species abundance is normally distributed. The model implies that resources affect species simultaneously. The `rad.lognormal()` command calculates lognormal models.
Zipf model	In the Zipf model there are two fitted parameters. The model implies that resources affect species in a sequential manner. The `rad.zipf()` command calculates Zipf models.
Mandelbrot model	The Mandelbrot model adds one parameter to the Zipf model. The model implies that resources affect species in a sequential manner. The `rad.zipfbrot()` command calculates Mandelbrot models.
Comparing models	The result of the `radfit()` command contains information about deviance and AIC values for the models, allowing you to select the 'best'. The AIC values and 'best' model are also displayed when you plot the result of `radfit()` that contains multiple samples. If you have replicate samples you can compare deviance between models using ANOVA.
Visualise RAD models with DD plots	Once you have your RAD model(s), which can be calculated with the `radfit()` command, you can visualise them graphically. The result of `radfit()` has plot() routines, the *lattice* package is used to display some models, for example the lattice system will display the 'best' model for each sample if `radfit()` is used on a dataset with multiple samples. Plotting a `radfit()` result for a single sample produces a single plot with the model fits overlaid for comparison. The `radlattice()` command will display the models in separate panels (using the *lattice* package) for a single sample.
Fisher's log-series	Fisher's log-series was an early attempt to model species abundances. One of the model parameters (*alpha*) can be used as an index of diversity, calculated via the `fisher.alpha()` command in the *vegan* package. The `fisherfit()` command will compute Fisher's log-series (using non-linear modelling). Use the `confint()` command in the *MASS* package to calculate confidence intervals for Fisher's log-series. The `as.fisher()` command in the *vegan* package allows you to convert abundance data into Fisher's log-series data.

Topic	Key Points
Preston's lognormal	Preston's lognormal model is an 'improvement' on Fisher's log-series, which implied infinite species richness. The frequency classes of the *x*-axis are collapsed and merged into wider bands, creating octaves of doubling size, 1, 2, 3–4, 5–8, 9–16 and so on. Furthermore, for each frequency half the species are transferred to the next highest octave.
	The `prestonfit()` and `prestondistr()` commands in the *vegan* package compute Preston's log-series. The commands have plotting routines allowing you to visualise the models.
	You can convert species count data to Preston octave data by using the `as.preston()` command.

11.5 Exercises

11.1 Rank abundance-dominance models derive from biological or statistical theories. The (biological) resource-partitioning models can be thought of as operating in two broad ways – what are they?

11.2 The Mandelbrot RAD model is a derivative of the Zipf model, two additional parameters being added – TRUE or FALSE?

11.3 The *DeVries* data is a matrix containing two samples; counts of tropical butterfly species in canopy or understorey habitat. Which is the 'best' model for each habitat?

11.4 The `radfit()` and `rad.xxxx()` commands assume that your data are count data. If you had cover data instead how might you modify the command(s)?

11.5 How many unseen (i.e. veiled) species are calculated to be in the *DeVries* data?

The answers to these exercises can be found in Appendix 1.

12. Similarity and cluster analysis

The themes of *similarity* and *clustering* are closely allied. You can think of one leading into the other – you start by assessing how similar community samples are to one another and then form groups based on the similarity. These groups (that is, clusters) can show you how related samples are to one another, especially if you draw them in some kind of graphical output.

12.1 Similarity and dissimilarity

When you look at sample similarity you are attempting to give some numerical value to how 'close' two samples are to one another based on their species composition.

If you have simple site lists you only have presence or absence data to work with. Your methods of assessing the community similarity will focus on how many shared species there are. If you have abundance data of any kind then your assessment of similarity can be more 'sensitive' because you have more 'sensitive' data, the abundance.

Measures of similarity are used in many analytical methods as a starting point – you saw them used in analysis of *beta diversity* for example in Section 10.7. Actually it is more common to use measures of *dissimilarity*. In a measure of similarity a large value indicates that two samples are close to one another, in a measure of dissimilarity a large value indicates that two samples are far apart.

12.1.1 Presence-absence data

If you have simple species lists as your data you obviously have presence-absence (i.e. binary) data. Your only way to assess the similarity of two samples is to examine the number of shared species.

Many methods of assessment exist, and they all use three basic quantities:

- A – the number of species in sample A.
- B – the number of species in sample B.
- J – the number of species shared by both samples.

Three commonly used measures of similarity are, Jaccard, Sørensen and Mountford indices. The Jaccard index uses the A, B and J quantities to produce an index that varies from 0–1 as shown in Figure 12.1.

$$\text{Jaccard} = \frac{J}{A + B - J}$$

Figure 12.1 The Jaccard similarity index for assessing sample similarity. A = # species in sample A, B = # species in sample B, J = # shared species. Values approaching 1 denote very similar samples.

The Sørensen index (also sometimes known as the Bray–Curtis index) uses the same quantities but in a slightly different way (Figure 12.2).

$$\text{Sorensen} = \frac{2J}{A + B}$$

Figure 12.2 The Sørensen index of similarity. A, B = # species in each sample, J = # shared species. This index is sometimes called the Bray–Curtis index.

The Mountford index is somewhat more complicated, an approximation to the formula is shown in Figure 12.3.

$$\text{Mountford} = \frac{2J}{2AB - J(A + B)}$$

Figure 12.3 The Mountford index of similarity for binary data. J = # shared species, A and B are # species in each sample.

It is easy enough to calculate these measures of similarity using Excel and in the following exercise you can have a go at working out similarity for three samples of data taken from the diet of hornbills in India.

Have a Go: Use Excel to calculate similarity between samples using presence-absence data

For this exercise you will need the *Hornbill fruits.xls* data. The data show three samples, one for each of three species of hornbill (frugivorous birds) from India (adapted from Datta & Rawat 2003). The diets of each species were examined and the results show the presence or absence of the various fruit species in their diets.

1. Open the spreadsheet *Hornbill fruits.xls*. Columns D:F show the presence or absence of the various fruit species. Go to cell C40 and type a label 'Richness'; you will calculate the number of fruit species for each hornbill shortly.

2. In cell D40 type a formula to calculate the number of fruit species found in the diet of the GH species (Great Hornbill), =SUM(D2:D38). Now copy this across into cells E40 and F40 so that you have values for all three hornbills.

3. Now you need to work to the number of shared species for each pair of samples. You'll need to make a grid for this. Start by copying the heading labels from cells D1:F1 to the clipboard. Paste the labels into cells D42:F42. Complete the grid by also clicking on cell C43. Use *Paste Special* (right click in the cell) and when the *Paste Special* menu appears select the Transpose button before clicking OK.

4. You should now have a grid with labels above and to the left. The diagonal represents the similarities between a sample and itself so you can leave this blank. You will calculate the number of shared species and fill in the lower triangle of the grid. Start by clicking in cell D44, which represents the similarity between GH and WH samples.

5. In cell D44 type a formula to calculate the number of shared species for the GH and WH samples: =SUMPRODUCT(D2:D38,E2:E38). The SUMPRODUCT function multiplies one column by the other, one row at a time, and then sums the results. You will only get a 1 if both columns contain a 1 because $1 \times 0 = 0$, $0 \times 1 = 0$ and $0 \times 0 = 0$.

6. Now go to cell D45 and type in a similar formula to step 5 to work out the shared species for the GH-OPH pair of samples: =SUMPRODUCT(D2:D38,F2:F38). You cannot simply copy the formula down from the cell above. You can click in the formula bar and then select and copy the actual formula, which you can then edit.

7. In cell E45 you need a formula to work out the shared species for the WH-OPH pair: =SUMPRODUCT(E2:E38,F2:F38).

8. Now you need to work out the Jaccard indices for each of the three pairs of samples. You can use the upper triangle of the grid. Start by clicking in cell E43 and type a formula to work out the Jaccard index for the GH-WH pairing: =D44/(D40+E40-D44).

9. In cell F43 type a formula to work out the Jaccard index for the GH-OPH pair: =D45/(D40+F40-D45).

10. In cell F44 type a formula to work out the Jaccard index for the WH-OPH pair: =E45/(E40+F40-E45).

You can see that the similarity for the WH-OPH pair is rather smaller than the other two.

It is awkward to calculate the matrix of similarities using Excel because you have to copy formulae down rows as well as across columns. This makes it impossible to simply copy and paste in the conventional way. When you have only a few samples then this is not too onerous a task but if you have a lot of samples it becomes more challenging.

It is not trivial to get Excel to produce a meaningful graphic to visualise the similarities and it is a lot simpler to use R to handle this (see Section 12.2).

Using R to calculate similarity indices

You can calculate similarities using R quite easily. The dist() command in the basic distribution of R will allow you to work out some measures of dissimilarity. More useful is the vegdist() command in the *vegan* package, which enables you to calculate several indices of dissimilarity using presence-absence data.

You usually start with a community sample that has rows as samples and columns as species. The vegdist() command creates a result that is a matrix of dissimilarities. To calculate Bray–Curtis (Sørensen) dissimilarities for presence-absence data you would type:

```
vegdist(x, method = "bray", binary = TRUE)
```

You set binary = TRUE to use presence-absence data. In the following exercise you can have a go at calculating dissimilarities using the diet of hornbills in India.

Have a Go: Use R to calculate dissimilarity indices for presence-absence data

For this exercise you'll need the *vegan* package and the *fruit* data, which is part of the *CERE.RData* file.

1. Start by preparing the *vegan* package:

```
> library(vegan)
```

2. Look at the *fruit* data (adapted from Datta & Rawat 2003), where you'll see three columns, one for each hornbill species:

```
> head(fruit)
           GH WH OPH
Acti.obov  0  1   0
Acti.angu  0  1   0
Alse.pedu  1  1   0
Amoo.wall  1  1   1
Apha.poly  1  1   0
Beil.assa  1  1   0
```

3. To compare the diets of the birds you will need to rotate the data so that the rows are the birds and the columns are the species of fruit in the diet – use the t() command, which produces a matrix:

```
> hf <- t(fruit)
> class(hf)
[1] "matrix"

> hf[,1:5]
     Acti.obov Acti.angu Alse.pedu Amoo.wall Apha.poly
GH           0         0         1         1         1
WH           1         1         1         1         1
OPH          0         0         0         1         0
```

4. Use the vegdist() command to work out the Jaccard index of dissimilarity:

```
> hf.jac <- vegdist(hf, method = "jaccard")
> hf.jac
           GH        WH
WH  0.6562500
OPH 0.6842105 0.8235294
```

5. The dist() command can also determine the Jaccard index, use it to work out the index of similarity:

```
> hf.bin <- 1 - dist(hf, method = "binary")
> hf.bin
           GH        WH
WH  0.3437500
OPH 0.3157895 0.1764706
```

The dist() and vegdist() commands always produce a measure of dissimilarity, where the larger the value the more different a pair of sample are to one another.

Both the `dist()` and `vegdist()` commands can calculate other dissimilarity indices using presence-absence data, Table 12.1 shows some of the indices that `vegdist()` can compute, along with the formulae.

Table 12.1 Dissimilarity indices calculated by the `vegdist()` command in the *vegan* package. A and B refer to number of species in each sample, J is the number of shared species.

Method	Calculation
`manhattan`	$A + B - 2J$
`euclidean`	$\sqrt{(A + B - 2J)}$
`canberra`	$(A + B - 2J)/(A + B - J)$
`bray`	$(A + B - 2J)/(A + B)$
`kulczynski`	$1 - (J/A + J/B)/2$
`jaccard`	$2b/(1 + b)$ This index is calculated from the Bray–Curtis index, b = Bray–Curtis index.
`gower`	$(A + B - 2J)/M$ M = the number of columns.
`altGower`	$(A + B - 2J)/(A + B - J)$
`horn`	$(A + B - 2J)/(A + B)$
`mountford`	Mountford index is defined as $M = 1/a$ where a is the parameter of Fisher's log-series assuming that the compared communities are samples from the same community. The index M is found as the positive root of equation $\exp(A * M) + \exp(B * M) = 1 + \exp((A + B - J) * M)$.
`raup`	The Raup–Crick dissimilarity is a probabilistic index based on presence-absence data. It is defined as $1 - \text{prob}(J)$, or based on the probability of observing at least J species in shared in compared communities.
`binomial`	$\log(2) \times (A + B - 2J)$

You will meet the `dist()` and `vegdist()` commands again shortly (Section 12.1.2) when you'll see how to determine sample similarity (or rather, dissimilarity) when you have abundance data.

Using indices of beta diversity as dissimilarities

Many of the indices of dissimilarity are used in calculations of *beta* diversity. You can use the `betadiver()` command (Section 10.1.1) in the *vegan* package to calculate a range of indices using presence-absence data.

You can see the range of indices that can be calculated by adding `help = TRUE` to the command like so:

```
> betadiver(help = TRUE)
```

These algorithms correspond to those reviewed in Koleff *et al.* (2003). Some are the same (e.g. 1 and 2, 8 and 9) and all give subtly different results. The most commonly used for *beta* diversity is that of Whittaker (1960), which corresponds to the `method = "w"`. In the following exercise you can have a go at calculating some of these dissimilarity indices for yourself.

Have a Go: Use indices of *beta* diversity as dissimilarity indices

For this exercise you'll need the *vegan* package and the *fruit* data, which is part of the *CERE.RData* file.

1. Start by preparing the *vegan* package:

```
> library(vegan)
```

2. The *fruit* data (adapted from Datta & Rawat 2003) are arranged in three columns, one for each species of hornbill. The rows represent the presence or absence of species of fruit in their diet:

```
> head(fruit)
          GH WH OPH
Acti.obov  0  1   0
Acti.angu  0  1   0
Alse.pedu  1  1   0
Amoo.wall  1  1   1
Apha.poly  1  1   0
Beil.assa  1  1   0
```

3. To compare the diets of the birds you will need to rotate the data so that the rows are the birds and the columns are the species of fruit in the diet – use the t() command, which produces a matrix:

```
> hf <- t(fruit)
> hf[,1:5]
    Acti.obov Acti.angu Alse.pedu Amoo.wall Apha.poly
GH          0         0         1         1         1
WH          1         1         1         1         1
OPH         0         0         0         1         0
```

4. Use the betadiver() command to work out a dissimilarity using the "g" method:

```
> betadiver(hf, method = "g")
            GH        WH
WH  0.6562500
OPH 0.6842105 0.8235294
```

5. Now use the "gl" method:

```
> betadiver(hf, method = "gl")
            GH        WH
WH  0.6976744
OPH 0.2400000 0.9000000
```

6. You can specify the number of the method (from the help = TRUE list), try #24, which corresponds to the "z" method:

```
> betadiver(hf, method = 24)
            GH        WH
WH  0.5737352
OPH 0.6040713 0.7655347
```

In addition to the methods computed by the `betadiver()` command you can compute your own index using the `designdist()` command in the *vegan* package, you'll see this in Section 12.1.3.

12.1.2 Abundance data

When you have abundance data, rather than simple presence or absence, you are able to determine the similarity (or rather, dissimilarity) more sensitively. There are many indices of dissimilarity – Figure 12.4 shows how the *Euclidean dissimilarity* is determined.

$$E_d = \sqrt{\sum \left(x_i - x_j\right)^2}$$

Figure 12.4 Euclidean dissimilarity. Terms x_i and x_j refer to the abundance of each species at sites i and j.

Another commonly used index is the Bray–Curtis dissimilarity; the formula is shown in Figure 12.5. There are several other commonly used indices; all use the terms x_i and x_j to represent the abundance of each species at sites i and j.

$$BC_d = \frac{\sum \left| x_i - x_j \right|}{\sum \left(x_i + x_j\right)}$$

Figure 12.5 Bray–Curtis dissimilarity. Terms x_i and x_j refer to the abundance of each species at sites i and j.

The Euclidean dissimilarity is one of the easiest indices to determine and it is easy to do this in Excel. In the following exercise you can have a go at calculating dissimilarities between from samples of forest trees from different areas of Brazil.

Have a Go: Use Excel to calculate Euclidean dissimilarities using abundance data

For this exercise you'll need the spreadsheet *Brazil tree density.xls*. This contains data on the densities of tree species from four types of forest area in Brazil (adapted from Periera 2003). Essentially the areas have experienced different levels of disturbance ranging from very little to completely cleared and regrown after periods of agriculture.

1. Open the spreadsheet and navigate to cell C55. Type a label to determine the species richness of the LD sample 'Richness'.

2. Now in cell D55 type a formula to work out the species richness of the LD sample: =COUNTIF(D2:D53,">0"). Note that you cannot simply add up the values or count cells (the sum is close to 100 and there are cells containing zero).

3. Copy the formula in cell D55 across to cells E55:G55 so that you have the richness for all four samples.

4. Now copy the cells D1:G1 to the clipboard. Paste them into cells D57:G57 to make a heading for the dissimilarity grid. Also use *Paste Special* to copy the labels into cells C58:C61, you can tick the *Transpose* box to do this from the *Paste Special* menu.

5. In cell D59 type a formula to work out the Euclidean distance between the LD-GF pair of samples: =SQRT(SUMXMY2(D2:D53,E2:E53)).

6. In cell D60 type a formula to work out the Euclidean distance between the LD-OR pair of samples: =SQRT(SUMXMY2(D2:D53,F2:F53)).

7. Complete the rest of the dissimilarity matrix/grid. You can save a lot of typing by copying the formula in its entirety (include the =) from one cell. Then paste into the appropriate cell and edit the formula to alter the column letters.

You can see readily that the lowest value is the distance between the LD-GF pair, these have had the least disturbance.

The Euclidean distance is the easiest to calculate using Excel because you can do it readily using a single formula. The Bray–Curtis distance is not intrinsically difficult to compute but is fiddly and needs intermediate steps. It is much easier to use R to conduct your calculations.

Using R to calculate dissimilarity indices

It is generally easier to determine the dissimilarities using R, rather than Excel. Furthermore, you will be able to use the results in additional analyses, such as *beta* diversity. It is also easy to take a dissimilarity matrix and create a graphic that shows the relationship between the samples – a *dendrogram*.

The dist() command in the basic distribution of R can determine several dissimilarity indices. The general form of the command is:

```
dist(x, method = "euclidean", diag = TRUE, upper = FALSE, p = 2)
```

The command can accept several instructions:

- x – a community dataset, usually a data.frame or a matrix. Rows represent the samples and the columns are the abundances of the individual species.
- method – the kind of index to calculate. The default is "euclidean", other options are "maximum", "manhattan", "canberra", "binary", and "minkowski".
- diag – by default the diagonal of the distance matrix is not shown, the diagonal will always consist of zeroes.
- upper – by default only the lower portion of the dissimilarity matrix is shown.
- p – the power of the Minkowski distance, the default is p = 2.

Note: Minkowski distance

If you set p = 1 for method = "minkowski" the result is equivalent to the Manhattan distance. If you set p = 2 the result is equivalent to the Euclidean distance. The Minkowski distance is defined as:

$$(\Sigma |x_i - -x_j|^p)^{1/p}$$

The `vegdist()` command in the *vegan* package calculates dissimilarities and uses similar instructions:

```
vegdist(x, method="bray", binary=FALSE, diag=FALSE, upper=FALSE)
```

The default dissimilarity for `vegdist()` is the Bray–Curtis index – other options are: `"manhattan"`, `"euclidean"`, `"canberra"`, `"kulczynski"`, `"jaccard"`, `"gower"`, `"altGower"`, `"morisita"`, `"horn"`, `"mountford"`, `"raup"`, `"binomial"` and `"chao"`. The Jaccard, Mountford and Raup–Crick indices are binary – determined from presence-absence data. The other indices use abundance data but their presence-absence alternatives can be calculated by setting `binary = TRUE`.

In Table 12.2 you can see the formulae for all the dissimilarity indices. In the table x_i and x_j represent the abundances of each species at sites i and j.

In the following exercise you can have a go at calculating dissimilarities between from samples of forest trees from different areas of Brazil.

Table 12.2 Dissimilarity indices calculated by the `vegdist()` command in the vegan package.

Method	Calculation
manhattan	$\Sigma\lvert x_i - x_j \rvert$
euclidean	$\sqrt{\Sigma(x_i - x_j)^2}$
canberra	$1/N_Z \times \Sigma[\lvert x_i - x_j\rvert/(x_i + x_j)]$ N_Z = number of non-zero entries.
bray	$\Sigma\lvert x_i - x_j\rvert/\Sigma(x_i + x_j)$
kulczynski	$1 - 0.5 \times (\Sigma\min(x_i, x_j)/\Sigma x_i + \Sigma\min(x_i, x_j)/\Sigma x_j)$
jaccard	$2b/(1 + b)$ This index is calculated from the Bray-Curtis index, b = Bray-Curtis index.
gower	$1/M \times \Sigma\lvert x_i - x_j\rvert/[\max(i, j) - \min(i, j)]$ M = the number of columns, $\max(i, j)$ is maximum number of species (min = the minimum).
altGower	$1/N_Z \times \Sigma\lvert x_i - x_j\rvert$ N_Z = number of non-zero entries.
morisita	$1 - 2\Sigma(x_i \times x_j)/[(\lambda_i + \lambda_j)\,\Sigma x_i\,\Sigma x_j]$ where $\lambda i = \Sigma x_i\,(x_i - 1)/[\Sigma x_i\,(\Sigma x_i - 1)]$ The index requires integer abundance values (e.g. count data).
horn	$1 - 2\Sigma(x_i \times x_j)/[(\lambda_i + \lambda_j)\,\Sigma x_i\,\Sigma x_j]$ where $\lambda i = \Sigma x_i\,2/(\Sigma x_i)2$
binomial	$\Sigma[x_i\log(x_i/n) + x_j\log(x_j/n) - n\log(1/2)]/n$ where $n = x_i + x_j$. The binomial index is derived from binomial deviance under null hypothesis that the two compared communities are equal. It should be able to handle variable sample sizes. The index does not have a fixed upper limit, but can vary among sites with no shared species.
chao	The Chao index tries to take into account the number of unseen species pairs. $1 - U_i \times U_j/(U_i + U_j - U_i \times U_j)$ where $U_i = C_i/N_i + (N_j - 1)/N_j * A1/(2*A2) * S_i/N_i$, and similarly for U_j. C_i is the total number of individuals in the species of site i that are shared with site j, N_i is the total number of individuals at site i, $A1$ (and $A2$) are the number of species occurring in site i that have only one (or two) individuals in site j, and S_i is the total number of individuals in the species present at site i that occur with only one individual in site j. The index requires integer values (e.g. count data).

Have a Go: Use R to calculate dissimilarities between samples with abundance data

You will need the *brasil* data for this exercise. This contains data on the densities of tree species from four types of forest area in Brazil (adapted from Periera 2003). Essentially the areas have experienced different levels of disturbance ranging from very little to completely cleared and regrown after periods of agriculture. The data are in the *CERE. RData* file. You'll also need the *vegan* package.

1. Start by preparing the *vegan* package:

    ```
    > library(vegan)
    ```

2. Look at the *brasil* data: the columns show the samples and the rows are the species:

    ```
    > head(brasil)
                LD   GF   OR NR
    Myra.urun 0.38 0.36 0.0  0
    Schi.bras 1.49 1.45 0.2  0
    Spon.tube 0.27 0.00 0.2  0
    Aspi.pyri 6.31 3.99 0.0  0
    Tabe.impe 2.94 2.36 0.0  0
    Ceib.glaz 0.21 1.27 0.0  0
    ```

3. Rotate the data so that the rows are the samples and the columns the species:

    ```
    > td <- t(brasil)
    > td[,1:5]
       Myra.urun Schi.bras Spon.tube Aspi.pyri Tabe.impe
    LD      0.38      1.49      0.27      6.31      2.94
    GF      0.36      1.45      0.00      3.99      2.36
    OR      0.00      0.20      0.20      0.00      0.00
    NR      0.00      0.00      0.00      0.00      0.00
    ```

4. Examine the Euclidean dissimilarities using the dist() command:

    ```
    > dist(td, method = "euclidean")
             LD        GF        OR
    GF 22.00358
    OR 46.19560 45.61583
    NR 78.76361 80.31157 77.50146
    ```

5. Use the vegdist() command to make a matrix showing the Bray-Curtis dissimilarities. Present the diagonals and the upper portion of the matrix:

    ```
    > vegdist(td, method = "bray", diag = TRUE, upper = TRUE)
              LD        GF        OR        NR
    LD 0.0000000 0.4116528 0.7319267 0.8837070
    GF 0.4116528 0.0000000 0.7353222 0.9272636
    OR 0.7319267 0.7353222 0.0000000 0.6846915
    NR 0.8837070 0.9272636 0.6846915 0.0000000
    ```

6. Now calculate the Gower index:

    ```
    > vegdist(td, method = "gower")
    ```

```
            LD          GF          OR
GF 0.5625369
OR 0.7001403 0.5612983
NR 0.6766782 0.4890798 0.2778097
```

7. Try calculating the Morisita index:

```
> vegdist(td, method = "morisita")
            LD          GF          OR
GF 0.1975058
OR 0.6490475 0.6164401
NR 0.9360652 0.9628954 0.7281461
Warning message:
In vegdist(td, method = "morisita") :
   results may be meaningless with non-integer data in method
morisita
```

You get a warning with the Morisita index because you can only calculate it using count data (integers). The Horn–Morisita variant however (method = "horn") is able to use non-integer data.

Tip: Accessing help entries

You can see details of the way the vegdist() command computes the various dissimilarities by accessing the help entry for the command. Use help(vegdist) if the *vegan* package is loaded. If the package is not loaded then use help(vegdist, package = vegan), of course the *vegan* package must be installed on your computer.

12.1.3 Which dissimilarity index to use?

As you have seen, there are quite a few indices of dissimilarity. Some of these indices have been 'invented' more than once and so the scientific literature knows them by several names. For instance, the Bray index is known also as Steinhaus, Czekanowski and Sørensen index. The quantitative version of Jaccard is also known as the Ružička index. The abbreviation "horn" for the Horn–Morisita index in the *vegan* package is misleading, since there is a separate Horn index.

Deciding which index you should use is a tricky process. The choice between a binary index and quantitative is simple enough – if you have presence-absence data then you use a binary index. However, this still leaves you with many options.

Gower, Bray–Curtis, Jaccard and Kulczynski indices are good in detecting underlying ecological gradients. Morisita, Horn–Morisita, binomial and Chao indices should be able to handle different sample sizes. Raup-Crick indices for presence-absence data should be able to handle unknown (and variable) sample sizes.

Euclidean and Manhattan dissimilarities are not good in gradient separation unless the community data have been standardised, which you can do with a separate command decostand() in the *vegan* package – this was used in the analysis of *beta* diversity (Chapter 10) and you will meet it again shortly.

Bray–Curtis and Jaccard indices are rank-order similar, and some other indices become identical or rank-order similar after some standardisations, especially with presence-absence transformation of equalising site totals with the decostand() command. The Jaccard index is metric, and probably should be preferred instead of the default Bray–Curtis which is semi-metric.

The *vegan* package provides two versions of the Gower index ("gower", "altGower") which differ in scaling: "gower" divides all distances by the number of observations (rows) and scales each column to unit range, but "altGower" omits double-zeros and divides by the number of pairs with at least one above-zero value, and does not scale columns.

You can use the decostand() command to add range standardisation to "altGower" (see next section). Although it has been recommended to remove double zeroes from data before calculating the index is often taken as the general feature of the Gower distances that they are left in place.

In the end the choice comes down to your own preference! If you are comparing to a previous study then you may wish to use the same index as that previous study for direct comparison. On the other hand it may be sensible to use more than one index and to compare the results – if your conclusions are the same for each result you can have more 'confidence' in your conclusions.

Standardising communities

Community data can of course be very variable: one sample may have numbers of individuals at several orders of magnitude greater than another. This can cause some problems – in the study of diversity partitioning for example, the *alpha*, *beta* and *gamma* components may not partition 'correctly' if there are large differences in abundances between samples. In Chapter 10 you saw how the decostand() command was able to carry out a standardisation and rescale the communities to overcome this issue.

The decostand() command can carry out various standardisation processes, the general form of the command is:

```
decostand(x, method, MARGIN, range.global, logbase = 2, na.rm=FALSE)
```

The command can accept various instructions:

- x – the community data, usually a data.frame or matrix object.
- method – the method of standardisation to use (see Table 12.3).
- MARGIN – set 1 for rows and 2 for columns of the data, each method has its own default.
- range.global – matrix from which the range is found in method = "range". This allows using same ranges across subsets of data. The dimensions of MARGIN must match with x.
- logbase – the base of the logarithm to use in method = "log".
- na.rm – if set to TRUE missing values in the data are ignored.

The command can conduct various methods. In Table 12.3 you can see these methods along with a brief explanation of how the standardisation is carried out.

Table 12.3 Methods of standardisation utilised by the decostand() command.

Method	Details
total	Divides by the row totals (i.e. MARGIN = 1).
max	Divides by maximum value in each column (i.e. MARGIN = 2).
freq	Divides by column totals and multiplies by number of non-zero items so that mean of non-zero entries is 1 (i.e. MARGIN = 2).
normalize	Makes row sums of squares equal to 1 (i.e. MARGIN = 1).
range	Standardises values in columns (i.e. MARGIN = 2) to the range 0–1 (if all values are constant they will become 0).
standardize	Scales columns to mean of 0 and variance of 1 (i.e. MARGIN = 2).

Table 12.3 Continued

Method	Details
pa	Converts values to presence-absence (i.e. 0 or 1).
chi.square	Divides by row sums (i.e. MARGIN = 1) and square root of column sums, then adjusts for square root of the grand total.
hellinger	Divides by the row totals and then takes the square root (i.e. MARGIN = 1).
log	Logarithmic transformation i.e. log_b (x) + 1, for values > 0, the default is logbase = 2.

These methods of standardisation can help overcome the difficulties caused by differences in samples sizes and species abundances across your samples. Euclidean and Manhattan distances in particular seem to benefit from standardisation. In the following exercise you can have a go at carrying out some standardisations for yourself.

Have a Go: Carry out sample standardisation on community data

For this exercise you'll need the *vegan* package. You will make the example data yourself using random values.

1. Start by preparing the *vegan* package:

    ```
    > library(vegan)
    ```

2. Set the random number generator and make 45 values. Split the values into five samples and assign labels to the rows and columns of the data matrix:

    ```
    > set.seed(1)
    > rv = runif(5*9, min = -1, max = 4)
    > rv = matrix(floor(10^rv), nrow = 5)
    > dimnames(rv) = list(LETTERS[1:5], letters[1:9])
    ```

3. Look at the values you just made. They are across three orders of magnitude:

    ```
    > rv
         a    b    c    d    e    f    g    h    i
    A    2 3104    1   30 4715    8   25  219 1272
    B    7 5289    0  387    1    0   99  935  171
    C   73  201  272 9110  181    8   29    0  821
    D 3475  139    8    7    0 2230    0  415   58
    E    1    0  706  771    2    5 1370   11   44
    ```

4. Standardise the community data by the column maxima:

    ```
    > print(decostand(rv, method = "max"), digits = 2)
            a      b      c       d       e      f     g     h     i
    A 0.00058 0.587 0.0014 0.00329 1.00000 0.0036 0.018 0.234 1.000
    B 0.00201 1.000 0.0000 0.04248 0.00021 0.0000 0.072 1.000 0.134
    C 0.02101 0.038 0.3853 1.00000 0.03839 0.0036 0.021 0.000 0.645
    D 1.00000 0.026 0.0113 0.00077 0.00000 1.0000 0.000 0.444 0.046
    E 0.00029 0.000 1.0000 0.08463 0.00042 0.0022 1.000 0.012 0.035
    attr(,"decostand")
    [1] "max"
    ```

5. Convert the data to presence-absence using the "pa" method:

```
> decostand(rv, method = "pa")
  a b c d e f g h i
A 1 1 1 1 1 1 1 1 1
B 1 1 0 1 1 0 1 1 1
C 1 1 1 1 1 1 1 0 1
D 1 1 1 1 0 1 0 1 1
E 1 0 1 1 1 1 1 1 1
attr(,"decostand")
[1] "pa"
```

6. Make the row sums of squares equal to unity:

```
> print(decostand(rv, method = "normalize"), digits = 2)
       a      b       c      d       e       f       g      h     i
A 0.00035 0.536 0.00017 0.0052 0.81422 0.00138 0.0043 0.0378 0.220
B 0.00130 0.982 0.00000 0.0718 0.00019 0.00000 0.0184 0.1735 0.032
C 0.00797 0.022 0.02971 0.9951 0.01977 0.00087 0.0032 0.0000 0.090
D 0.83684 0.033 0.00193 0.0017 0.00000 0.53702 0.0000 0.0999 0.014
E 0.00058 0.000 0.40953 0.4472 0.00116 0.00290 0.7947 0.0064 0.026
attr(,"decostand")
[1] "normalize"
```

7. Use a logarithmic transformation of the data:

```
> print(decostand(rv, method = "log", logbase = 2), digits = 4)
      a       b        c       d      e       f       g      h      i
A  2.000 12.600   1.000   5.907 13.20  4.000   5.644  8.775 11.313
B  3.807 13.369   0.000   9.596  1.00  0.000   7.629 10.869  8.418
C  7.190  8.651   9.087  14.153  8.50  4.000   5.858  0.000 10.681
D 12.763  8.119   4.000   3.807  0.00 12.123   0.000  9.697  6.858
E  1.000  0.000  10.464  10.591  2.00  3.322  11.420  4.459  6.459
attr(,"decostand")
[1] "log"
```

8. Now rescale the species columns to zero mean and unit variance:

```
> sptrans <- decostand(rv, "standardize")
> print(sptrans, digits = 3)
       a      b      c      d      e      f      g      h      i
A -0.459  0.573 -0.639 -0.514  1.788 -0.444 -0.469 -0.251  1.454
B -0.456  1.496 -0.642 -0.423 -0.468 -0.452 -0.345  1.605 -0.550
C -0.413 -0.653  0.243  1.783 -0.382 -0.444 -0.462 -0.819  0.633
D  1.789 -0.679 -0.616 -0.520 -0.469  1.789 -0.510  0.257 -0.756
E -0.460 -0.738  1.655 -0.326 -0.468 -0.447  1.785 -0.791 -0.781
attr(,"scaled:center")
   a    b    c    d    e    f    g    h    i
 712 1747  197 2061  980  450  305  316  473
attr(,"scaled:scale")
    a    b    c    d    e    f    g    h    i
 1545 2367  307 3953 2089  995  597  386  549
attr(,"decostand")
[1] "standardize"
```

The decostand() command produces a result that it a matrix, which you can deal with as appropriate for your subsequent analytical tasks. The "standardize" method produces a more complicated result than the other methods, but it is still essentially a matrix.

Tip: Double standardisation

The `wisconsin()` command in the *vegan* package is a shortcut to the Wisconsin double standardisation procedure in which species are standardised by maxima and then by site totals. In other words the command runs `decostand(x, method = "max", MARGIN = 2)` then `decostand(x, method = "total", MARGIN = 1)`.

Design your own dissimilarity index

The dissimilarity indices that you can calculate using the `vegdist()` command in the *vegan* package are probably adequate for most tasks. If you have presence-absence data you can also use the `betadiver()` command to compute other indices (generally used in *beta* diversity), as you saw in Sections 10.1.1 and 12.1.1.

You can also compute any other index by using the `designdist()` command in the *vegan* package. The command allows you to use either presence-absence or abundance data, so you can make virtually any index. The command uses three basic measures, *A*, *B* and *J*, which can be combined as binary terms (for presence-absence) or in other ways (for quantitative data). You can also specify *N* for the number of sites (rows) and *P* for the number of species (columns). The general form of the command is as follows:

```
designdist(x, method, terms, name)
```

You need to supply the 'formula' as a value (in quotes) to the `method` instruction. How the terms A, B and J (as well as N and P) are combined depends on the value of the `terms` instruction:

- `"quadratic"`: $J = \Sigma(xy)$, $A = \Sigma(x^2)$, $B = \Sigma(y^2)$.
- `"minimum"`: $J = \Sigma pmin(x, y)$, $A = \Sigma x$, $B = \Sigma y$.
- `"binary"`: J = shared species, A = species in *x*, B = species in *y*.

The terms are used across pairs of rows (*x* and *y*) to produce a result similar to that of the `dist()` command. You can assign a name to your index by using the `name` instruction. In the following exercise you can have a go at making some dissimilarity indices for yourself.

Have a Go: Design your own dissimilarities

You will need the *brasil* data for this exercise. This contains data on the densities of tree species from four types of forest area in Brazil (adapted from Periera 2003). Essentially the areas have experienced different levels of disturbance ranging from very little to completely cleared and regrown after periods of agriculture. The data are in the *CERE. RData* file. You'll also need the *vegan* package.

1. Start by preparing the *vegan* package:

   ```
   > library(vegan)
   ```

2. Look at the *brasil* data, the columns show the samples and the rows are the species:

   ```
   > head(brasil)
                LD   GF   OR NR
   Myra.urun 0.38 0.36 0.0  0
   Schi.bras 1.49 1.45 0.2  0
   ```

```
Spon.tube 0.27 0.00 0.2  0
Aspi.pyri 6.31 3.99 0.0  0
Tabe.impe 2.94 2.36 0.0  0
Ceib.glaz 0.21 1.27 0.0  0
```

3. Rotate the data so that the rows are the samples and the columns the species:

```
> td <- t(brasil)
> td[,1:5]
    Myra.urun  Schi.bras  Spon.tube  Aspi.pyri  Tabe.impe
LD      0.38       1.49       0.27       6.31       2.94
GF      0.36       1.45       0.00       3.99       2.36
OR      0.00       0.20       0.20       0.00       0.00
NR      0.00       0.00       0.00       0.00       0.00
```

4. Make a simple Jaccard index using "binary" terms:

```
> designdist(td, method = "(A+B-2*J)/(A+B-J)", terms = "binary",
           name = "Jaccard")
          LD         GF         OR
GF 0.3265306
OR 0.6078431 0.5384615
NR 0.8163265 0.7777778 0.6800000
```

5. Use the same method but alter the terms to "minimum" to work out the Ružička index:

```
> designdist(td, method = "(A+B-2*J)/(A+B-J)", terms = "minimum",
           name = "Ružička")
          LD         GF         OR
GF 0.5832210
OR 0.8452167 0.8474763
NR 0.9382637 0.9622593 0.8128390
```

6. Now alter the terms to "quadratic" to work out a dissimilarity ratio:

```
> designdist(td, method = "(A+B-2*J)/(A+B-J)", terms = "quadratic",
           name = "Dissimilarity Ratio")
          LD         GF         OR
GF 0.4431076
OR 0.7997574 0.7765815
NR 0.9677035 0.9814919 0.8459096
```

7. Use the number of species to work out the Raup–Crick index. You'll need to use the phyper() command for this, which uses the hypergeometric distribution:

```
> designdist(td, method = "1-phyper(J-1, A, P-A, B)",
           terms = "binary", name = "Raup-Crick")
           LD         GF         OR
GF 0.11378336
OR 0.96729153 0.07325862
NR 0.83749192 0.24356157 0.01415359
```

8. Finally, look at the Arrhenius dissimilarity:

```
> designdist(td, "(log(A+B-J)-log(A+B)+log(2))/log(2)",
           terms = "binary", name = "Arrhenius")
```

```
            LD          GF          OR
GF 0.2571578
OR 0.5226782 0.4525122
NR 0.7567288 0.7104934 0.5994621
```

The Arrhenius dissimilarity is illustrated in the help entry for the `designdist()` command. It is not strictly reasonable to use it here because the samples are not from the same areas (although they've been standardised).

Once you have a dissimilarity result you can use it in a variety of analyses as if you had calculated it using the `dist()` or `vegdist()` commands. You've already seen how to use dissimilarities in analysis of *beta* diversity (Section 10.7). Later you will see how to use dissimilarities in ordination but first you will see how to carry out cluster analysis.

12.2   Cluster analysis

Samples that are most similar to one another (in their species composition) form clusters. When you carry out cluster analysis you are generally trying to work out how your samples break down into clusters and how the samples and clusters relate to one another.

There are two main approaches to clustering:

- *Hierarchical* – you split your data according to how similar the samples are. The most similar samples will be 'near' to one another. As samples become more dissimilar they are positioned further away. The result is hierarchical, you can create clusters at different hierarchical levels.
- *Partitioning* – you decide how many clusters there should be and partition the data by assigning each sample to a particular cluster.

In both approaches you will want to visualise the clusters and be able to assign each sample to a particular cluster.

R is able to handle clustering in various ways. The basic distribution of R comes with commands for hierarchical clustering and partitioning. In addition, the *cluster* package, which comes with the basic distribution of R, is able to handle several other methods.

In most cases you start by making a dissimilarity matrix from your community data and then go on to make the clusters. There are various ways to visualise the clusters and to examine cluster membership, as you will see shortly.

12.2.1 Hierarchical clustering

In an hierarchical clustering approach your samples end up being ordered in such a way that the most similar are close together and the least similar are far apart. The starting point for most methods of hierarchical clustering is the dissimilarity matrix. R can calculate a range of dissimilarities, as you have seen in Section 12.1.

The mainstay of hierarchical clustering is the `hclust()` command in conjunction with a matrix of dissimilarities. You can make your dissimilarities using the `dist()` command or the `vegdist()` command (in the *vegan* package). Once you have a matrix of dissimilarities you can use the `hclust()` command to make a cluster result object. The general form of the command is like so:

```
hclust(d, method)
```

Here d is a dissimilarity matrix and method is the kind of agglomeration method to use (you supply a name, in quotes). There are several choices of method: "ward", "single", "complete", "average", "mcquitty", "median" and "centroid".

The result is an object that contains various elements, which can be used to visualise the relationships between the samples. The result of the hclust() command has a class "hclust", for which there is a specific plotting method, accessed via the plot() command. In the following exercise you can have a go at making a cluster object and visualising it as a dendrogram.

Have a Go: Make a hierarchical clustering result and a dendrogram

You will need the *brasil* data for this exercise. This contains data on the densities of tree species from four types of forest area in Brazil (adapted from Periera 2003). Essentially the areas have experienced different levels of disturbance ranging from very little to completely cleared and regrown after periods of agriculture. The data are in the *CERE.RData* file. You'll also need the *vegan* package.

1. Start by preparing the *vegan* package:

    ```
    > library(vegan)
    ```

2. The *brasil* data are arranged with species as rows so rotate the data to make the rows the samples:

    ```
    > td = t(brasil)
    > td[,1:5]
         Myra.urun  Schi.bras  Spon.tube  Aspi.pyri  Tabe.impe
    LD      0.38       1.49       0.27       6.31       2.94
    GF      0.36       1.45       0.00       3.99       2.36
    OR      0.00       0.20       0.20       0.00       0.00
    NR      0.00       0.00       0.00       0.00       0.00
    ```

3. Make a Bray–Curtis dissimilarity matrix using the vegdist() command:

    ```
    > td.eu = vegdist(td, method = "bray")
    ```

4. Now use the Bray–Curtis dissimilarity matrix as the basis for a hierarchical cluster:

    ```
    > td.cl = hclust(td.eu, method = "complete")
    ```

5. The result of the hclust() command contains several elements, look at these with the str() command:

    ```
    > str(td.cl)
    List of 7
     $ merge      : int [1:3, 1:2] -1 -3 1 -2 -4 2
     $ height     : num [1:3] 0.412 0.685 0.927
     $ order      : int [1:4] 1 2 3 4
     $ labels     : chr [1:4] "LD" "GF" "OR" "NR"
     $ method     : chr "complete"
     $ call       : language hclust(d = td.eu, method = "complete")
     $ dist.method: chr "bray"
     - attr(*, "class")= chr "hclust"
    ```

6. Make a dendrogram of the result using the `plot()` command. Your result should resemble Figure 12.6:

```
plot(td.cl)
```

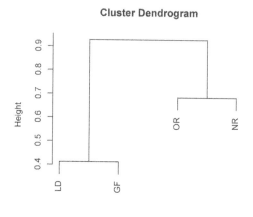

Figure 12.6 Hierarchical cluster relationship between samples of forest trees in Brazil via Bray–Curtis dissimilarity. LD = little disturbed, GF = grazed fragment, OR = old regrowth, NR = new regrowth.

This is a simple situation since Figure 12.6 only shows four samples. However, it serves to illustrate how the dendrogram is a useful tool for looking at sample relationships.

In the preceding exercise you were dealing with a situation with only four samples and you could probably spot the relationships without the dendrogram. However, when you have a more complicated situation the dendrogram becomes much more useful.

Splitting a dendrogram

When you have the result from the `hclust()` command the dendrogram shows the relationships between the samples. The samples are not really divided into definite clusters because at the end of each branch is a single sample. However, you are dealing with a hierarchical structure and so it is possible to 'step up' the hierarchy to make clusters by joining the various branches.

The `cutree()` command allows you to split a hierarchical cluster result and form clusters; you can specify a particular number of clusters or a level of dissimilarity. The general form of the command is:

```
cutree(tree, k = NULL, h = NULL)
```

In the `cutree()` command, `tree` is the result of the `hclust()` command, `k` is the number of required clusters and `h` = the cut height you require (you specify either `k` or `h`).

You can label the ends of the branches of your dendrogram with the groups to which they belong. In addition, you can highlight the groups by adding rectangles around them. In the following exercise you can have a go at making a dendrogram and splitting it into clusters.

Have a Go: Split and label a complicated dendrogram

You will need the *gb.biol* data for this exercise, these show abundances of ground beetles from 18 samples across three habitat types. The data are found in the *CERE.RData* file.

1. Start by looking at the sample names for the *gb.biol* data. You'll see that there are six replicates from each of three habitat types. The samples relate to *Grass, Edge* and *Wood* habitat types:

```
> rownames(gb.biol)
 [1] "E1" "E2" "E3" "E4" "E5" "E6" "G1" "G2" "G3" "G4" "G5" "G6"
[13] "W1" "W2" "W3" "W4" "W5" "W6"
```

2. Use the `dist()` command to make a Euclidean dissimilarity matrix:

```
> gb.eu <- dist(gb.biol, method = "euclidean")
```

3. Now create an hierarchical cluster object:

```
> gb.clus <- hclust(gb.eu)
```

4. Create a dendrogram to show the relationships between the samples, your dendrogram should resemble Figure 12.7:

```
> plot(gb.clus)
```

Cluster Dendrogram

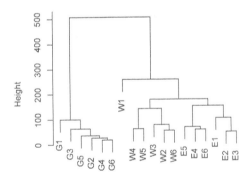

gb.eu
hclust (*, "complete")

Figure 12.7 Hierarchical cluster relationship between samples of ground beetles in three habitat types: Edge, Grass and Wood.

5. Look at Figure 12.7; you can see that there are at least three groups, as you'd expect, but that one sample appears slightly awry. Use the `cutree()` command to view the group membership if there were only three groups:

```
> cutree(gb.clus, k = 3)
E1 E2 E3 E4 E5 E6 G1 G2 G3 G4 G5 G6 W1 W2 W3 W4 W5 W6
 1  1  1  1  1  1  2  2  2  2  2  2  3  1  1  1  1  1
```

6. Now look at the group membership using several numbers of groups to help you decide the 'best' value for *k*:

```
> cutree(gb.clus, k = 1:5)
   1 2 3 4 5
E1 1 1 1 1 1
E2 1 1 1 1 1
E3 1 1 1 1 1
E4 1 1 1 1 2
E5 1 1 1 1 2
E6 1 1 1 1 2
G1 1 2 2 2 3
G2 1 2 2 2 3
G3 1 2 2 2 3
G4 1 2 2 2 3
G5 1 2 2 2 3
G6 1 2 2 2 3
W1 1 1 3 3 4
W2 1 1 1 4 5
W3 1 1 1 4 5
W4 1 1 1 4 5
W5 1 1 1 4 5
W6 1 1 1 4 5
```

7. From step 6 it looks like setting k = 4 would be the most sensible option. Make a result that splits the original tree into four groups:

```
> gb.ct <- cutree(gb.clus, k = 4)
```

8. Redraw the dendrogram but assign the group numbers to the ends of the branches:

```
> plot(gb.clus, labels = as.character(gb.ct))
```

9. The dendrogram you made in step 8 was not very helpful because you cannot see which samples relate to which groups. Make some custom labels and redraw the dendrogram using the new labels, your new tree should resemble Figure 12.8:

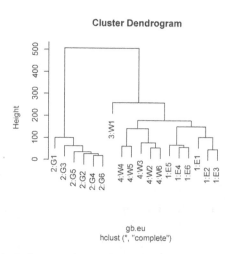

Cluster Dendrogram

gb.eu
hclust (*, "complete")

Figure 12.8 Hierarchical cluster relationship between samples of ground beetles in three habitat types: Edge, Grass and Wood. Tree cut into k = 4 groups.

```
> gb.lab <- paste(as.character(gb.ct), names(gb.ct), sep = ":")
> plot(gb.clus, labels = gb.lab)
```

10. Highlight the four clusters on the dendrogram:

```
> rect.hclust(gb.clus, k = 4)
```

11. Now redraw the tree and highlight the clusters that result from cutting at a height of 150:

```
> plot(gb.clus)
> rect.hclust(gb.clus, h = 150)
```

The cutree() and rect.hclust() commands are helpful in being able to help you to split your dendrogram into meaningful clusters.

Note: Identify clusters on a dendrogram interactively

You can use the identify() command to highlight clusters on a dendrogram using the mouse. You supply the name of the cluster result and the MAXCLUSTER instruction (an integer value, usually the same as the number of clusters you want). Then when you click on the dendrogram the cluster identified at the particular cut height is highlighted with a rectangle. For example: identify(gb.clus, MAX = 4).

Using the *cluster* package

The *cluster* package is installed as part of the main distribution of R but it is not loaded automatically. The package contains a range of routines for cluster analysis. You can carry out hierarchical clustering in two main ways: agglomeratively (in a similar fashion to the hclust() command) and divisively.

Conducting agglomerative clustering

The agnes() command in the *cluster* package carries out hierarchical clustering by agglomerative clustering. The command can carry out the dissimilarity calculations as part of the command – meaning that you do not have to start from a dissimilarity matrix. The *cluster* package has its own command for calculating dissimilarity indices, the daisy() command. The only indices it can calculate are Euclidean, Manhattan and Gower so you will probably use the vegdist() command in the *vegan* package to create dissimilarities.

The agnes() command has the following general form:

```
agnes(x, diss, metric, stand, method)
```

The command accepts various instructions:

- x – the input data, which will be a community dataset or a dissimilarity result.
- diss – if you set diss = TRUE then the assumption is that the data is a dissimilarity result.
- metric – the calculation method for the dissimilarity, currently only "euclidean", "manhattan" or "gower" are accepted.
- stand – if TRUE, the measurements in x are standardised before calculating the dissimilarities. This is done by subtracting the mean value and dividing by the mean absolute deviation for each column (species).

- method – the method of assembling the hierarchy: `"average"` ([unweighted pair-]group average method, UPGMA), `"single"` (single linkage), `"complete"` (complete linkage), `"ward"` (Ward's method), `"weighted"` (weighted average linkage) and its generalisation `"flexible"`, which uses a constant version of the Lance–Williams formula. The default is `"average"`.

The result of an `agnes()` command is an object that contains various components in a similar way to the `hclust()` command result. However, it also contains an agglomeration coefficient. This can be used to compare datasets, but only if they are of similar size.

The result of the `agnes()` command has its own plotting routine so you can create dendrograms. You can cut the dendrogram into clusters in much the same way as you saw earlier for the `hclust()` result but you need to convert the result into an `hclust()` result first using the `as.hclust()` command. In the following exercise you can have a go at using `agnes()` to create a cluster object and then explore the hierarchy and dendrogram.

Have a Go: Use agglomerative nesting

You will need the *ant* data for this exercise, this is found in the *CERE.RData* file. The data show the abundances of various species of ant from sites in Northern Australia (Hoffmann 2003). The sites have different burning/grazing regimes and underlying soil type.

1. Start by ensuring that the *cluster* package is loaded and ready:

```
> library(cluster)
```

2. Look at the *ant* data, which contains columns representing the various samples:

```
> tail(ant)
          E2b  E3b  L2b  L3b  Ub  E2r  E3r  L2r  L3r  Ur
Para.spB    0    0    0    0   1   33   20   60   16  39
Para.spC    0    0    0    0   0    0    0    1    0   0
Poly.inco   2    0    2    0   2    1    0    1    0   0
Poly.prom   0    0    1    0   0    0    0    4    1   0
Poly.seni   2    0   12    3  11    1   11   15    4   1
Poly.spD    1    0    0    0   0    0    0    0    0   0
```

3. Rotate the data so that the rows are the samples and the columns are the species:

```
> af = t(ant)
```

4. Use the Gower metric and the Average joining method to make an `agnes()` result:

```
> af.ag <- agnes(af, metric = "gower", method = "average")
> af.ag
Call:      agnes(x = af, metric = "gower", method = "average")
Agglomerative coefficient:  0.6152038
Order of objects:
 [1] E2b E3b L2b L3b Ub  E2r Ur  E3r L3r L2r
Height (summary):
   Min.  1st Qu.  Median    Mean  3rd Qu.    Max.
  49.77    64.42  181.70  169.00   217.80  352.00

Available components:
[1] "order"     "height"     "ac"         "merge"     "diss"
[6] "call"      "method"     "order.lab"  "data"
```

5. Use the `plot()` command to show the dendrogram. Yours should resemble Figure 12.9:

```
> plot(af.ag, which.plots = 2, main = "", las = 1)
```

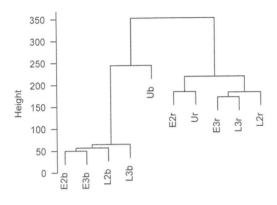

af
Agglomerative Coefficient = 0.62

Figure 12.9 Agglomerative nesting for ants in Northern Australia. Sites have different burning regimes and soils.

6. It looks like there are two main groups, with the *Ub* sample (unburnt, brown soil) being something of an outlier. Convert the result into a `hclust()` object using the `as.hclust()` command:

```
> af.clus <- as.hclust(af.ag)
```

7. Use the `cutree()` command to explore the groups:

```
> af.ct = cutree(af.clus, k = 3)
> af.ct
E2b E3b L2b L3b  Ub E2r E3r L2r L3r  Ur
  1   1   1   1   2   3   3   3   3   3

> af.lab = paste(as.character(af.ct), names(af.ct), sep = ":")
> plot(af.clus, labels = af.lab)
```

The `as.hclust()` command is useful here as it allows you to convert the `agnes()` result into a `hclust()` result, which makes it easier to manage the subsequent tree cutting.

Conducting divisive clustering

So far you've seen agglomeration used as the way to build up an hierarchy. The `diana()` command in the *cluster* package creates the hierarchy in the opposite manner. The data are successively split into smaller elements until each element contains a single sample. The method involves taking everything as one large group and then looking for the item with the largest dissimilarity. This forms a splinter group and the process carries on by reallocating samples to one group or the other based on the dissimilarity.

The `diana()` command has a general form like so:

```
diana(x, diss, metric, stand)
```

You can supply a community sample or a dissimilarity matrix, in the latter case you set `diss = TRUE`. The `metric` is applied to the community data (ignored if it is a dissimilarity matrix), you can specify `"euclidean"` or `"manhattan"`. If `stand = TRUE` the community data is standardised (this is done by subtracting the mean value and dividing by the mean absolute deviation for each column).

The result contains various elements, which you can utilise, and are used in the plotting method to make a dendrogram. You can also convert the result to a `hclust()` object and hence use the `cutree()` command. In the following exercise you can have a go at divisive clustering for yourself.

Have a Go: Carry out divisive hierarchical clustering

You will need the ground beetle data and the ant data for this exercise. Both are found in the *CERE.RData* file. You'll also need the *vegan* and *cluster* packages. The ground beetle data show abundances of beetles for three habitat types (six replicates for each). The ant data show abundances of ants from two soil types and several burning regimes (Hoffmann 2003).

1. Start by preparing the *cluster* and *vegan* packages:

```
> library(cluster)
> library(vegan)
```

2. Look at the row names of the *gb.biol* data; you'll see that they show the site names:

```
> rownames(gb.biol)
 [1] "E1" "E2" "E3" "E4" "E5" "E6" "G1" "G2" "G3" "G4" "G5" "G6"
[13] "W1" "W2" "W3" "W4" "W5" "W6"
```

3. Make a divisive clustering result using the Manhattan dissimilarity – standardise the data since there are large variations in species abundances:

```
> gb.di <- diana(gb.biol, metric = "manhattan", stand = TRUE)
```

4. Look at the result object, which contains several elements, and pick out the divisive coefficient:

```
> names(gb.di)
[1] "order"     "height"    "dc"        "merge"     "diss"
[6] "call"      "order.lab" "data"

> gb.di$dc
[1] 0.7442595
```

5. Now plot the dendrogram, which should resemble Figure 12.10a:

```
> pltree(gb.di, las = 1, main = "(a) Ground beetles")
```

```
> mtext(paste("Divisive coefficient",
        format(gb.di$dc, digits = 3),
        sep = " "), side = 1)
```

6. Look at the *ant* data, which shows that the species are the columns. You will need to rotate the data so that the samples are the rows:

```
> af <- t(ant)
```

7. Make a dissimilarity matrix using the Horn–Morista metric. Use the `decostand()` command to standardise the samples first:

```
> af.bc <- vegdist(decostand(af, method = "max"), method = "horn"
```

8. Now make a divisive clustering result using the dissimilarities from step 7:

```
> af.di <- diana(af.bc, diss = TRUE)
```

9. View the divisive coefficient:

```
> af.di$dc
[1] 0.4418091
```

10. Now prepare the dendrogram – yours should resemble Figure 12.10b:

```
> pltree(af.di, las = 1, main = "(b) Ants")
> mtext(paste"Divisive coefficient",
        format(af.di$dc, digits = 3),
        sep = " "), side = 1)
```

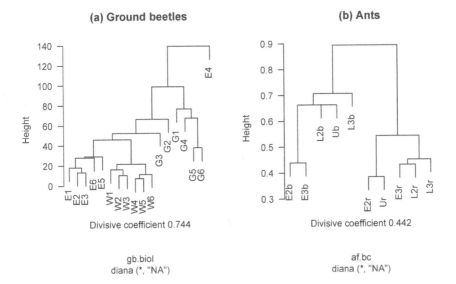

(a) Ground beetles

Divisive coefficient 0.744

gb.biol
diana (*, "NA")

(b) Ants

Divisive coefficient 0.442

af.bc
diana (*, "NA")

Figure 12.10 Divisive hierarchical clustering: (a) ground beetles from three habitat types in UK and (b) ants and fire regime from Northern Australia.

The method of creating the dissimilarity has a potentially large impact on the dendrogram. You could try different methods and see if your conclusions are the same.

Note: Splitting the plot window

It is possible to split the plot window into parts, allowing multiple plots. The `par()` command can do with via the `mfrow` instruction. You specify the number of rows and columns you want, e.g. `par(mfrow = c(1, 2))` produces one row and two columns. When you are finished you should restore the default `mfrow = c(1, 1)`, closing the graphics window will achieve this.

Agglomerative coefficients

An agglomerative coefficient is a way to measure the clustering structure of a hierarchical clustering. Essentially each observation is compared to the first cluster it is merged with to get a dissimilarity. This is then divided by the dissimilarity to the final step of the merger. The average of all these values is subtracted from unity to give a coefficient (which varies from 0 to 1).

The coefficient will increase simply by increasing number of observations so its usefulness is limited if datasets are of very different sizes. The `coef.hclust()` command in the *cluster* package will determine the agglomerative coefficient for the results of hierarchical clustering such as obtained by `hclust()`, `agnes()` or `diana()` commands.

Testing significance of clusters

It is possible to examine a dendrogram and its hierarchical structure and determine the statistical significance of the clusters. The *pvclust* package carries out hierarchical clustering and then applies bootstrapping procedures to determine significance of the various 'splits' in the dendrogram (and therefore the clusters).

The main command is `pvclust()`, and its general form is:

```
pvclust(data, method.hclust, method.dist, nboot)
```

You specify various instructions:

- `data` – the community dataset. This must be arranged with samples as columns and species as rows, which is unlike most community data.
- `method.hclust` – the agglomerative method used in hierarchical clustering. This should be one of `"average"` (the default), `"ward"`, `"single"`, `"complete"`, `"mcquitty"`, `"median"` or `"centroid"`. These are the same as the `hclust()` command.
- `method.dist` – the distance measure to be used. This should be one of `"correlation"` (the default), `"uncentered"`, `"abscor"` or those which are allowed the `dist()` command (i.e. `"euclidean"`, `"maximum"`, `"manhattan"`, `"canberra"`, `"binary"` or `"minkowski"`).
- `nboot` – the number of bootstrap replications, the default is 1000.

Notice that you cannot specify a dissimilarity matrix as the input data; this limits the dissimilarities that you can use a little but there is still plenty of choice.

The command can take quite a while to complete its operations so you need to be patient with large datasets! The result contains a variety of components, which you can utilise afterwards. The *pvclust* package contains various helper commands to help you analyse the result; in the following exercise you can have a go at analysis of an hierarchical clustering for yourself.

Have a Go: Examine significance of hierarchical clustering

You will need the ground beetle data for this exercise – these are found in the *CERE. RData* file. The data represent 18 samples of beetles from three different habitat types in the UK. You'll also need the *pvclust* package. If you do not have the package installed then you'll need to run the command `install.packages("pvclust")`.

1. Start by ensuring the *pvclust* package is loaded.

   ```
   > library(pvclust)
   ```

2. The beetle data are in the *gb.biol* data object. You can see that the samples form the rows and the columns are the species.

   ```
   > rownames(gb.biol)
    [1] "E1" "E2" "E3" "E4" "E5" "E6" "G1" "G2" "G3" "G4" "G5" "G6"
   [13] "W1" "W2" "W3" "W4" "W5" "W6"
   ```

3. The `pvclust()` command requires the data to be rotated so that the samples are the columns. However, you can do this 'on the fly' so use the `pvclust()` command to make and test an hierarchical clustering:

   ```
   > gb.clp <- pvclust(t(gb.biol))
   ```

4. The command in step 3 will take some while to complete. You used all the default settings to make the result, which you should view:

   ```
   > gb.clp

   Cluster method: average
   Distance       : correlation

   Estimates on edges:

         au     bp se.au se.bp      v       c  pchi
    1 0.989 0.522 0.002 0.005 -1.170  1.114 0.001
    2 0.973 0.568 0.005 0.005 -1.052  0.880 0.057
    3 0.985 0.759 0.003 0.005 -1.435  0.732 0.618
    4 0.987 0.858 0.003 0.004 -1.654  0.582 0.543
    5 0.949 0.429 0.008 0.005 -0.729  0.909 0.599
    6 0.952 0.393 0.008 0.005 -0.696  0.967 0.000
    7 0.994 0.488 0.001 0.005 -1.232  1.262 0.847
    8 0.937 0.588 0.010 0.005 -0.876  0.654 0.035
    9 0.934 0.388 0.010 0.005 -0.610  0.894 0.706
   10 0.885 0.237 0.017 0.004 -0.243  0.957 0.544
   11 0.940 0.526 0.009 0.005 -0.809  0.745 0.238
   12 0.976 0.468 0.004 0.005 -0.951  1.032 0.945
   13 0.596 0.174 0.036 0.004  0.348  0.592 0.909
   14 0.626 0.091 0.042 0.003  0.508  0.829 0.717
   15 0.462 0.401 0.031 0.005  0.173  0.077 0.882
   16 0.263 0.522 0.025 0.005  0.289 -0.344 0.027
   17 1.000 1.000 0.000 0.000  0.000  0.000 0.000
   ```

5. The result contains various components:

   ```
   > names(gb.clp)
   [1] "hclust" "edges"  "count"  "msfit"  "nboot"  "r"      "store"
   ```

6. The "edges" component contains information that you are most interested in:

```
> names(gb.clp$edges)
[1] "au"     "bp"     "se.au" "se.bp" "v"      "c"      "pchi"
```

7. The "edges" result holds the approximate unbiased *p*-value (au, high is good), the bootstrap estimate (bp), the standard errors of the au and bp, the signed distance (v) and curvature of boundary (c) as well as a chi square *p*-value based on asymptotic theory. The au value is generally of most interest:

```
> gb.clp$edges$au
 [1]  0.9888111 0.9733284 0.9848670 0.9873021 0.9492591 0.9518328
 [7]  0.9936959 0.9370284 0.9337776 0.8850043 0.9399309 0.9763517
[13]  0.5962537 0.6259027 0.4617401 0.2633094 1.0000000
```

8. Use the plot() command to see a dendrogram and associated statistics. Your dendrogram should resemble Figure 12.11:

```
>plot(gb.clp)
```

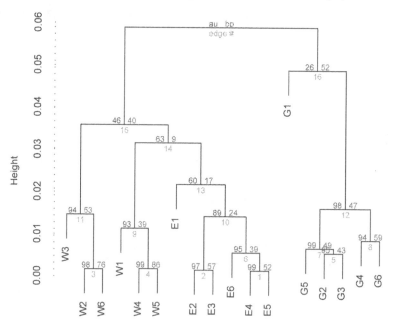

Cluster dendrogram with AU/BP values (%)

Distance: correlation
Cluster method: average

Figure 12.11 Hierarchical clustering of ground beetle communities with approximate unbiased *p*-values (expressed as 1 − *p*).

9. The *p*-values are represented as $1 - p$ so a significant cluster will have a value of ≥ 0.95. On the dendrogram these values are multiplied by 100. You can highlight clusters with high values using the `lines()` command:

```
> lines(gb.clp, alpha = 0.95, pv = "au", type = "geq")
```

10. In step 9 you highlighted significant clusters with bars on the dendrogram. The `type` instruction allows you to specify which clusters are highlighted relative to the `alpha`. You can specify one of `"geq"`, `"leq"`, `"gt"` or `"lt"`. You can also highlight clusters according to bp values by specifying `pv = "bp"`. The `pvrect()` command allows you to highlight the clusters with rectangles in a similar way:

```
> pvrect(gb.clp, pv = "au", border = "blue")
```

11. Use the `pvpick()` command to view the significant clusters:

```
> pvpick(gb.clp)
$clusters
$clusters[[1]]
[1] "E2" "E3"

$clusters[[2]]
[1] "W2" "W6"

$clusters[[3]]
[1] "W4" "W5"

$clusters[[4]]
[1] "E4" "E5" "E6"

$clusters[[5]]
[1] "G2" "G3" "G4" "G5" "G6"


$edges
[1]  2  3  4  6 12
```

12. The `msplot()` command can draw the curve-fitting results for one or more edges. Look at the significant edges and one non-significant one; your graph should resemble Figure 12.12:

```
> msplot(gb.clp, edges = c(1:4,12,16))
```

13. You can also examine the standard errors in a plot via the `seplot()` command. If you specify `identify = TRUE` as part of the command you can click on the graph and the edge number will appear (press the Esc key to stop):

```
> seplot(gb.clp, type = "au", identify = TRUE)
```

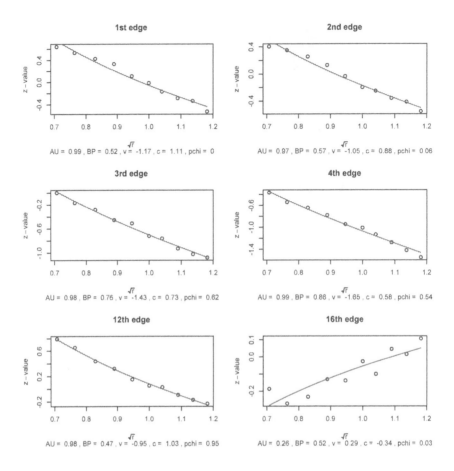

Figure 12.12 Curve-fitting for hierarchical clustering of ground beetle data.

Tip: Examining species clusters

Most often you are examining samples of communities and assessing the similarity (or otherwise) of the samples according to their species composition. There is no reason why you cannot look at clustering the species instead. If your data are arranged with rows as sites you can rearrange easily using the t() command; this simply rotates the data.

Using hierarchical clustering to create species tables

You generally arrange your community data in one of two ways: with samples as columns or with rows as columns. If you've used a Pivot Table to assemble your data (which is a good idea) the species and samples will be arranged simply in alphabetical order.

Most of the analytical commands in R require the data to be arranged with the samples as the rows but most people prefer to view their data with the species as the rows. Switching between the two 'views' is not a problem – you can simply use the t() command to rotate your data.

The *vegan* package contains a command, vegemite(), which creates a very compact data table. It is designed to allow you to see the species as the rows and the sites as the columns. A key feature of the `vegemite()` command is that you can use the result of a hierarchical cluster to reorder the species and sites into a more 'meaningful' order. The general form of the command is like so:

```
vegemite(x, use, scale, zero=".")
```

You specify the community dataset (where rows are the sites) and any `hclust()` cluster result (you can also use ordination results produced by commands in the *vegan* package). The compact table that the `vegemite()` command produces uses only a single digit for abundance data. This means that many datasets will need to be rescaled – the `scale` instruction carries this out (it actually calls the `coverscale()` command internally).

You specify the cover scale you require from the following options: `"Braun.Blanquet"`, `"Domin"`, `"Hult"`, `"Hill"`, `"fix"`, or `"log"`. The first three are fairly standard methods of converting percentage data into a smaller scale. The `"Hill"` method takes its values from the TWINSPAN program, and uses the default cut-levels from that program as a scale. In all these cases values <1 are shown as + and 10 is represented by X.

The `"fix"` scale simply truncates values between 1 and 9 to integers, while smaller values are represented by + and larger by X. The `"log"` takes the maximum abundance and then splits the range into range 0–9, small values are shown by +.

You can alter the character that is displayed in lieu of 0 by using the `zero` instruction; the default is a period.

In the following exercise you can have a go at making compact tables using hierarchical clustering to reorder the species and samples.

Have a Go: Use hierarchical clustering to reorder a compact species table

You will need the *cluster* and *vegan* packages for this exercise. You'll also need the *pqd* data, which comes with the *CERE.RData* file. The data represents plant species abundance (using a Domin scale). There are ten samples, that is five each from two locations.

1. Start by preparing the *cluster* and *vegan* packages:

    ```
    > library(cluster)
    > library(vegan)
    ```

2. Look at the *pqd* data, it is in biological recording format so you'll need to assemble a community dataset by cross-tabulation with the `xtabs()` command:

    ```
    > pqd.biol <- xtabs(domin ~ obs + abbr, data = pqd)
    ```

3. If you prepare a `vegemite()` table from the 'raw' data the species and sites are ordered in the same way as the original data (alphabetically):

    ```
    > pqd.vt <- vegemite(pqd.biol)

                  11111luuuuu
                  ooooopppppp
                  wwwwwppppp
    ```

```
           eeeeeeeeee
           rrrrrrrrrr
           1234512345
Ach.mil 333331..1.
Agr.cap .2654.....
Alc.vul 1.........
Ant.odo 41432.....
Bel.per 3112211132
Bot.lun ..1.......
Cam.rot ....2.....
Car.car 21111.....
Car.fla 21322.....
Cer.fon .111..111.
Cir.vul 1222....1.
Dac.glo .3.33.....
Des.ces .3343.....
Fes.ovi 8676553.32
Gal.sax 11.31.....
Lol.per .....2.222
Lot.cor ..1.1.....
Luz.cam 42.21.....
Nar.str 32.12.....
Pla.lan .1213.....
Pla.maj 11.1142233
Poa.ann .....87758
Poa.tri .....53473
Pot.ere 31332.....
Pru.vul 33332...23
Ran.rep 3143355545
Rhy.squ 64354.....
Rum.ace 1..212....
Tar.off .....1.11.
Tri.rep 3443355555
Ver.ser 2....21112
Vio.riv ....1.....
    sites species
       10      32
```

4. The table in step 3 is not very useful. It would be better to reorder the data according to an hierarchical clustering result. Make an `hclust()` result to use in the reordering:

```
> pqd.bc <- vegdist(pqd.biol, method = "bray")
> pqd.cl <- hclust(pqd.bc)
```

5. Now make a new species table using the `hclust()` result to reorder the data:

```
> pqd.vt <- vegemite(pqd.biol, use = pqd.cl)
```

```
           uuuuulllll
           pppppooooo
           pppppwwwww
           eeeeeeeeee
           rrrrrrrrrr
           4231513524
```

```
Tar.off 1.11......
Poa.tri 73453.....
Poa.ann 57788.....
Lol.per 2.222.....
Ver.ser 111222....
Pla.maj 322431.111
Ran.rep 4555534313
Tri.rep 5555534343
Cer.fon 111...1.11
Bel.per 3111231212
Alc.vul .....1....
Fes.ovi 33.5287566
Pru.vul 2...333233
Bot.lun ......1...
Rum.ace ...2.1.1.2
Ach.mil 1..1.33333
Cir.vul 1....12.22
Lot.cor ......11..
Ant.odo .....44213
Car.car .....21111
Nar.str .....3.221
Luz.cam .....4.122
Car.fla .....23212
Pot.ere .....33213
Rhy.squ .....63445
Cam.rot ........2..
Vio.riv ........1..
Pla.lan ......2311
Agr.cap ......6425
Des.ces ......3334
Gal.sax .....1.113
Dac.glo .......333
 sites species
    10      32
```

6. The `vegemite()` result contains two components, look at these:

    ```
    > names(pqd.vt)
    [1] "sites"   "species"
    > pqd.vt$sites
     [1]  9  7  8  6 10  1  3  5  2  4
    > pqd.vt$species
     [1] 29 23 22 16 31 21 26 30 10  5  3 14 25  6 28  1 11 17  4  8
    [21] 19 18  9 24 27  7 32 20  2 13 15 12
    ```

7. If you want to use a different kind of hierarchical clustering result then you need to convert the result. Use the `diana()` command to make a divisive hierarchical cluster result:

    ```
    > pqd.di <- diana(pqd.biol)
    > pqd.cl <- as.hclust(pqd.di)
    ```

8. Now you can use the `diana()` result to reorder the data table:

    ```
    > pqd.vt <- vegemite(pqd.biol, use = pqd.cl)
    ```

```
        11111uuuuu
        oooooppppp
        wwwwwppppp
        eeeeeeeeee
        rrrrrrrrrr
        1245312354
Alc.vul 1.........
Luz.cam 4221......
Nar.str 3212......
Car.car 21111.....
Gal.sax 1131......
Rhy.squ 64543.....
Dac.glo .333......
Ant.odo 41324.....
Pot.ere 31323.....
Car.fla 21223.....
Des.ces .3433.....
Ach.mil 333331...1
Agr.cap .2546.....
Rum.ace 1.21.2....
Pla.lan .1132.....
Cir.vul 122.2....1
Cam.rot ...2......
Vio.riv ...1......
Fes.ovi 8665753.23
Lot.cor ...11.....
Pru.vul 33323...32
Bot.lun ....1.....
Bel.per 3122111123
Cer.fon .11.1.11.1
Tri.rep 3433455555
Ran.rep 3133455554
Ver.ser 2....21121
Pla.maj 1111.42233
Poa.ann .....87785
Tar.off .....1.1.1
Poa.tri .....53437
Lol.per .....2.222
 sites species
   10      32
```

The results of the hierarchical clustering result obviously affect the order of the sites and species. The compact nature of the table makes it easier to get an overview than looking at the original 'raw' data.

Note: Compact species tables and reordering

When you reorder a compact species table using an hierarchical clustering result you generally see the data take a particular pattern. In general the species that occur at most of the sites will be in the middle. Species that tend to occur at few sites tend to be at the top and bottom of the table, the data is rearranged so that you end up with a diamond pattern.

You'll see the compact species tables used again using the results of ordinations in Section 14.4.3.

12.2.2 Clustering by partitioning

Methods of clustering by partitioning produce clusters in a non-hierarchical way. In these partitioning methods you decide how many clusters there ought to be and the data are then 'assigned' to clusters according to 'how far' they are from the centre of the cluster.

There are several methods for clustering by partitioning available in R. The *cluster* package contains three methods and the *stats* package (one of the core elements of R) another.

K-means

In k-means clustering data is arranged around various cluster centres. The idea is to minimise the sums of squares of the distances between the data and the centre. The `kmeans()` command carries out the analysis. Its general form is:

```
kmeans(x, centers, iter.max, nstart, algorithm)
```

You can supply various instructions to the command:

- `x` – the data, which must be a numeric `matrix` or a `data.frame`.
- `centers` – usually you give a numeric value for the number of centres you require.
- `iter.max` – the maximum number of iterations allowed, the default is 10.
- `nstart` – a value giving the number of random starts to use. The default is 1 but larger values can be helpful if there are a lot of clusters.
- `algorithm` – the method of assigning the data to the clusters, you choose one of: `"Hartigan-Wong"` (the default), `"Lloyd"`, `"Forgy"` or `"MacQueen"`.

In the simplest situation you have two variables – you can visualise this in a scatter plot. The k-means analysis randomly assigns centres for the number of cluster you require. The data are then assigned to a cluster (the nearest) and the sums of squares are calculated. The centres are then shifted to 'better' positions and the process repeated.

When you have three variables the situation becomes more complicated – you can still visualise the situation, though, as a 3-D scatter plot. When you have more than three variables you cannot visualise the situation but the underlying process is the same.

K-means analysis can be carried out on community data but it is hard to produce a meaningful visual result. You can, however, assign samples to a particular cluster and tabulate your results using the method. In the following exercise you can have a go at k-means clustering for a fairly simple example, where you have only two clusters.

Have a Go: Use k-means analysis

You will need the *pqd* plant data for this exercise. The data are in the *CERE.RData* file. The data show plant abundance (in Domin scale) for ten samples taken from quadrats at adjacent sites.

1. The *pqd* data are in biological recording format so you'll need to cross-tabulate to get a community dataset to work with:

```
> pqd.biol <- xtabs(domin ~ obs + abbr, data = pqd)
```

2. Carry out a k-means analysis – there are two main sites so you are expecting that the samples will form two clusters:

```
> pqd.k <- kmeans(pqd.biol, centers = 2)
```

3. Look at the result, which contains several components:

```
> pqd.k
K-means clustering with 2 clusters of sizes 5, 5

Cluster means:
  Ach.mil Agr.cap Alc.vul Ant.odo Bel.per Bot.lun Cam.rot Car.car
1     3.0     3.4     0.2     2.8     1.8     0.2     0.4     1.2
2     0.4     0.0     0.0     0.0     1.6     0.0     0.0     0.0
  Car.fla Cer.fon Cir.vul Dac.glo Des.ces Fes.ovi Gal.sax Lol.per
1       2     0.6     1.4     1.8     2.6     6.4     1.2     0.0
2       0     0.6     0.2     0.0     0.0     2.6     0.0     1.6
  Lot.cor Luz.cam Nar.str Pla.lan Pla.maj Poa.ann Poa.tri Pot.ere
1     0.4     1.8     1.6     1.4     0.8       0     0.0     2.4
2     0.0     0.0     0.0     0.0     2.8       7     4.4     0.0
  Pru.vul Ran.rep Rhy.squ Rum.ace Tar.off Tri.rep Ver.ser Vio.riv
1     2.8     2.8     4.4     0.8     0.0     3.4     0.4     0.2
2     1.0     4.8     0.0     0.4     0.6     5.0     1.4     0.0

Clustering vector:
lower1 lower2 lower3 lower4 lower5 upper1 upper2 upper3 upper4
     1      1      1      1      1      2      2      2      2
upper5
     2

Within cluster sum of squares by cluster:
[1] 118.0  57.2
 (between_SS / total_SS =  71.8 %)

Available components:

[1] "cluster"      "centers"      "totss"       "withinss"
[5] "tot.withinss" "betweenss"    "size"
```

4. Make a simple table showing how many samples are in each cluster:

```
> table(pqd.k$cluster)

1 2
5 5
```

5. Now make a more complicated table to show the sample names and the clusters:

```
> table(rownames(pqd.biol), pqd.k$cluster)

        1 2
lower1  1 0
lower2  1 0
lower3  1 0
lower4  1 0
lower5  1 0
upper1  0 1
upper2  0 1
upper3  0 1
upper4  0 1
upper5  0 1
```

You can see from the table in step 5 that the samples fall nicely into the two clusters.

It is not easy to produce a plot that shows the clusters convincingly using the k-means method. You can plot two columns of the data, one species against another, but this only shows a small part of the 'result'. You'll see more about visualising cluster results using the `clusplot()` command shortly.

Other methods of partition clustering 'collapse' the data into a two-dimensional form, allowing a visual representation of the clusters. The `pam()` command in the *cluster* package for example, is able to produce a plot, as you will see next.

Partitioning around medoids (PAM)

A more robust form of k-means analysis is called *partitioning around medoids* and is implemented by the `pam()` command in the *cluster* package. Unlike the `kmeans()` command `pam()` starts with a dissimilarity matrix. The aim is to minimise the dissimilarity between the data and the centres of the clusters.

The `pam()` command has the following general form:

```
pam(x, k, diss, metric, stand)
```

You can supply several instructions:

- `x` – the data to use; this can be a community dataset or a dissimilarity matrix.
- `k` – the number of clusters to create.
- `diss` – if TRUE then the input data are treated as a dissimilarity matrix.
- `metric` – the metric to use to make the dissimilarity matrix; you can choose `"euclidean"` or `"manhattan"`.
- `stand` – if TRUE the data are standardised before analysis. This is done by subtracting the mean value and dividing by the mean absolute deviation for each column (species).

The `pam()` command produces a result with several components and which can be plotted. In the following exercise you can have a go at using the `pam()` command and exploring the results for yourself.

Have a Go: Use partitioning around medoids for cluster analysis

You will need the *ant* data for this exercise, which can be found in the *CERE.RData* file. The data show the abundances of various species of ant from sites in Northern Australia (Hoffmann 2003). The sites have different burning/grazing regimes and underlying soil type. You'll also need the *cluster* package – this should be installed already as it comes as part of the basic distribution of R.

1. Start by preparing the *cluster* package:

```
> library(cluster)
```

2. The *ant* data are arranged with samples as columns so rotate the data:

```
> af <- t(ant)
```

3. Make a dissimilarity matrix using the Euclidean metric. The `pam()` command can do this but it is more useful to make your own dissimilarity as `pam()` can only use two different metrics:

```
> af.eu <- dist(af)
```

4. Carry out a *pam* analysis using the dissimilarity matrix – there are two main soil types so you are expecting that the samples will form two clusters. The command will recognise that the data is a dissimilarity but use the `diss` instruction as a reminder:

```
> af.pam <- pam(pqd.eu, k = 2, diss = TRUE)
```

5. Look at some basic information about the result:

```
> af.pam
Medoids:
     ID
[1,] "2" "E3b"
[2,] "7" "E3r"
Clustering vector:
E2b E3b L2b L3b  Ub E2r E3r L2r L3r  Ur
  1   1   1   1   1   2   2   2   2   2
Objective function:
   build     swap
116.0127 113.1744

Available components:
[1] "medoids"    "id.med"     "clustering" "objective"
[5] "isolation"  "clusinfo"   "silinfo"    "diss"
[9] "call"
```

6. Use the `silhouette()` command to get some details about the clusters – the nearer the width to unity the better the cluster. Negative values indicate that a sample could be misplaced:

```
> silhouette(af.pam)
    cluster neighbor sil_width
L3b      1         2 0.7077503
E3b      1         2 0.6984494
E2b      1         2 0.6714635
L2b      1         2 0.6680857
Ub       1         2 0.4053516
L3r      2         1 0.4457791
Ur       2         1 0.4327786
L2r      2         1 0.4228036
E3r      2         1 0.4223321
E2r      2         1 0.3999806
attr(,"Ordered")
[1] TRUE
attr(,"call")
pam(x = af.eu, k = 2)
attr(,"class")
[1] "silhouette"
```

7. The `summary()` command also provides details of the result but you can inspect individual components – look at the cluster membership and the cluster information:

```
> af.pam$clustering
E2b E3b L2b L3b  Ub E2r E3r L2r L3r  Ur
  1   1   1   1   1   2   2   2   2   2

> af.pam$clusinfo
     size max_diss  av_diss diameter separation
[1,]    5 250.0920  81.03624 258.2073   290.7043
[2,]    5 189.7999 145.31246 245.2550   290.7043
```

8. Look at the silhouette data as a graphic. Your plot should resemble Figure 12.13:

```
> plot(af.pam, which.plots = 2)
```

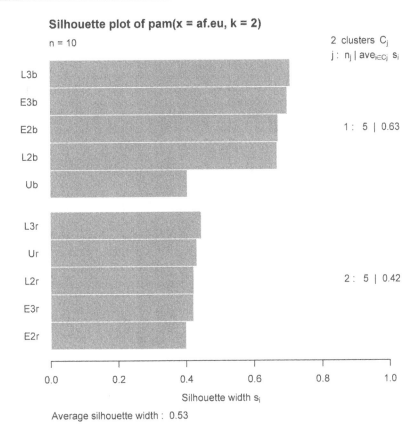

Figure 12.13 PAM cluster statistics for samples of ant communities from Northern Australia.

9. The command you used in step 8 can be used to visualise the clusters by using `which.plots = 1`. However, there is also a `clusplot()` command so use that to make a plot that resembles Figure 12.14:

```
> clusplot(af.pam)
```

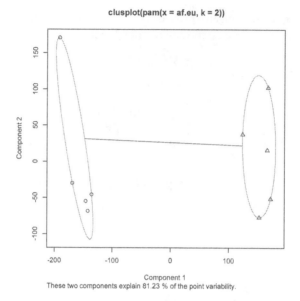

Figure 12.14 PAM clustering for samples of ant species from Northern Australia.

The plot you made in step 9 shows the clusters as well as an ellipse around each cluster and lines connecting them. In this case of course the plot is fairly simple as there are only two clusters.

Fuzzy analysis

In a 'fuzzy' analysis each observation is spread out across more than one cluster. The aim is to minimise the dissimilarity between each observation and the cluster(s). The 'fuzziness' of the algorithm can be altered, leading to more or less leeway on cluster membership.

The `fanny()` command in the *cluster* package carries out fuzzy clustering. The general form of the command is:

```
fanny(x, k, diss, memb.exp, metric, stand)
```

You can supply several instructions to the command:

- x – the data to use, which can be a community dataset or a dissimilarity matrix.
- k – the number of clusters to create.
- diss – if TRUE then the input data are treated as a dissimilarity matrix.
- memb.exp – a number (>1) that is used as the 'fuzziness' exponent. Generally larger values lead to more fuzziness. The default value is 2.
- metric – the metric to sue for calculating the dissimilarity, one of: "euclidean", "manhattan" or "SqEuclidean".
- stand – if TRUE the data are standardised before analysis. This is done by subtracting the mean value and dividing by the mean absolute deviation for each column (species).

The `fanny()` command produces a result that contains several components and can be plotted. In the following exercise you can have a go at a fuzzy cluster analysis for yourself.

Have a Go: Carry out fuzzy cluster analysis

You will need the ground beetle data for this exercise – these are found in the *CERE. RData* file. The data represent 18 samples of beetles from three different habitat types in the UK. You'll also need the *vegan* and *cluster* packages.

1. Start by preparing the *vegan* and *cluster* packages:

    ```
    > library(vegan)
    > library(cluster)
    ```

2. Use the `vegdist()` command to make a Bray–Curtis dissimilarity matrix:

    ```
    > gb.bc <- vegdist(gb.biol, method = "bray")
    ```

3. Now make a fuzzy cluster result using three clusters:

    ```
    > gb.fan <- fanny(gb.bc, k = 3)
    ```

4. You can use `summary()` on the result but for now simply look at the group memberships:

    ```
    > gb.fan$cluster
    E1 E2 E3 E4 E5 E6 G1 G2 G3 G4 G5 G6 W1 W2 W3 W4 W5 W6
     1  1  1  1  1  1  2  2  2  2  2  2  1  3  3  3  3  3
    ```

5. The sample *W1* looks like it might be in the wrong cluster – view the silhouette information:

    ```
    > silhouette(gb.fan)
       cluster neighbor   sil_width
    E6       1        3   0.33573802
    E1       1        3   0.33012560
    E4       1        3   0.25042396
    E5       1        3   0.20563930
    E2       1        3   0.13014213
    E3       1        3   0.06559154
    W1       1        3  -0.01627251
    G2       2        1   0.69730404
    G6       2        1   0.69485373
    G4       2        1   0.69402449
    G5       2        1   0.67428470
    G3       2        1   0.60959431
    G1       2        1   0.55026932
    W6       3        1   0.44364032
    W2       3        1   0.36125448
    W3       3        1   0.33880798
    W4       3        1   0.31075679
    W5       3        1   0.28094924
    ```

6. Try the analysis again but use a larger value for the 'fuzziness' exponent:

    ```
    > gb.fan <- fanny(gb.bc, k = 3, memb.exp = 4)
    ```

7. Look at the cluster membership again:

    ```
    > gb.fan$cluster
    E1 E2 E3 E4 E5 E6 G1 G2 G3 G4 G5 G6 W1 W2 W3 W4 W5 W6
     1  1  1  1  1  1  2  2  2  2  2  2  3  3  3  3  3  3
    ```

8. Visualise the clusters using the `clusplot()` command, your graphic should resemble Figure 12.15:

    ```
    > clusplot(gb.fan)
    ```

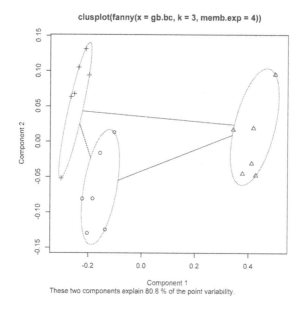

Figure 12.15 Fuzzy cluster analysis of ground beetle communities from Southern England.

The plot shows the clusters separating quite clearly.

Large clusters

A variant of the partitioning around medoids method (as implemented by the pam() command) is clara(), which uses slightly simpler algorithms to determine the clustering. This can save on computer memory and processing time. The command is part of the *cluster* package and has the following general form:

```
clara(x, k, metric, stand, samples)
```

You can supply various instructions to the command:

- x – the data to use, a community dataset.
- k – the number of clusters to create.
- metric – the metric to use to make the dissimilarity matrix; you can choose "euclidean" or "manhattan".
- stand – if TRUE the data are standardised before analysis. This is done by subtracting the mean value and dividing by the mean absolute deviation for each column (species).
- samples – the number of random samples to use, the default is 5.

The clara() command cannot operate on a dissimilarity matrix and you are therefore limited to the two metric built-in to the command (Euclidean and Manhattan).

The command uses broadly the same approach as the pam() command but the simpler random sampling approach makes processing more efficient. In practice a 'large' dataset is very large indeed and on most modern computers several thousand observations is not a problem. In the following example you can have a go at a clara() clustering analysis.

Have a Go: Use a large cluster approach in a partitioning cluster analysis

You will need the *golf* data for this exercise – the data are in the *CERE.RData* file and show plant abundance (as percentage cover) for 71 samples and 122 species in total. Samples were taken from sites at UK golf courses. You'll also need the *cluster* and *vegan* packages.

1. Start by preparing the *cluster* and *vegan* packages:

   ```
   > library(cluster)
   > library(vegan)
   ```

2. Look at the golf data using the `attributes()` command to give a quick list of the sites and species:

   ```
   > attributes(golf)
   $names
     [1] "Ach.pta" "Agr.can" "Agr.cap" "Agr.eup" "Agr.sto" "Agr.vin"
     [7] "Alo.pra" "Ana.ten" "Ant.odo" "Arr.ela" "Ath.fil" "Bar.are"
    [13] "Bel.per" "Bet.pen" "Bet.pub" "Bri.med" "Cal.vul" "Car.dem"
    [19] "Car.fla" "Car.hir" "Car.hos" "Car.lep" "Car.nig" "Car.pil"
    [25] "Car.pra" "Car.pul" "Carpanc" "Carpant" "Cen.nig" "Cer.fon"
    [31] "Cir.arv" "Cir.pal" "Cra.mon" "Cyn.cri" "Dac.glo" "Dac.pra"
    [37] "Dan.dec" "Des.ces" "Des.fle" "Dry.dil" "Dry.fil" "Ele.pal"
    [43] "Ele.qui" "Ely.rep" "Epi.cil" "Epi.hir" "Epi.obs" "Epi.par"
    [49] "Equ.arv" "Equ.flu" "Equ.pal" "Eri.ang" "Fes.fil" "Fes.ovi"
    [55] "Fes.rub" "Fra.exc" "Gal.apa" "Gal.pal" "Gal.sax" "Gly.dec"
    [61] "Her.sph" "Hie.sub" "Hol.lan" "Hol.mol" "Hyd.vul" "Hyp.rad"
    [67] "Hyp.tet" "Jun.acu" "Jun.art" "Jun.con" "Jun.eff" "Jun.inf"
    [73] "Jun.sur" "Lat.pra" "Lol.per" "Lot.cor" "Lot.ped" "Luz.cam"
    [79] "Men.aqu" "Men.tri" "Myo.cae" "Nar.str" "Phl.pra" "Pil.off"
    [85] "Pla.lan" "Poa.hum" "Poa.tri" "Pot.ang" "Pot.ans" "Pot.ere"
    [91] "Pot.rep" "Pru.sp." "Pru.vul" "Pul.dys" "Que.rob" "Ran.acr"
    [97] "Ran.fla" "Ran.rep" "Rub.fru" "Rum.ace" "Sag.pro" "Sch.aru"
   [103] "Sch.pra" "Scu.gal" "Sen.aqu" "Sen.jac" "Ser.tin" "Sil.flo"
   [109] "Sol.dul" "Sta.syl" "Ste.gra" "Suc.pra" "Tar.sp." "Tri.dub"
   [115] "Tri.Pal" "Tri.pra" "Tri.rep" "Ule.gal" "Val.dio" "Vic.cra"
   [121] "Vic.sat" "Vio.riv"

   $class
   [1] "data.frame"

   $row.names
     [1] "Q10" "Q11" "Q12" "Q13" "Q14" "Q15" "Q16" "Q17" "Q18" "Q19"
    [11] "Q20" "Q21" "Q22" "Q23" "Q24" "Q25" "Q26" "Q27" "Q28" "Q29"
    [21] "Q30" "Q31" "Q32" "Q33" "Q34" "Q35" "Q36" "Q37" "Q38" "Q39"
    [31] "Q40" "Q41" "Q42" "Q43" "Q44" "Q45" "Q49" "Q50" "Q51" "Q52"
    [41] "Q53" "Q54" "Q55" "Q56" "Q57" "Q58" "Q59" "Q60" "Q61" "Q62"
    [51] "Q63" "Q64" "Q65" "Q66" "Q67" "Q68" "Q69" "Q7 " "Q70" "Q71"
    [61] "Q72" "Q73" "Q74" "Q75" "Q76" "Q77" "Q78" "Q79" "Q8 " "Q80"
    [71] "Q9 "
   ```

3. The *golf.hab* data object holds information about the habitat types; look at that to make a guess about the number of clusters required:

   ```
   > levels(golf.hab$Habitat)
   [1] "Acid grassland"    "Flush"
   [3] "Grassland"         "Heathy grassland"
   ```

```
    [5] "Mire"                      "Mire edge"
    [7] "Species-rich grassland"    "Transitional flush/scrub"
```

4. Now make a `clara()` object using three clusters – this seems reasonable given the types of habitat sampled (see step 3):

```
> golf.cla = clara(golf, k = 3)
```

5. The result contains various components:

```
> names(golf.cla)
  [1] "sample"     "medoids"    "i.med"      "clustering"
  [5] "objective"  "clusinfo"   "diss"       "call"
  [9] "silinfo"    "data"
```

6. You can use `summary()` on the result but for now simply look at the cluster membership:

```
> golf.cla$clustering
Q10 Q11 Q12 Q13 Q14 Q15 Q16 Q17 Q18 Q19 Q20 Q21 Q22 Q23 Q24 Q25
  1   1   1   1   1   2   2   1   1   3   2   1   1   1   1   1
Q26 Q27 Q28 Q29 Q30 Q31 Q32 Q33 Q34 Q35 Q36 Q37 Q38 Q39 Q40 Q41
  1   3   1   1   1   1   2   1   1   1   1   2   3   2   3
Q42 Q43 Q44 Q45 Q49 Q50 Q51 Q52 Q53 Q54 Q55 Q56 Q57 Q58 Q59 Q60
  2   2   2   2   2   2   2   2   2   2   2   2   2   3   2   2
Q61 Q62 Q63 Q64 Q65 Q66 Q67 Q68 Q69 Q7  Q70 Q71 Q72 Q73 Q74 Q75
  2   2   3   2   2   2   2   3   3   1   3   2   2   3   3   2
Q76 Q77 Q78 Q79 Q8  Q80 Q9
  2   2   2   2   1   2   1
```

7. Make a silhouette plot of the result, which should resemble Figure 12.16:

```
> plot(golf.cla, which.plots = 2)
```

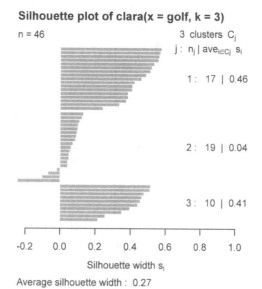

Silhouette plot of clara(x = golf, k = 3)

n = 46   3 clusters C_j
                                          j : n_j | ave$_{i \in C_j}$ s_i

                                          1 : 17 | 0.46

                                          2 : 19 | 0.04

                                          3 : 10 | 0.41

    -0.2   0.0   0.2   0.4   0.6   0.8   1.0

              Silhouette width s_i

        Average silhouette width : 0.27

Figure 12.16 Cluster silhouette plot showing sample cluster statistics for golf course plant communities.

8. Now try a plot of the clusters:

```
> clusplot(golf.cla)
Error in princomp.default(x, scores = TRUE, cor = ncol(x) != 2) :
  'princomp' can only be used with more units than variables
```

9. The `clusplot()` command in step 8 fails! This generally happens if you run a cluster analysis on the 'raw' data. In the case of `clara()` you have no choice but you can get around the problem by making a Euclidean dissimilarity matrix and then plot that, your plot should resemble Figure 12.17:

```
> golf.eu = dist(golf)
> clusplot(golf.eu, clus = golf.cla$clustering, diss = TRUE)
```

CLUSPLOT(golf.eu)

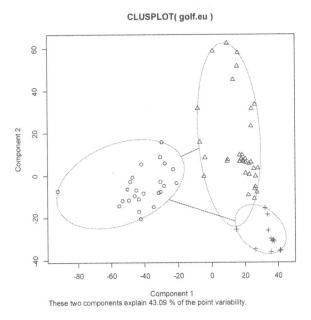

Component 1
These two components explain 43.09 % of the point variability.

Figure 12.17 Clustering based on Euclidean dissimilarity using clara() command in *cluster* package.

The `clusplot()` command is quite general and you are able to use it to show the clusters from the Euclidean dissimilarity and 'point' to the clustering membership.

In practice you are only likely to use the `clara()` command when you have extremely large datasets. The advice is, try the `pam()` command first and resort to `clara()` only if this gives any problems!

Visualising partition cluster results

You've seen how you can use the `clusplot()` command in the *cluster* package to visualise clusters. The command is quite flexible and you can alter the way you display the clusters in various 'useful' ways. Generally you simply use the `clusplot()` command on the result of a `pam()` or `fanny()` command and the defaults are perfectly acceptable. However, you can press the command into service for other cluster results – you saw an example of this in the preceding section with the `clara()` command.

Here you'll see two other uses of the `clusplot()` command: one is plotting the result of a `kmeans()` command, the other is adding custom labels to the plot to help you identify the individual samples.

Plotting a k-means result

K-means analysis is carried out by the `kmeans()` command. The results of this command do not have a 'dedicated' plotting routine, which means you need to find another way. You can use the `clusplot()` command from the *cluster* package to plot a `kmeans()` result, but you have to represent the clusters by using a dissimilarity matrix. In the following exercise you can have a go at representing a k-means analysis for yourself.

Have a Go: Make a visual representation of a k-means analysis

You will need the *pqd* plant data for this exercise. The data are in the *CERE.RData* file. The data show plant abundance (in Domin scale) for ten samples taken from quadrats at adjacent sites.

1. The *pqd* data need to be rearranged because they are in biological recording format:

    ```
    > pqd.biol <- xtabs(domin ~ obs + abbr, data = pqd)
    ```

2. Now make a k-means result using two clusters:

    ```
    > pqd.k <- kmeans(pqd.biol, centers = 2)
    ```

3. View the group memberships:

    ```
    > pqd.k$cluster
    lower1 lower2 lower3 lower4 lower5 upper1 upper2 upper3 upper4
         1      1      1      1      1      2      2      2      2
    upper5
         2
    ```

4. In order to view the clusters you will need to make a dissimilarity matrix. Use a Euclidean metric to make a dissimilarity result:

    ```
    > pqd.eu <- dist(pqd.biol)
    ```

5. Now you can use the `clusplot()` command to view the clusters by plotting the dissimilarity matrix and using the group membership from the `kmeans()` result. Your plot should resemble Figure 12.18:

    ```
    > clusplot(pqd.eu, clus = pqd.k$cluster, diss = TRUE)
    ```

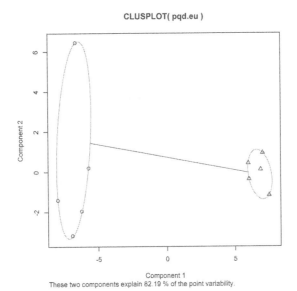

Figure 12.18 K-means clusters for upland plant communities from two sites, using Euclidean dissimilarity.

Using different dissimilarity metrics will affect the general look of your plot but you have little option if you want to use the `kmeans()` command.

Using a dissimilarity matrix is more or less your only choice if you want to use the `kmeans()` command for k-means analysis. A better option would probably be to use the `pam()` command instead, which uses dissimilarity matrices from the outset.

The `clusplot()` command has various options and you'll find out more in the following section.

Adding custom labels to a cluster plot

The `clusplot()` command can be 'customised' – allowing you to tweak the visualisation of your cluster results. The command has the following general form:

```
clusplot(x, clus, diss, lines, shade, color,
          labels, col.p, col.txt, col.clus)
```

There are various instructions you can give to the command:

- `x` – the data to plot; usually this will be a dissimilarity matrix but you can specify a result object directly (e.g. a `pam()` result).
- `clus` – the cluster membership.
- `diss` – if TRUE, `x` is treated as a dissimilarity matrix.
- `lines` – how to join the cluster ellipses: 0 = no lines, 1 = join the centres of the clusters, 2 = join the boundaries of the clusters. If two ellipses overlap no line is drawn.
- `shade` – if TRUE the ellipses are shaded (using cross hatch) by density (number of points divided by area).
- `color` – if TRUE the ellipses are given different colours according to their density.

- `labels` – how to handle labels: 0 = no labels (the default), 1 = points and ellipses can be labelled by clicking with the mouse, 2 = all points and ellipses are labelled, 3 = only the points are labelled, 4 = only the ellipses are labelled, 5 = the ellipses are labelled and the points can be labelled by clicking with the mouse.
- `col.p` – the colour of the points.
- `col.txt` – the colour of any labels, by default this is the same as the point colour.
- `col.clus` – the colours to use for the ellipses. If `color = FALSE` only one colour will be used.

There are other instructions that could be used; use `help(clusplot.default, package = cluster)` to see the help entry for the command. You can also use some standard graphical instructions, e.g. `xlim, ylim, cex`.

By using these instructions you can subtly alter the appearance of your plot. One of the more useful instructions is `labels`, which allows you to identify individual samples on the plot. You can also use your own set of labels and in the following exercise you can have a go at using custom labels on a cluster plot for yourself.

Have a Go: Add custom labels to a cluster plot

You will need the *golf* data for this exercise – the data are in the *CERE.RData* file and show plant abundance (as percentage cover) for 71 samples and 122 species in total. Samples were taken from sites at UK golf courses. You'll also need the *cluster* and *vegan* packages.

1. Start by preparing the *cluster* and *vegan* packages:

    ```
    > library(cluster)
    > library(vegan)
    ```

2. Look at the sample names for the *golf* plant data:

    ```
    > rownames(golf)
     [1] "Q10" "Q11" "Q12" "Q13" "Q14" "Q15" "Q16" "Q17" "Q18" "Q19"
    [11] "Q20" "Q21" "Q22" "Q23" "Q24" "Q25" "Q26" "Q27" "Q28" "Q29"
    [21] "Q30" "Q31" "Q32" "Q33" "Q34" "Q35" "Q36" "Q37" "Q38" "Q39"
    [31] "Q40" "Q41" "Q42" "Q43" "Q44" "Q45" "Q49" "Q50" "Q51" "Q52"
    [41] "Q53" "Q54" "Q55" "Q56" "Q57" "Q58" "Q59" "Q60" "Q61" "Q62"
    [51] "Q63" "Q64" "Q65" "Q66" "Q67" "Q68" "Q69" "Q7 " "Q70" "Q71"
    [61] "Q72" "Q73" "Q74" "Q75" "Q76" "Q77" "Q78" "Q79" "Q8 " "Q80"
    [71] "Q9 "
    ```

3. Each quadrat sample comes from a particular habitat type; look at the *golf.hab* data object to see which samples match to which habitats:

    ```
    > table(golf.hab)
                            HabType
    Habitat                 AG  F  G HG  M Me SRG TFS
      Acid grassland        32  0  0  0  0  0   0   0
      Flush                  0  8  0  0  0  0   0   0
      Grassland              0  0  1  0  0  0   0   0
      Heathy grassland       0  0  0  6  0  0   0   0
      Mire                   0  0  0  0  3  0   0   0
      Mire edge              0  0  0  0  0  3   0   0
      Species-rich grassland 0  0  0  0  0  0   3   0
      Transitional flush/scrub 0  0  0  0  0  0   0  15
    ```

4. You will use the *HabType* variable as labels for the eventual clusters but first you need to make a dissimilarity matrix – use the Bray–Curtis metric:

```
> golf.bc <- vegdist(golf, method = "bray")
```

5. Make a pam() cluster result using three clusters:

```
> golf.pam <- pam(golf.bc, k = 3)
```

6. View the cluster membership:

```
> golf.pam$clustering
Q10 Q11 Q12 Q13 Q14 Q15 Q16 Q17 Q18 Q19 Q20 Q21 Q22 Q23 Q24 Q25
  1   1   1   1   1   2   1   1   1   3   1   1   1   1   1   1
Q26 Q27 Q28 Q29 Q30 Q31 Q32 Q33 Q34 Q35 Q36 Q37 Q38 Q39 Q40 Q41
  1   3   1   3   1   1   1   1   1   1   1   1   1   3   3   3
Q42 Q43 Q44 Q45 Q49 Q50 Q51 Q52 Q53 Q54 Q55 Q56 Q57 Q58 Q59 Q60
  3   3   3   3   3   3   3   3   2   2   2   3   3   3   3   3
Q61 Q62 Q63 Q64 Q65 Q66 Q67 Q68 Q69 Q7  Q70 Q71 Q72 Q73 Q74 Q75
  3   3   3   3   2   2   2   3   3   1   3   3   3   3   3   3
Q76 Q77 Q78 Q79 Q8  Q80 Q9
  3   3   3   3   1   3   1
```

7. Now assign the required labels (the habitats types) to the dissimilarity matrix:

```
> attr(golf.bc, "Labels") <- golf.hab$HabType
```

8. You can now plot the results and use the labels from the dissimilarity matrix as custom labels. Make a plot now and use some of the clusplot() instructions to alter the appearance; your plot should resemble Figure 12.19:

```
> clusplot(golf.pam, labels = 3, lty = 2, col.clus = 1,
           cex = 0.5, lines = 1, col.p = "black",
           main = "Bray-Curtis and PAM", las = 1)
```

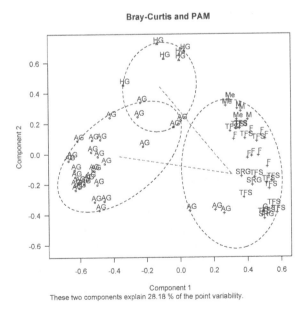

Figure 12.19 A PAM clustering using Bray–Curtis dissimilarity for plant communities at UK golf courses.

In step 8 you used a range of instructions to modify the plot appearance: the `lty` instruction made the lines dashed, `col.clus` set the colour of the ellipses (all to black because color = FALSE), cex made the points smaller (this does not work with the labels), `lines` made the ellipse joining lines point to the centres (you cannot alter the colour), `col.p` set the points to black, `main` set the title and `las` rotated the axis labels to 'all horizontal'.

The labels overlap somewhat – it is virtually impossible to resize these labels but you can try labelling the points interactively by using `labels = 5`. Then click with the mouse – the label appears to the side of the point where the mouse is so you can alter the position to minimise overlap.

Note: Plotting cluster results with `clusplot()`

When you use the `clusplot()` command to plot a cluster result the command looks for the original dissimilarity matrix. If this is missing you get an error. The labels of the dissimilarity matrix are used for the labels of the plot.

12.3 Summary

Topic	Key Points
Similarity and dissimilarity	The more closely the species composition of two samples becomes the higher the similarity. Conversely the dissimilarity decreases as species composition matches more closely.
	Many analyses use dissimilarities rather than similarities.
	You can determine similarity using presence-absence data or with abundance information.
Presence-absence data	Similarity can be determined using three quantities; the number of species in each sample (A and B) and the number of species common to both (J).
	Common algorithms are Jaccard, Sørensen (Bray–Curtis) and Mountford.
	The `dist()` command can determine dissimilarity for the Jaccard metric. The `vegdist()` command in the *vegan* package uses a wide range of metrics. Setting binary = TRUE for `vegdist()` will compute the binary (i.e. presence-absence) version of a metric.
Beta diversity	Measures of *beta* diversity can be used as the basis for dissimilarity. The `betadiver()` command in the *vegan* package can use a wide range of metrics based on presence-absence data.
Abundance data	If you have abundance data then the various metrics use two quantities; x_i and x_j to represent the various species abundance at sites i and j.
	There are many metrics, e.g. Euclidean, Bray–Curtis, Manhattan, Gower.
	The `dist()` command can calculate some metrics whilst the `vegdist()` command in the *vegan* package can calculate others.
	The Euclidean metric is easily computed in Excel but other metrics are more difficult, especially with many samples.

Normalisation and standardisation	Some metrics of dissimilarity are sensitive to differences in abundance. The `decostand()` command in the *vegan* package can normalise data, which is especially useful for Euclidean and Manhattan metrics.
	Wisconsin double standardisation can be carried out with the `wisconsin()` command.
Distance metrics	There are many dissimilarity metrics – those not explicitly used in the `dist()`, `betadiver()` or `vegdist()` commands can be 'created' using the `designdist()` command in the *vegan* package. You can use presence-absence or abundance data.
Hierarchical clustering	In hierarchical clustering data are rearranged in such a way as to form 'layers' of clustering. This forms a kind of family tree, which can be visualised with a dendrogram. The ends of the branches show samples that are most similar to one another.
	The `hclust()` command takes a dissimilarity matrix and forms an hierarchical clustering result, which can be plotted as a dendrogram via `plot()`.
	The `cutree()` command can cut a dendrogram into groups (or by height).
Agglomerative nesting	A form of hierarchical clustering carried out by the `agnes()` command in the *cluster* package. You can convert the result to an `hclust()` result via `as.hclust()`, allowing the `cutree()` command to be used.
Divisive clustering	A form of hierarchical clustering carried out by the `diana()` command in the *cluster* package. The result can be converted to an `hclust()` result via the `as.hclust()` command.
Cluster significance	The *pvclust* package contains a command, `pvclust()`, which can assess the significance of hierarchical cluster results. The command uses bootstrapping to randomise groups and clusters in its assessment.
	You can plot the result and also highlight significant clusters with the `pvrect()` command. The `pvpick()` command will show a text summary of the significant clusters.
Compact ordered community tables	You can use the result of an hierarchical clustering to reorder the samples and species in a compact table, allowing a summary of the community displayed in a meaningful manner. The `vegemite()` command in the *vegan* package creates compact tables.
Partitioning	Clustering by partitioning is done in various ways by assigning samples to groups that cluster around centres (medoids). You have to select how many groups you want.
K-means	K-means analysis is a form of partitioning.
	The `kmeans()` command can carry out k-means analysis but the results are hard to visualise.
	The `pam()` command in the cluster package carries out k-means (partitioning around medoids) analysis. The results of `pam()` can be visualised as a kind of scatter plot showing groups, using the `clusplot()` command.
	The `silhouette()` command can provide details about groups from a `pam()` result.
Fuzzy set analysis	The `fanny()` command in the cluster package can carry out fuzzy analysis. You can specify how 'fuzzy' you want the analysis to be using the `memb.exp` instruction. The result of a `fanny()` analysis can be visualised using the `clusplot()` command.

Topic	Key Points
Large datasets	If processing time/power is an issue then you can use the `clara()` command, which uses a simpler algorithm (than the `pam()` command) to carry out partitioning around medoids.
Visualising partitioning results	The `clusplot()` command in the cluster package can visualise results of partitioning around medoids clustering.
	You can use results from k-means analysis by assigning group membership and using a dissimilarity matrix.
	The `clusplot()` can be customised in many ways to help visualise results.

12.4 Exercises

12.1 Similarity is a measure of species composition so you can use measures of *beta* diversity to explore similarity – TRUE or FALSE?

12.2 Which of the following metrics of dissimilarity cannot be used by the `vegdist()` command in the *vegan* package?

A) Jaccard
B) Sørensen
C) Euclidean
D) Kulczynski
E) Binary

12.3 Hierarchical clustering creates a kind of family tree (a dendrogram) allowing you to visualise relationships between samples according to a dissimilarity metric. What are the main methods of producing hierarchical clustering results?

12.4 Look at the ant data, which shows abundances of ants at ten sites from Northern Australia. Using a hierarchical clustering with a Euclidean distance metric, are any clusters significant?

12.5 In clustering by partitioning methods you have to specify the number of clusters you want at the start of the analysis – TRUE or FALSE?

The answers to these exercises can be found in Appendix 1.

13. Association analysis: identifying communities

So far the examples you've seen have focused on the samples rather than the species. It has been somewhat implicit that each sample has been from a particular community. This chapter takes a different approach and looks to examine the species explicitly.

Species that tend to be found together are likely to be from the same community. Species that tend to be found apart would be from different communities. The larger the area you sample from, the greater the chance that you collect data from more than one community.

Association analysis is about identifying those species that tend to live together (a positive association) and also determining the strength (and possibly statistical significance) of the association. Your analyses will also show those species that tend to live apart, negative association, and so belong to different communities.

Association analysis is not so different from a dissimilarity or similarity measure in that you end up with a matrix of values that represent how 'associated' pairs of species are; there are two main ways you might proceed:

- *Area approach* – you take random samples looking at the presence-absence of the species. By taking many samples you can build up a picture of the different communities that are in your sample area.
- *Transect approach* – you sample across some gradient of some kind, usually where you perceive that there are changes in the community across the gradient.

There is not really a lot of difference between these two approaches; a more important consideration is the method of data collection itself. The larger the sampling unit the greater the chance that you will sample from more than one community. So, lots of small samples are the order of the day. In Section 13.1 you'll see the area approach demonstrated using presence-absence data of plants in small quadrats, whilst in Section 13.2 you'll see the transect approach demonstrated using abundance data.

The mainstay of association analysis is the *chi squared* test, which uses frequency data. This means that whatever measure of abundance you have used in your sampling will be 'converted' to presence-absence data; the frequency of co-occurrence is the important element.

Association analysis can also be used to give an idea of 'indicator species'. An indicator species would be one that is significantly positively associated with a particular community and significantly negatively associated with other communities. You'll see this kind of analysis in Section 13.4.

13.1   Area approach to identifying communities

In any sampling area you may have more than one community, and the larger the area becomes the greater the chance that there are more communities. If you use very small sampling units then each unit is most likely to contain species from a single community and less likely to be in a 'border', with species from two (or more) communities. This means that if you collect a lot of sample units you will be able to pick out those species that tend to live together (i.e. co-occur in sample units) and those that live apart.

For the association analysis you are interested in the number of sampling units that species have in common. The following example shows some presence-absence data for three species (called simply a, b and c) at 12 sites (A to L):

```
  A B C D E F G H I J K L
a 0 0 0 1 1 1 1 1 0 0 1 0 0
b 1 1 0 1 1 0 0 0 1 1 0 0
c 0 1 1 1 1 1 0 0 1 1 0 0
```

By inspection you can work out how many times each species was found co-occurring with the others:

```
    a b
  b 3
  c 4 5
```

So, you can see that a and b were found together three times, a and c four times and b and c five times. You want to know how many times these co-occurrences would have happened if the species were randomly associated with one another. To do this you need to know how many times each species was found and how many samples there were in total. You can easily add up the rows to work out the frequency for each species:

```
  a b c
  5 6 7
```

The total number of samples is 12. Now you can determine the 'expected' value for each pairing. The probability of finding species a is 5/12. The probability of finding species b is 6/12. The probability of finding both together is the product of these two probabilities. You want the number of times rather than the probability so you multiply by the total number of samples to give: $5/12 \times 6/12 \times 12 = 5 \times 6 \div 12 = 2.5$. The process is repeated for each pair:

```
            a        b
  b 2.500000
  c 2.916667  3.500000
```

In general terms this is:

Expected = SpA × SpB ÷ total no. samples

Now you need to assess how close this expected value is to the actual number of co-occurrences. You use the chi squared formula:

$$X^2 = (\text{Obs} - \text{Exp})^2 \div \text{Exp}$$

In the formula Obs is the number of times the two species co-occurred and Exp is the calculated expected value for the co-occurrence. For your overall sample you simply add together all the individual chi squared values and determine the statistical significance by comparing to a table of critical values, taking into account the number of species involved (the degrees of freedom).

In the vast majority of cases you will get a statistically significant result because it is very unlikely that the species you sample are randomly associated – they will tend to form communities. What you can now do is to examine the species pair by pair and look to see how strong the association is between them. This is subtly different from the approach above.

You start by looking at the number of times two species co-occur. You also need to know how many times each species was found and the total number of sampling units. With this information you can work out how many sampling units contained the first species but not the second, how many units contained the second species only, and how many contained neither species. You can make a simple contingency table, the following shows the species b and c from the previous example:

```
          c
 b      + -  Sum
   +    5 1   6
   -    2 4   6
   Sum  7 5  12
```

This contingency table can now be used in a new chi squared analysis to examine the association between the two species (it comes out at $p = 0.242$).

You can use the chi squared approach to make a dissimilarity matrix. You can start from your co-occurrence matrix. The chi squared formula makes all values positive (by squaring Obs – Exp) as you were only interested in the magnitude of the association. If you use the square root of the formula you can calculate the *Pearson residuals*:

$$\text{Pearson resid} = \text{Obs} - \text{Exp} \div \sqrt{\text{Exp}}$$

These residuals will be positive if the species positively associate with one another and negative if they negatively associate. To use the Pearson residuals as a dissimilarity measure you need to rescale the values to get rid of negatives, then subtract each value from the maximum. Thus, the largest of the Pearson residuals becomes the smallest – the least dissimilar.

13.1.1 Using Excel for area-based association analysis

You can carry out association analysis fairly easily using Excel but it becomes quite tedious to copy formulae for pairwise comparisons when you have more than a few species. You cannot easily produce a visual representation of your results using Excel, since the hierarchical cluster dendrogram is not really part of the Excel armoury. In the following exercise you can have a go at making a simple spreadsheet using a small dataset so that you can see the principles involved.

Have a Go: Use Excel for association analysis

You'll need the spreadsheet *Moorland example.xls* for this exercise. The data are a subset of a larger dataset and show the presence-absence of some species of plants on Dartmoor in England.

1. Open the *Moorland example.xls* spreadsheet. There are two worksheets: the *Data only* sheet contains presence-absence information for seven plant species on Dartmoor based on 20 sample quadrats (25 cm square). The other worksheet, *Completed*

version, contains the completed analysis. If you need to check your work (or simply cheat), then look here. For now you can start by working out the number of sample quadrats that each species is found in. Go to cell W1 and type a heading 'Total'.

2. In cell W2 type a formula to work out the number of times *Western Gorse* was found; =COUNT(C2:V2) will do nicely. Copy the formula down the column.

3. Make a row of column totals to show the number of species in each quadrat. Start with a label in cell B9, 'Total'. Then in C9 use, =COUNT(C2:C8). Copy this formula across the row.

4. In cell W9 you want the total number of quadrats used, which is 20. You can either type the value or count the number of column totals, =COUNT(C9:V9).

5. Highlight the common names in cells B2:B8 and copy to the clipboard. Now paste the names into cell B11. Now also click in cell C18 and use *Home > Paste > Paste Special*. Select the *Transpose* button and put the names in cells C18:I18.

6. Highlight cells C18:I18 and use the *Home > Format > Format Cells* button. From the menu that appears use the *Alignment* tab and set the text *Orientation* to 90°.

7. The cells above C18:I18 are going to be your co-occurrence matrix. Start by clicking in cell C12 and type a formula to work out the number of quadrats that *Ling Heather* and *Western Gorse* occur together, =SUMPRODUCT(C2:V2,C3:V3).

8. Now you need to fill out the rest of the matrix. Click in cell C12 and edit the formula to read, =SUMPRODUCT(C$2:V$2,C3:V3). The $ will 'fix' the 2nd row and allow you to copy the formula down the rest of the column. However, you cannot use this trick to go across the next column so you will need to type the formula afresh.

9. In cell C13 type a formula to determine the co-occurrence of *Ling Heather* and *Bell Heather*, =SUMPRODUCT(C$3:V$3,C4:V4). Note the $, which allows you to copy this down the column. Fill out the rest of the matrix using similar formula (you can check your results in the *Completed version* worksheet).

10. Now prepare a matrix for the expected values. Start by copying the species names from cells B11:B17 and paste them into cells B20:26. Then use *Paste Special* to place the names in cells C27:I27, which can then be formatted to rotate them by 90°.

11. Each expected value is the total for species A × species B ÷ total quadrats. So in cell C21 type a formula to work out the expected value for Gorse and Ling, =W$2*W3/$W$9. Note the $, which allow you to copy down the row.

12. In cell D22 type a formula for the expected value between *Ling* and *Bell Heather*, =W$3*W4/$W$9. You can now fill in the rest of the matrix with the expected values for all the species pairs.

13. You now need to make a matrix for the chi squared values. Start in cell B29 with the species names (B29:B35), these need to go in cells C36:I36 as well.

14. Click in cell C30 and type a formula to calculate the chi squared value for *Gorse* and *Ling*, =IF(C12=0,0,(C12-C21)^2/C21). Notice that you have to take into account possible zero values. However, this formula will copy down and across without needing $, so fill in the rest of the matrix by copying the formula into the other cells.

15. Now go to cell G31 and type a heading for the final result, 'Chi-squared' will be fine. Under this type headings 'Total', 'Critical' and 'p-value' into the cells L32: L34.

16. In cell N32 type a formula to work out the total chi squared value, which is the sum of the individual chi squared values, =SUM(C29:I35).

17. In cell N33 type a formula to see the critical value for chi squared with degrees of freedom equal to the number of species (7), =CHIINV(0.05, 7).

18. In cell N34 type a formula to work out the statistical significance of the result, =CHIDIST(N32,7).

19. Next you should work out the significance of individual pairs of species. Type the name of a species into cell D38, 'Ling Heather'. Now in cell B41 type 'Tormentil'. These are the two species you will examine.

20. In Cells D39:F39 type some labels, '+', '-', 'Tot'. You might need to enter the + and – by prefixing with a ' so that Excel treats them as labels (depends on your version of Excel). Enter similar labels in cells C40:C42.

21. Now in cell D40 enter the number of quadrats in which both species co-occurred (so, in the position ++ in the table), =D15. In cell F40 enter the number of total occurrences of *Tormentil*, =W6. In cell D42 enter the number of occurrences of *Ling*, =W3. In cell F42 enter the total number of quadrats placed, =W9.

22. The rest of the table (D40:F42) can be filled in by simple subtraction, for example: if there are eight *Tormentil* in total and five co-occurrences with *Ling*, the missing value (for number of occurrences of *Ling* without *Tormentil*) must be $8 - 5 = 3$. Use this principle to fill in the rest of the table (you can check your results in the *Completed version* worksheet).

23. Now work out the expected values in cells D44:E45. In cell D44 the value will correspond to =F40*D42/F42. This is the row total * the column total/grand total. Use this approach for the rest of the matrix.

24. You now need to make the chi squared values in cells G44:H45. Start in cell G44 and work out the value using =(D40-D44)^2/D44. This corresponds to (Observed - Expected)^2 /Expected; use this approach to fill out the rest of the matrix.

25. Click in cell F47 and type a label, 'ChiSq'. Under this add to more labels, 'Crit' and 'P-val'. Now click in cell G47 and type a formula to work out the total chi squared from the sum of the individual values, =SUM(G44:H45).

26. In cell G48 type a formula to work out the critical value of chi square for one degree of freedom, =CHIINV(0.05,1).

27. In cell G49 type a value to determine the probability of the result (the statistical significance), =CHIDIST(G47,1). The result is 0.09, which is not significant.

You can investigate other species pairs using this approach. The results are not significant, but this is hardly surprising since the data are only part of the original. You generally need quite a lot of data for this kind of analysis: 80–100 quadrats would be more or less a minimum.

In the preceding exercise you looked at part of a dataset and did not find any statistically significant results. You could alter the spreadsheet and calculate the Pearson residuals instead of (or in addition to) the chi squared values. These values could be used to explore the hierarchy and perhaps to construct a dendrogram. However, you'd have to do this manually as there is no easy way to get Excel to do this for you.

13.1.2 Using R for areas-based association analysis

The R program can be pressed into service for association analysis. The `chisq.test()` command is part of the basic installation of R but you need to carry out a bit of manipulation before you get to that point. The *vegan* package is useful here as the `designdist()` command can be used for co-occurrence and expected values calculations.

In the following exercise you can have a go at carrying out an association analysis for yourself. You'll use the same data as in the preceding exercise but this time you can use the 'full version', which contains 100 quadrat samples and 15 species.

Have a Go: Use R for area-based association analysis

You'll need the *moor.pa* data for this exercise, which is found in the *CERE.RData* file. The data are composed of presence-absence data for 15 plant species in 100 quadrats. The data were collected from a valley-bog in Dartmoor in the UK. You will also need the *vegan* package.

1. Look at the data for the Dartmoor plant data: there are two items:

```
> ls(pattern = "moor")
[1] "moor.nam" "moor.pa"
```

2. The *moor.pa* data contains the presence-absence data::

```
> moor.pa[1:4, 1:10]
                 X1 X2 X3 X4 X5 X6 X7 X8 X9 X10
Ulex galii        0  0  1  0  0  0  0  0  0   0
Calluna vulgaris  0  0  0  0  0  0  0  0  0   0
Erica cinerea     0  0  0  0  0  0  0  0  0   0
Erica tetralix    0  1  1  0  0  0  0  0  1   0
```

3. The *moor.nam* data holds the common names for the species (the row names are the same) and an abbreviated name. These additional names can be useful as labels on plots:

```
> names(moor.nam)
[1] "Common" "Abbr"
```

4. Start by making a co-occurrence matrix using the `designdist()` command. You'll need the *vegan* package:

```
> library(vegan)
> moor.co = designdist(moor.pa, method = "J", terms = "binary")
```

5. The co-occurrence matrix can be viewed simply by typing its name but the display is rather unwieldy. Make another matrix for the expected values, the `designdist()` command will do the job:

```
> moor.exp =designdist(moor.pa,method ="A*B/P",terms="binary")
```

6. Now calculate chi squared values using the co-occurrence and expected values:

```
> moor.csq = (moor.co - moor.exp)^2 / moor.exp
```

7. You can see the total chi squared value from the sum of the individual values:

```
> sum(moor.csq)
[1] 106.1635
```

8. Use the total chi squared value to work out the statistical significance:

```
> pchisq(sum(moor.csq), df = nrow(moor.pa), lower.tail = FALSE)
[1] 8.757557e-16
```

9. Now work towards making a dissimilarity matrix so you can draw an hierarchical cluster dendrogram, begin by working out the Pearson residuals:

```
> moor.resid = (moor.co - moor.exp) / sqrt(moor.exp)
```

10. Rescale the Pearson residuals, first make all values positive by 'shifting' the baseline, then subtracting all the values from the maximum:

```
> moor.pr = moor.resid + abs(min(moor.resid))
> moor.pr = max(moor.pr) - moor.pr
```

11. Now the Pearson residuals have been rescaled you can use the values to make a hclust() result, which you can then plot. Use the *Common* names in the *moor.nam* data as custom labels for your plot, which should resemble Figure 13.1:

```
> moor.hc = hclust(moor.pr)
> plot(moor.hc, labels = moor.nam$Common)
```

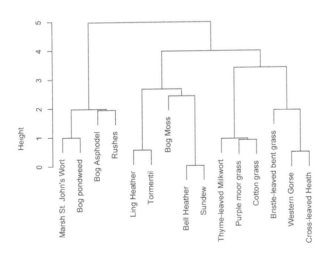

Figure 13.1 Moorland plant association. Distances based on chi squared dissimilarity.

12. The *moor.pa* data contains the presence-absence data, look at the row names, which hold the species names:

```
> rownames(moor.pa)
 [1] "Ulex galii"              "Calluna vulgaris"
 [3] "Erica cinerea"           "Erica tetralix"
 [5] "Potentilla erecta"       "Molina caerulea"
 [7] "Agrostis curtsii"        "Eriopherum spp."
 [9] "Narthecium ossifragum"   "Polygala serpyllifolia"
[11] "Drossera rotundifolia"   "Sphagnum spp."
[13] "Hypericum elodes"        "Potamogeton polygonifolius"
[15] "Juncus spp."
```

13. Look at the association between *Calluna vulgaris* and *Erica cinerea*. Start by setting the names to variables, use the number from the `rownames()` command from step 12:

```
> a = rownames(moor.pa)[2]
> b = rownames(moor.pa)[3]
> a ; b
[1] "Calluna vulgaris"
[1] "Erica cinerea"
```

14. Get the value for the number of co-occurrences, you'll have to 'convert' the co-occurrence matrix (which is really a `dist` object) into a 'proper' matrix:

```
> COOC = as.matrix(moor.co)[a,b]
> COOC
[1] 4
```

15. Now get the values for the species occurrence:

```
> SpA = rowSums(moor.pa > 0)[a]
> SpB = rowSums(moor.pa > 0)[b]
> SpA ; SpB
Calluna vulgaris
              27
Erica cinerea
               5
```

16. Determine the total number of quadrats from the original data:

```
> TOT = ncol(moor.pa) ; TOT
[1] 100
```

17. Now you can fill in the rest of the pairwise species occurrence contingency table by various subtractions:

```
> noA = TOT - SpA
> noB = TOT - SpB
> AnoB = SpA - COOC
> BnoA = SpB - COOC
> nowt = TOT - COOC - AnoB - BnoA
```

18. Make a `matrix` object using the values required in the contingency table:

```
> mat = matrix(c(COOC, BnoA, AnoB, nowt), ncol = 2,
         dimnames = list(Sp.A = c("+", "-"), Sp.B = c("+", "-")))
> names(dimnames(mat)) = c(a,b)
```

19. Look at the contingency table, this contains values for co-occurrence, frequency of each species alone and the number of 'empty' quadrats:

```
> mat
                  Erica cinerea
Calluna vulgaris +  -
               + 4 23
               - 1 72
```

20. You can view the marginal totals with the addmargins() command:

```
> addmargins(mat)
                  Erica cinerea
Calluna vulgaris  +   -   Sum
               +  4   23  27
               -  1   72  73
             Sum  5   95  100
```

21. Now carry out a chi squared test, the Yates' correction will be applied by default:

```
> moor.cs = chisq.test(mat)
Warning message:
In chisq.test(mat) : Chi-squared approximation may be incorrect
```

22. The chi square test gives a warning: view the result and look also at the expected values:

```
> moor.cs

    Pearson's Chi-squared test with Yates' continuity
    correction

data:  mat
X-squared = 4.9374, df = 1, p-value = 0.02628

> moor.cs$exp
                  Erica cinerea
Calluna vulgaris     +       -
               +  1.35    25.65
               -  3.65    69.35
```

In this case you see a significant association but the result should be treated with a little caution, the low expected value leading to the warning.

Note: R code for species associations

The R code illustrated in the exercise is available as several custom commands as part of the CERE.RData file. The dist_chi() command takes a community dataset and calculates the overall species associations. The result of the dist_chi() command has its own class "distchi", which can be plotted (there is also a summary method). The species_assoc() command takes a community dataset and carries out a pairwise chi squared test on a pair of species (which you specify).

Note: Empty spreadsheet cells and R

Some datasets contain empty cells, which may represent missing values, and R converts them to NA when the data are imported via the read.csv() command. However, sometimes the missing values are really zero, this can happen in presence-absence data because you are most likely to enter 1s and not bother with the 0s.

The *gdata* package contains a useful command, NAToUnknown(), which can convert NA items to 0 (or any value).

13.2   Transect approach to identifying communities

The transect-based approach to identifying communities is not much different from the area-based one. The principles are exactly the same – you are looking for co-occurrences of species so that you can see the positive and negative associations and so split the species into community groups.

The difference from the area-based approach is that you are expecting to see a change in community as you go across your gradient. You can carry out this kind of analysis using Excel but if you are dealing with abundance data or you have a lot of species, the process becomes very tedious and repetitive.

When you have presence-absence data (as ones and zeros or blanks) you can use =SUMPRODUCT to determine species co-occurrence. When you have abundance data there is no 'shortcut' formula to determine species co-occurrence. You have several choices:

- Work out co-occurrence manually for each pair of species.
- Copy the data to a new spreadsheet using a formula that converts any number to 1 (e.g. IF(A1 > 0, 1, "")).
- If your data are arranged in biological recording format make a Pivot Table using *Count* of data rather than *Sum*.

Even if you are able to convert the data to presence-absence the processes become unwieldy later on and you will still be unable to make a dendrogram. You can see where this is heading – it is easier to use R right from the outset. In the following exercise you can have a go at a transect-based association analysis using some abundance data.

Have a Go: Use R for transect-based association analysis

You will need the *ridgeT* data for this exercise. The data show plant abundances, measured as frequencies (using point quadrats) at 79 sample locations. The samples were taken from ten transect stations (there were eight replicated transects). The transect covered an area on a beach edge moving back away from the shore – thus showing a successional gradient (less disturbed away from the shore). The data are part of the *CERE.RData* file. You'll also need the *vegan* package.

1. Look at the data, which are arranged with species as rows and samples as columns:

```
> ridgeT[5:10,12:20]
```

	T13	T14	T15	T16	T17	T18	T19	T20	T22
Glaucum flavum	0	0	0	0	0	0	0	0	0
Tripleurospermum maritimum	0	0	0	0	0	0	0	0	0

Aquilea millefolia	5	2	5	3	4	0	0	2	3
Cerastium sp.	0	0	0	0	0	0	0	0	0
Heracleum spondylum	1	0	0	0	0	0	0	0	2
Hypochaeris radicata	1	3	2	1	0	0	0	0	0

2. There is 'support' for the *ridgeT* data, look at the *ridge.nam* object, which contains alternative names as well as the variable *Type*:

```
> head(ridge.nam)
```

	Common	Type	Abbr
Cochlearia danica	Danish Scurvy Grass	P	Coc.dan
Elymus farctus	Sand Couch Grass	P	Ely.far
Erodium lebelii	Sticky stork's-bill	P	Ero.leb
Euphorbia paralias	Sea Spurge	P	Eup.par
Glaucum flavum	Yellow Horned Poppy	P	Gla.fla
Tripleurospermum maritimum	Sea Mayweed	P	Tri.mar

3. The information in the *ridge.nam* data can be helpful for visualising results. Look at the *Type* variable: it contains five levels, these correspond to the sort of species, *H* = Hardy generalists, *MC* = Meadow Community, *MS* = Maritime Specialist, *P* = Pioneer and *SC* = Shrub Community:

```
> levels(ridge.nam$Type)
[1] "H"  "MC" "MS" "P"  "SC"
```

4. Start the association analysis by making a co-occurrence matrix, you'll need the *vegan* package:

```
> library(vegan)
> ridge.co = designdist(ridgeT, method = "J", terms = "binary")
```

5. Now calculate the expected values and use these to determine the chi squared values and Pearson residuals:

```
> ridge.exp =designdist(ridgeT,method ="A*B/P",terms="binary")
> ridge.csq = (ridge.co - ridge.exp)^2 / ridge.exp
> ridge.resid = (ridge.co - ridge.exp) / sqrt(ridge.exp)
```

6. Look at the sum of the chi squared values and work out the statistical significance:

```
> sum(ridge.csq)
[1] 1464.984
> pchisq(sum(ridge.csq), df = nrow(ridgeT), lower.tail = FALSE)
[1] 1.134069e-271
```

7. In order to use the Pearson residuals as a dissimilarity measure you'll need to rescale the values:

```
> ridge.pr = ridge.resid + abs(min(ridge.resid))
> ridge.pr = max(ridge.pr) - ridge.pr
```

8. Now the Pearson residuals are resealed you can form a hierarchical cluster object:

```
> ridge.hc = hclust(ridge.pr)
```

9. Make some custom labels that combine the common name and the *Type* variable – the paste() command can join items:

```
> labs = paste(ridge.nam$Common, ridge.nam$Type, sep = "-")
```

10. Now plot the dendrogram and use the custom labels, your result should resemble Figure 13.2:

```
> plot(ridge.hc, labels = labs)
```

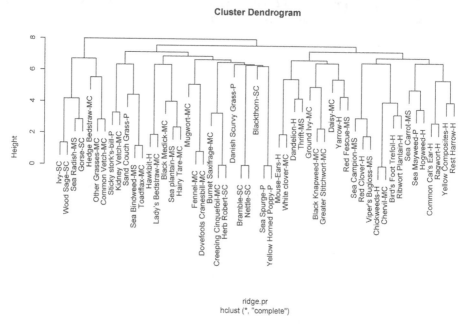

Figure 13.2 Plant communities across a shingle ridge. Dissimilarities using chi squared.

11. With many species the plot can be hard to read, you can resize it using the mouse to rescale the plot window.

12. You can carry out a pair by pair species association using the same approach as in the preceding exercise. This would involve taking the co-occurrence matrix and making a 2 × 2 contingency table to show: number of samples containing both species, number containing species *a* but not species *b*, number containing *b* but not *a*, and number of samples with neither species. You then carry out a chi squared test on the contingency table. You can look back at the preceding exercise to see how to carry this out but the *CERE.RData* file contains a custom function that will perform the necessary calculations for you.

13. Start by looking at the species available:

```
> rownames(ridgeT)
 [1] "Cochlearia danica"        "Elymus farctus"
 [3] "Erodium lebelii"          "Euphorbia paralias"
 [5] "Glaucum flavum"           "Tripleurospermum maritimum"
 [7] "Aquilea millefolia"       "Cerastium sp."
 [9] "Heracleum spondylum"      "Hypochaeris radicata"
...
```

14. There are 53 species (only 10 are shown in step 13). Look at species 4 and 5, these are *Yellow-horned poppy* and *Sea spurge*. The species_assoc() command in the *CERE. RData* file will carry out the analysis:

```
> ridge.as = species_assoc(4,5,ridgeT)

    Pearson's Chi-squared test with Yates' continuity
    correction

data:  mat
X-squared = 9.2486, df = 1, p-value = 0.002357

Warning message:
In chisq.test(mat) : Chi-squared approximation may be incorrect
```

15. Look at the result object you just made in step 14. There are several components:

```
> names(ridge.as)
[1] "spA"       "spB"       "chi.sq"    "co.occur"
```

16. Look at the expected values – these are part of the $chi.sq element, which is the result of a chisq.test(), you'll see that there are some low values here, which is why you got the warning in step 14:

```
> ridge.as$chi$exp
                      Glaucum flavum
Euphorbia paralias       +              -
                  + 0.02531646   0.9746835
                  - 1.97468354  76.0253165
```

17. The $co.occur element contains the co-occurrence contingency table, view it along with the marginal totals:

```
> addmargins(ridge.as$co.occur)
                      Glaucum flavum
Euphorbia paralias  +   -    Sum
                  +  1   0    1
                  -  1  77   78
                Sum  2  77   79
```

The plot in step 10 does show that there is some pattern forming but it is hard to see. You probably need more data to make the pattern more discernible but even so, you can start to see the shrubs and meadow species starting to split apart.

A big issue with this kind of analysis is that you need relatively high numbers of samples. You need to collect enough data so that expected values are greater than 5. In practice this means 100–200 samples or more.

13.3   Using alternative dissimilarity measures for identifying communities

Using the chi squared approach is not the only way to set about identifying the various communities in your samples. The chi squared method allows you to put a level of significance to species associations but you can use other dissimilarity measures. These will allow you to carry out hierarchical clustering and so view the data splitting into community clusters.

R assumes that your community data are arranged with species as columns and samples as rows. If your aim is to identify the species clusters then you need to switch this around; the t() command will do this easily. Once you have your data in the appropriate arrangement you can carry out hierarchical clustering using the methods outlined in Section 12.2.1.

It does not matter if your data are presence-absence or are some kind of abundance meas-ure, you will be able to use dissimilarity measures appropriate to either. In the following exercise you can have a go at using presence-absence data and some dissimilarity measure to explore the same moorland data you looked at earlier – this will form a useful comparison.

Have a Go: Use dissimilarity with presence-absence data for community identification

You'll need the *moor.pa* data for this exercise, which is found in the *CERE.RData* file. The data are composed of presence-absence data for 15 plant species in 100 quadrats. The data were collected from a valley-bog in Dartmoor in the UK. You will also need the *vegan* package.

1. Start by preparing the *vegan* package:

   ```
   > library(vegan)
   ```

2. Because the data are presence-absence you can only use the 'binary' type of dis-similarity measures. Start with the dist() command:

   ```
   > moor.bin = dist(moor.pa, method = "binary")
   ```

3. Now make a hierarchical cluster object and draw the dendrogram – use the com-mon names for the labels (from the *moor.nam* object), your dendrogram should resemble Figure 13.3:

   ```
   > moor.bin = dist(moor.pa, method = "binary")
   > moor.hc = hclust(moor.bin)
   > plot(moor.hc, labels = moor.nam$Common)
   ```

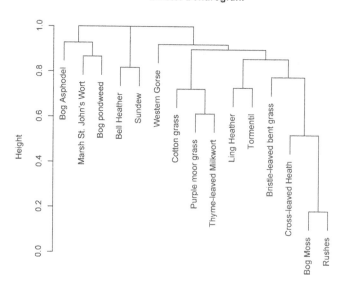

Cluster Dendrogram

moor.bin
hclust (*, "complete")

Figure 13.3 Plant species from a valley bog on Dartmoor. Based on Jaccard dissimilarity.

4. The dendrogram you made in step 3 shows a reasonable hierarchy but try another dissimilarity measure for comparison. Use the Raup–Crick measure and produce a dendrogram, yours should resemble Figure 13.4:

```
> moor.ra = vegdist(moor.pa, method = "raup", binary = TRUE)
> moor.hc = hclust(moor.ra)
> plot(moor.hc, labels = moor.nam$Common)
```

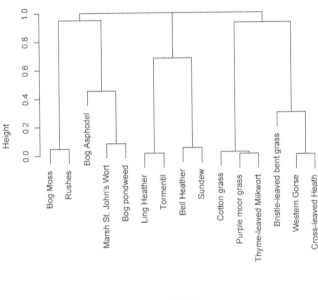

Figure 13.4 Plant species from a valley bog on Dartmoor. Based on Raup–Crick dissimilarity.

Compare the two dendrograms to the one you made earlier using the chi squared approach (Figure 13.1). They all show similar, but slightly different patterns.

The dissimilarity measure you use will affect the final dendrogram, since each measure will give you subtly different results. In general the more samples you have the closer the patterns will be. It is always worth taking several approaches and comparing the final dendrograms. Hopefully you will be able to spot similarities between the results, which can strengthen your conclusions. In the Dartmoor plant communities, for example, all three dendrograms pick out similar groupings – use the rect.hclust() command to cut the dendrogram into groups. If the groups end up broadly the same you can be more confident of the community groupings. If you use k = 3 for the three dendrograms from the moorland data you'll see that the chi squared and the Raup–Crick communities are very similar.

If you have abundance data you'd expect the dissimilarity measure to be more 'accurate' than for presence-absence data. In the following exercise you can have a go at comparing hierarchical dendrograms using the shingle ridge data that you used earlier.

Have a Go: Use dissimilarity with abundance data for community identification

You will need the *ridgeT* data for this exercise. The data show plant abundances, measured as frequencies (using point quadrats) at 79 sample locations. The samples were taken from ten transect stations (there were eight replicated transects). The transect covered an area on a beach edge moving back away from the shore – thus showing a successional gradient (less disturbed away from the shore). The data are part of the *CERE.RData* file. You'll also need the *vegan* package.

1. Start by preparing the *vegan* package:

   ```
   > library(vegan)
   ```

2. Try using the Bray–Curtis dissimilarity:

   ```
   > ridge.bc = vegdist(ridgeT, method = "bray")
   ```

3. Before making the dendrogram, make some custom labels:

   ```
   > labs = paste(ridge.nam$Common, ridge.nam$Type, sep = "-")
   ```

4. Now take the dissimilarity matrix, make an hierarchical cluster object and plot it; your dendrogram should resemble Figure 13.5:

   ```
   > ridge.hc = hclust(ridge.bc)
   > plot(ridge.hc, labels = labs)
   ```

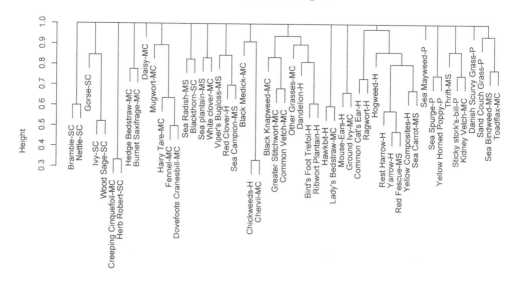

Figure 13.5 Plant species across a shingle ridge successional gradient. Based on Bray–Curtis dissimilarity.

5. Try the Kulczynski dissimilarity and make another dendrogram; yours should resemble Figure 13.6:

```
> ridge.ku = vegdist(ridgeT, "kulczynski")
> ridge.hc = hclust(ridge.ku)
> plot(ridge.hc, labels = labs)
```

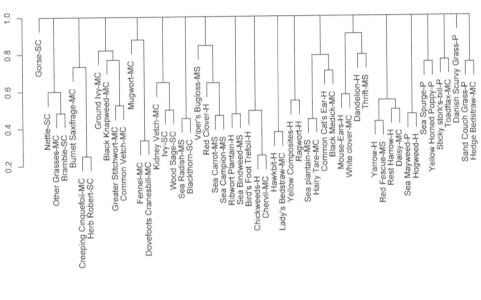

Cluster Dendrogram

ridge.ku
hclust (*, "complete")

Figure 13.6 Plant species across a shingle ridge successional gradient. Based on Kulczynski dissimilarity.

Compare these dendrograms to the one you made using the chi squared approach (Figure 13.2). The overall patterns are similar, but the Bray–Curtis and Kulczynski measures produce dendrograms with a 'flat' top.

The problem with the *ridge* data is that there are not enough samples and too many zero values. Some of the species are not very abundant and these have a large effect on the results. This highlights a fundamental problem with this kind of analysis, which is that you need to collect your data in an appropriate manner and get lots of it!

Note: Community identification and cluster significance

You could use the `pvclust()` command from the *pvclust* package to explore the significance of the groupings in a hierarchical clustering (see Section 12.2.1).

13.4 Indicator species

So far you've seen how to use species co-occurrence to identify which species associate positively with one another, and so form communities. The process also identifies which species disassociate with one another and so form separate communities. The chi squared approach can also be used to examine the associations between species and samples (i.e. habitats or sites). If you find a species positively associated with one habitat but negatively associated with other habitats then you have potentially found an *indicator species*.

Indicator species can be useful because you only have to record the presence of the species in a habitat to recognise the community that you are dealing with. Many conservation efforts use indicator species as monitoring tools for example – so it can be helpful to identify which species are potential indicators.

The 'classic' method of using the chi squared test is to construct a contingency table that shows the frequency of observations across two ranges of categories. In the case of indicators the categories would be the species and the samples (i.e. habitats or sites):

	Habitat-A	Habitat-B	Habitat-C	Habitat-D	Sum
Species-a	23	16	6	2	47
Species-b	97	8	42	9	156
Species-c	5	12	43	51	111
Species-d	0	63	12	11	86
Sum	125	99	103	73	400

The analysis requires that the data are in the form of frequencies – this means counts of individuals if you are dealing with animals. If you are dealing with plants then you might have to be more 'creative' – percentage cover is a form of frequency so is 'acceptable' as long as you don't limit the values that can occur in any one cell of your contingency table. If you standardise your data then your results will be misleading so always deal with the 'raw' counts or frequencies.

The expected values for each cell of your contingency table are calculated quite simply:

Exp = row total × column total ÷ grand total

For the example shown here the occurrence of species *a* at site *A* is 23. There are 47 of species *a* in total and in site *A* 125 individuals were found (of all species). The grand total is 400. The probability of finding species *a* is 47 ÷ 400, and the probability of finding something in site *A* is 125 ÷ 400. The probability of finding species *a* at site *A* is therefore the product of these two. You want the actual number likely to be found, rather than a probability, so multiply by the grand total to give: 47 ÷ 400 × 125 ÷ 400 × 400. After a bit of cancelling you get: 47 × 125 ÷ 400 = 14.7.

Once you have all the expected values you can work out chi squared:

$X^2 = (Obs – Exp)^2 ÷ Exp$

For the purposes of indicator analysis it would better to use the Pearson residuals, though, because they are positive for positive associations and negative for negative associations. The residuals are found by taking the square root of the formula:

Pearson resid = $(Obs – Exp) ÷ \sqrt{Exp}$

Once you have your Pearson residuals you can look to see if any of the species can be regarded as indicators:

	Habitat-A	Habitat-B	Habitat-C	Habitat-D
Species-a	2.17	1.28	-1.754	-2.25
Species-b	6.91	-4.93	0.289	-3.65
Species-c	-5.04	-2.95	2.697	6.83
Species-d	-5.18	9.04	-2.156	-1.19

The Pearson residuals can be considered to be approximately normally distributed and a value of ~2 would be considered as significant. In the example here you can see that there are several values above the threshold. However, to be an indicator species you would need to see one positive value and the rest negative, for any one species. Species c shows values that are all above the threshold but there are two positive and two negative values.

You can carry out chi squared indicator analysis quite easily using Excel – there is a fair amount of copying and pasting involved but since you do not have any triangular matrices to work out these operations are easy enough. In the following exercise you can have a go at an indicator analysis using a small dataset showing polypore fungi communities in Caribbean mangrove swamps.

Have a Go: Use Excel for indicator species analysis

You'll need the *Mangrove Fungi.csv* file for this exercise. The file is a comma separated values (CSV) file, which you can read using Excel. The data shows the occurrence of several species of fungi in three different kinds of mangrove swamp (Gilbert & Sousa 2002).

1. Open the *Mangrove Fungi.csv* file using Excel. From the *Open* window you may have to alter the *filetype* box (just above the *Open* button) to display *Text Files* or *All Files*. Alternatively you can find the file using the File Explorer, CSV files are likely to open with Excel as default (but you can also right-click and select *Open with*).

2. Click in cell E1 and type a label for the totals, 'Total' will do nicely. Now click in cell E2 and type a formula to calculate the row sum, that is the number of occurrences of the *Phe.swi* species: =SUM(B2:D2). Copy this formula down the column to fill cells E2:E10 with the row totals.

3. Click in cell A11 and type a label for the column (site) totals, 'Total' will be fine. Now in cell B11 type a formula to calculate the column total: =SUM(B2:B10). Copy the formula across the row to fill cells B11:E11. You should now have the row totals, column totals and a grand total (in cell E11).

4. Click in cell B1 and then highlight the three column totals B1:D1. Copy these to the clipboard and paste the result into cell G1. The new labels will be used for the expected values, which you will calculate next.

5. It can be helpful to make the row and column totals bold, so that you can differentiate them from the main data. So, highlight them and click the bold button from the *Home* menu (in the *Font* section). Now click in cell G2 and type a formula to work out the expected value for the *Phe.swi* species in the *Black* mangrove habitat: =E2*B11/E11.

6. The formula you typed in step 5 takes the row total × column total ÷ grand total. You need to copy this across all columns and rows but it will need modification otherwise the cell references will not work correctly. Fix the grand total reference to $E$11. Now the column totals are all in row 11 so fix the row to B$11. The column totals are all in column E so fix that to $E2. The final formula should read: =$E2*B$11/$E$11.

7. Copy your formula from step 6 across the rows, then down the columns to fill all the expected values. You may want to format the cells to an appropriate number of decimal places (0 or 1) to make them display nicely.

8. Now copy the site names to the clipboard and paste them into cell K1 – these will make the headings for the Pearson residuals. Go to cell K2 and type a formula to calculate the Pearson residual for the *Phe.swi* species in the *Black* mangrove habitat: =(B2-G2)/SQRT(G2). The formula does not need any modification so you can copy it to all the other cells in the Pearson residuals matrix.

9. Now you have the Pearson residuals for all the species and site combinations. You can inspect them so see if there are any 'interesting' results. It can be helpful to highlight the significant results (i.e. those > 2 or < -2). So, highlight all the Pearson residuals (cells K2:M10) and use conditional formatting. Use *Home > Conditional Formatting > New Rule* to bring up a dialogue box.

10. From the *New Formatting Rule* window select the *Rule Type: Format only cells that contain*. Then edit the rule description so that you get *Cell Value not between -2 and 2*. Now click the *Format* button and select the style of formatting – colour red and bold font are adequate. The final window will look somewhat like Figure 13.7:

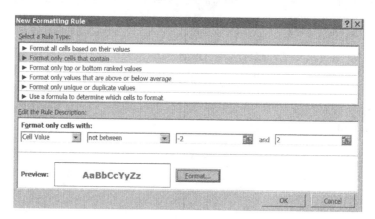

Figure 13.7 Conditional formatting to highlight significant Pearson residual results.

11. Once you have set the conditional formatting in step 10 you can click the OK button and you will see the significant Pearson residuals highlighted, which makes it easier to see.

12. You can see that the three topmost species are each significantly associated positively with one type of mangrove and negatively with the others. The result needs to be treated with caution though, as there are several expected values of < 5.

13. You can make a chart of the results – a bar chart is simplest and most effective. There are two ways to go: show the habitats as the categories or species as categories. To show habitats as categories is easiest as each data series will be a row from your Pearson residuals table. Click on the spreadsheet away from any data and start to build a bar chart from *Insert > Column > 2-D Column*. The species will each be a separate data series and the *Horizontal (Category) Axis Labels* will be the site names.

14. Since you have positive and negative bars the axis labels will overlap the data in the middle of the plot. You can alter this from the Format Axis dialogue window. Click once on the chart, then use the *Chart Tools > Layout > Axis button*. Select *Primary Horizontal Axis* and then *More Primary Horizontal Axis Options*. From the Axis Options section you'll see the *Axis labels* set to *Next to Axis*, alter this to *Low* and the labels appear at the bottom.

15. You'll need a bit of fiddling to set axis titles and generally make the graph appear acceptable, hopefully your final graph will resemble Figure 13.8:

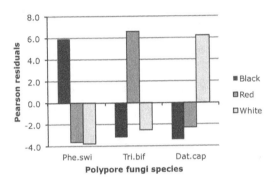

Figure 13.8 Polypore fungi species indicators at three mangrove types in the Caribbean.

16. You can use a similar process (steps 13–15) to make a chart that shows the species as the categories. In this case the columns of the Pearson residuals are the data series and the species names form the horizontal category labels. If your data are not in a contiguous block you'll have to use the control key to select the individual cells you want. It might be simpler to copy the data to the clipboard and use *Paste Special* to construct the chart data as a separate block (paste the values only). In any event you'll need to do some fiddling to get your chart to appear something like Figure 13.9:

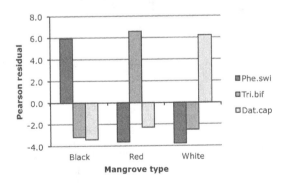

Figure 13.9 Mangrove types and polypore fungi as indicator species in the Caribbean.

When you have a lot of sites or a lot of species you need to exercise some restraint with your graphics – try not to overcrowd a graph. It might be better to make several graphs rather than squeezing everything in one.

If you use R you can use the `chisq.test()` command to carry out the calculations; the command gives the expected values and the Pearson residuals as part of the result. With a bit of manipulation you can extract the significant results, which you can plot in a similar manner to the Excel charts you made in the preceding exercise. In the following exercise you can have a go at carrying out an indicator analysis using ground beetle data from three habitat types.

Have a Go: Use R for indicator species analysis

You'll need the *gb.biol* data for this exercise – it is part of the *CERE.RData* file. The data show the abundances of ground beetles at three habitat types, *Wood*, *Grass* and *Edge*. Each habitat has six replicate samples. You'll also need the *gb.site* data, which contains the *Habitat* grouping variable.

1. Start by looking at the row names of the gb.biol data, there are 18 samples, which is six replicates for each habitat:

```
> rownames(gb.biol)
 [1] "E1" "E2" "E3" "E4" "E5" "E6" "G1" "G2" "G3" "G4" "G5" "G6"
[13] "W1" "W2" "W3" "W4" "W5" "W6"
```

2. Look at the *Habitat* variable of the *gb.site* data, this matches up with the original samples:

```
> gb.site$Habitat
 [1] Edge  Edge  Edge  Edge  Edge  Edge  Grass Grass Grass Grass
[11] Grass Grass Wood  Wood  Wood  Wood  Wood  Wood
Levels: Edge Grass Wood
```

3. You need to amalgamate the samples by *Habitat* to produce a new data item. Use the `rowSum()` command to achieve this:

```
> gb.grp = rowSum(gb.biol, group = gb.site$Habitat)
> gb.grp[,1:6]
      Aba.par Acu.dub Ago.afr Ago.ful Ago.mue Ago.vid
Edge     2146       0       0       0       3       0
Grass     709       3      20       3       0       2
Wood     2013       0       0       0       0       0
```

4. Use the `chisq.test()` command to carry out a chi squared test on your grouped data:

```
> gb.cs = chisq.test(gb.grp)
Warning message:
In chisq.test(gb.grp) : Chi-squared approximation may be incorrect
```

5. The result from step 4 contains the expected values and the Pearson residuals. Look at these then make a new object to hold the residuals:

```
> names(gb.cs)
[1] "statistic" "parameter" "p.value"   "method"    "data.name"
[6] "observed"  "expected"  "residuals"

> PR = gb.cs$residuals
```

6. You can view the results by typing the name but you'll see all the values and not just the significant ones. To see only those results where the *Edge* habitat is significant use a which() command like so:

```
> PR[, which(abs(PR[1,]) > 2)]
            Aba.par    Ago.afr    Bem.big    Bem.man    Cal.fus    Cli.fos
Edge     4.092740  -2.841032  -3.192145  -4.170376  -2.841032  -2.979700
Grass   -2.034203   9.506136  11.182116  18.561500   9.506136   9.970119
Wood    -2.706151  -2.963855  -3.629967  -7.107075  -2.963855  -3.108518
            Lei.ful    Lei.ruf    Neb.bre    Not.big    Poe.cup
Edge     3.385070  -7.245067  -3.209243  -3.0466677  -2.682207
Grass   -1.429568  -5.450846  -6.097483   0.2020969  17.636905
Wood    -2.389537  10.205860   6.724140   2.7995061  -7.980425
            Pte.mad    Pte.mel    Pte.nige
Edge     2.4886677   5.124711  -5.4475749
Grass   -5.1617001  -1.531811   0.8567046
Wood     0.7025095  -3.995918   4.7092925
```

7. The command in step 6 can be 'refined' by adding a statement for each row:

```
> PR[, which(abs(PR[1,]) > 2 & abs(PR[2,]) > 2 & abs(PR[3,]) > 2)]
            Aba.par    Ago.afr    Bem.big    Bem.man    Cal.fus    Cli.fos
Edge     4.092740  -2.841032  -3.192145  -4.170376  -2.841032  -2.979700
Grass   -2.034203   9.506136  11.182116  18.561500   9.506136   9.970119
Wood    -2.706151  -2.963855  -3.629967  -7.107075  -2.963855  -3.108518
            Lei.ruf    Neb.bre    Poe.cup
Edge    -7.245067  -3.209243  -2.682207
Grass   -5.450846  -6.097483  17.636905
Wood    10.205860   6.724140  -7.980425
```

8. Now from step 7 you can see only those species that have significant Pearson residuals for all three habitats. Use the up arrow to get the command back on the entry line and edit it to save the result to a named object, which you can then use for plotting:

```
> PRS = PR[, which(abs(PR[1,]) > 2 & abs(PR[2,]) > 2 & abs(PR[3,]) > 2)]
```

9. Use the barplot() command to make a plot showing the results grouped by species, your chart should resemble Figure 13.10:

```
> barplot(PRS, beside = TRUE, las = 2, legend = TRUE,
         args.legend = list(bty = "n", x = "topleft"))
> abline(h = c(-2,2), lty = 3)
> abline(h = 0)
> title(ylab = "Pearson residuals")
```

10. If you select a single row of the result you can show indicators for a single habitat. If you select a single column you can show a single species associations for all the habitats. You can also use the t() command to rotate the result so that your barplot() is grouped by habitat rather than species. Try splitting the plot window into sections and making multiple plots. Your result should resemble Figure 13.11:

```
> op <- par(mfrow = c(2,2)) # split the window 2 x 2
> barplot(PRS[1,], las = 2, cex.names = 0.8) # plot first row
> title(main = "Edge habitat", ylab = "Pearson residuals")
> abline(h = c(-2, 2), lty = 3) # add horizontal lines at the significance
```

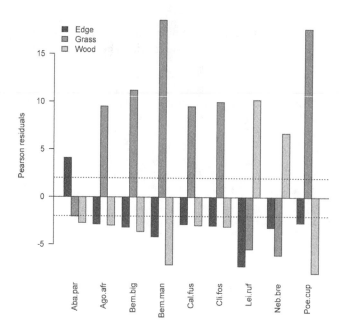

Figure 13.10 Ground beetle indicator species for three habitat types in southern England.

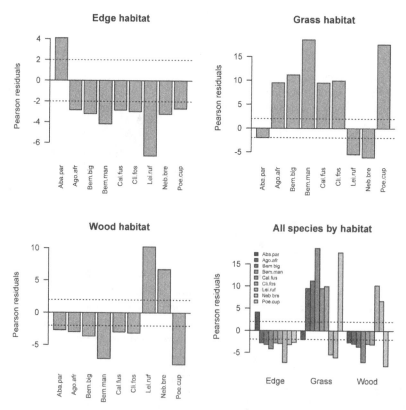

Figure 13.11 Ground beetle indicator species for three habitat types in southern England.

```
> abline(h = 0) # add a line across the middle
> title(ylab = "Pearson residuals") # add a title

> barplot(PRS[2,], las = 2, cex.names = 0.8)
> title(main = "Grass habitat", ylab = "Pearson residuals")
> abline(h = c(-2, 2), lty = 3)
> abline(h = 0)

> barplot(PRS[3,], las = 2, cex.names = 0.8)
> title(main = "Wood habitat", ylab = "Pearson residuals")
> abline(h = c(-2, 2), lty = 3)
> abline(h = 0)

## Rotate the data to group by habitat
> barplot(t(PRS), beside = TRUE, las = 1, legend = TRUE,
        args.legend = list(bty = "n", x = "topleft", cex = 0.6))
> title(main = "All species by habitat", ylab = "Pearson residuals")
> abline(h = c(-2, 2), lty = 3)
> abline(h = 0)

> par(op) # reset the window to single (the current plot is unaffected)
```

Using multiple plots in one window can be useful, allowing you to present multiple charts in a compact manner.

In the preceding exercise you can see the various indicator species fairly clearly. Of course the result should be treated with some caution, as some expected values were < 5. You must also be careful in interpreting the result of indicators. Look at the abundance of the *Aba.par* species for example:

```
Edge Grass  Wood
2146  709  2013
```

Its Pearson residuals show it to be significantly associated with the *Edge* habitat and disassociated with the *Grass* and *Wood* habitats:

```
Edge Grass  Wood
   4   -2    -3
```

This species is still to be found in substantial numbers in habitats that it is significantly disassociated with! You can perhaps think of the indicator statistic as showing the preference of the species; you still need to exercise some common sense.

Note: R commands for indicator species analysis

When you have only a few habitat types you can use the which() command to help select the significant Pearson residuals from your chi squared result, as was demonstrated in the exercise. When you have more habitats this could become unwieldy. The species_ind() command in *CERE.RData* file will carry out the chi squared indicator analysis for you and return a result that contains only the significant results. You can use the command on data where the species are rows or columns by setting the rows instruction (the default is "species" as rows).

13.5 Summary

Topic	Key Points
Species associations	If you sample from a large area you are more likely to collect species from more than one community. Association analysis is one way to identify communities from your samples. The more often species are found together the more likely it is that they are from the same community; they are positively associated. Species that tend to be found in separate samples are likely to be from different communities (i.e. they are negatively associated).
Chi squared	The chi squared statistic is used to examine associations. You use count data to determine the frequency of co-occurrence. You calculate expected values from the data and use the deviation from the observed values to help you calculate a chi squared statistic.
	Pearson residuals are derived from chi squared values, positive residuals relate to positive associations, negative residuals relate to negative associations.
	The `designdist()` command in the *vegan* package can compute co-occurrence matrices and expected values.
Area-based analysis	In an area-based approach you sample over a geographical area and identify the communities in that area.
	You use the Pearson residuals as the basis for creating a dissimilarity matrix, which you can visualise with a dendrogram using the `hclust()` and `plot()` commands.
	The custom command `dist_chi()` in *CERE.RData* will compute the expected values and Pearson residuals. The command has plot and summary methods, allowing you to make and display a dendrogram.
Transect-based analysis	In a transect-based approach you sample along a gradient, either environmental or successional.
	The analysis is the same as for the area-based approach.
Pairwise associations	You can use chi squared tests to examine the statistical significance of pairwise co-occurrence. The Yates' correction is usually applied. The `chisq.test()` command can carry out the test and apply the correction.
	The custom command `species_assoc()` in *CERE.RData* will carry out a chi squared tests on two species from a dataset.
Alternative dissimilarity metrics	Any measure of dissimilarity could be used with a hierarchical clustering analysis to explore species groupings.
	For most R commands you'll need to set the species as the rows to make them the 'target' of the analysis. The `t()` command can rotate a dataset.
	You can use the `pvclust()` command in the *pvclust* package to examine cluster significance using a variety of dissimilarity metrics.
Indicator species	The chi squared approach can be used to look at indicator species. An ideal indicator would be one that has significant positive association with one community and negative associations with others.

To explore data for indicator species you make a contingency table showing species frequency in the various samples (usually amalgamated by habitat) and carry out a 'classic' chi squared test. This can be easily done in Excel or with the `chisq.test()` command in R.

The custom command `species_ind()` in *CERE.RData* will carry out the analysis and return only species with significant associations.

Even when you find a species with the required positive and negative associations it can still be found in the 'wrong' communities in substantial numbers.

13.6 Exercises

13.1 When carrying out chi squared analysis of species co-occurrence for identifying communities why would you use Pearson residuals rather than 'regular' chi squared values?

13.2 Which of the following statements does **not** apply to the area-based community identification approach?

A) You want to keep sampling units small.
B) You must use a random sampling approach.
C) You only need presence-absence data.
D) You can use abundance data.

13.3 In a transect-based approach you use are expecting there to be changes in communities along the transect(s) that you sample – TRUE or FALSE?

13.4 Here is a contingency table that shows the co-occurrence of two species:

	B-present	B-absent
A-present	150	40
A-absent	50	25

Is there a significant association between these two species?

13.5 In indicator species analysis a species would be taken as a good indicator if it had a significant positive association with a habitat – TRUE or FALSE?

The answers to these exercises can be found in Appendix 1.

14. Ordination

Ordination refers to a variety of methods that take complicated data items and reduce them to a more manageable form, thus revealing patterns and relationships that were not evident on first inspection. The term ordination is a broad one: you can think of it as 'putting things in order' – another term often used in this context is *multivariate analysis*.

If you have a simple environmental condition, let's say pH, you could examine the growth response of various plant species and arrange them along an axis of pH. Each species is likely to have an optimum pH at which it grows and you can make a 1-D graph to show how the plants relate to one another. This is a simple ordination.

If you have a second environmental condition you could plot the species responses on a second axis to produce a 2-D scatter plot. This second axis would help to separate the species and permit you to see groupings that weren't evident from the 1-D plot. If the relationship between the two environmental variables was *orthogonal* (i.e. there was low correlation between the variables) then the pattern is maximised – you would gain very little additional information if the two axes were highly correlated with one another.

If you add a third variable you might be able to make a 3-D plot but beyond that you have a problem – you simply haven't got the physical space to represent the data. This is what ordination does – it takes the multi-dimensional data and 'squashes' it into fewer dimensions (usually 2). This allows you to see the results in a form that you can comprehend (such as a 2-D scatter plot).

Commonly in community ecology you have several samples and wish to explore the relationships between the samples in terms of the species composition (the communities); you therefore have multivariate data with many variables (the species being the variables). The various methods of ordination take the samples (the sites) and reorder them according to the species composition. Some (but not all) methods also allow you to visualise the species and how they relate to the samples.

It is possible to utilise environmental data in ordination – commonly you wish to use information about the environment to help 'align' the data – this is called constraining. As you'll see later, it is possible to see if the constraining variables are statistically significant. Most of the methods of ordination are not like traditional hypothesis tests – more often you use them as exploratory methods. You can think of the methods of ordination falling into two major camps:

- Indirect gradient analysis.
- Direct gradient analysis.

With *indirect* gradient analysis you are starting with just the species composition in various samples. Any environmental gradients must be inferred after the main analysis. With *direct* gradient analysis you already have environmental data and the methods of analysis use these data to help sort out the patterns in the data.

There are several methods of ordination in common use, and in the next section you'll get an idea of their relative merits. For most of these you'll use the R program because the routines are simply too complicated to be easily carried out using Excel. The basic distribution of R can carry out several types of ordination and there are additional packages that can conduct others. The *vegan* package is especially useful.

14.1 Methods of ordination

All the main methods of ordination use a dissimilarity matrix of some kind – some methods allow you a choice over the distance measure that is used, others do not. The main methods considered here are:

- Polar ordination (Bray–Curtis ordination).
- Principal co-ordinates analysis (metric or classical multi-dimensional scaling).
- Non-metric multi-dimensional scaling.
- Principal components analysis.
- Correspondence analysis (reciprocal averaging).
- Detrended correspondence analysis.
- Canonical correspondence analysis.
- Redundancy analysis.

That is quite a choice but you can whittle down your selection easily – sometimes with a bit of trial and error.

14.1.1 Overview of ordination methods

Polar ordination (PO) is a method for indirect gradient analysis. It is not especially computer intensive and is a method that you can carry out using your spreadsheet. It is not often used because of advances in computing but remains a useful method, not least because it can help you to understand the general principals of ordination. However, if you have a lot of data the method can be a bit tedious. The method uses a distance matrix and aims to maximise the distance between samples using this matrix – you use endpoints (poles) of an axis that maximises separation.

Principal co-ordinates analysis (PCoA, PCO) is also called classical multi-dimensional scaling (MDS). It is used for indirect gradient analysis. It is similar to polar ordination and aims to maximise the linear correlation between the distances in the dissimilarity matrix and the distances in a low-dimension space. If you use the Euclidean distance metric the results are the same as principal components analysis (PCA). A potential problem with the method is that samples along a gradient can appear arched in the result. This can make interpretation difficult.

Non-metric multi-dimensional scaling (NMDS) is similar to PCO (metric multi-dimensional scaling) but uses ranks to carry out the correlation between the distance matrix and the low-dimensional space. This helps to reduce the arch effect and can make it easier to interpret the results. This method can be used with presence-absence data (since you convert the data to dissimilarity) and works well when you have species with very large differences in abundance.

Principal components analysis (PCA) is based on a Euclidean dissimilarity between samples. Essentially the method rotates the data such that the maximum variance (in the dissimilarity matrix) is projected onto an axis. A second axis is constructed by projecting another variance but this is orthogonal to the first. The result is that the first axis explains the most variance and the second axis explains a little less. Additional axes are usually calculated with diminishing explanatory power. This is a very popular method but it does have problems with community data – the arch effect is very pronounced (due to correlations) and can lead to horseshoe patterns, which can be hard to interpret.

Correspondence analysis (CA or RA) is also known as reciprocal averaging, because one algorithm for finding the solution involves the repeated averaging of sample scores and species scores. A score is a position along an axis. The method maximises the correspondence between species scores and sample scores. For the first axis, species scores and sample scores are assigned such that the correlation between the two is maximised. The second axis also maximises the correlation between species scores and sample scores, but is constrained to be orthogonal to the previous axis. There are some issues with the method, in addition to the arch; the axis extremes can be compressed.

Detrended correspondence analysis (DCA) is a method of removing the arch effect of a correspondence analysis (Hill & Gauch 1980). Essentially it is a correspondence analysis with an extra step or two. The method involves splitting the main axis into segments – the samples in each segment are then centred so that their mean on the second axis is zero. This process is repeated for different segments until the 'best' solution is found. The compression of the ends of the gradients is corrected by nonlinear rescaling. Rescaling shifts sample scores along each axis such that the average width is equal to unity.

Canonical correspondence analysis (CCA) is a cross between correspondence analysis and multiple regression. You use various environmental factors as explanatory variables. The method maximises the correlation between species scores and sample scores but the sample scores are constrained to be linear combinations of the explanatory variables. If a combination of environmental variables is strongly related to species composition, this method will create an axis from these variables that makes the species response curves most distinct. The method uses relative abundance, that is species composition, rather than absolute abundance.

Redundancy analysis (RDA) is similar to canonical correspondence analysis in that explanatory variables are used to constrain the ordination. In contrast to CCA this method uses absolute abundance. The upshot is that if you have a gradient along which all species are positively correlated, RDA will detect such a gradient while CCA may not.

14.1.2 Choosing an appropriate method

To help you choose the most appropriate method of ordination for your data you need to work out what you want to do!

- PO – indirect, simple. This method is not used very much except for demonstrating ordination to students! It is useful because you can undertake the analysis using Excel but this is only easy if you use the Euclidean dissimilarity and do not have many samples. You can use the ordination axes as response variables in further analyses, perhaps to correlate to an environmental variable.
- PCO (MDS) – indirect. You can use most kinds of data with this method, including presence-absence. The results can suffer from the arch effect, which can make

interpretation difficult. The method does not produce species scores but you can use additional methods to get them.

- NMDS – indirect. Like MDS you can use presence-absence data. The analysis minimises arch effect so is a good general choice. You can use a range of dissimilarity metrics, which adds to its usefulness. The method does not produce species scores but you can use additional methods to get them.
- PCA – indirect. This method provides sample and species scores, with species being represented with biplot arrows. PCA suffers from the horseshoe and arch effects, which can make interpretation difficult. It can be a good method when there are lots of correlations between species but it is falling out of favour for community analysis. PCA creates compound variables as axes, which can be used in further analyses, like correlation.
- CA – indirect. This method can be useful when you want to show relationships and are not bothered about axes showing compound variables. The method suffers from the arch effect, which can make interpretation difficult. You get sample and species scores from CA, which can be used in further analyses.
- DCA – indirect. This method is allied to CA and removes the arch effect, making it a useful choice when you see arch effects in CA. The axes are in units of *beta* diversity rather than compound variables.
- CCA – direct. This method focuses on species' relative abundance (composition) and uses environmental factors to constrain the ordination. You can use permutation tests to assess the significance of environmental variables and this makes it a common choice for direct gradient analysis. You get sample, species and environmental scores with CCA. The environmental variables are generally plotted as biplot arrows.
- RDA – direct. This method can be thought of as a linear spin-off of CCA, which focuses on species abundance (rather than composition). It can detect gradients where CCA will not, so it is a good complement to CCA. Like CCA you can use permutation tests to assess the significance of environmental variables.

Generally you'll either have environmental data or not. If you do have the data then you will use either CCA or RDA most of the time. You might also decide to use an indirect method and then look to correlate the results with an environmental factor.

If you do not have environmental data then you'll have to use a method of indirect gradient analysis. If you have presence-absence data then you need to use a method that starts with a dissimilarity matrix. Sometimes the arch effect is not too pronounced – generally the approach is to 'suck it and see', trying various methods until you get a result that aids your interpretation. This is quite unlike traditional hypothesis-based approaches!

Some methods of analysis produce sample scores and not species scores (a score being the position along an axis of ordination). However, it is often possible to calculate the species scores, which you can add to your results.

14.2   Indirect gradient analysis

When you use indirect gradient analysis you are using only the species by sample data. If there are any environmental data such data are used afterwards – mainly as an aid to interpretation. When you carry out the ordination you are looking to determine what the important gradients are by using changes in species composition between samples. Even if you do not have any information about the environmental gradient(s) the analytical methods will help you to see that they exist.

In most cases you can take the results of your ordination and add the environmental information to your plots. Of course the original ordination did not 'take into account' the variables, but you can see how the variables 'map' onto the ordination you have.

14.2.1 Polar ordination

Mostly you'll be using the R program to carry out ordination because the analyses are just too complex to use Excel. However, the method of polar ordination (also called Bray–Curtis ordination) is one that you **can** undertake using your spreadsheet.

The method begins with a dissimilarity matrix. Your choice of metric will influence the result, every metric will 'lose information' about the system. Most people have their favourite but a good approach is to use several metrics and see if your interpretation is the same (the rankindex() command in the *vegan* package can be used to help decide).

Once you have a dissimilarity matrix you select the two samples with the largest dissimilarity to be your first axis. You then calculate scores for this axis based on the dissimilarities of each sample compared to the ends of the axis (the poles). The second axis scores are calculated in a similar manner. However, there is a bit of trial and error because you want an axis orthogonal to the first. The usual method is to select a new axis based on a large dissimilarity and work out the scores. You can then look at the correlation to the first axis. Repeat for several candidate axes until you find one with a low correlation. You can then plot the samples using the axis scores to produce a scatter plot.

The best way to visualise what is happening is to try it for yourself. In the following exercise you can have a go at a polar ordination using some data on bryophytes on trees in North Carolina.

Have a Go: Use Excel for polar ordination

You'll need the Excel file *Polar ordination data.xls* for this exercise. The data show the importance scores (a measure of relative abundance) for various species of bryophyte on trees in a forest in North Carolina (Palmer 1986). Each sample is an average of ten trees at three sites. Some trees were found in more than one site.

1. Open the Excel data file *Polar ordination data.xls*. You'll see the data arranged in biological recording format with a column *Bryophyte* for species name, a column for *Tree* (the site or sample name) and a column for the importance score, *Imp*. There is also a column containing the species abbreviated names.

2. Make a Pivot Table of the data; you need to end up with the *Column Labels* as the tree names (*Tree*) and the *Row Labels* as the species names (*Bryophyte*). The data in the table (*Values*) should be the importance scores (*Imp*). You do not want the row or column grand totals to turn these off using the *Pivot Table Tools > Design > Grand Totals* button. Now also turn blank cells into 0 values using the *Pivot Table Tools > Options > Options* button. There is a completed Pivot Table already made for you in the *Completed Pivot* worksheet but try it for yourself as practice.

3. Copy the data from the Pivot Table (cells A4:J29) to the clipboard and use *Paste Special* to place a copy of the data in a new worksheet. You can use the *Completed Pivot*

worksheet if you prefer. You can operate on a Pivot Table but it is generally better to extract the data as plain data in case you modify the Pivot Table later.

4. Now copy the tree names in cells B1:J1 to the clipboard. Paste the names into cells B28:J28 to form the columns for the dissimilarity matrix. Use Paste Special to copy the names down the A column in cells A29:A37. You'll need to tick the *Transpose* box to do this.

5. Click in cell B29. The dissimilarity between BN2 and itself will be 0 so simply type that into the cell. Now fill in zeros for the rest of the diagonal, which represents the dissimilarities between samples and themselves.

6. You will now calculate Euclidean dissimilarities for all the pairs of species. However, you only need to fill out the lower triangle of the matrix at the moment. Use the mouse to select the entire matrix (cells B29:J37) and format them to show integer values (from the *Home* menu you can alter the cell formatting or simply right-click). Click in cell B30 and type a formula to calculate the Euclidean distance between samples BN2 and LT1: =SQRT(SUMXMY2(B2:B26,C2:C26)). The result should be 42.

7. With a triangular matrix you won't be able to add $ to 'fix' appropriate rows or columns but you can copy the basic formula to the clipboard and simply edit the results. Click in cell B31 and type a formula to work out the Euclidean distance between BN2 and LT2: =SQRT(SUMXMY2(B2:B26,D2:D26)). The result should be 43.

8. Carry on down the column and fill in the dissimilarities between BN2 and the other samples. The B2:B26 part will be the same for all but you'll need to change the second half of the formula. The results should be: 55, 57, 46, 67, 50 and 46.

9. Now click in cell C31 and type a formula to work out the dissimilarity between LT1 and LT2. These are columns C and D: =SQRT(SUMXMY2(C$2:C$26,D$2:D$26)).

10. Carry on down the column and then fill out the rest of the dissimilarity matrix. Remember that you are always dealing with rows 2:26 but the columns will alter. This is a tedious process but consider it character building. There is a completed Euclidean dissimilarity matrix in the *Completed Dissimilarity* worksheet so that you can check your results.

11. The calculations to come are greatly easier if you also complete the top part of the dissimilarity matrix. You can do this most easily by setting the values to equal to their lower-triangle counterparts. So, in cell C29 type =B30 for example.

12. When you have completed the dissimilarity matrix you need to work out the axis scores for the polar ordination. Start by identifying the largest dissimilarity, as this will separate the samples the most. The largest value is 76, which represents the dissimilarity between PT3 and PO2. So your first PO axis will be based on the dissimilarity between these two samples. Make some labels for the axis scores; start in cell A39 and type 'i'. In cell A40 type 'j'. Now copy the names of the samples from cells A29:A37 to the clipboard and paste them into cells A42:A50. In cell B41 type 'Ax1' as a reminder that this is the first Axis.

13. You are going to calculate how far along the axis each sample lies; the axis represents the distance between PT3 and PO2 so type these as labels in cells B39:B40. The

distance that each sample lies along the axis is determined by the formula shown in Figure 14.1:

$$\text{Axis}_1 \text{ score} = \frac{D^2 + D_i^2 - D_j^2}{2D}$$

Figure 14.1 Calculating an axis score for polar ordination. D is largest dissimilarity. D_i is dissimilarity between sample i and axis pole 1. D_j is dissimilarity between sample j and axis pole 2.

14. Click in cell B42 and type a formula to work out the axis score for the BN2 sample along the new axis using the formula in Figure 14.1. In this case D will be 76 (cell F35), Di will be 67 (cell H29) and Dj will be 57 (cell F29). Your formula should be: =($F$35^2+H29^2-F29^2)/2/$F$35.

15. By adding the $ to the cell representing the largest dissimilarity you can now copy the formula down the column. Notice that the value for PT3 is 0, because this is at the start of the axis. The other end is PO2 and its value is 76, which is the maximum dissimilarity. The other values are between these extremes. Notice that the values are already starting to show an ordination (the pines are low values and the broad-leaved trees are high values, birch is in the middle).

16. You could plot these values along a single axis but it would be better to make a second axis so that you can make a 2-D scatter plot. You need to identify another dissimilarity, that does not involve the first two samples, and repeat the process. Try setting i to be QA1 and j to be PE3. The dissimilarity for this pair is 65.

17. Add labels in cells C39:C41 for the axis poles; 'QA1', 'PE3' and 'Ax2' will be fine. Now in cell C42 type a formula to work out the axis score for sample BN1: =($E$36^2+I29^2-E29^2)/2/$E$36. Copy this down the column to make scores for all the samples relative to this axis. Note that QA1 will be 0 and PE3 will be 65 because these are the poles. However, you also get values outside of this (67 and –2), which is fine.

18. You want to check to see if this axis is orthogonal to the first axis so in cell B51 type a label 'r=' for the correlation coefficient. In cell C51 type a formula to determine the correlation: =CORREL($B$42:$B$50,C42:C50). The result (–0.996) shows that these are highly correlated and therefore not orthogonal.

19. Try other candidate axes and fill in more columns (D:J) using the same principles. Try: PT1-LT2, QA1-PT1, QR1-BN2, QA1-LT2, QR1-LT2, QA1-QR1 and LT1-QR1. These are more or less in descending order of dissimilarity. Once you have determined the axis scores copy the correlation formula in cell C51 across the row to work out correlation of your candidate axes to the first axis.

20. You can see from the completed axis scores matrix that Ax6 and Ax7 have low correlation to the first axis (Ax1). Either of these would make a good candidate for the secondary axis – ideally you want to minimise the correlation and maximise the dissimilarity. In this case Ax6 makes the slightly better choice. Use the Ax1 and Ax6 axis scores to make a scatter plot. With some tinkering you should be able to make a chart that resembles Figure 14.2.

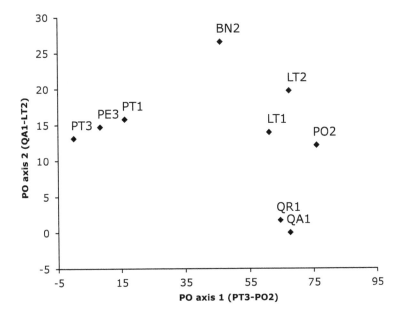

Figure 14.2 Polar ordination of bryophyte species on trees in North Carolina. BN = *Betula nigra*, LT = *Liriodendron tulipifera*, PE = *Pinus echinata*, PO = *Platanus occidentalis*, PT = *Pinus taeda*, QR= *Quercus rubra*, QA= *Quercus alba*.

The completed chart shows the relationship between the tree species (in terms of their bryophyte communities) quite clearly.

Tip: Scatter plot chart labels

Excel usually doesn't allow you to label data points with anything other than a numerical value. The *XY chart labeler* is an add-in for Excel that allows you to add custom labels to individual points on scatter plots.

Even though there were only nine samples and 25 species, undertaking the polar ordination was quite an ordeal.

14.2.2 Multi-dimensional (metric) scaling (principal co-ordinates analysis)

Classical MDS is also known as metric multi-dimensional scaling or principal co-ordinates analysis (PCO). It is similar to polar ordination in that it uses a dissimilarity matrix as a starting point. The method maximises the linear correlation between the distances in the dissimilarity matrix and the low-dimension space of the final result. You can think of it somewhat as like taking a cloud of points and shining a torch through them to produce a shadow. The shadow is the low-dimensional space.

The `cmdscale()` command can carry out PCO; this command is in the *stats* package, which comes as part of the basic distribution of R. The command has the following general form:

```
cmdscale(d, k = 2, eig = FALSE)
```

You must give the data as a dissimilarity matrix. You also specify the number of dimensions you want the result to be; the default is 2, which makes a 2-D scatter plot. You can also specify eig = TRUE, in which case the eigenvalues of the result are shown as well as the goodness of fit. The more dimensions you specify (up to the number of samples minus 1) the closer to unity the goodness of fit will be. Generally you'll use k = 2 to start and then see how good the fit is. If the fit is not substantially better with k = 3 then you can be moderately happy with your solution.

The result of the cmdscale() command can be used to make a scatter plot. If eig = FALSE (the default) the result is simply a matrix of axis scores, otherwise the axis scores are part of the $points component. You can add text labels to the points (or instead of) using the text() command. The usual 'running order' is:

1. Make a dissimilarity matrix using the dist() or vegdist() command.
2. Use the cmdscale() command to carry out the MDS.
3. Plot the result displaying an 'empty' plot (that is, make the co-ordinate system but do not display the actual points).
4. Add text labels to the plot to show the sample names.
5. View the goodness of fit of the result via the $GOF component.
6. Add an extra dimension to the result using k = 3 and view the $GOF.

In the following exercise you can have a go at a MDS for yourself. You can use the bryophyte data from the polar ordination exercise so that you can compare results.

Have a Go: Carry out classical (metric) MDS

You'll need the *moss* data for this exercise; these are part of the *CS4E.Rdata* file. The data show the importance scores (a measure of relative abundance) for various species of bryophyte on trees in a forest in North Carolina (Palmer 1986). Each sample is an average of ten trees at three sites. Some trees were found in more than one site.

1. Start by looking at the *moss* data. You'll see that the columns are the samples and the rows are the species:

```
> head(moss)
          BN2    LT1    LT2 PE3    PO2 PT1   PT3   QA1   QR1
Amb.ser   1.0    0.0    5.0   0    0.0   0     0   3.2   2.3
Ano.att   0.9   17.6   26.4   0   41.2   0     0  27.3  22.4
Ano.min   2.1    1.0    5.4   0    9.4   0     0   2.4   0.6
Ano.ros   0.0    0.0    1.4   0    4.7   0     0  14.0  13.6
Bra.acu   0.0    3.1    0.0   0    0.0   0     0   5.3   2.9
Bra.oxy   0.9    1.8    0.8   0    1.6   0     0   1.7   0.5
```

2. You need to rotate the data so that the rows are the samples and then make a Euclidean dissimilarity matrix:

```
> moss.biol = t(moss)
> moss.eu = dist(moss.biol, method = "euclidean")
```

3. In step 2 you could have made the dissimilarity in 'one go' using `t(moss)` instead of your rotated data but it is 'useful' to have a copy of the data arranged in the appropriate way. Now use the dissimilarity result in the `cmdscale()` command to make a multi-dimensional scaling result (i.e. a PCO or MDS):

```
> moss.pco = cmdscale(moss.eu, k = 2)
> moss.pco
            [,1]        [,2]
BN2   -0.4446475  34.8789419
LT1  -18.1046921   2.8738551
LT2  -22.5535658  -0.1146282
PE3   39.2108396  -2.0936131
PO2  -29.1878891 -13.1531653
PT1   31.1197817   3.1753791
PT3   45.2687047 -13.5379996
QA1  -24.0843294  -7.6439190
QR1  -21.2242021  -4.3848508
```

4. This simple result contains only the two axis scores. Make a plot of the result; display the sample names instead of simple points, your result should resemble Figure 14.3:

```
> plot(moss.pco, type = "n", asp = 1, xlab = "PCoA1", ylab = "PCoA2")
> text(moss.pco, labels = rownames(moss.pco))
```

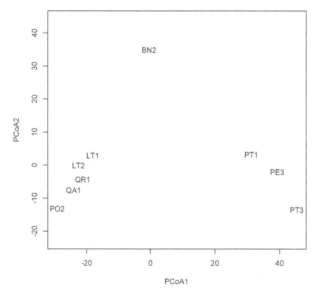

Figure 14.3 Classical multidimensional scaling (Euclidean dissimilarity) of bryophyte species on trees in North Carolina. BN = *Betula nigra*, LT = *Liriodendron tulipifera*, PE = *Pinus echinata*, PO = *Platanus occidentalis*, PT = *Pinus taeda*, QR= *Quercus rubra*, QA= *Quercus alba*.

Here you labelled the axes 'PCoA' as a reminder that MDS is a synonym (you could of course have used 'MDS'). You can compare the result of the MDS to the result of the PO that you carried out earlier (Figure 14.2). The pattern is broadly the same although there are evidently some minor differences.

One of the problems with MDS (PCO) is that samples that lie on a gradient can be displayed as an arch rather than as a straight line. This can make plots hard to interpret. In the following exercise you can have a go at another MDS using some transect data that should show a gradient (pH). You can also explore the goodness of fit by setting eig = TRUE.

Have a Go: View the arch effect in a PCO

You'll need the *hsere.biol* data for this exercise; they are found in the *CERE.RData* file. The data show the abundance of some plant species across a gradient of pH in a hydrosere. There are ten transect samples and the data shows the cumulative frequency for five replicates. You'll also need the *vegan* package.

1. Start by making sure the *vegan* package is ready:

    ```
    > library(vegan)
    ```

2. Look at the data and see that the samples are the columns and species the rows:

    ```
    > head(hsere.biol)
                             T1    T2    T3   T4   T5   T6    T7   T8   T9  T10
    Carex rostrata          216   140   119   98   96   65     9    4   10    0
    Fillipendula ulmaria     38    41     9    1    0    0     0    0    0    0
    Caltha palustris         22    10    15    0    0    0     0    0    0    0
    Bryophyte                48    43    39   65   53   38   100   53   67   15
    Galium palustre          55    59     1    7    4    1     0    0    0    0
    Valeriana officianalis    0     0     2    0    0    0     0    0    0    0
    ```

3. You'll need to rotate the data to make an appropriate dissimilarity matrix; use the Bray–Curtis metric from the vegdist() command:

    ```
    > hsere.bc = vegdist(t(hsere.biol), method = "bray")
    ```

4. Now you can carry out the PCO. Set eig = TRUE to get additional results, including the goodness of fit:

    ```
    > hsere.pco = cmdscale(hsere.bc, eig = TRUE)
    > names(hsere.pco)
    [1] "points"    "eig"    "x"      "ac"     "GOF"
    ```

5. The axis scores are in the $points component. Use these to make a scatter plot and add the sample names in lieu of points. Your plot should resemble Figure 14.4:

    ```
    > plot(hsere.pco$points, type = "n", asp = 1, xlab = "MDS1",
           ylab = "MDS2")
    > text(hsere.pco$points, labels = rownames(hsere.pco$points))
    ```

6. Now return to the PCO result and look at the goodness of fit:

    ```
    > hsere.pco$GOF
    [1] 0.7964238 0.7964238
    ```

7. Make a new PCO result using k = 3 dimensions and examine the goodness of fit again:

    ```
    > cmdscale(hsere.bc, eig = TRUE, k = 3)$GOF
    [1] 0.867574 0.867574
    ```

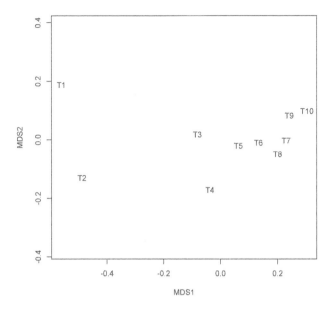

Figure 14.4 Principal co-ordinates analysis (Bray–Curtis dissimilarity) of plant species in a bog hydrosere.

The goodness of fit increases as you add more dimensions, until you've added every-thing and get unity. Here you can see that the basic 2-D solution gives a good fit of 80%, whereas the 3-D solution gives 87%.

   Look at the plot (Figure 14.4) and see how there is a bit of an arch (the beginnings of a U shape). In this case the arch is not very pronounced. Try the same process on the *ridge. biol* data and see a more pronounced effect.

The arch effect can distort a 'gradient' and make interpretation difficult. However, it is still fairly common to try to use the axis scores as a variable in a correlation. You can extract the scores from the result or use the `scores()` command in the *vegan* package. In the follow-ing exercise you can use the result of a MDS and pH data.

Have a Go: Use axis scores from MDS in a correlation

You'll need the *hsere.biol* data for this exercise; they are found in the *CERE.RData* file. The data show the abundance of some plant species across a gradient of pH in a hydrosere. There are ten transect samples and the data shows the cumulative frequency for five replicates. You'll also need the *vegan* package.

1.   Start by making sure the *vegan* package is ready:

```
> library(vegan)
```

2.   The data are arranged with samples as columns so start by rotating the data. Then you can make a Bray–Curtis dissimilarity and carry out a basic MDS:

```
> hsere.bc = vegdist(t(hsere.biol), method = "bray")
> hsere.pco = cmdscale(hsere.bc)
```

3. Look at the axis scores using the `scores()` command in the *vegan* package:

```
> scores(hsere.pco)
             Dim1          Dim2
T1   -0.55871648   0.18724723
T2   -0.48709708  -0.13061449
T3   -0.08141886   0.01795278
T4   -0.03765352  -0.17146209
T5    0.06220023  -0.02061298
T6    0.13251581  -0.01036445
T7    0.22962298  -0.00276169
T8    0.19974872  -0.04979448
T9    0.23999389   0.08272150
T10   0.30080431   0.09768868
```

4. Look at the pH data in the *hsere.ph* data object:

```
> hsere.ph
      T1   T2   T3   T4   T5   T6  T7   T8   T9 T10
pH 7.12 6.46 6.26 5.76 5.14 5.02 4.7 4.3 4.22 4.2
```

5. Because the pH data are in a `data.frame` with multiple columns you'll need to extract the values as a single element. You can use `as.numeric()` or `t()` commands; the former returns a vector and the latter a `matrix`:

```
> as.numeric(hsere.ph)
 [1] 7.12 6.46 6.26 5.76 5.14 5.02 4.70 4.30 4.22 4.20
> t(hsere.ph)
       pH
T1   7.12
T2   6.46
T3   6.26
T4   5.76
T5   5.14
T6   5.02
T7   4.70
T8   4.30
T9   4.22
T10  4.20
```

6. Carry out a correlation test between the axis scores and the pH data:

```
> cor.test(scores(hsere.pco)[,1], t(hsere.ph))

  Pearson's product-moment correlation

data:  scores(hsere.pco)[, 1] and t(hsere.ph)
t = -8.0971, df = 8, p-value = 4.003e-05
alternative hypothesis: true correlation is not equal to 0
95 percent confidence interval:
 -0.9870054 -0.7752407
sample estimates:
     cor
```

7. Look at the other axis:

```
> cor(scores(hsere.pco)[,2], as.numeric(hsere.ph))
[1] -0.01165334
```

> You can see that there is a very good correlation with the first axis and a very poor one
> with the second axis – of course this demonstrates the effect of orthogonality.

If you have environmental data you can use PCO/MDS in a redundancy analysis or par-
tially constrained PCO. You'll see how to do this later (Section 14.3.2).

The arch effect comes about partly because the method maximises linear correlation.
One way to reduce the arch effect is to use a non-metric method, utilising ranks rather than
absolute values – you will see this shortly (Section 14.2.3) but before that you'll see how to
get species scores from a PCO.

Species scores from an MDS

The `cmdscale()` command computes sample scores only as part of the analysis. The whole
idea of multi-dimensional scaling is to produce the sample scores but you can determine
species scores using the `wascores()` command in the *vegan* package. The `wascores()`
command calculates weighted average scores using the samples scores from the ordination and the original data. To get species scores you proceed as follows:

1. Make a PCO result from your data using a dissimilarity of your choice:

    ```
    > hsere.bc = vegdist(t(hsere.biol), method = "bray")
    > hsere.pco = cmdscale(hsere.bc, eig = TRUE)
    ```

2. Now make a result object holding the species scores using the `wascores()`
    command:

    ```
    > hsere.wa = wascores(hsere.pco$points, hsere)
    ```

3. Plot the results on a blank plot:

    ```
    > plot(hsere.pco$points, type = "n", xlab = "MDS1", ylab = "MDS2")
    ```

4. Add the site scores as points or text:

    ```
    > text(hsere.pco$points, labels = rownames(hsere.pco$points))
    ```

5. Add the species scores as text or simple points:

    ```
    > points(hsere.wa, pch = "+", col = "darkgreen")
    ```

This process works with any ordination that produces sample scores but some methods
standardise the data, so you need to know what process of standardisation has been carried out before calculating the species scores. You'll see this process in the next section on
non-metric multi-dimensional scaling.

Note: Weighted MDS

The *vegan* package contains a command `wcmdscale()`, which can carry out weighted
MDS. This allows you to downweight rare species so that they have a smaller effect on the
analysis. Setting all the weights to the same value gives the same results as regular MDS as
carried out by the `cmdscale()` command.

Note: Conditioning variables and MDS

You can partial out the effects of one or more variables in a MDS (PCO) analysis using the `capscale()` command in the *vegan* package. This is essentially a redundancy analysis of a classical MDS (see Section 14.3.2).

14.2.3 Non-metric multi-dimensional scaling

NMDS is a way to overcome some of the problems of the arch effect. You might think of it as analogous to non-parametric statistical tests, such as Spearman's rank test compared with Pearson's product moment. Instead of linear correlation between dissimilarities and the low-dimensional result NMDS uses a measure of 'stress', which is the mismatch between the rank order of the dissimilarities and the low-dimensional solution. The minimisation of the 'stress' is the target of the method – the more dimensions you select the smaller will be the final stress. Usually you choose two dimensions and then compare the changes in stress with additional dimensions to help decide on the 'explanatory power' of your solution.

The method is considered to be quite robust and is one of the more popular methods of ordination. There are commands to conduct NMDS in the basic distribution of R – they are found in the *MASS* package, which comes with R but is not loaded by default. The main command is `isoMDS()`, which implements Kruskal's NMDS. The `sammon()` command carries out another version (Sammon's NMDS). Both commands have a similar form and both require a dissimilarity matrix as a starting point:

```
isoMDS(d, k = 2)
sammon(d, k = 2)
```

You must supply, d, a dissimilarity and the number of dimensions for the final result; the default is k = 2. The results of both commands is an object that contains a `$points` and a `$stress` component. You can plot the `$points` to make a scatter plot and of course can use the axis scores as a variable in a correlation.

In the following exercise you can have a go at a NMDS using some plant data from a hydrosere succession.

Have a Go: Conduct non-metric multi-dimensional scaling

You'll need the *hsere.biol* data for this exercise; they are found in the *CERE.RData* file. The data show the abundance of some plant species across a gradient of pH in a hydrosere. There are ten transect samples and the data shows the cumulative frequency for five replicates.

1. The *hsere* data are arranged with species as rows so begin by rotating the data so that the samples form the rows:

   ```
   > hsere = t(hsere.biol)
   ```

2. Now make a Euclidean dissimilarity matrix and use this to carry out NMDS:

   ```
   > hsere.eu = dist(hsere, method = "euclidean")
   > hsere.nmds = MASS::isoMDS(hsere.eu)
   initial  value 8.858713
   ```

```
iter   5 value 5.034676
iter  10 value 4.246873
iter  15 value 4.007225
final  value 3.690918
converged
```

3. The result shows the axis scores and final stress of the model:

```
> hsere.nmds
$points
            [,1]          [,2]
T1   -227.80200  -21.1792652
T2   -155.16675   43.7759924
T3    -32.37222  -56.0001348
T4    -27.68303    0.3373198
T5     18.17286  -52.4649619
T6     32.07896   57.8673889
T7    121.26515  -45.8779972
T8     50.84830   32.2091144
T9    108.13871   -7.9563339
T10   112.52002   49.2888774

$stress
[1] 3.690918
```

4. Use the axis scores to make a scatter plot – show the points as sample names using the text() command. Your plot should resemble Figure 14.5:

```
> plot(hsere.nmds$points, type = "n", xlab = "NMDS1",
      ylab = "NMDS2", asp = 1)
> text(hsere.nmds$points, labels = rownames(hsere.nmds$points))
```

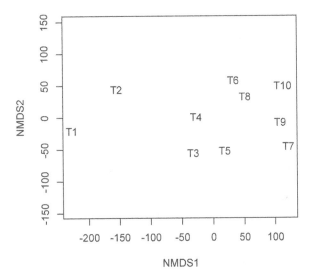

Figure 14.5 Kruskal's non-metric multidimensional scaling. Plant abundance in a bog hydrosere.

5. Note that you set asp = 1 to maintain a 1:1 aspect ratio for the final plot in step 4. Look at the correlation between the axis 1 scores and pH (from *hsere.ph*):

```
> cor(scores(hsere.nmds)[,1], t(hsere.ph), method = "spearman")
            pH
[1,] -0.9272727
```

6. Look at how the stress changes as you add more dimensions:

```
> for(i in 1:5) print(isoMDS(hsere.eu, k = i, trace = FALSE)$stress)
[1] 10.29016
[1]  3.690918
[1]  1.534728
[1]  0.1690944
[1]  0.007746814
```

There is a big drop in stress from k = 1 to k = 2. The two-dimensional solution seems adequate but there is an even better drop from k = 3 to k = 4.

The preceding exercise used the isoMDS() command but the principles are exactly the same for the sammon() command (with subtly different results). In both cases you only get the sample scores but it is possible to get species scores by using the metaMDS() command from the *vegan* package.

Get species scores from NMDS

The MDS and NMDS analyses do not compute species scores. However, it would be useful to be able to do so. The *vegan* package contains a command, metaMDS(), that carries out NMDS and calculates the species scores as well. The general form of the command is like so:

```
metaMDS(comm, distance = "bray", k = 2)
```

You supply the community data (rows as samples) rather than a dissimilarity matrix – the command will make the dissimilarity using the Bray–Curtis metric by default. You can specify other metrics – those used by the vegdist() command simply by providing the name. You also set the number of dimensions you require; k = 2 by default. A side-benefit of the metaMDS() command is that you can use some additional plotting commands. In the following exercise you can have a go at a NMDS for yourself and explore some of the options for displaying the results.

Have a Go: Get species scores for a NMDS and plot the results

You will need the *ridge.biol* data for this exercise; the data are in the *CERE.RData* file. The data show plant abundance for species along a transect from a psammosere. There are ten samples and each shows frequency from eight replicates. You'll also need the *vegan* package.

1. Start by ensuring the *vegan* package is ready:

```
> library(vegan)
```

2. The *ridge.biol* data are arranged with species as rows so you'll need to rearrange the data before you start. There are also some environmental data in the *ridge.env* object. Rotate both so that the samples are the rows:

```
> ridge = t(ridge.biol)
> env = t(ridge.env)
```

3. The metaMDS() command can use a variety of dissimilarity metrics. Have a look to see which one might be the 'best' with the rankindex() command:

```
> rankindex(env, ridge)
      euc        man        gow        bra        kul
0.3546772  0.4363756  0.5287220  0.3009223  0.3677207
```

4. A good dissimilarity index for multi-dimensional scaling should have a high rank-order similarity with gradient separation. The rankindex() command compares the metrics from the vegdist() command against gradient separation using rank correlation coefficients. You can see from step 3 that the Gower metric is likely to be a good metric so use this in the metaMDS() command to make a NMDS result:

```
> ridge.nmds = metaMDS(ridge, distance = "gower")
Wisconsin double standardization
Run 0 stress 0.06073532
Run 1 stress 0.0607357
... procrustes: rmse 0.0003135526  max resid 0.0005886017
*** Solution reached
```

5. Look at the changes in stress for different dimensions; use k = 1 to 5:

```
> for(i in 1:5) print(metaMDS(ridge, distance = "gower", k = i,
                    trace = FALSE)$stress *100)
[1] 14.2528
[1]  6.073532
[1]  1.62488
[1]  0.1204245
[1]  0.06485988
```

6. From step 5 it looks like a 3- or 4-D solution would be better than the 2-D solution so recalculate the result using k = 3:

```
> ridge.nmds = metaMDS(ridge, distance = "gower", k = 3)
Wisconsin double standardization
Run 0 stress 0.0162488
Run 1 stress 0.01627280
... procrustes: rmse 0.003890412  max resid 0.005785606
*** Solution reached
```

7. There is a plot() command associated with the metaMDS results – this allows you some additional control. Start with a basic plot of the sites only, your plot should resemble Figure 14.6:

```
> plot(ridge.nmds, type = "t", display = "sites")
```

8. In step 7 you plotted the data as text (the default shows the sample or species names) and chose to display only the sites. Now show the species as well as the sites as simple points, your result should resemble Figure 14.7:

```
> op = plot(ridge.nmds, type = "p", display = c("site", "species"))
```

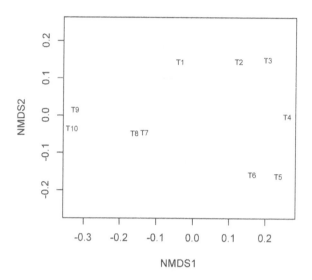

Figure 14.6 NMDS plot of sample scores for psammosere plant abundances.

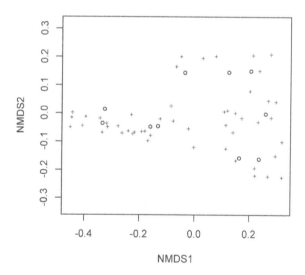

Figure 14.7 NMDS plot for psammosere plant abundances. Open circles are sites, + are species.

9. The plot you made in step 8 is somewhat congested now that you have displayed the species scores. Notice that you assigned the plot to a named object (*op*). You can use this to help interactively label the plot. Use the identify() command to add some labels to species. You can use abbreviated species names from the ridge.nam object. Once you enter the command you'll be able to click on points and the name will appear. When you are finished press the *esc* key:

```
> identify(op, what = "species", cex = 0.8, col = "red",
        labels = ridge.nam$Abbr)
```

If you click slightly to the left of a point the label will appear to the left, if you click to the right it appears to the right. Similarly if you click above or below the label appears above or below. The command here makes the labels red, and slightly smaller than "standard", your plot should resemble Figure 14.8:

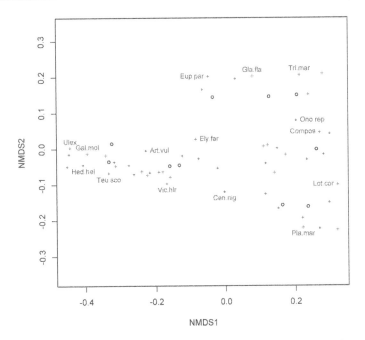

Figure 14.8 NMDS plot for psammosere plant abundances. Open circles are sites, + are species. Some species have been labelled using the interactive identify() command.

10. You can see the axis (site) scores using the scores() command or by accessing the $points component of the result:

```
> scores(ridge.nmds)
          NMDS1          NMDS2          NMDS3
T1    -0.03616464    0.143555642    0.05743401
T2     0.12475715    0.144925098    0.06942493
T3     0.20470658    0.149456182   -0.06879748
T4     0.25952324   -0.002142783   -0.02933284
T5     0.23554777   -0.162255407   -0.04976017
T6     0.16305980   -0.158034476    0.09255079
T7    -0.13247529   -0.045959929    0.07187495
T8    -0.16006535   -0.048100300   -0.14814423
T9    -0.32527748    0.014263578   -0.10380939
T10   -0.33361177   -0.035707606    0.10855942

> ridge.nmds$points
...
```

Try a k=4 model and you'll find that the algorithm does not produce a 'solution'.

You can get species scores independently using the `wascores()` command. The command needs the site scores from the ordination and the species data. However, the `metaMDS()` command has standardised the data as part of the analysis. You need to follow these steps:

1. Examine the NMDS result to see the standardisation:

    ```
    > ridge.nmds$data
    [1] "wisconsin(ridge)"
    ```

2. Use the `wascores()` command with the sites scores and the standardised data, you need to set `expand = TRUE`, which matches the (weighted) variance of the species and sample scores:

    ```
    > wascores(ridge.nmds$po, wisconsin(ridge), expand = TRUE)
    ```

You can also use the `wascores()` command to get weighted average scores for the species in relation to any environmental variables. You supply the appropriate environmental data and the species data in the command:

```
> wascores(env, ridge)
                                Soil     Light    Humid      Wind
Cochlearia danica           0.230000 16231.25 27.76500 3.9710000
Elymus farctus              8.308065 14713.84 29.27097 0.9204839
Erodium lebelii             2.666667 15637.67 28.86667 3.5033333
Euphorbia paralias          0.000000 16250.00 28.30000 5.1300000
Glaucum flavum              0.075000 16250.00 27.62500 3.7200000
Tripleurospermum maritimum  1.700000 16875.00 27.35000 2.5650000
```

As an additional feature you can examine the eigenvalues of the environmental variables using the `eigengrad()` command – this gives the relative 'importance' of the variables:

```
> eigengrad(env, ridge)
     Soil      Light      Humid       Wind
0.4541225  0.3845023  0.1170576  0.4462910
```

If you have environmental information you can use the axis sample scores in a correlation as you saw in the preceding section on MDS/PCO. However, it is more likely that you will use it in some form of direct gradient analysis such as CCA or redundancy analysis (see Section 14.3).

14.2.4 Principal components analysis

PCA is a way to take a 'cloud' of data and project it onto a new low-dimensional space. This is done in such a way as to maximise the variance explained by the first axis. The second axis is projected in a similar manner but ensuring that it is *orthogonal* to the first axis. The process continues with subsequent axes, this inevitably means that the 'variance explained' is less each time. There are usually as many PCA axes as there are samples in the original data. Most often the first two axes are the most important. The eigenvalues from the analysis represent the variance extracted by each axis, and are usually expressed as a percentage of the sum of all the eigenvalues (in other words, the total variance).

Essentially PCA creates new vectors that maximise their linear correlation with all the other variables. The solution is similar to PCO, when conducted using a Euclidean distance metric. PCA returns sample and species scores. The species scores are usually represented graphically as arrows, implying increasing abundance in the direction of the arrow. This linear approach does, however, lead to a problem – the samples are often distorted into an arch, which can be quite pronounced and horseshoe-shaped. This can make interpretation difficult and make it hard to correlate environmental variables with PCA axis scores.

The `prcomp()` command for PCA

The basic distribution of R contains the `prcomp()` command, which carries out PCA. It uses singular value decomposition (SVD) of the data matrix to determine the PCA solution. The general form of the command is like so:

```
prcomp(x, center = TRUE, scale. = FALSE)
```

You supply the data (as a `data.frame` or `matrix`) with samples as rows and species as columns. The `prcomp()` command allows centring and rescaling (to unit variance) of variables – this is useful if they have been measured in different units but for community data you'll usually have species in the same units.

   The result of the `prcomp()` command contains various elements that you can use, such as the variance explained by each PCA axis and the scores of the samples and species. There are `summary()` and `print()` routines as well as a `plot()` command. There is also a special plotting command `biplot()`, which displays the samples as points and the species as arrows. In the following exercise you can have a go at a PCA using data from a study of tree bryophytes.

Have a Go: Use principal components analysis

You'll need the *moss* data for this exercise; these are part of the *CS4E.Rdata* file. The data show the importance scores (a measure of relative abundance) for various species of bryophyte on trees in a forest in North Carolina (Palmer 1986). Each sample is an average of 10 trees at three sites. Some trees were found in more than one site.

1. Look at the *moss* data; they are arranged with species as rows:

```
> head(moss)
         BN2  LT1  LT2 PE3    PO2 PT1 PT3   QA1   QR1
Amb.ser  1.0  0.0  5.0   0    0.0   0   0   3.2   2.3
Ano.att  0.9 17.6 26.4   0   41.2   0   0  27.3  22.4
Ano.min  2.1  1.0  5.4   0    9.4   0   0   2.4   0.6
Ano.ros  0.0  0.0  1.4   0    4.7   0   0  14.0  13.6
Bra.acu  0.0  3.1  0.0   0    0.0   0   0   5.3   2.9
Bra.oxy  0.9  1.8  0.8   0    1.6   0   0   1.7   0.5
```

2. Make a new data object from the moss data; rotate the original so that the new data have the rows as samples:

```
> moss.biol = t(moss)
```

3. Now carry out a PCA on the data with the `prcomp()` command:

```
> moss.pca = prcomp(moss.biol)
> names(moss.pca)
[1] "sdev"    "rotation" "center"   "scale"   "x"
```

4. The result contains the explained variance for each PCA axis as the standard deviation in the $sdev component. View this and then look at a screeplot() of the variance; your plot should resemble Figure 14.9:

```
> moss.pca$sdev^2
 [1] 9.092852e+02 2.091523e+02 5.012740e+01 4.046632e+01
 [5] 2.431393e+01 6.625639e+00 4.798909e+00 2.311628e+00
 [9] 1.993057e-30

> screeplot(moss.pca)
```

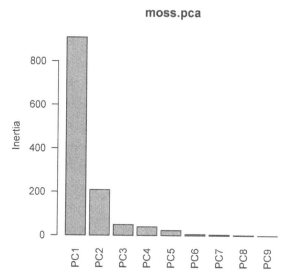

moss.pca

Figure 14.9 Variance explained by PCA axes for tree bryophyte data.

5. Use the summary() command to view the importance of the various PCA axes:

```
> summary(moss.pca)
Importance of components:
                           PC1      PC2     PC3      PC4     PC5
Standard deviation      30.1544 14.4621 7.0801 6.36131 4.9309
Proportion of Variance   0.7291  0.1677 0.0402 0.03245 0.0195
Cumulative Proportion    0.7291  0.8968 0.9370 0.96949 0.9890
                           PC6      PC7     PC8      PC9
Standard deviation      2.57403 2.19064 1.52040 1.412e-15
Proportion of Variance  0.00531 0.00385 0.00185 0.000e+00
Cumulative Proportion   0.99430 0.99815 1.00000 1.000e+00
```

6. You can plot the proportion of the variance but will need to extract the appropriate components, the result is identical to that from step 4 but with different y-axis scaling:

```
> barplot(summary(moss.pca)$importance[2,], las = 2)
> title(ylab = "Proportion of variance", xlab = "PCA axis")
```

7. Try a `biplot()` to view the sample scores and the species as arrows. The command can accept various graphical parameters, if you give two values these apply to species then scores. Your graph should resemble Figure 14.10:

```
> biplot(moss.pca, col = c("blue", "darkgreen"), cex = c(1, 0.8))
```

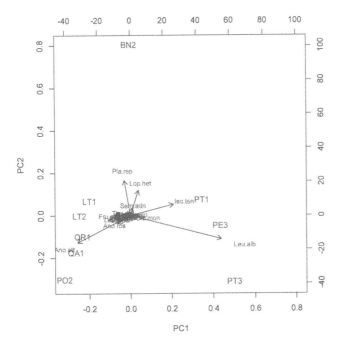

Figure 14.10 Bryophytes and tree species in North Carolina. PCA bi-plot.

8. You can get sample (site) scores using the `scores()` command:

```
> scores(moss.pca, choices = 1:2, display = "sites")
              PC1          PC2
BN2    -0.4446475   34.8789419
LT1   -18.1046921    2.8738551
LT2   -22.5535658   -0.1146282
PE3    39.2108396   -2.0936131
PO2   -29.1878891  -13.1531653
PT1    31.1197817    3.1753791
PT3    45.2687047  -13.5379996
QA1   -24.0843294   -7.6439190
QR1   -21.2242021   -4.3848508
```

9. Use the `scores()` command to view the species (column) scores:

```
> scores(moss.pca, choices = 1:2, display = "species")
                PC1            PC2
Amb.ser  -0.03286895   -0.0001433745
Ano.att  -0.44676881   -0.4601104326
Ano.min  -0.06677371   -0.0413068730
```

```
Ano.ros  -0.10923248  -0.1366403589
Bra.acu  -0.03372452  -0.0264878042
Bra.oxy  -0.02052265   0.0001435151
Bry.ill  -0.04253102   0.0070905281
Cam.his  -0.01791163   0.0039579047
Cla.par  -0.06571515  -0.0701570942
...
```

The `scores()` command will give all the axis scores unless you specify the `choices` instruction.

The sample scores are held in the `$x` component and the species scores in the `$rotation` component. You can use these scores to construct your own plot but the species scores will need to be rescaled to fit in the plot space. The *vegan* package makes it easier to construct custom plots, as you will see shortly. Before that though, you should consider (briefly) an alternative PCA command from the basic R distribution.

The `princomp()` command for PCA

An alternative to the `prcomp()` command is `princomp()`. They are computationally quite different and in most instances `prcomp()` is the command of choice. The `princomp()` command uses a different approach (eigenvalue analysis of the correlation or covariance matrix), which is generally considered to be less accurate than SVD. In addition, `princomp()` requires that your data have more rows (samples) than columns (species or variables) – this is often not the case with community data. The general form of the command is:

```
princomp(x, cor = FALSE)
```

The `princomp()` command allows you to specify if you want to use the correlation matrix (`cor = TRUE`) or the covariance matrix (`cor = FALSE`); the latter option (the default) is what you should use for community data.

The `princomp()` command has special `summary()` and `print()` routines and the `biplot()` command can be used to make the standard bi-plot, showing species as arrows. However, customisation of the graphical output is not always trivial; for this reason it is useful to consider the *vegan* package, where the `rda()` command can carry out PCA, as you will see next.

Using the vegan package for PCA

The `rda()` command in the *vegan* package can carry out PCA, using a method similar to that of the `prcomp()` command. The *vegan* package has several plotting routines that make customising PCA plots a lot easier. The `rda()` command rescales the species scores so that they 'fit' in the same scale as the sample scores, which makes things simpler.

To use the `rda()` command for PCA you simply specify the community data in the command:

```
rda(comm)
```

The result of the `rda()` command contains several components but it is generally easier to use one of the 'helper' commands to access them. In the following exercise you can have a go at a PCA using the `rda()` command to explore some plant community data.

Have a Go: Explore PCA results using the vegan package

You will need the *psa* data for this exercise; these data show the abundance of plant species at several sites in the UK. The abundance is an average Domin score from five quadrats. The data are in the *CERE.RData* file and are in the form of biological records. You'll also need the *vegan* package.

1. Start by preparing the *vegan* package:

```
> library(vegan)
```

2. Look at the data, which are in biological recording format:

```
> head(psa)
   Site              Species Qty
1  ML1 Achillea millefolium   6
2  ML1      Centaurea nigra    4
3  ML1   Lathyrus pratensis    5
4  ML1 Leucanthemum vulgare    5
5  ML1    Lotus corniculatus   7
6  ML1   Plantago lanceolata   5
```

3. Use cross-tabulation to rearrange the data into an appropriate data object with rows as sites and columns as species:

```
> psa.biol = xtabs(Qty ~ Site + Species, data = psa)
```

4. Now use the rda() command to carry out a PCA:

```
> psa.pca = rda(psa.biol)
```

5. Use the scores() command to access the site scores for PCA axes 1 and 2:

```
> scores(psa.pca, choices = 1:2, "sites")
            PC1         PC2
ML1 -2.8376924   2.4726691
ML2 -1.8925630  -2.1161538
MU1 -2.2339608  -1.9208739
MU2 -2.5739986   3.5197994
PL2  1.3225987  -1.3196237
PU2 -0.3510521  -2.5180412
SL1  0.9717673  -2.3032552
SL2  1.5615621   1.1093255
SU1  3.1272889   0.9418854
SU2  2.9060501   2.1342683
```

6. Look at the species scores:

```
> scores(psa.pca, choices = 1:2, "species")
              PC1           PC2
Achlm -0.969234174   0.104955585
Agpdp  0.053465888   0.037981887
Agrsc -0.403514198  -0.629691284
Agrss -0.024194704  -0.259924865
Anths  0.266867221   0.161201615
Arctm  0.024874859   0.018268654
Arrhe  1.408852600   0.574444204
Bdnsc  0.090568175  -0.090364453
...
```

7. Get the eigenvalues for the axes using the `eigenvals()` command:

```
> eigenvals(psa.pca)
      PC1       PC2       PC3       PC4       PC5       PC6       PC7
77.86321  55.31192  31.18960  25.81939  17.53097  14.17354  11.94598
      PC8       PC9
 4.79615   4.00925
```

8. The eigenvalues show the proportion of variance explained – get the proportion directly from the `summary()`:

```
> summary(psa.pca)$cont  # Just proportion explained
Importance of components:
                          PC1      PC2      PC3      PC4      PC5
Eigenvalue            77.8632  55.3119  31.1896  25.8194  17.53097
Proportion Explained   0.3209   0.2280   0.1285   0.1064   0.07225
Cumulative Proportion  0.3209   0.5489   0.6774   0.7838   0.85606
                          PC6      PC7      PC8      PC9
Eigenvalue            14.17354 11.94598 4.79615 4.00925
Proportion Explained   0.05841  0.04923 0.01977 0.01652
Cumulative Proportion  0.91448  0.96371 0.98348 1.00000
```

9. The `screeplot()` command can visualise the explained variance; you can show bars (the default) or lines. Use this to show the variance as lines, your result should resemble Figure 14.11:

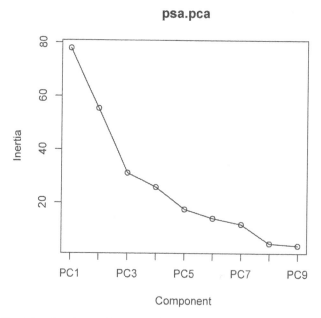

psa.pca

Figure 14.11 Explained variance for PCA of plant abundance data.

10. The `plot()` command can display site and/or species scores but does not produce arrows for the species. The 'long' species names also lead to congested plots so

make some abbreviated names for the species and use them to build a plot. Your final plot should resemble Figure 14.12:

```
> psa.nam = abbreviate(colnames(psa.biol), minlength = 5)
> plot(psa.pca, display = "sites", type = "text")
> text(psa.pca, display = "species", labels = psa.nam, col = "red", cex = 0.6)
> arrows(0,0, psa.sc$species[,1], psa.sc$species[,2],
        length = 0.05, col = "red", lwd = 0.5)
> abline(h = 0, lty = 3)
> abline(v = 0 , lty = 3)
```

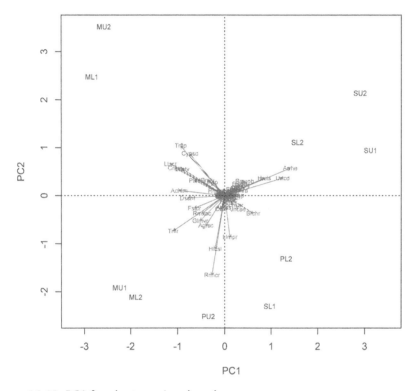

Figure 14.12 PCA for plant species abundance.

11. In step 10 you made a basic plot and displayed the sites only as text. Then you added the species names from their abbreviations, making the labels red and small. The `arrows()` command added the arrows for the species. Finally the `abline()` command added the lines passing through the origin.

12. The results of the `rda()` command and `plot()` produce a graphic that can be used interactively. This allows you to produce a plot of points, which is less congested. You can then click on individual points to make labels. Try this out now; your final plot should resemble Figure 14.13. Start by making a plot that shows the sites as text and the species as small points:

```
> op = plot(psa.pca, type = "n")
> text(psa.pca, display = "sites")
> points(psa.pca, display = "species", pch = ".", col = "red")
```

13. Now use the `identify()` command to add labels to some of the points. You'll add the arrows afterwards. The labels will appear to the side of the point where you click. When you've added a few labels press the Esc key to end the process. The 'identity' of the species will be stored so you can make the arrows:

```
> sp = identify(op, "species", col = "red", cex = 0.6,
   labels = psa.nam)
> arrows(0,0, psa.sc$species[sp,1], psa.sc$species[sp,2],
         col = "red", lwd = 0.5, length = 0.1)
```

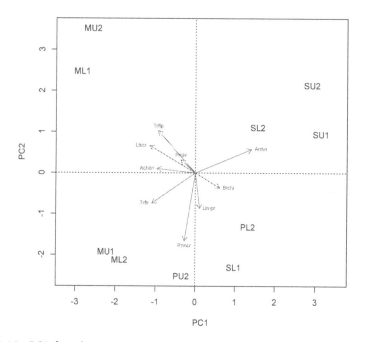

Figure 14.13 PCA for plant species abundance; produced interactively to select species bi-plot arrows.

You can make a bi-plot directly using the `biplot()` command but the species names will be very long. It is not easy to alter the display directly and it's easier to make the species abbreviations right at the start. The `biplot()` result can also be used interactively.

The ordination plots produced with the *vegan* package can be used interactively and there are also additional ways to produce your results – you can select individual species (or sites) for example. However, this is not helpful for PCA results as you cannot easily make the arrows. You'll see more examples of plotting in the section on correspondence analysis (Section 14.2.5).

Tip: Zooming-in on ordination plots

Ordination plots are often quite congested. Bi-plots of PCA results are often very 'busy' towards the origin. You can zoom in to view a portion of the plot but you can only do this by redrawing the plot. Use the `xlim` and `ylim` instructions to specify the start and end points of the axes, e.g. `xlim = c(-1, 1)`. Note that you always have to specify both starting and ending values.

Producing specific bi-plot arrows is a challenge. The `biplot()` command does not accept the `select` instruction, which you will find out about in the section about correspondence analysis (Section 14.2.5). The answer is to determine the row numbers of the species (or sites) that you want to display. You can use the `colnames()` command to see the species names or, if you have made abbreviated names you can 'interrogate' the result to get the number of the species you are interested in:

```
> psa.nam = abbreviate(colnames(psa.biol), minlength = 5)
> head(psa.nam)
  Achillea millefolium Aegopodium podagraris   Agrostis capillaris
              "Achlm"              "Agpdp"               "Agrsc"
  Agrostis stolonifera Anthriscus sylvestris        Arctium minus
              "Agrss"              "Anths"               "Arctm"

> which(psa.nam == "Trflr")
Trifolium repens
              73
```

Once you have the number(s) of the species you can use them with `text()`, `points()` or `arrows()` commands. The arrowheads and labels tend to overlap so a good compromise is to make your arrows and then add the labels using the `identify()` command:

```
> op = plot(psa.pca, type = "n")
> arrows(0,0, psa.sc$species[73,1], psa.sc$species[73,2],
         length = 0.1)
> identify(op, "species", labels = psa.nam)
```

In general PCA plots are the most 'difficult' to deal with. The `rda()` command makes things easier because the species scores are scaled 'nicely', allowing you to add the bi-plot arrows without fiddling with scales. Fortunately PCA is not the most useful method of ordination for community data!

PCA scores and environmental data

Because of the horseshoe effect it can be difficult to interpret PCA results. You can easily extract axis scores and can attempt correlation with environmental variables. In cases where the arch or horseshoe are pronounced you can use a non-parametric correlation. For example Figure 14.14 shows a PCA for plant data collected across a psammosere successional gradient:

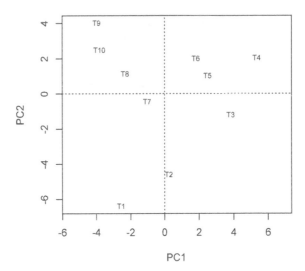

Figure 14.14 PCA of plant data from a psammosere successional gradient, showing a pronounced horseshoe effect.

The cor.test() command can carry out a Spearman rank correlation test for any axis:

```
> cor.test(scores(ridge.pca,2,"sites"), env$Soil, method = "spearman")

  Spearman's rank correlation rho

data:  scores(ridge.pca, 2, "sites") and env$Soil
S = 40, p-value = 0.01592
alternative hypothesis: true rho is not equal to 0
sample estimates:
      rho
0.7575758
```

The horseshoe makes interpretation difficult – you'd expect PCA axis 1 to be the most important axis yet the correlation between axis 2 and the environmental variable is significant. Presumably there is another gradient that is not easily discernible.

14.2.5 Correspondence analysis (reciprocal averaging)

CA is also known as reciprocal averaging – one algorithm for finding the solution involves the repeated averaging of sample scores and species scores. CA maximises the correspondence between species scores and sample scores. The first axis is made by assigning species and sample scores so that the weighted correlation (weight being species abundance) between them is maximised. You cannot rearrange the rows and columns in any way that makes the correlation better.

Subsequent axes are made in a similar manner except that they are forced (constrained) to be orthogonal to the first axis. The eigenvalues of the axes is equivalent to the correlation between species and sample scores. Species scores are often standardised to unit variance and zero mean; in which case the eigenvalues represent the variance in sample scores. The sum of the eigenvalues is called the inertia.

Unlike PCA the species are represented by centres rather than arrows. You can think of the species score as being at the peak of its abundance. The method can result in an arch effect and the ends of axes can also be somewhat compressed. This can lead to difficulties in interpretation and led to the invention of detrended correspondence analysis (DCA), which you will meet in the next section (14.2.6).

The *MASS* package comes as part of the basic distribution of R but is not loaded by default. It contains a command, `corresp()`, which will carry out correspondence analysis. The *vegan* package has the `cca()` command, which will also conduct correspondence analysis (as well as constrained analysis, see Section 14.3.1).

Of the two commands, the `cca()` command in the *vegan* package is by far the easier to manage. The results are easier to extract and plot and there are additional features that make interpretation easier than the `corresp()` command.

Using MASS for CA

The `corresp()` command in the *MASS* package can carry out CA. The general form of the command is like so:

```
corresp(x, nf = 1)
```

You must supply the data, usually as a date frame or matrix, with rows as samples and columns as sites. The default is to produce a single axis so usually you want to set `nf = 2` to produce two CA axes for plotting.

The `corresp()` command produces a result with several components, in particular the `$rscore` and `$cscore` components, which hold the sample and species scores respectively. There are `print()` and `plot()` routines associated with the results – the `print()` command produces a bi-plot. In the following exercise you can have a go at using the `corresp()` command to make a CA analysis of some plant community data.

Have a Go: Use the MASS package for correspondence analysis

You will need the *psa* data for this exercise; these data show the abundance of plant species at several sites in the UK. The abundance is an average Domin score from five quadrats. The data are in the *CERE.RData* file and are in the form of biological records. You'll also need the *MASS* package (which is part of the basic distribution of R).

1. Make sure the *MASS* package is loaded:

   ```
   > library(MASS)
   ```

2. The data are in biological recording format so you'll need to use cross-tabulation to construct a community data matrix with rows as samples and species as columns. Use the `abbreviate()` command to make the species names shorter:

   ```
   > head(psa)
       Site              Species  Qty
   1   ML1   Achillea millefolium    6
   2   ML1        Centaurea nigra    4
   3   ML1     Lathyrus pratensis    5
   4   ML1   Leucanthemum vulgare    5
   ```

```
5   ML1    Lotus corniculatus     7
6   ML1    Plantago lanceolata    5

> comm = xtabs(Qty ~ Site + abbreviate(Species, minlength = 5),
               data = psa)
> comm[1:5,1:9]
       abbreviate(Species, minlength = 5)
Site  Achlm  Agpdp  Agrsc  Agrss  Anths  Arctm  Arrhe  Bdnsc  Brchr
  ML1     6      0    0.0      0      0      0      0    0.0      0
  ML2     3      0    8.0      0      0      0      0    0.0      0
  MU1     5      0    8.0      0      0      0      0    0.0      0
  MU2     4      0    5.6      0      0      0      0    0.0      0
  PL2     0      0    8.0      0      0      0      4    3.2      9
```

3. Now make a CA result. Use nf = 2 (number of factors) to create two CA axes:

```
> mod = corresp(comm, nf = 2)
Warning message:
In corresp.matrix(x, ...) : negative or non-integer entries in table
```

4. You get a warning about non-integer values in step 3 but do not worry about this. Look at the result:

```
> names(mod)
[1] "cor" "rscore" "cscore" "Freq"
```

5. There is a print() routine associated with the result (you can access it by simply typing the name of the result), but the output is quite lengthy. View the sample scores in the $rscore component:

```
> mod$rscore
             [,1]          [,2]
ML1   -1.0940206    0.82828785
ML2   -0.6863088   -0.15521074
MU1   -0.6984149   -0.05949684
MU2   -1.1083597    0.95972268
PL2    0.4654786   -2.17490438
PU2   -0.2849861   -0.88488892
SL1    0.3126034   -1.03534045
SL2    1.4934307    1.35203432
SU1    1.2367184    0.11601535
SU2    1.7685848    1.16669525
```

6. The species scores are in the $cscore component:

```
> head(mod$cscore)
             [,1]          [,2]
Achlm  -0.9476265    0.3435649
Agpdp   1.7183079    1.8887449
Agrsc  -0.1460886   -0.5561127
Agrss  -0.1763278   -1.0046847
Anths   1.8600502    1.2104718
Arctm   2.0348941    1.6298327
```

7. The default plotting routine produces a bi-plot with site and species scores on different scales. It is not easy to customise this, so have a go at making your own bi-plot. Start by working out the scales the x and y axes will need to be:

```
> arx = range(c(mod$cscore[,1], mod$rscore[,1]))
> ary = range(c(mod$cscore[,2], mod$rscore[,2]))
> arx ; ary
[1]  -1.268908  2.034894
[1]  -3.038266  1.888745
```

8. Now make a blank plot to set the co-ordinate system. Then add text for the sites and the species scores separately – your plot should resemble Figure 14.15:

```
> plot(mod$rscores, xlim = arx * 1.2, ylim = ary * 1.2,
        xlab = "CA1", ylab = "CA2")
> text(mod$rs, labels = rownames(mod$rscore), cex = 0.9)
> text(mod$cscore, cex = 0.7, col = "red",
        labels = rownames(mod$cscore))
```

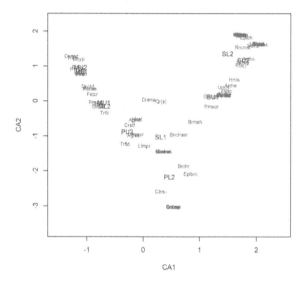

Figure 14.15 Correspondence analysis of plant abundance data.

You can access the site and sample scores but there are not any 'helper' commands to make things easier.

The `corresp()` command conducts the CA easily enough but was never designed with community ecology in mind. The `cca()` command in the *vegan* package, however, is aimed at community data and makes dealing with results a lot easier. This is the subject of the following section.

Using the vegan package for CA

The `cca()` command in the *vegan* package carries out correspondence analysis; it can also handle constrained (canonical) analysis (see Section 14.3.1). To carry out CA you simply provide the command with the name of the community data, where the rows are the samples (sites) and the columns are the species.

The result contains several components – there are helper commands to facilitate access. There are also `print()`, `summary()` and `plot()` routines. The `plot()` command in particular makes customising plots much easier than in the `corresp()` command that is in the *MASS* package. In the following exercise you can have a go at CA using the `cca()` command – you'll use the same data as the preceding exercise so you can compare.

Have a Go: Use the vegan package for correspondence analysis

You will need the *psa* data for this exercise; these data show the abundance of plant species at several sites in the UK. The abundance is an average Domin score from five quadrats. The data are in the *CERE.RData* file and are in the form of biological records. You'll also need the *vegan* package.

1. Start by ensuring the *vegan* package is loaded and ready:

   ```
   > library(vegan)
   ```

2. Look at the data – you'll see that they are arranged in recording format:

   ```
   > head(psa)
     Site              Species  Qty
   1 ML1  Achillea millefolium    6
   2 ML1       Centaurea nigra    4
   3 ML1     Lathyrus pratensis    5
   4 ML1  Leucanthemum vulgare    5
   5 ML1     Lotus corniculatus    7
   6 ML1    Plantago lanceolata    5
   ```

3. Rearrange the data into a community dataset by using cross-tabulation via the `xtabs()` command:

   ```
   > psa.biol = xtabs(Qty ~ Site + Species, data = psa)
   ```

4. Your new data now contain rows for each sample and columns for the abundances of the species. Carry out a correspondence analysis using the `cca()` command in the *vegan* package:

   ```
   > psa.ca = cca(psa.biol)
   ```

5. You can view the results of the analysis using `summary()` or `print()` commands; the latter is the same as typing the result name but allows you to specify the number of significant figures displayed:

   ```
   > print(psa.ca, digits = 3)
   Call: cca(X = psa.biol)

                 Inertia Rank
   Total            2.94
   Unconstrained    2.94    9
   Inertia is mean squared contingency coefficient

   Eigenvalues for unconstrained axes:
   ```

```
    CA1    CA2    CA3    CA4    CA5    CA6    CA7    CA8    CA9
  0.755  0.512  0.414  0.357  0.266  0.211  0.187  0.160  0.076
```

6. The `summary()` command produces a result with several components – you can use the `cont` component to look at the contribution of the eigenvalues for example:

```
> names(summary(psa.ca))
 [1] "species"     "sites"      "call"       "tot.chi"
 [5] "unconst.chi" "cont"       "scaling"    "digits"
 [9] "inertia"     "method"

> summary(psa.ca)$cont
Importance of components:
                         CA1    CA2    CA3    CA4    CA5     CA6
Eigenvalue            0.7554 0.5124 0.4142 0.3572 0.2663 0.21101
Proportion Explained  0.2570 0.1743 0.1409 0.1215 0.0906 0.07178
Cumulative Proportion 0.2570 0.4313 0.5722 0.6937 0.7843 0.85606
                         CA7     CA8     CA9
Eigenvalue            0.18687 0.16030 0.07596
Proportion Explained  0.06357 0.05453 0.02584
Cumulative Proportion 0.91963 0.97416 1.00000
```

7. Get the eigenvalues directly using the `eigenvals()` command. The sum is the same as the `unconst.chi` component of the `summary()` result:

```
> eigenvals(psa.ca)
      CA1        CA2        CA3        CA4        CA5        CA6
0.7553847  0.5124232  0.4141913  0.3571938  0.2663379  0.2110118
      CA7        CA8        CA9
0.1868727  0.1603037  0.0759594

> summary(psa.ca)$unconst.chi
[1] 2.939678
```

8. Use the `scores()` command to get sites or species scores – try the site scores now:

```
> scores(psa.ca, display = "sites")
            CA1          CA2
ML1  -1.0940206   0.82828785
ML2  -0.6863088  -0.15521074
MU1  -0.6984149  -0.05949684
MU2  -1.1083597   0.95972268
PL2   0.4654786  -2.17490438
PU2  -0.2849861  -0.88488892
SL1   0.3126034  -1.03534045
SL2   1.4934307   1.35203432
SU1   1.2367184   0.11601535
SU2   1.7685848   1.16669525
```

9. The species names are 'full' scientific and will clutter the display if you try a plot (Figure 14.16):

```
> plot(psa.ca)
```

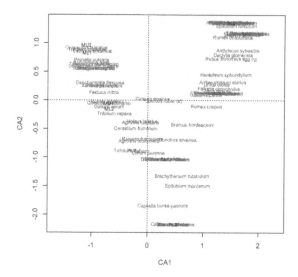

Figure 14.16 Correspondence analysis of plant abundance data showing full scientific names of species. Abbreviated names are more useful to reduce congestion in ordination plots.

10. Use the `abbreviate()` command to make shorter species names, which you can use as labels in plotting:

```
> psa.nam = abbreviate(colnames(psa.biol), minlength = 5)
```

11. Make a plot using labels for the site names but only indicate the species with points; your plot should resemble Figure 14.17:

```
> plot(psa.ca, type = "n")
> text(psa.ca, display = "sites")
> points(psa.ca, display = "species", col = "red", cex = 0.7, pch = "+")
```

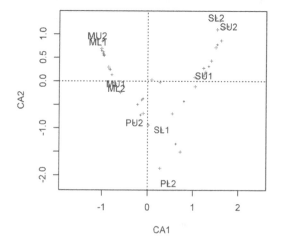

Figure 14.17 Correspondence analysis of plant species data plotted to show site labels and species as unlabelled points.

12. You can add species names to the plot using the `text()` command but things are still congested. Try it for yourself (plot not shown):

```
> plot(psa.ca, type = "n")
> text(psa.ca, display = "sites")
> text(psa.ca, display = "species", col = "blue", cex = 0.7, labels = psa.nam)
```

13. The *vegan* package allows you to identify components of a correspondence analysis result – try redrawing the plot from step 11 (you'll need to name the plot) and then add some species names using the mouse (press the *Esc* key when done labelling); your plot will resemble Figure 14.18 (depending on which species you labelled):

```
> op = plot(psa.ca, type = "n")
> text(psa.ca, display = "sites")
> points(psa.ca, display = "species", col = "red", cex = 0.7, pch = "+")
> identify(op, what = "species", labels = psa.nam, col = "blue")
```

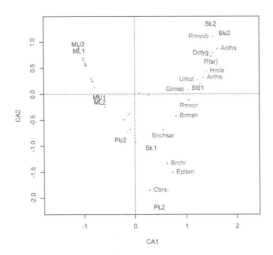

Figure 14.18 Correspondence analysis of plant species data plotted to show site labels and species as unlabelled points. Some species have been labelled explicitly using the `identify()` command.

14. You can select explicit species (or sites) to add to your plots by using the `select` instruction in the `text()` or `points()` command. If you use the abbreviated names you'll need to specify the species name(s) in abbreviated form. If you simply give the row numbers then this is not necessary. Try the following to produce a plot showing only selected species; your plot should resemble Figure 14.19:

```
> plot(psa.ca, type = "n")
> text(psa.ca, display = "sites", cex = 0.8)

> text(psa.ca, display = "species", col = "blue", cex = 0.7,
       labels = psa.nam, select = c("Urtcd"))

> text(psa.ca, display = "species", col = "blue", cex = 0.7,
       labels = psa.nam, select = psa.nam[1:5])
```

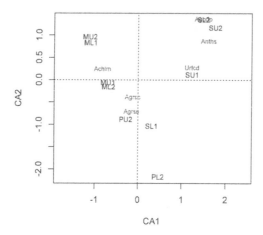

Figure 14.19 Correspondence analysis of plant species data plotted to show site labels. Only some species are shown.

The more sites and species you've got the more congested and complex your plots will be!

Tip: Zoom in on ordination plots

Even with abbreviated species names the resulting plots can be quite congested. You can zoom in to focus on a section of a plot but you can only do this by redrawing the plot and specifying the section you wish to view. You use the `xlim` and `ylim` instructions in the `plot()` command to specify the limits of the *x* and *y* axes. Each instruction needs a starting and ending value e.g. `xlim = c(-1, 1)`.

You can see from the example used in this section that correspondence analysis can produce results that are quite arched. This can lead to difficulties in interpretation, which led to the advent of detrended correspondence analysis. This is the subject of the following section.

14.2.6 Detrended correspondence analysis

DCA eliminates the arch effect that results from correspondence analysis. The most common method of achieving this is to split the first axis into segments. The samples in each segment are then centred to have a zero mean on the second axis. The process continues by assigning different starting points for the segments and centring again. Other axes are detrended in a similar manner.

The sample scores are rescaled using non-linear rescaling; this corrects the compression that usually occurs at the ends of the 'arch' in CA. The sample scores are shifted along each axis so that the average width is unity. This rescaling means that the axes are scaled in units of *beta* diversity.

The `decorana()` command in the *vegan* package carries out DCA – the command has `print()`, `summary()` and `plot()` routines associated with it. The `plot()` command works much like for the `cca()` command that you saw in the preceding section. This allows you to customise your plots fairly easily. In the following exercise you can have a go at a DCA using the plant data from the preceding section on correspondence analysis – you can then see the effect of the detrending for yourself.

Have a Go: Carry out detrended correspondence analysis

You will need the *psa* data for this exercise; these data show the abundance of plant species at several sites in the UK. The abundance is an average Domin score from five quadrats. The data are in the *CERE.RData* file and are in the form of biological records. You'll also need the *vegan* package.

1. Start by ensuring the *vegan* package is loaded and ready:

   ```
   > library(vegan)
   ```

2. The data are in biological recording format so you'll need to rearrange them with samples as rows and species as columns:

   ```
   > names(psa)
   [1] "Site"    "Species" "Qty"
   > psa.biol = xtabs(Qty ~ Site + Species, data = psa)
   ```

3. The species names are 'long' so it will be useful to make abbreviated versions:

   ```
   > psa.nam = abbreviate(colnames(psa.biol), minlength = 5)
   ```

4. Now carry out a DCA using the decorana() command:

   ```
   > psa.dca = decorana(psa.biol)
   ```

5. The decorana() command calculates four axes. You can see the result using print() or summary() commands but it is easier to get the scores for sites or species with the scores() command:

   ```
   > scores(psa.dca, display = "sites", choices = 1:2)
             DCA1         DCA2
   ML1  -1.9387870  -0.04592691
   ML2  -1.1155124   0.01433617
   MU1  -1.1434691   0.04318351
   MU2  -1.9652579  -0.01250189
   PL2   1.0358345   1.40613715
   PU2  -0.2705119  -0.42713187
   SL1   0.8074944  -0.84388521
   SL2   2.2837902   0.61782652
   SU1   1.9532815  -0.22335776
   SU2   2.6750034  -0.19266165
   ```

6. Get the eigenvalues for the axes from the result component, evals:

   ```
   > psa.dca$evals
        DCA1       DCA2       DCA3       DCA4
   0.7469290  0.2904719  0.2032709  0.1848151
   ```

7. Make a plot to show the site names with species as points – your plot should resemble Figure 14.20:

   ```
   > plot(psa.dca, type = "n")
   > text(psa.dca, display = "sites")
   > points(psa.dca, display = "species", col = "red", pch = "+")
   ```

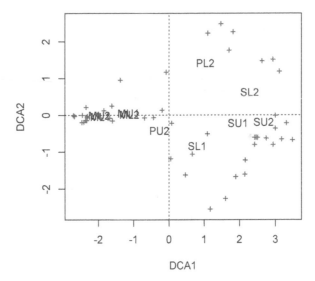

Figure 14.20 Detrended correspondence analysis of plant abundance data.

8. Redraw the plot but this time name it and then use the `identify()` command to label some of the species (press *Esc* to finish); your plot will resemble Figure 14.21, depending which species you identified:

```
> op = plot(psa.dca, type = "n")
> text(psa.dca, display = "sites")
> points(psa.dca, display = "species", pch = "+", col = "red")
> identify(op, "species", labels = psa.nam, cex = 0.7, col = "red")
```

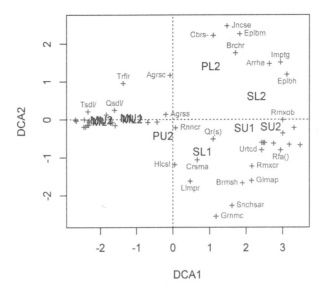

Figure 14.21 Detrended correspondence analysis of plant abundance data. Selected species have been labelled using the `identify()` command.

9. Zoom into one portion of the plot to help resolve the congestion – your figure will resemble Figure 14.22:

```
> op = plot(psa.dca, type = "n", xlim = c(-3, -1), ylim = c(-1, 1))
> text(psa.dca, display = "sites")
> points(psa.dca, display = "species", pch = "+", col = "red")
> identify(op, "species", labels = psa.nam, cex = 0.7, col = "red")
```

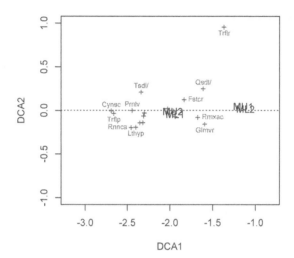

Figure 14.22 Detrended correspondence analysis of plant abundance data focusing on one area of the ordination space. Selected species have been labelled using the `identify()` command.

The zoomed-in plot from step 9 still shows some congestion – you can play around with different axis values until you get the most useful plot for your purpose.

Linking DCA and environmental gradients

Since the DCA axes are linked to *beta* diversity diversity it is possible to plot the sample scores against an environmental gradient. This can show if the species 'view' the gradient differently from the way you measure it. Higher slopes indicate areas of high *beta* diversity. In the following exercise you can have a go at visualising the link between DCA sample scores and environmental gradients.

Have a Go: Link DCA scores to environmental gradients

You will need the *ridge.biol* data for this exercise; the data are in the *CERE.RData* file. The data show plant abundance for species along a transect from a psammosere. There are ten samples and each shows frequency from eight replicates. You'll also need the *vegan* package.

1. Start by ensuring the *vegan* package is ready:

```
> library(vegan)
```

2. The *ridge.biol* data are arranged with species as rows so you'll need to rearrange the data before you start. There are also some environmental data in the *ridge.env* object. Rotate both so that the samples are the rows:

```
> ridge = t(ridge.biol)
> env = t(ridge.env)
```

3. Now make a DCA result object. Plot the result as simple points; your plot should resemble Figure 14.23:

```
> ridge.dca = decorana(ridge)
> plot(ridge.dca, type = "p")
```

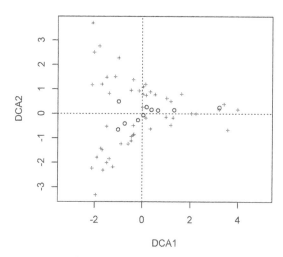

Figure 14.23 Detrended correspondence analysis of psammosere plant community data.

4. The sample scores are obtained using the `scores()` command, you'll need the first column for the DCA axis 1 scores:

```
> scores(ridge.dca)[,1]
        T1           T2           T3           T4           T5
3.20771359   1.31790105   0.64328055   0.37534910   0.17896114
        T6           T7           T8           T9          T10
0.04389717  -0.17610547  -0.73289905  -1.01208887  -0.98187265
```

5. The environmental data are in a matrix (the result of the `t()` command), and you'll be using columns 1 and 4:

```
> env
      Soil   Light   Humid   Wind
T1     0.0   16250    28.3   5.13
T2     0.1   16250    27.4   3.25
T3     3.3   17500    27.3   1.88
```

```
T4     3.8   16625    28.6   1.13
T5     3.6   15875    26.6   0.75
T6     6.1   16625    26.9   0.63
T7     8.0   14413    30.0   0.25
T8    14.4   14368    51.4   0.00
T9    13.7   14064    25.2   0.25
T10   14.1   11381    25.5   0.25
```

6. Use the environmental data as the x-axis and use the sample scores as the y-axis data. You can visualise these using the plot() command. Start by splitting the plot window into two sections (one row and two columns):

```
> opar = par(mfrow = c(1,2))
```

7. Now make a plot for the *wind* data:

```
> plot(env[,4], scores(ridge.dca)[,1],
       ylab = "DCA sample scores", xlab = "Wind speed")
```

8. Add a locally weighted polynomial regression – the lowess() command calculates the regression and the lines() command adds it to the plot (Figure 14.24a):

9. Make a plot for the sample scores against the *soil* data:

```
> plot(env[,1], scores(ridge.dca)[,1],
       ylab = "DCA sample scores", xlab = "Soil depth")
```

10. Add a locally weighted polynomial regression line to make a plot like Figure 14.24b:

```
> lines(lowess(env[,1], scores(ridge.dca)[,1]))
```

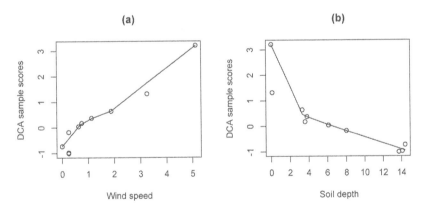

Figure 14.24 DCA sample scores for psammosere plants and environmental variables.

11. Reset the graphic window back to the original settings (i.e. with only a single plot). The current plot is unaffected:

```
> par(opar)
```

You can see from Figure 14.24 that there is a clear break in the regression line for the soil data. It looks like the communities in the shallow soils have higher *beta* diversity.

You can add environmental information to ordination plots in several ways: you'll see more in Section 14.4.1.

14.3 Direct gradient analysis

If you have environmental information of some kind you can use it at the start of your analysis rather than afterwards – this is direct gradient analysis. You can think of this as a form of regression where you are linking species composition to the environmental variables that you've got. The direct gradient analysis allows you to test hypotheses about the relationship between species composition and the measured variables that you have – usually carried out by permutation tests.

As well as being able to test relationships between species composition and environmental variables you can 'partial-out' variables. This means that you can remove the effects of one or more variables from the analysis – which is sometimes called *conditioning*.

In this section you'll see two main methods of direct gradient analysis, using three commands from the *vegan* package:

- `cca()` – CCA is an extended form of CA.
- `rda()` – redundancy analysis (RDA) is a spin-off from CCA using slightly different methods.
- `capscale()` – constrained analysis of principal co-ordinates (CAP) is a form of RDA based on PCO; it allows dissimilarity measures other than the Euclidean metric to be used (which is the basis of RDA).

In all three cases you can partial-out variables from your analysis (i.e. apply conditioning variables). The methods of CCA and RDA allow for a measure of model-building, that is, you can add environmental variables to your analysis and so build a model that is 'most effective' in explaining the variability in your data.

14.3.1 Canonical correspondence analysis

Canonical (constrained) correspondence analysis (CCA) is a combination of regular correspondence analysis and linear regression. The analysis maximises the correlation between species scores and sample scores. However, in CCA the sample scores are constrained to be linear combinations of the environmental variables. You can also think of CCA as a way to find the best dispersion of the species scores – relating species composition to combinations of environmental variables. Subsequent axes are created in a similar manner but are orthogonal.

You end up with as many constrained axes as there are explanatory variables and the total inertia explained is the sum of the eigenvalues of these constrained axes. Unconstrained axes can be thought of as residual (i.e. unexplained). The total inertia from the result is the sum of the eigenvalues of both constrained and unconstrained axes. So, you can gain a measure of how well species composition is explained by the variables by comparing the eigenvalues of the constrained axes to the total of all eigenvalues.

The `cca()` command in the *vegan* package carries out CCA. In this command the data are chi squared transformed and then subjected to weighted linear regression using the constraining variables. The fitted values are then used in a correspondence analysis (using singular value decomposition). The general form of the command is like so:

```
cca(formula, data)
```

You supply a formula that describes the response on the left of a ~ and the predictors (constraining variables) on the right of the ~. You can optionally give the name of the data object (usually a `data.frame` or `matrix`) that 'holds' the predictor variables. If you want to use a conditioning variable to 'partial-out' you can use `Condition(var)` to name the variable (`var`) in the formula.

The `cca()` command produces a complicated result with many components, but fortunately there are `print()`, `summary()` and `plot()` commands associated with the command. The `plot()` command allows you to see the samples and species as well as the constraining variables (shown as biplot arrows). There are also some 'helper' commands, e.g. `scores()`, that can extract parts of the result in a meaningful way.

Basic canonical correspondence analysis

The `cca()` command uses a formula to describe the analytical model. Usually you'll have your biological data in one dataset, as a `matrix` or `data.frame` with samples as rows and species as columns. You'll then have a separate dataset containing the environmental variables. The formula syntax allows you to specify complicated models easily and also permits you to alter the model quickly and simply. In the following exercise you can have a go at a CCA using data from a psammosere succession.

Have a Go: Carry out canonical correspondence analysis

You will need the *ridge.biol* data for this exercise; the data are in the *CERE.RData* file. The data show plant abundance for species along a transect from a psammosere. There are 10 samples and each shows frequency from 8 replicates. You'll also need the *vegan* package.

1. Start by ensuring the *vegan* package is ready:

    ```
    > library(vegan)
    ```

2. The *ridge.biol* data are arranged with species as rows so you'll need to rearrange the data before you start. There are also some environmental data in the *ridge.env* object. Rotate both so that the samples are the rows:

    ```
    > ridge = as.data.frame(t(ridge.biol))
    > env = as.data.frame(t(ridge.env))
    ```

3. Now look at the variables in the environmental data:

    ```
    > names(env)
    [1] "Soil"  "Light" "Humid" "Wind"
    ```

4. Use all the variables in the env data as constraining variables in a CCA:

    ```
    > ridge.cca = cca(ridge, env)
    ```

5. The `scores()` command can get you the scores for samples ("`sites`"), species ("`species`") or constraining variables ("`bp`"):

    ```
    > scores(ridge.cca, display = "sites")
             CCA1        CCA2
    T1  -4.2176951   7.1181887
    T2  -1.4827868   0.9975761
    ```

```
T3  -0.7918630  -0.3838182
T4  -0.6413930  -0.8510676
T5  -0.3685498  -0.7327989
T6  -0.1377318  -0.6848219
T7   0.4007082  -0.1664721
T8   0.9911504   0.4733635
T9   1.3620394   0.6889541
T10  1.3750469   0.9426872
> scores(ridge.cca, display = "bp")
              CCA1         CCA2
Soil     0.9488892   0.28258715
Light   -0.7944369  -0.45323598
Humid    0.2071015   0.08786683
Wind    -0.8973366   0.40979670
```

6. Use the `inertcomp()` command to compare the inertia of the constrained and unconstrained variables:

```
> inertcomp(ridge.cca, display = "sites")
            CCA           CA
T1   0.14956922  0.165340647
T2   0.17976522  0.067419746
T3   0.06464694  0.081106790
T4   0.04599451  0.069766535
T5   0.07523393  0.046755205
T6   0.04894603  0.058866209
T7   0.02691362  0.104516483
T8   0.12475212  0.001979242
T9   0.12654133  0.028485140
T10  0.18043567  0.035103985
```

7. The ratio of the CCA inertia to the total inertia is a measure of goodness of fit. You can use the values from step 6 or use the `goodness()` command to calculate the values:

```
> goodness(ridge.cca, choices = 1:4, statistic = "explained",
           display = "sites")
              CCA1        CCA2       CCA3       CCA4
T1   1.742262e-01  0.4743353  0.4744421  0.4749588
T2   4.577727e-01  0.7151474  0.7164586  0.7272498
T3   3.370804e-01  0.3380781  0.3428472  0.4435354
T4   1.929517e-01  0.3961080  0.3965840  0.3973229
T5   6.429199e-02  0.4956983  0.5249179  0.6167265
T6   4.406122e-05  0.3783179  0.3784241  0.4539933
T7   5.788874e-02  0.1261640  0.1303305  0.2047752
T8   2.977630e-01  0.3609044  0.9590011  0.9843824
T9   5.496145e-01  0.6242958  0.6405154  0.8162563
T10  4.366324e-01  0.7229466  0.8011433  0.8371345
```

8. Run a permutation test on the model – you can only test the overall model because you did not specify the environmental variables explicitly in the `cca()` command:

```
> anova(ridge.cca, test = "F")
Permutation test for cca under reduced model
```

```
Model: cca(X = ridge, Y = env)
         Df  Chisq      F N.Perm Pr(>F)
Model     4 1.0228 1.9391    199  0.015 *
Residual  5 0.6593
---
Signif. codes:  0 '***' 0.001 '**' 0.01 '*' 0.05 '.' 0.1 ' ' 1
```

9. Now make a plot of the result – start by showing the sites and constraining variables, then add the species as points, your plot should resemble Figure 14.25:

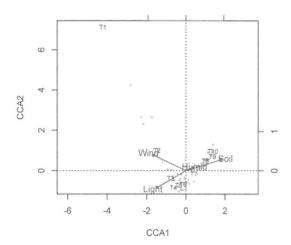

Figure 14.25 Canonical correspondence analysis of plant communities in a psammosere.

You can use the identify() command to label points as you've seen in other examples, however you'd need to assign a name to your plot in order to do this.

In the preceding exercise you used all the available environmental variables in the analysis. This is not usually recommended as it can lead to serious over-fitting and a confusing
outcome. It is better to be more selective and to pick only 'meaningful' variables. You'll
see how to undertake model-building to produce the 'most effective' model shortly
but before that you can have a go at using the formula syntax to select some variables
from the alternatives; this also allows you to carry out permutation tests on individual
variables.

Have a Go: Use permutation tests to explore a CCA

You'll need the *fw.biol* data for this exercise – these data are the abundance of various
freshwater invertebrates at 18 different samples. The samples themselves are in three
main groupings, which relate to the kind of site they were collected from. The *fw.env*
data contain the same sample names and show results for 15 different environmental
variables. Both datasets are in the *CERE.RData* file. You'll also need the *vegan* package.

1. Start by preparing the *vegan* package:

```
> library(vegan)
```

2. Have a look at the data you'll be using. There are 18 samples and 15 environmental variables:

```
> rownames(fw.biol)
 [1] "R1"  "R2"  "R3"  "R4"  "R5"  "R6"  "S7"  "S8"  "S9"  "S10"
[11] "S11" "S12" "D13" "D14" "D15" "D16" "D17" "D18"

> names(fw.env)
 [1] "DO2"    "TDS"    "cond"   "temp"   "pH"     "width"
 [7] "depth"  "pebble" "gravel" "silt"   "clay"   "turbid"
[13] "litter" "shade"  "poach"
```

3. Use *TDS, pH* and *DO2* variables in a CCA:

```
> fw.cca = cca(fw.biol ~ TDS + pH + DO2, data = fw.env)
```

4. Now carry out a permutation test on the model terms – your results will look slightly different as this is a randomisation process:

```
> anova(fw.cca, by = "terms")
Permutation test for cca under reduced model
Terms added sequentially (first to last)

Model: cca(formula = fw.biol ~ TDS + pH + DO2, data = fw.env)
         Df  Chisq     F  N.Perm Pr(>F)
TDS       1  0.1007 0.5815    99   0.81
pH        1  0.1346 0.7771    99   0.66
DO2       1  0.1431 0.8261    99   0.45
Residual 14  2.4250
```

5. You can use `scores()`, `goodness()` and `inertcomp()` commands on the result as you did in the preceding exercise. However, this time use the `eigengrad()` command to look at the strength of each environmental variable:

```
> eigengrad(fw.env, fw.biol)
       DO2        TDS       cond       temp         pH      width
0.13808483 0.10071600 0.10223115 0.22536764 0.14901178 0.10418564
     depth     pebble     gravel       silt       clay     turbid
0.19365753 0.04563151 0.45241748 0.42516169 0.29986697 0.25315756
    litter      shade      poach
0.31136363 0.09526314 0.07208707
```

6. Now plot the result – show the site names and the environmental variables (as biplot arrows) but display the species as simple points to avoid congestion. Your plot should resemble Figure 14.26:

```
> plot(fw.cca, type = "n")
> text(fw.cca, display = "bp", col = "blue")
> points(fw.cca, display = "species", col = "red", pch = "+", cex = 0.5)
> text(fw.cca, display = "sites", cex = 0.8)
```

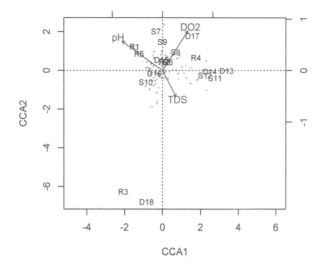

Figure 14.26 Canonical correspondence analysis of freshwater invertebrates from three different habitat types (R = river, S = stream, D = ditch). Model is not significant (F 3 = 0.728, permutation test).

You can see that the current model is not significant (from step 4). By editing the `cca()` command in step 3 you can explore the effects of the different environmental variables.

The model you made in the preceding exercise is not a significant one. You can explore the effects of different variables by editing the model or you can take a model-building approach to select the 'best' variables to add to your model – you'll see how to do this next.

Tip: Selecting all variables for a model

The `.` character acts as a wildcard and selects all variables if you use it in a formula. For example in the command `cca(fw.biol ~ ., data = fw.env)`, the `.` selects all the variables in the *fw.env* data to use in the CCA.

Model-building in canonical correspondence analysis

When you have more than a few environmental variables it is useful to employ a model-building process to create a 'best' or 'minimum adequate' model for your CCA. You can use the `add1()` command to help you achieve this. The `add1()` command takes the available variables and adds them to an existing model one at a time. The result shows you the effect of adding each term and you can then modify the model as appropriate. The usual way to proceed is as an iterative process like so:

1. Make a 'full' model containing all the available terms. This will be used to tell the `add1()` command what the potential candidate variables are (the `scope`). Save the result to a variable.

2. Make a 'blank' model, that is one without any terms except an intercept.
3. Run the `add1()` command on the blank model and use the full model to select the candidate variables.
4. Select a term (variable) to add to your model from the `add1()` result and modify the blank model by replacing the intercept.
5. Run the `add1()` command again.
6. If there are potentially significant terms then add the 'best' to the model.
7. Repeat steps 5 and 6 until there are no more significant terms.

You do not necessarily have to include all the significant terms, as some may be more biologically relevant than others. This process can save you a lot of time, especially if you have a lot of environmental variables. In the following exercise you can have a go at model-building using the freshwater data from the preceding exercise.

Have a Go: Use model-building to make a CCA

You'll need the *fw.biol* data for this exercise – these data are the abundance of various freshwater invertebrates at 18 different samples. The samples themselves are in three main groupings, which relate to the kind of site they were collected from. The *fw.env* data contain the same sample names and show results for 15 different environmental variables. Both datasets are in the *CERE.RData* file. You'll also need the *vegan* package.

1. Start by preparing the *vegan* package:

```
> library(vegan)
```

2. Have a look at the data you'll be using; there are 18 samples and 15 environmental variables:

```
> rownames(fw.biol)
 [1] "R1"  "R2"  "R3"  "R4"  "R5"  "R6"  "S7"  "S8"  "S9"  "S10"
[11] "S11" "S12" "D13" "D14" "D15" "D16" "D17" "D18"

> names(fw.env)
 [1] "DO2"    "TDS"    "cond"   "temp"   "pH"     "width"
 [7] "depth"  "pebble" "gravel" "silt"   "clay"   "turbid"
[13] "litter" "shade"  "poach"
```

3. Use all the variables to make a new result object – this will be used as the `scope` in the `add1()` command:

```
> fwm = cca(fw.biol ~ ., data = fw.env)
```

4. Now make a blank model (containing an intercept only) to use as the base for the model-building:

```
> fw.cca = cca(fw.biol ~ 1, data = fw.env)
```

5. Now run the `add1()` command to see if any of the variables is likely to significantly affect the model:

```
> add1(fw.cca, scope = formula(fwm), test = "permutation")
```

```
           Df    AIC      F N.Perm Pr(>F)
<none>        127.92
DO2        1 129.01 0.8289     99  0.580
TDS        1 129.26 0.5962     99  0.780
cond       1 129.25 0.6055     99  0.880
temp       1 128.41 1.3987    199  0.110
pH         1 128.94 0.8982     99  0.420
width      1 129.24 0.6176     99  0.710
depth      1 128.63 1.1873     99  0.160
pebble     1 129.63 0.2647     99  0.980
gravel     1 126.75 3.0790    199  0.095 .
silt       1 126.96 2.8603    199  0.080 .
clay       1 127.89 1.9164    199  0.145
turbid     1 128.22 1.5883    199  0.055 .
litter     1 127.80 1.9991    199  0.025 *
shade      1 129.30 0.5628     99  0.850
poach      1 129.45 0.4223     99  0.990
---
Signif. codes:  0 '***' 0.001 '**' 0.01 '*' 0.05 '.' 0.1 ' ' 1
```

6. From step 5 you can see that the *litter* variable is significant (from a permutation test, your values might be slightly different). The *litter* variable has the lowest AIC value so edit the *fw.cca* model, replacing the intercept with the variable:

```
> fw.cca = cca(fw.biol ~ litter, data = fw.env)
```

7. Run the add1() command again:

```
> add1(fw.cca, scope = formula(fwm), test = "permutation")
           Df    AIC      F N.Perm Pr(>F)
<none>        127.80
DO2        1 129.03 0.6552     99  0.800
TDS        1 129.05 0.6401     99  0.900
cond       1 129.04 0.6466     99  0.900
temp       1 128.00 1.5778    199  0.060 .
pH         1 128.56 1.0690     99  0.350
width      1 128.93 0.7498     99  0.670
depth      1 128.83 0.8371     99  0.550
pebble     1 129.38 0.3593     99  0.870
gravel     1 126.20 3.3194    199  0.050 *
silt       1 126.30 3.2259    199  0.050 *
clay       1 127.13 2.4048    199  0.055 .
turbid     1 127.90 1.6739    199  0.030 *
shade      1 128.22 1.3780    199  0.125
poach      1 129.10 0.5962     99  0.890
---
Signif. codes:  0 '***' 0.001 '**' 0.01 '*' 0.05 '.' 0.1 ' ' 1
```

8. This time, step 7 shows you that the gravel variable has the lowest AIC value and is significant so add this to the model:

```
> fw.cca = cca(fw.biol ~ litter + gravel, data = fw.env)
```

9. Run add1() again, remember that you can use the up arrow to recall previous commands. This time add the *temp* variable:

```
> add1(fw.cca, scope = formula(fwm), test = "permutation")
> fw.cca = cca(fw.biol ~ litter + gravel + temp, data = fw.env)
```

10. At this point the `add1()` command shows that none of the remaining variables are worth adding so run a permutation test on the existing model terms:

```
> anova(fw.cca, by = "terms")
Permutation test for cca under reduced model
Terms added sequentially (first to last)

Model: cca(formula = fw.biol ~ litter + gravel + temp, data = fw.env)
          Df  Chisq      F N.Perm Pr(>F)
litter     1 0.3114 2.4176     99   0.02 *
gravel     1 0.4515 3.5061     99   0.09 .
temp       1 0.2375 1.8438     99   0.07 .
Residual  14 1.8030
---
Signif. codes:  0 '***' 0.001 '**' 0.01 '*' 0.05 '.' 0.1 ' ' 1
```

11. Make a plot of the model, yours should resemble Figure 14.27:

```
> plot(fw.cca)
```

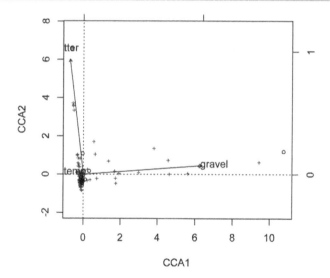

Figure 14.27 Canonical correspondence analysis of freshwater invertebrates from three different habitat types (R = river, S = stream, D = ditch). Model is significant (F 3 = 2.589, permutation test).

Your p-values might be slightly different, the permutation test relies on randomly rearranging the data. However, the AIC values should be the same.

Note: add1() command instructions

The `add1()` command can be used with different values for the number of permutations used. The `pstep` instruction selects the number of permutations for each step (default `pstep = 100`). The `perm.max` instruction set the maximum number of permutations (default `perm.max = 200`). Setting larger values might give 'more reliable' results but will take longer to process.

The AIC values are used to help you decide upon the 'best' variables so that you end up with the most 'effective' or 'minimum adequate' model. However, you may want to explore how other variables affect the species composition. The model-building process is a tool to help you, not a means in itself.

14.3.2 Redundancy analysis

RDA is similar to canonical correspondence analysis in that the analysis maximises the correlation between species scores and sample scores. However, in RDA the data are chi squared transformed and then subjected to weighted linear regression using the constraining variables. The fitted values are then used in a correspondence analysis (using non-weighted singular value decomposition). The `rda()` command in the *vegan* package carries out RDA – the general form of the command is like so:

```
rda(formula, data, scale = FALSE)
```

You supply a formula that describes the response on the left of a ~ and the predictors (constraining variables) on the right of the ~. You can optionally give the name of the data object (usually a `data.frame` or `matrix`) that 'holds' the predictor variables. If you want to use a conditioning variable to 'partial-out' you can use `Condition(var)` to name the variable (`var`) in the formula. You can also use `scale = TRUE` to rescale all species to unit variance – usually you'd only do this if the species were measured using different scales of measurement.

There are `print()`, `summary()` and `plot()` commands associated with the command. The `plot()` command allows you to see the samples and species as well as the constraining variables (shown as biplot arrows). There are also some 'helper' commands, e.g. `scores()`, that can extract parts of the result in a meaningful way.

There are two possible advantages of RDA over CCA:

- If there is an environmental gradient and the species are all (positively) correlated RDA will detect the gradient where CCA may not.
- The variance explained is really variance and not merely inertia, so it is more readily interpreted.

Using the `rda()` command is virtually identical to using `cca()`; you can use the `add1()` command to carry out model-building and the same helper commands are available as for `cca()`. In the following exercise you can have a go at a redundancy analysis using conditioning to partial-out some variables.

Have a Go: Carry out partial redundancy analysis

You'll need the *bryo.biol* dataset for this exercise – the data are abundance of various bryophyte species from churchyard sites in central England. There are 30 sites altogether but each site has several samples taken from various substrates. The *bryo.env* data shows various environmental variables (50 altogether): some are continuous variables, some are binomial (dummy variables, taking a value of 0 or 1) and some are categorical. The data are in the *CERE.RData* file. You'll also need the *vegan* package.

1. Start by preparing the *vegan* package:

```
> library(vegan)
```

2. Look at the environmental variables available in the data:

```
> names(bryo.env)
 [1] "Site"       "Area"       "Perim"      "Age"        "Easting"
 [6] "Northing"   "Slope"      "South"      "West"       "Aspect"
[11] "Northness"  "Eastness"   "CQ"         "Dir"        "N"
[16] "NE"         "E"          "SE"         "S"          "SW"
[21] "W"          "NW"         "Shade"      "Water"      "Livi_wood"
[26] "Expo_root"  "Deca_stum"  "Wood_situ"  "Natu_rock"  "Maso"
[31] "Dry_ston"   "Tarm"       "Conc"       "Basi_grav"  "Acid_grav"
[36] "Bare_grou"  "Broa_gras"  "Fine_gras"  "Disc_gras"  "Roof_fall"
[41] "Habitat"    "Rural_res"  "Ind"        "Urb_res"    "Agric"
[46] "post_Ind"   "Urb_gre"    "Urb"        "Rural"      "Use"
```

3. Carry out a redundancy analysis and remove the effect of site area (i.e. carry out pRDA); use *Northness, Water* and *Rural* as the variables in the analysis:

```
> bryo.rda = rda(bryo.biol ~ Northness + Water + Rural + Condition(Area),
          data = bryo.env)
```

4. The result object is very extensive with many components; you can use `scores()`, `inertcomp()` and `goodness()` commands to help interrogate the result as well as the `summary()` command. However, try the `print()` command in this case:

```
> print(bryo.rda)
Call: rda(formula = bryo.biol ~ Northness + Water + Rural
+ Condition(Area), data = bryo.env)

                 Inertia  Proportion  Rank
Total           36.506941   1.000000
Conditional      0.171970   0.004711      1
Constrained      1.749770   0.047930      3
Unconstrained   34.585201   0.947360     51
Inertia is variance

Eigenvalues for constrained axes:
  RDA1   RDA2   RDA3
1.0913 0.3607 0.2978

Eigenvalues for unconstrained axes:
  PC1   PC2   PC3   PC4   PC5   PC6   PC7   PC8
6.110 4.049 2.921 1.766 1.617 1.604 1.522 1.249
(Showed only 8 of all 51 unconstrained eigenvalues)
```

5. The `eigenvals()` command will give you the eigenvalues for constrained and unconstrained axes, use the command to get the constrained values and the proportions explained:

```
> eigenvals(bryo.rda)[1:3]
     RDA1         RDA2        RDA3
1.0913222 0.3606842 0.2977639

> eigenvals(bryo.rda)[1:3]/sum(eigenvals(bryo.rda)[1:3])
```

```
     RDA1      RDA2      RDA3
0.6236945 0.2061323 0.1701731
```

6. Use the `vif.cca()` command to look at the variance inflation factors:

```
> vif.cca(bryo.rda)
     Area Northness     Water     Rural
 1.140283  1.037825  1.139568  1.227685
```

7. From step 6 you can see that none of the variables has a high inflation value (>10) so try a permutation test to see the significance of the variables:

```
> anova(bryo.rda, by = "terms")

Permutation test for rda under reduced model
Terms added sequentially (first to last)

Model: rda(formula = bryo.biol ~ Northness + Water + Rural +
Condition(Area), data = bryo.env)
          Df    Var     F N.Perm Pr(>F)
Northness  1  0.757 4.1156    99   0.01 **
Water      1  0.692 3.7627    99   0.01 **
Rural      1  0.300 1.6332    99   0.04 *
Residual 188 34.585
---
Signif. codes:  0 '***' 0.001 '**' 0.01 '*' 0.05 '.' 0.1 ' ' 1
```

8. Look at the basic plot of the result with the `plot()` command, your plot should resemble Figure 14.28:

```
> plot(bryo.rda)
```

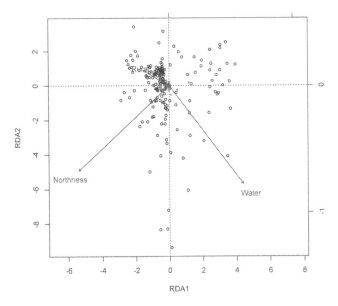

Figure 14.28 Partial redundancy analysis for bryophyte communities in UK Midlands churchyards. Model is significant (F3 = 3.171, permutation test).

9. The result of the `rda()` command can be plotted much like that for the `cca()` command and you can use the `identify()` command for example (but the plot needs a name). Try building a plot resembling Figure 14.28 (but with a few subtle alterations) by separate commands:

```
> plot(bryo.rda, type = "n")
> points(bryo.rda, display = "sites", pch = 16, cex = 0.5)
> text(bryo.rda, display = "bp", col = "blue", cex = 0.6)
> points(bryo.rda, display = "species", col = "red", pch = "+", cex = 0.6)
```

10. The current model is significant but you may want to explore it further. Set-up a maximum model to use as the scope in the `addl()` command and then have a look at the effects of adding other variables:

```
> bryom = rda(bryo.biol ~ . + Condition(Area), data = bryo.env)
> addl(bryo.rda, scope = formula(bryom), test = "permutation")
```

The result of the `addl()` command is not shown (it is quite extensive). If you try it you'll see readily that there are several variables that might be added to the model.

Note: The `plot()` command and ordination results in the vegan package

The `plot()` command in the *vegan* package allows great flexibility when making graphs of ordination results (those obtained with commands from the *vegan* package). In general you can choose to `display` the sample scores, species scores and biplot arrows (where appropriate). To do this you choose one or more of "sites", "species" or "bp".

The general RDA approach is to use a Euclidean distance metric when analysing the data. It is possible to carry out RDA with other dissimilarity metrics as you will see shortly.

Constrained analysis of principal co-ordinates

Allied to RDA is constrained analysis of principal co-ordinates (CAP). The `capscale()` command in the *vegan* package can carry out the analysis. Essentially the command carries out MDS and then goes on to perform the RDA. The general form of the command is like so:

```
capscale(formula, data, distance = "euclidean", dfun = vegdist)
```

You supply a formula to the command where the left of the ~ is the community data or a dissimilarity matrix. To the right of the ~ you give the environmental variables; you can use `Condition()` to partial-out a variable. If you provide a community dataset you must specify the dissimilarity metric you wish to use – this should be one available via the `vegdist()` command. Alternatively you can give a different dissimilarity metric and provide the name of the command that can produce it using the `dfun` instruction.

The result of the `capscale()` command is a complicated object with many components but you can use the various helper commands as well as `print()`, `summary()` and `plot()`. In the following exercise you can have a go at using CAP for yourself using the psammosere plant data.

Have a Go: Carry out constrained analysis of principal co-ordinates

You will need the *ridge.biol* data for this exercise; the data are in the *CERE.RData* file. The data show plant abundance for species along a transect from a psammosere. There are ten samples and each shows frequency from eight replicates. You'll also need the *vegan* package.

1. Start by ensuring the *vegan* package is ready:

   ```
   > library(vegan)
   ```

2. The *ridge.biol* data are arranged with species as rows so you'll need to rearrange the data before you start. There are also some environmental data in the *ridge.env* object. Rotate both so that the samples are the rows:

   ```
   > ridge = as.data.frame(t(ridge.biol))
   > env = as.data.frame(t(ridge.env))
   ```

3. Now look at the available environmental variables:

   ```
   > names(env)
   [1] "Soil"  "Light" "Humid" "Wind"
   ```

4. Use the default distance (Euclidean) with `capscale()` to produce a pRDA result – use *Light* as a conditioning variable:

   ```
   > ridge.cap = capscale(ridge ~ Soil + Wind + Condition(Light),
                          data = env)
   ```

5. Now use the same formula but use the Bray–Curtis metric:

   ```
   > ridge.cap1 = capscale(ridge ~ Soil + Wind + Condition(Light),
                           data = env, distance = "bray")
   ```

6. Try the Manhattan metric – make a dissimilarity matrix first and use this in the formula:

   ```
   > ridge.man = dist(ridge, method = "manhattan")
   > ridge.cap2 = capscale(ridge.man ~ Soil + Wind + Condition(Light),
                           data = env)
   ```

7. Now use the Manhattan metric again but this time specify it within the `capscale()` command:

   ```
   > ridge.cap3 = capscale(ridge ~ Soil + Wind + Condition(Light), data = env,
                           distance = "manhattan", dfun = dist)
   ```

8. Try the `inertcomp()` command – you'll find that you can only analyse site results:

   ```
   > inertcomp(ridge.cap)
   Error in inertcomp(ridge.cap) : cannot analyse species with 'capscale'
   > inertcomp(ridge.cap, display = "sites")
            pCCA        CCA        CA
   T1    10.332946 156.028254 10.11171
   T2    10.332946  24.701184 20.41276
   T3    57.856555   3.910265 52.44780
   ```

```
T4    20.539458   12.818527 80.18246
T5     3.598351   40.318624 25.23525
T6    20.539458   18.723039 27.15853
T7    10.496220   21.652721 57.40308
T8    11.545684   28.830645 18.02099
T9    19.945105   16.697884 28.11447
T10 193.006611    6.252661 23.09692
```

9. The `inertcomp()` and `goodness()` commands can only analyse site axes but you can still get species scores:

```
> scores(ridge.cap, display = "species")
```

10. The results can be plotted and you can access sites, species and biplot arrows. Try looking at the four results for comparison. Split the plot window into four sections; your plot should resemble Figure 14.29:

```
> opar = par(mfrow = c(2,2)) # Split the plot window into 4 parts
> plot(ridge.cap)
> title(main = "Euclidean")
> plot(ridge.cap1)
> title(main = "Bray-Curtis")
> plot(ridge.cap2)
> title(main = "Manhattan (from matrix)")
> plot(ridge.cap3)
> title(main = "Manhattan (specified)")
> par(opar) # Reset the plot window
```

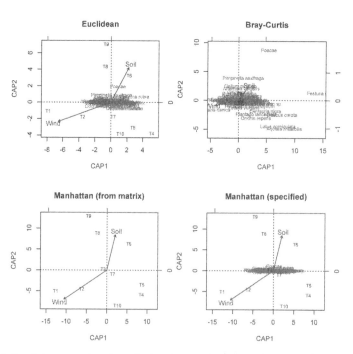

Figure 14.29 Partial constrained analysis of co-ordinates (a form of RDA) for plant psammosere communities. Analyses via `capscale()` command.

11. You can use the `anova()` command to carry out a permutation test on the various models:

```
> anova(ridge.cap1)
Permutation test for capscale under reduced model

Model: capscale(formula = ridge ~ Soil + Wind + Condition(Light),
data = env, distance = "bray")
          Df      Var     F N.Perm Pr(>F)
Model      2 0.84243 4.6028    199  0.005 **
Residual   6 0.54908
```

Notice how in Figure 14.29 the plot that resulted from using the Manhattan matrix does not show the species scores. If you want the species scores then specify the dissimilarity metric and the corresponding command that will produce it as part of the `capscale()` command, as in step 7.

14.4 Using ordination results

You've seen how ordination can help you to 'put things in order' – the various methods of analysis are intended to help you make sense of changes in species composition. The end result is generally the ordination plot – a 2-D representation of the samples (and sometimes species or environmental data) that allow you to see patterns that were not evident before the analysis.

In many ways the process of ordination is the beginning rather than the end – you still have to make some kind of sense out of the result. A good ordination result will allow you plenty of useful interpretation but you might also want to use the ordination scores (for sites or species) as the starting point for some other analysis.

Adding extra information to your ordination plots is especially useful as the ordination plot makes the most easily interpretable result. In the following section you'll see how to add grouping information to your ordination plots and also how to add additional environmental data to your plots.

14.4.1 Adding environmental information

The ordination plot is usually the 'main event' of your analysis – you simply cannot interpret reams of species scores (but see Section 14.4.3). If you can add extra information to the ordination plot you can aid interpretation but beware of over-filling your plots with data, your task is to make patterns clearer, not make them more confusing.

There are two main ways you can add data to your ordination plots:

- Identifying logical groups.
- Adding environmental data.

Identifying groups on ordination plots

It is often useful to be able to identify certain logical groups on your ordination plots. This makes it easier to see how these groups are arranged relative to one another in the ordination space. The *vegan* package makes it easy to identify groups – there are three commands:

- ordihull() – this command draws a line between the outermost points in each group, thus creating a polygon (a hull).
- ordiellipse() – this command draws an ellipse around the points that make up a group.
- ordispider() – this command joins the centre of each group to each point in that group, thus creating a series of 'spider webs'.

The general form of these commands is like so:

```
ordixxx(ord, groups, display = "sites", show.groups,
        label = FALSE, ...)
```

Each command requires the ordination result (ord), which can be any that is generated by the *vegan* package, and the grouping variable (groups). You'll probably have the grouping variable in the same dataset as any environmental variables. You can also specify groups based on species rather than sites but most commonly you'll stick to display = "sites". By using the show.groups instruction you can specify one or more groups to display rather than all groups. By default the groups are not labelled, you can add labels using label = TRUE.

There are other instructions you can pass to the commands. Some are general ones relating to the lines drawn, e.g. col, lty and lwd to control colour, line type and line width.

There are many ways to use these commands. In the following exercise you can have a go at using the commands for yourself to get a flavour of the possibilities.

Have a Go: Use grouping information on ordination plots

You'll need the *gb.biol* data for this exercise – it is part of the *CERE.RData* file. The data show the abundances of ground beetles at three habitat types, *Wood*, *Grass* and *Edge*. Each habitat has six replicate samples. You'll also need the *gb.site* data, which contains the *Habitat* grouping variable. The *vegan* package is also required.

1. Start by ensuring that the *vegan* package is loaded and ready:

    ```
    > library(vegan)
    ```

2. Make a DCA result for the ground beetle data:

    ```
    > gb.dca = decorana(gb.biol)
    ```

3. Make a basic plot of the result – display the samples only, using text labels. Your plot should resemble Figure 14.30:

    ```
    > plot(gb.dca, display = "sites", type = "t")
    ```

4. You can just about make out the groupings in Figure 14.30 because the points are labelled with the sample names, which are logical (they correspond to the habitat types). However, if you displayed points only you wouldn't have a clue! Plot the points alone and then add a spider to join the centre of each group to each point in that group. Your plot should resemble Figure 14.31:

    ```
    > plot(gb.dca, display = "sites", type = "p")
    > ordispider(gb.dca, groups = gb.site$Habitat, label = TRUE,
                 cex = 0.7, lty = 2)
    ```

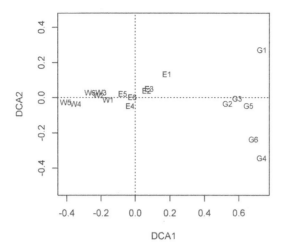

Figure 14.30 DCA result showing ground beetle communities for three habitat types. Sites are labelled with sample names.

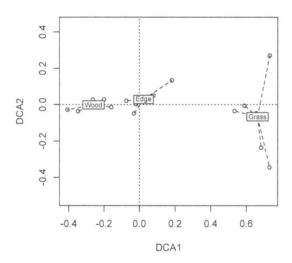

Figure 14.31 DCA result showing ground beetle communities for three habitat types. Habitat groups are identified using the ordispider() command.

5. Now try an ellipse – the ordiellipse() command will add to the plot, so redraw the points to make something resembling Figure 14.32:

```
> plot(gb.dca, display = "sites", type = "p")
> ordiellipse(gb.dca, groups = gb.site$Habitat, label = TRUE, lwd = 1.5)
```

6. Now redraw the plot and use the ordihull() command to make a polygon around each group. Your plot should resemble Figure 14.33:

```
> plot(gb.dca, display = "sites", type = "p")
> ordihull(gb.dca, groups = gb.site$Habitat, label = TRUE,
           lty = 3, lwd = 0.5, cex = 0.7)
```

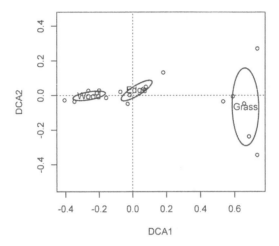

Figure 14.32 DCA result showing ground beetle communities for three habitat types. Habitat groups are identified using the `ordiellipse()` command.

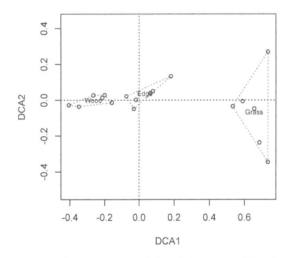

Figure 14.33 DCA result showing ground beetle communities for three habitat types. Habitat groups are identified using the `ordihull()` command.

7. The `ordiellipse()` command can produce ellipses based on standard deviation of the points or the standard error of the weighted averages via the `kind` instruction; you specify `kind = "sd"` (the default) or `"se"`. You can optionally specify a confidence interval. Make an NMDS result and then plot it. Add an ellipse for the *Edge* group and then add a 95% confidence interval to produce a plot resembling Figure 14.34:

```
> plot(gb.nmds)
> ordiellipse(gb.nmds, groups = gb.site$Habitat, label = TRUE,
        show.groups = "Edge")
```

```
> ordiellipse(gb.nmds, groups = gb.site$Habitat, label = TRUE,
              show.groups = "Edge", kind = "sd", conf = 0.95, lty = 2)
```

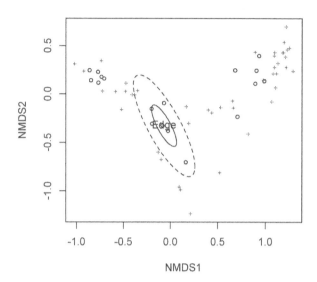

Figure 14.34 NMDS of ground beetle communities in three habitat types. The 'edge' group are highlighted with 95% confidence interval (std. deviation).

8. The `ordihull()` command returns results, which you can utilise separately. You can produce the plotting co-ordinates without drawing on the plot:

```
> oh = ordihull(gb.dca, groups = gb.site$Habitat, draw = "none")
> oh
$Edge
            DCA1         DCA2          DCA3         DCA4
E4 -0.03037186 -0.04859590  0.151261466  0.03044433
E5 -0.07412194  0.02083155  0.009707801 -0.00898395
E1  0.18138813  0.13271807 -0.107797270 -0.04146416
E4 -0.03037186 -0.04859590  0.151261466  0.03044433

$Grass
           DCA1         DCA2         DCA3         DCA4
G4 0.7334490 -0.34507441  0.10874864 0.071529737
G2 0.5330316 -0.03512119 -0.11004338 0.217467725
G1 0.7343998  0.27006693  0.06988733 0.002578648
G4 0.7334490 -0.34507441  0.10874864 0.071529737

$Wood
           DCA1         DCA2          DCA3         DCA4
W4 -0.3473194 -0.03620598 -0.017573049 -0.009647351
W5 -0.4067104 -0.02786967 -0.008866406  0.000990355
W6 -0.2635463  0.02698606  0.035104765  0.047653105
```

```
W3 -0.1997046  0.02813660  0.104459918  0.094857532
W1 -0.1581917 -0.01393944 -0.073710945 -0.075284043
W4 -0.3473194 -0.03620598 -0.017573049 -0.009647351

attr(,"class")
[1] "ordihull"
```

9. The `ordiellipse()` command also produces a result that you can use. The result is more extensive than for `ordihull()` so use the `summary()` command to get an overview:

```
> oe = ordiellipse(gb.dca, groups = gb.site$Habitat, draw = "none")
> summary(oe)
              Edge          Grass          Wood
DCA1 0.023519818   0.66428093 -0.259368140
DCA2 0.027967712  -0.06598190 -0.002081734
Area 0.009626618   0.05533217  0.006955489
```

Adding environmental information to ordination plots

If you have used a direct gradient analysis, such as CCA or RDA, then you'll already have environmental data that you can use – typically you will plot the biplot arrows representing those variables. However, you may wish to add extra environmental data that were not included in the original analysis. If you have carried out an indirect gradient analysis then you will not have included any environmental data in any case.

There are two main ways to add environmental information to your ordination plots:

- Bi-plot arrows representing gradients.
- Contours showing an environmental gradient.

You can use the `envfit()` command in the *vegan* package to fit environmental factors onto an ordination. The `ordisurf()` command uses general additive modelling to fit a smooth surface (using thinplate splines). The upshot is that you end up with a 'contour map', which is applied over your ordination plot. The `ordisurf()` command is part of the *vegan* package but requires the *mgcv* package for operation.

The general form of the `envfit()` command is like so:

```
envfit(ord, env, permutations = 999, display = "sites")
```

You need to supply the result of an ordination (`ord`) and the environmental variables (usually in a `data.frame` or `matrix`). You can also supply the environmental variables as a `formula` where the ordination is on the left of the ~ and the variables are on the right (you also supply the location of the variables with the `data` instruction).

The general form of the `ordisurf()` command is like so:

```
ordisurf(x, y, add = FALSE, nlevels = 10, levels)
```

You supply the result of an ordination (or the scores), `x` and the environmental variable to be plotted, `y`. By default the command will make a new plot showing the sites from the ordination and the environmental surface – you can add the contours to an existing plot with `add = TRUE`. The command will 'split' your chosen variable into ten levels by default;

you can alter this via the `nlevels` instruction. You can also provide explicit levels for the contours via the `levels` instruction.

Both the `envfit()` and `ordisurf()` commands will accept various graphical instructions such as `col` and `lty` for example (colour and line type). In the following exercise you can have a go at adding environmental data to an ordination result for yourself to get a feel for some of the options.

Have a Go: Add environmental data to ordination plots

You will need the *ridge.biol* data for this exercise; the data are in the *CERE.RData* file. The data show plant abundance for species along a transect from a psammosere. There are ten samples and each shows frequency from eight replicates. There is also a *ridge.env* dataset, which contains some environmental information. You'll also need the *vegan* and *mgcv* packages.

1. Start by ensuring that the *vegan* and *mgcv* packaged are loaded. The `ordisurf()` command will load *mgcv* for you as required but it is as well to do this at the beginning if you know it will be needed:

   ```
   > library(vegan)
   > library(mgcv)
   ```

2. The data are arranged with samples as columns for both the biological and environmental data so rotate both using the `t()` command – the results will be matrix objects so coerce them into data frames like so:

   ```
   > ridge = as.data.frame(t(ridge.biol))
   > env = as.data.frame(t(ridge.env))
   ```

3. Look at the environmental data – there are four variables:

   ```
   > env
        Soil Light Humid Wind
   T1    0.0 16250  28.3 5.13
   T2    0.1 16250  27.4 3.25
   T3    3.3 17500  27.3 1.88
   T4    3.8 16625  28.6 1.13
   T5    3.6 15875  26.6 0.75
   T6    6.1 16625  26.9 0.63
   T7    8.0 14413  30.0 0.25
   T8   14.4 14368  51.4 0.00
   T9   13.7 14064  25.2 0.25
   T10  14.1 11381  25.5 0.25
   ```

4. Make an ordination using NMDS (an indirect gradient method):

   ```
   > ridge.nmds = metaMDS(ridge)
   Wisconsin double standardization
   Run 0 stress 0.0523968
   Run 1 stress 0.05239665
   ```

```
... New best solution
... procrustes: rmse 0.0006489827  max resid 0.001284218
*** Solution reached
```

5. Now use the `envfit()` command to fit the environmental variables to the ordination result:

```
> ridge.fit = envfit(ridge.nmds, env)
> ridge.fit

***VECTORS

          NMDS1     NMDS2      r2 Pr(>r)
Soil    0.93984   0.34160 0.9222  0.001 ***
Light  -0.69709  -0.71698 0.7521  0.013 *
Humid   0.96421   0.26512 0.1145  0.783
Wind   -0.92595   0.37764 0.9679  0.001 ***
---
Signif. codes:  0 '***' 0.001 '**' 0.01 '*' 0.05 '.' 0.1 ' ' 1
P values based on 999 permutations.
```

6. You could use a formula, which allows you to specify the variables more explicitly:

```
> envfit(ridge.nmds ~ Soil + Wind, data = env)

***VECTORS

          NMDS1     NMDS2      r2 Pr(>r)
Soil   0.93984   0.34160 0.9222  0.003 **
Wind  -0.92595   0.37764 0.9679  0.002 **
---
Signif. codes:  0 '***' 0.001 '**' 0.01 '*' 0.05 '.' 0.1 ' ' 1
P values based on 999 permutations.
```

7. Now plot the NMDS result using points only for species and sites; your plot will resemble Figure 14.35a:

```
> plot(ridge.nmds, type = "p")
```

8. Add the environmental data from the ridge.fit result you calculated in step 5. Note that when you plot the `envfit()` result the data are added to the existing plot. You could specify this explicitly using `add = TRUE` to remind yourself; your plot should resemble Figure 14.35b:

```
> plot(ridge.fit, add = TRUE)
```

9. Now redraw the NMDS plot and add the environmental data from `envfit()` again, but this time select only variables that are significant ($p < 0.05$). Your plot should resemble Figure 14.35c:

```
> plot(ridge.nmds, type = "p")
> plot(ridge.fit, p.max = 0.05, col = "red")
```

10. Add a surface contour plot to the current plot (from step 9) using the `ordisurf()` command. This should resemble Figure 14.35d:

```
> ordisurf(ridge.nmds, env$Wind, add = TRUE, col = "blue")
```

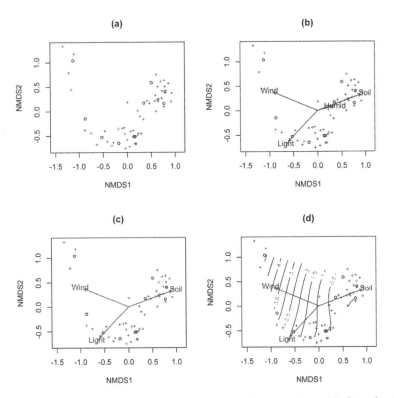

Figure 14.35 NMDS of plant psammosere successional communities, (a) plain plot, (b) all environmental factors, (c) only significant environmental factors ($p < 0.05$), (d) significant environmental factors plus environmental surface showing wind speeds.

If you specify add = FALSE with the ordisurf() command you'll end up with a plot of your sites and the environmental surface overlaid.

14.4.2 Comparing ordination results

There are times when you want to compare the results of two ordinations, to see which one is 'better' or indeed if they are really so different after all. The reality is that you cannot easily compare ordination results from different methods. You can, however, compare results based on a method that employs dissimilarity matrices. For example NMDS uses a dissimilarity matrix as a starting point. You can compare the results obtained from different dissimilarity metrics.

If you are using a direct gradient analysis then you are more likely to be interested in the effects of adding different environmental variables – you can use AIC values and the add1() command to help you do this (see Section 14.3).

The procrustes() command in the *vegan* package allows you to compare two ordination results. It does a similar job to the mantel() test that you met in Section 10.8 0 (comparing *beta* diversity). The difference is that the procrustes() command can take the results from an ordination whilst mantel() requires a 'square' matrix (e.g. a dissimilarity matrix).

Essentially the procrustes() command rotates one ordination so that it is maximally similar to a second ordination – minimising the sums of squares along the way. The command protest() carries out a significance test on the result by permutation.

In the following exercise you can have a go at procrustean rotation to evaluate differences in NMDS solutions using different dissimilarity metrics.

Have a Go: Use `procrustes()` to evaluate differences in NMDS solutions

You will need the *ridge.biol* data for this exercise; the data are in the *CERE.RData* file. The data show plant abundance for species along a transect from a psammosere. There are ten samples and each shows frequency from eight replicates. You'll also need the *vegan* package.

1. Start by making sure the *vegan* package is loaded and ready:

   ```
   > library(vegan)
   ```

2. The data are arranged with samples as columns so use the `t()` command to rotate so that rows are samples:

   ```
   > ridge = as.data.frame(t(ridge.biol))
   ```

3. Use the `vegdist()` command to make three dissimilarity matrices using three different metrics:

   ```
   > ridge.bc = vegdist(ridge, method = "bray")
   > ridge.eu = vegdist(ridge, method = "euclidean")
   > ridge.go = vegdist(ridge, method = "gower")
   ```

4. The processes use randomisation so to mimic the results here set your random number generator (don't do this on 'real' data):

   ```
   > set.seed(16)
   ```

5. Use the `monoMDS()` command to make three NMDS results, one for each of the dissimilarities:

   ```
   > nmds1 = monoMDS(ridge.bc, k = 2)
   > nmds2 = monoMDS(ridge.eu, k = 2)
   > nmds3 = monoMDS(ridge.go, k = 2)
   ```

6. The `procrustes()` command performs the rotations necessary and calculates various results – compare the *nmds1* and *nmds2* results:

   ```
   > pc12 = procrustes(nmds1, nmds2)
   > pc12

   Call:
   procrustes(X = nmds1, Y = nmds2)

   Procrustes sum of squares:
   2.308
   ```

7. You'll need the `protest()` command to see if the correlation between NMDS solutions is significant:

   ```
   > protest(nmds1, nmds2)

   Call:
   protest(X = nmds1, Y = nmds2)

   Correlation in a symmetric Procrustes rotation:  0.877
   Significance:  0.001
   Based on 1000 permutations.
   ```

8. The `procrustes()` result has a plotting routine – this can produce two sorts of results, a comparison of the ordination solutions and a summary of the residuals. Try this now to produce something resembling the top row of Figure 14.36:

```
> plot(pc12, kind = 1, main = "Bray vs. Euclidean")
> plot(pc12, kind = 2)
```

9. Now compare the NMDS solutions from the Bray–Curtis and Gower dissimilarities:

```
> protest(nmds1, nmds3)

Call:
protest(X = nmds1, Y = nmds3)

Correlation in a symmetric Procrustes rotation:  0.572
Significance:  0.088
Based on 1000 permutations.
```

10. Run the `procrustes()` command to generate the rotation comparison and then plot the results to give something resembling the bottom row of Figure 14.36:

```
> pc13 = procrustes(nmds1, nmds3)
> plot(pc13, kind = 1, main = "Bray vs. Gower")
> plot(pc13, kind = 2)
```

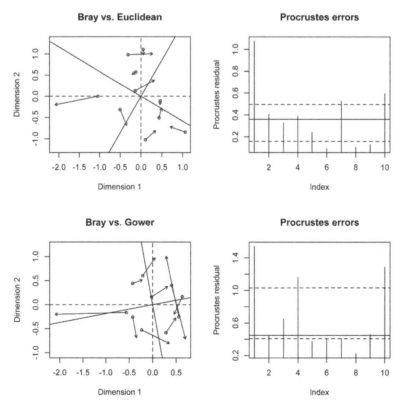

Figure 14.36 Procrustes rotation of NMDS solutions comparing dissimilarity metrics. Left column shows comparison of solutions, right column shows corresponding procrustes residuals.

> You will get different results each time you run the analysis because the `protest()` command uses a permutation test. The `set.seed()` command in step 3 merely helps 'standardise' your computer by setting the random number generator. If you omit the command you'll get different results.

Tip: Permutations and the `protest()` command

You can set the number of permutations used in the `protest()` command via the `permutations` instruction, the default is `permutations = 999`. If you use larger values you may get 'better' results but you'll also take longer to calculate them.

14.4.3 Compact ordered community tables

When you are dealing with community data it can be useful to get an overview. One way to do this is with a table showing the sites and species. Before you undertake any analysis it is likely that you can only arrange the data in alphabetical order or sampling order. This is not especially useful.

You can use the results of an ordination to help reorder a community data table in a more meaningful fashion – the `vegemite()` command in the *vegan* package is able to do this for you. You met this command briefly in Section 12.2.1, where it was used with hierarchical clustering results. The command produces a compact table, which is one where the species abundance is represented by a single digit/character and zero entries are shown with a period (you can alter this). The general form of the command is like so:

```
vegemite(x, use, scale, zero=".", maxabund)
```

You supply the community data, x, which are then tabulated using the order specified by `use`. Generally the use instruction will point to an ordination result, in which case the scores are used to reorder the sites and species. You can use a vector of values to give an explicit order, but this will only affect the sites. You can also use the results from a hierarchical clustering result.

The `scale` instruction determines how the abundance values are treated – they will be shown as a single character so any dataset with abundances >9 will need to be rescaled. The options are: `"Braun.Blanquet"`, `"Domin"`, `"Hult"`, `"Hill"`, `"fix"`, and `"log"`. If you use the `"log"` option then you can optionally set `maxabund` (the default takes the maximum abundance from the data). The `"fix"` option is for 10-point scales (values are truncated to integers, values of 10 or larger are shown as x and values <1 as +).

In the following exercise you can have a go at creating a compact community table for yourself:

Have a Go: Create compact ordered community tables

You'll need the *moss* data for this exercise; the data show abundances of bryophytes from various trees in Duke forest, North Carolina (Palmer 1986). The data are in the *CERE.RData* file. You will also need the *vegan* package.

1. Start by preparing the *vegan* package:

```
> library(vegan)
```

2. The data are arranged with sites as columns so rearrange to make sites the rows using the `t()` command:

```
> mosses = t(moss)
```

3. Make a compact table without using any form or reordering. Use the Hult scale (0–5) for this:

```
> vegemite(mosses, scale = "Hult")

        BLLPPPPQQ
        NTTEOTTAR
        212321311
Amb.ser 1.1....11
Ano.att 134.4..43
Ano.min 111.2..11
Ano.ros ..1.1..33
Bra.acu .1.....11
Bra.oxy 111.1..11
Bry.ill 111.1..11
Cam.his 11.....11
Cla.par .12.2..11
Dic.mon 1..2.12..
Dic.sco ...1.2...
Ent.sed .11.1....
Fru.ebo 132.2..22
Hap.tri .11.1..11
Iso.ten 311414311
Leu.alb 1..4.45..
Leu.jul 112.2..22
Lop.het 3112.2..1
Pla.rep 412.11.11
Por.pla .11.1..11
Rad.com .1..1..11
Rad.obc .11.1..11
Sem.adn 221212111
The.asp .1.....11
Thu.del 131.1..11
  sites species
      9       25
scale:  Hult
```

4. As you can see from step 3, the data are simply arranged alphabetically. Make an ordination result to use in reordering the data – an NMDS based on the Bray–Curtis dissimilarity metric will do for a demonstration:

```
> mosses.nmds = metaMDS(mosses, distance = "bray", k = 2, trace = 0)
```

5. Now use the NMDS result to reorder the community table:

```
> vegemite(mosses, use = mosses.nmds, scale = "Hult")

        PPPBLLQPQ
        TETNTTROA
        331221121
```

```
Dic.sco .12......
Leu.alb 5441.....
Dic.mon 2211.....
Iso.ten 344311111
Lop.het .223111..
Sem.adn 122212111
Pla.rep ..1421111
Cam.his ...1.11.1
Amb.ser ...11.1.1
Bry.ill ...111111
Bra.oxy ...111111
Ano.min ...111121
Fru.ebo ...123222
Thu.del ...113111
Leu.jul ...121222
Ano.att ...143344
Ent.sed ....11.1.
Cla.par ....21121
The.asp .....11.1
Rad.obc ....11111
Por.pla ....11111
Rad.com .....1111
Hap.tri ....11111
Bra.acu .....11.1
Ano.ros ....1.313
  sites species
      9     25
scale: Hult
```

6. You can see that there is a pattern, species at the top tend to be found primarily in the sites to the left, and species at the bottom tend to be found at the sites on the right.

7. Just for comparison try making a Bray–Curtis dissimilarity matrix and then a hierarchical cluster result to use for the reordering:

```
> mosses.bc = vegdist(mosses, method = "bray")
> mosses.cl = hclust(mosses.bc, method = "complete")
> vegemite(mosses, use = mosses.cl, scale = "Hult")

        LQQLPBPPP
        TARTONTET
        111222331
The.asp 111......
Thu.del 311111...
Bra.acu 111......
Rad.com 111.1....
Por.pla 11111....
Bry.ill 111111...
Cam.his 111..1...
Rad.obc 11111....
Ano.ros .3311....
Hap.tri 11111....
```

```
Fru.ebo 322221...
Ent.sed 1..11....
Bra.oxy 111111...
Leu.jul 122221...
Ano.att 343441...
Amb.ser .111.1...
Cla.par 11122....
Ano.min 111121...
Pla.rep 111214..1
Sem.adn 211112122
Lop.het 1.11.3.22
Iso.ten 111113344
Dic.mon .....1221
Leu.alb .....1544
Dic.sco ......12
  sites species
      9     25
  scale:  Hult
```

Try using other scales – the "log" scale is especially useful as it gives a good range of values.

Tip: Using scales with different community datasets

If you have different communities to compare then you need to be careful to use appropriate scales in the vegemite() command. The "log" scale is useful as you can set the maxabund instruction to any value – thus "standardising" community tables. Alternatively you can use the decostand() command to standardise your community data beforehand.

14.4.4 Alternatives to ordination

Ordination can be thought of simply as a way to take species composition from various samples and rearrange those samples into some order along several axes. The resulting axes can be viewed as a 2-D plot. In indirect gradient analysis the axes can reveal something of the species' responses to environmental gradients. In direct gradient analyses you are using existing environmental data from the beginning to help you arrange the samples in order.

There are other ways to look at your samples as alternatives to methods of ordination. You can think of ordination as similar to analysis of *beta* diversity (see Chapter 10), which is essentially concerned with changes in species composition between sampling units. For example other methods of analysis might include:

- Mantel tests (Section 10.8) – for comparison of two matrices. Usually you will convert your data into a dissimilarity matrix for the analysis.
- Permutational MANOVA (Section 10.5.2 and 10.7) – for analysis of species composition in relation to one or more variables.
- Multi-response permutation procedure (Section 10.5.3) – for analysis of species composition in relation to grouping variables.

- Clustering and similarity (Chapter 12) – for using species composition to identify 'natural groupings'; hierarchical clustering methods (Section 12.2.1) are most closely allied to indirect gradient analysis.
- Analysis of similarities – for testing differences in species composition between sampling units (Section 10.6.2 and 10.7). This method is allied to NMDS in that it uses rank order of dissimilarity values.

Ordination is simply one approach to looking at community composition. The alternatives outlined above give plenty of additional analytical methods to keep you busy.

14.5 Summary

Topic	Key Points
What is ordination?	Ordination is the term given to a variety of multivariate analyses. The main aim of ordination is to take complicated community data and simplify them, producing a low-dimensional result. The results are generally viewed in a 2-D plot, akin to a scatter plot.
	The samples (and sometimes species) are arranged along the axes of the ordination result, giving rise to the name, ordination (putting things in order). The axes are constructed so that they are orthogonal.
Types of ordination: • Direct gradient • Indirect gradient	There are two main approaches to ordination: In indirect gradient analysis environmental information is not used but gradients can be inferred from the results. In direct gradient analysis the environmental information is used at the start and helps to 'constrain' the ordination.
	With direct gradient analysis you can test hypotheses about the effects of environmental variables and species composition.
PO (Bray–Curtis)	Polar ordination is a form of indirect gradient analysis. A dissimilarity matrix is used to arrange samples (not species) along each axis (ordering from one pole to the other, hence the name).
	You can carry out polar ordination using Excel but it becomes difficult with large datasets and dissimilarity metrics other then Euclidean.
PCO/PCoA/MDS	Principal co-ordinate analysis or classical (metric) multi-dimensional scaling is an indirect gradient analysis. If you use the Euclidean distance metric the results are the same as Principal components analysis. You can use presence-absence data. A potential problem is the arch effect, which can make interpretation difficult.
	The `cmdscale()` command can carry out MDS, which produces sample scores. You can partial-out one or more variables with the `Condition()` instruction.
	The `wascores()` command in the *vegan* package can calculate species scores from an MDS result.
	Weighted MDS can be carried out using the `wcmdscale()` command, allowing you to downweight rare species.
NMDS/nMDS	Non-metric multi-dimensional scaling is an indirect gradient analysis. It is similar to MDS but uses ranks (so is non-linear), which helps to reduce the arch effect. You can use any metric of dissimilarity. Presence-absence data can be used.

The `isoMDS()` and `sammon()` commands in the *MASS* package can carry out NMDS, giving sample scores only.

The `metaMDS()` command in the *vegan* package can get sample and species scores for NMDS. It also has a plotting method, which helps visualise results.

PCA   Principal components analysis is an indirect gradient analysis. You specify the community data rather than a dissimilarity. The method produces sample and species scores; species being represented in plots by biplot arrows. The results often show an arch or horseshoe effect, which can make interpretation difficult.

The `prcomp()` and `princomp()` commands carry out PCA but the former uses SVD and is generally better. The `biplot()` command produces a visual representation of the results.

The `rda()` command in the *vegan* package can carry out PCA. The results are easier to deal with as there are various helper commands.

CA/RA   Correspondence analysis or reciprocal averaging is an indirect gradient analysis. You specify the community data rather than a dissimilarity. The results can suffer from the arch effect, which can make interpretation difficult.

The `corresp()` command in the *MASS* package can perform CA and has a dedicated plotting routine.

The `cca()` command in the *vegan* package has additional helper functions, including plotting methods. You can also partial-out one or more variables using the `Condition()` instruction.

DCA   Detrended correspondence analysis is an indirect gradient analysis. This is a spin-off from CA where the arch effect is removed by segmenting the axes and rescaling the scores. The axes are in units relating to *beta* diversity.

The `decorana()` command in the *vegan* package carries out DCA. There are various helper commands including plotting routines.

CCA   Canonical correspondence analysis is a direct gradient analysis. Environmental variables are used to constrain the ordination. You can also partial out one or more variables.

The `cca()` command in the *vegan* package carries out CCA. You can partial-out variables using the `Condition()` instruction. The `cca()` has various helper commands, including routines for plotting.

The results can be analysed with a permutation test using `anova()` to assess significance.

RDA   Redundancy analysis is a direct gradient analysis. It is a spin-off from CCA. In contrast to CCA this method uses absolute abundance. The upshot is that if you have a gradient along which all species are positively correlated, RDA will detect such a gradient while CCA will not.

The `rda()` command in the *vegan* package will carry out RDA.

CAP   Constrained analysis of principal co-ordinates is a direct gradient analysis. It is akin to RDA and essentially you carry out a PCO and then an RDA (so you get species and sample scores as well as scores for the constraining variables). You can use any metric to make a dissimilarity.

Topic	Key Points
	The `capscale()` command in the *vegan* package can carry out CAP. You can partial-out variables and the `capscale()` result has various helper commands, including plotting routines.
Model-building	In CCA and RDA you can use a model-building approach to help select the 'best' environmental variables. In the *vegan* package you start with a blank model (containing an intercept only) and use the add1() command. This gives details about AIC values, allowing you to pick the best single variable to add to your model. You run the process iteratively and build your model stepwise.
Visualising results	Many of the ordination commands in the *vegan* package have dedicated plotting routines to help visualise results. You can choose to display sample or species scores (and sometimes environmental scores) and can also select one or more sites or species to add to the plot. You can also add `text()` or `points()` to existing plots. The `identify()` command can be used to interactively label ordination plots.
Adding information	You can use commands in the *vegan* package to help identify groupings on your ordination plots, e.g. the `ordihull()`, `ordiellipse()` and `ordispider()` commands.
	You can compute information about environmental variables to add to ordination plots in two main ways: The `envfit()` command can calculate scores for environmental variables, that you can add as biplot arrows. The `ordisurf()` command can plot environmental 'contours' onto an ordination plot.
Comparison and diagnostics	You can compare ordination results using the `procrustes()` command in the *vegan* package. This examines how much rotation one solution has to undergo to correlate maximally with another. The `protest()` command can put a significance to the difference. This is especially useful for NMDS, where you can compare ordination using different dissimilarity metrics.
	For constrained ordinations there are several diagnostic commands in the *vegan* package: `inertcomp()` looks at inertia, whilst `goodness()` looks at goodness of fit for sites or species. The `vif.cca()` command looks at variance inflation factors.
Alternatives to ordination: • Mantel test • MANOVA • Analysis of similarities • Multi-response permutation	Ordination may not be the only way to explore a situation, alternatives are: Mantel tests look at correlations between dissimilarity matrices; the `mantel()` command in the *vegan* package can do this. The *ade4* package has `mantel.rtest()` and `mantel.randtest()` commands. Cluster analysis can look at patterns in species composition. Analysis of *beta* diversity is another way to explore changes in species composition from sample to sample. The *vegan* package provides various ways to do this including: `adonis()`, `mrpp()` and `anosim()` commands.

14.6 Exercises

14.1 When constructing axes in ordinations you need to make sure that they are *orthogonal*. What does this mean and why is it important?

14.2 The *hsere.biol* data contains plant abundances at ten samples in hydrosere communities. You want to carry out a NMDS analysis. You have pH data (*hsere.ph*) and think that pH is an important gradient. Rank the following dissimilarities in decreasing order of their ability to separate the data: Bray–Curtis, Euclidean, Gower, Horn–Morisita. What is the final stress achieved from the NMDS?

14.3 In a direct gradient analysis you can partial-out one or more variables by specifying the variable(s) in the formula. Which of the following would be a correct formula:

```
A) response ~ light + condition(ph)
B) response ~ light + partial(ph)
C) response ~ light - Condition(pH)
D) response ~ light + Condition(ph)
E) response ~ light - partial.out(ph)
```

14.4 In model-building for CCA or RDA you use the add1() command to look at the various variables. The 'best' variable is the one with the highest AIC value – TRUE or FALSE?

14.5 The *fw.biol* data contains abundances of freshwater invertebrates in 18 samples. The corresponding environmental data are in *fw.env*. Carry out an NMDS then decide which of *DO2* (dissolved oxygen) or *cond* (conductivity) has the best correlation to the ordination.

The answers to these exercises can be found in Appendix 1.

Appendices

Here you'll find the answers to the end-of-chapter exercises (Appendix 1) and a section on the custom R commands that have been produced to go with the book (Appendix 2).

Appendix 1: Answers to exercises

The answers to the end-of-chapter exercises are set out chapter by chapter.

Chapter 1

1.1 The main topics of community ecology are: diversity, similarity, clustering, association and ordination. You could add indicator species as a spin-off from association analysis (which itself could be thought of as a spin-off of similarity/clustering).

1.2 Diversity is measured at different scales; *alpha* diversity is the diversity of a single sampling unit (e.g. a quadrat). *Gamma* diversity is the diversity of all your samples together (think landscape diversity). *Beta* diversity is the diversity between samples. The three are related like so: *Alpha* × *Beta* = *Gamma*, although sometimes the relationship can be additive.

1.3 Association analysis can be used to identify communities by examining which species tend to be found together in your samples. You can also use it for identifying indicator species.

1.4 FALSE. Indirect gradient analysis only infers environmental gradients after the fact. To test hypotheses about the environment you would use direct gradient analysis. It is possible to use indirect analysis results and link to environmental variables but this is done as a secondary step after the main ordination.

1.5 A null hypothesis is not simply the opposite of your hypothesis but states that no relationship occurs. In this case H0 would be something like 'there is no correlation between species richness and soil moisture'.

Chapter 2

2.1 Excel is most useful as a data management tool. You can use it as a simple database to manage and manipulate data. You can also carry out preliminary summaries and produce graphs. You can undertake a range of statistical analyses using Excel but many of the analyses in community ecology are simply too complicated to make it worthwhile.

2.2 The *Analysis ToolPak* is usually found in the *Data > Data Analysis* button on the ribbon of the later versions of Excel.

2.3 FALSE. The R program is entirely free and is a huge *Open Source* project.

2.4 FALSE. R can carry out simple statistics as well as complicated ones with equal ease.

Chapter 3

3.1 The minimum you ought to include in a biological record is: who, what, where, when.

3.2 The sort of columns you use will depend largely on the subject but as a general rule each column should be unique. For example have a column for the species name and another for the abundance as well as one for the site name. Your data can be easily rearranged later using a Pivot Table.

3.3 FALSE. Although you may well want to tell your supervisor what you were up to and when, it is more important to know the date because much biological data is time-specific. If you were looking at bats for example then the time of year might be particularly important to their activity (and so the abundances you record).

Chapter 4

4.1 C. The `1_data` item is not a valid name in R because it starts with a number.

4.2 FALSE. You can get ZIP files from the R website but it is generally simpler to get and install them directly from R via the `install.packages()` command.

4.3 TRUE. You can filter data using more than one column but you'll need to select the items to filter for each one.

4.4 The *Report Filter* allows you to apply a filter to your results. The *Column Labels* box is for the column variables. The *Row Labels* box is for row variables and the *Values* box is where you place the variables to use as the main data of the table.

4.5 To get data from Excel to R you first open Excel. Save the data as a CSV file. Then open R and use the `read.csv()` command to import the data.

Chapter 5

5.1 Species richness is the simplest kind of diversity measure. You can use richness as a response variable in many of the classical hypothesis looking at differences or correlations for example. You can also examine the species richness at different scales, i.e. in the study of *beta* diversity.

5.2 B. Population studies are least useful. In theory you could use population estimates for all the species in a habitat or sampling area, but generally you simply record the abundances (or simple presence). Correlation is often used to link environmental factors to species richness or ordination scores. Differences tests can be used in various ways to compare diversity indices or *beta* diversity for example. Regression is an extension of correlation (just more complicated). Analysis of covariance could be used in various ways – think of it as a mixture of regression and a differences test.

5.3 In short, clustering can be hierarchical clustering or clustering by partitioning. In hierarchical clustering samples are split by their similarity (or dissimilarity) to form a kind of family tree (a dendrogram). Partitioning methods form 'clumps' by assigning samples to groups based around cluster centres (called medoids).

5.4 Ordination methods can be split into two broad themes, indirect gradient analysis and direct gradient analysis. In the former, environmental data are not used but gradients are inferred later. In the latter, the environmental data are used to help order the samples and species.

5.5 TRUE. The association analysis looks to put a value on the 'closeness' of things (such as species). This value can be used like any other similarity or dissimilarity measure as the basis for cluster analysis – either by hierarchical or partitioning methods.

Chapter 6

6.1 A bar chart or box-whisker plot would be most suitable for this task. The bar chart is a lot easier to prepare in Excel and you could do this via a Pivot Chart. Each bar would represent a habitat and the height of the bar would be the number of species.

6.2 It is possible that you would end up with duplicate abbreviations. If you use a Lookup table you can see your list of abbreviations and can manually edit any clashes. You can still use the text formula to make the original abbreviations.

6.3 A Pivot allows you to rearrange your data with species as rows in the table. You can then easily see 'duplicates', either misspellings or added spaces. You can then go back to the original data and use a Filter to view the errant items. You might still have the odd species misspelt but consistency is more important.

6.4 To start with you can look at the names of the columns:

```
> names(psa)
[1] "Site"    "Species"    "Qty"
```

Then you need to use the xtabs() command to cross-classify the data. The *Qty* column is the "data" variable. The *Species* needs to be the rows so that should be specified before the *Site* variable, which will be the columns. The command you need would be:

```
> xtabs(Qty ~ Species + Site, data = psa)
                        Site
Species                 ML1   ML2   MU1   MU2   PL2   PU2   SL1
   Achillea millefolium  6.0   3.0   5.0   4.0   0.0   3.2   0.0
   Aegopodium podagraris 0.0   0.0   0.0   0.0   0.0   0.0   0.0

   Agrostis capillaris   0.0   8.0   8.0   5.6   8.0   0.0   5.0

   Agrostis stolonifera  0.0   0.0   0.0   0.0   0.0   5.0   0.0

   Anthriscus sylvestris 0.0   0.0   0.0   0.0   0.0   0.0   0.0
```

6.5 There are two main ways you can do this. The first way is to get the species richness from the *psa* data and then use the barplot() command:

```
> sr = colSums(table(psa$Species, psa$Site))
ML1 ML2 MU1 MU2 PL2 PU2 SL1 SL2 SU1 SU2
 15  16  21  14  13  11  16  24  27  26
> barplot(sr)
```

The second way is to use the cross-classified result you made in question 6.4:

```
> psa.biol = xtabs(Qty ~ Species + Site, data = psa
> sr = colSums(psa.biol > 0)
```

```
ML1 ML2 MU1 MU2 PL2 PU2 SL1 SL2 SU1 SU2
 15  16  21  14  13  11  16  24  27  26
> barplot(sr)
```

Chapter 7

7.1 First of all you need to get the species richness (save the result to an object called *rich*). There are two ways, using the rowSums() or specnumber() commands (the latter from the *vegan* package):>

```
> rich = rowSums(gb.biol > 0)
E1 E2 E3 E4 E5 E6 G1 G2 G3 G4 G5 G6 W1 W2 W3 W4 W5 W6
17 14 15 25 21 17 28 22 18 28 26 24 12 11 11 12 12 10

> rich = specnumber(gb.biol)
```

You can use ANOVA for the test as the data are normally distributed – try using: tapply(rich, gb.site$Habitat, shapiro.test) to carry out a normality test on the three samples. Now run the ANOVA and a post-hoc test:

```
> gb.aov = aov(rich ~ Habitat, data = gb.site)
> summary(gb.aov)
            Df  Sum Sq  Mean Sq F value    Pr(>F)
Habitat      2  507.44   253.72  23.277  2.519e-05 ***
Residuals   15  163.50    10.90

---
Signif. codes:  0 '***' 0.001 '**' 0.01 '*' 0.05 '.' 0.1 ' ' 1

> TukeyHSD(gb.aov)
  Tukey multiple comparisons of means
  95% family-wise confidence level

Fit: aov(formula = rich ~ Habitat, data = gb.site)

$Habitat
                  diff          lwr        upr     p adj
Grass-Edge   6.166667     1.215550  11.117784  0.0144428
Wood-Edge   -6.833333   -11.784450  -1.882216  0.0071810
Wood-Grass -13.000000  -17.951117  -8.048883  0.0000164
```

Visualise the results using a box-whisker plot:

```
> boxplot(rich ~ Habitat, data = gb.site)
```

7.2 FALSE. The general pattern is a logarithmic one, which is linear, but there may also be non-linear relationships. The fitspecaccum() command in the *vegan* package can fit various models to your species richness data.

7.3 The differences are 0 and 0.18. You need to use the rarefy() command in the *vegan* package to carry out rarefaction. Start by getting the basic species richness as measured:

```
> rowSums(DeVries > 0)
canopy  under
    56     65
```

Now you need to know the smallest total abundance so that you can rarefy the communities 'down' to the lowest:

```
> rowSums(DeVries)
canopy under
  5774   5922
```

Now carry out the rarefaction:

```
> rarefy(DeVries, sample = 5774)
   canopy      under
56.00000  64.82439
```

So, the difference in this case is very small – you could have predicted that by looking at the total abundance for each sample, the values are close.

7.4 C. The "logit" model is not used in fitspecaccum(), the actual model is "logis".

7.5 The ACE method gives the closest estimates (and the smallest standard error). You have two separate samples in the *DeVries* data (*canopy* and *under*) so you have to use an abundance-based method of estimation. The estimateR() command in the *vegan* package will conduct the necessary calculations:

```
> estimateR(DeVries)
               canopy       under
S.obs     56.000000  65.00000
S.chao1   58.500000  75.50000
se.chao1   4.882888  31.10868
S.ACE     58.700553  69.49063
se.ACE     3.675424   3.67565
```

Chapter 8

8.1 The main difference is this treatment of the proportions of the species. In Simpson's index the proportion of each species is squared. In the Shannon index the proportions are multiplied by their logarithm.

8.2 Sample evenness is usually determined by dividing the *effective species* by the *species richness*. This is sometimes called E-evenness, exp(H)/richness. The alternative is J-evenness, which is Shannon entropy divided by the logarithm of the richness, H/log(richness). The logarithm should match that used to determine H (usually the natural log). In R you can use the specnumber() command (from *vegan*) to get the species richness and the diversity() command (also from *vegan*) to get the Shannon entropy:

```
> exp(diversity(DeVries, index = "shannon"))/specnumber(DeVries)
   canopy      under
0.2503988  0.3036880

> diversity(DeVries, index = "shannon") /
log(specnumber(DeVries))
   canopy      under
0.6560051  0.7145080
```

You could also use the diversityresult() command in the *BiodiversityR* package:

```
> diversityresult(DeVries, index = "Jevenness", method = "s")
> diversityresult(DeVries, index = "Eevenness", method = "s")
```

8.3 It depends what you mean by 'flatlines'! If the plot is more or less flat across the entire profile this indicates that the sample is very even, with species being present in more or less equal abundance. If the curve falls sharply and flattens out it indicates that there are dominant species – the faster the curve falls away the more uneven the community is.

8.4 FALSE. The scales for Rényi do relate to other diversities but you need to convert to Hill numbers by taking the exponent of the Rényi result (although at scale = 1, Rényi = Shannon). By contrast the Tsallis entropy scales are directly related to other diversity indices and at scale = 0 the Tsallis entropy equals species richness.

8.5 The scale at which sample evenness is at a minimum is q^* and is considered an important quality of a diversity profile. You can use the eventstar() command in the *vegan* package to determine q^* like so:

```
> eventstar(DeVries)
            qstar         Estar       Hstar      Dstar
canopy  0.68474940  0.111300568  5.0625792  20.615670
under   0.60784225  0.093201905  6.9298194  28.455174
```

The command also calculates evenness (*Estar*), Tsallis entropy (*Hstar*) and effective species (*Dstar*) at the scale value of q^*.

Chapter 9

9.1 FALSE. Tsallis and Rényi plots show entropy over several scales, which helps to take into account rare species.

9.2 TRUE. If you have unreplicated data, and therefore only have one sample for each site, then you would need a special version of the t-test (for Simpson's or Shannon indices). However, if you have replicated data then you could use a t-test. However, your replicates would have to be 'meaningful'. Having said that, a bootstrapping approach would be a more 'robust' alternative.

9.3 Since you have unreplicated data you cannot use a regular hypothesis test, such as a t-test or ANOVA. You could use a version of the t test for Shannon entropy for each pair of samples and then adjust the p-values to take into account the multiple tests (e.g. a Bonferroni correction). You could also use bootstrapping to get confidence intervals and determine p-values (by simulation or via a z-score), but you would still need to adjust the p-values for multiple tests. An alternative to a hypothesis test is to calculate the confidence intervals for the four samples and to present the results graphically.

9.4 A, B C and D are all true. Bootstrapping is a randomisation process and the distribution of the data is not important. You can compare two unreplicated samples or use replicated data. Any index of diversity could be compared so Rényi entropy could be used (but you would have to compare one scale at a time).

9.5 TRUE. The lowess() command can produce locally weighted polynomial regressions. These lines are sometimes called scatter plot smoothers because they follow the contours of the points rather than being straight.

Chapter 10

10.1 For species turnover you need a, the number of species shared between two samples. You also need b and c, which are the number of species unique to each sample respectively. These quantities are used in various ways, depending on the metric: Whittaker's metric is $(b + c)/(2 * a + b + c)$.

10.2 You need to use effective species. For Shannon entropy this would be exp(H) for example.

10.3 A. This is the wrong way around – you normalise data so that total *alpha* diversity does not exceed the *gamma* diversity.

10.4 TRUE. A dissimilarity matrix is a measure of differences in species composition and therefore of *beta* diversity.

10.5 There are two main ways you can proceed here: you can use a Mantel test or additive diversity partitioning. The easiest method is to use the adonis() command, which allows use of continuous variables:

```
> adonis(gb.biol ~ Max.Ht, data = gb.site)

Call:
adonis(formula = gb.biol ~ Max.Ht, data = gb.site)

          Df SumsOfSqs MeanSqs F.Model      R2  Pr(>F)
Max.Ht     1   0.66348 0.66348  12.326 0.43515  0.004 **
Residuals 16   0.86124 0.05383         0.56485
Total     17   1.52472                 1.00000
---
Signif. codes: 0 '***' 0.001 '**' 0.01 '*' 0.05 '.' 0.1 ' ' 1
```

Alternatively you can use a Mantel test. You first need to make dissimilarity matrices for the community data and the vegetation height. You'll find that the vegdist() command will be fine for the community data but not for the height (it needs at least two variables); you'll need the dist() command instead. Select the dissimilarity metric(s) of your choice then run the mantel() command like so:

```
> gb.bc = vegdist(gb.biol, method = "bray")
> ht.eu = dist(gb.site$Max.Ht, method = "euclidean")
> mantel(gb.bc, ht.eu)

Mantel statistic based on Pearson's product-moment correlation

Call:
mantel(xdis = gb.bc, ydis = ht.eu)

Mantel statistic r: 0.3616
      Significance: 0.003

Empirical upper confidence limits of r:
  90%    95%   97.5%    99%
0.0826 0.1164 0.2613 0.3024

Based on 999 permutations
```

Chapter 11

11.1 Resource-partitioning models can be thought of as operating over ecological time (e.g. broken stick), or evolutionary time (e.g. niche preemption model).

11.2 FALSE. The Mandelbrot model is a derivative of Zipf but there is only one additional parameter in the model (called β).

11.3 You can approach this in two ways. The simplest way is to make a `radfit()` result and look at the deviance or AIC values. However, the data are a matrix and the `radfit()` command needs a `data.frame` so you must convert first:

```
> radfit(as.data.frame(DeVries))

Deviance for RAD models:

               canopy     under
Null          4706.45   3363.52
Preemption    2223.89    588.82
Lognormal      244.20    552.36
Zipf           430.61   1296.77
Mandelbrot     386.46    588.82
```

So, the lognormal model is best for both habitats because the deviance is lowest. If you took a `summary()` of the result you'd see the AIC values too. You can also look at each habitat separately and use `radlattice()` to view the result – the AIC values are given on the plot. For the first sample you would use the following:

```
> radlattice(radfit(DeVries[1,]))
```

11.4 You can use the `family` instruction to specify other distributions when using the `radfit()` or `rad.xxxx()` commands. The default is `family = poisson`. For cover data `family = Gamma` would be a good alternative.

11.5 You can compute the unseen (veiled) species using the `veiledspec()` command in the *vegan* package. The starting point are the `prestonfit()` or `prestondistr()` commands, which use slightly different fitting methods. The *DeVries* data is a matrix but in any event the best way is to use `apply()` to create a result containing all samples first, then use `sapply()` to get the veiled species from that:

```
> DV1 = apply(DeVries, MARGIN = 1, FUN = prestonfit)
> sapply(DV1, veiledspec)
                    canopy        under
Extrapolated     62.745778    71.062599
Observed         56.000000    65.000000
Veiled            6.745778     6.062599
```

If you replace the `FUN` in the first command with `prestondistr` you will calculate the veiled species for the alternative method of computation:

```
> DV2 = apply(DeVries, MARGIN = 1, prestondistr)
> sapply(DV2, veiledspec)[3,]
  canopy      under
1.713444   2.163121
```

Chapter 12

12.1 TRUE. *Beta* diversity is a measure of how species composition changes between samples so can be used for analysis of similarity. The `betadiver()` command for example can create a dissimilarity matrix based on one of many metrics based on presence-absence data.

12.2 E. The `binary` instruction to the `vegdist()` command will use presence-absence versions of the other metrics, but it is not itself a metric. The Sørensen metric is the same as Bray–Curtis. The Jaccard metric is equivalent to the binomial metric specified in the `dist()` command.

12.3 The `hclust()` command in the basic R distribution carries out an *agglomeration* method by assigning and reassigning samples to clusters. The `agnes()` command in the *cluster* package will also carry out agglomerative nesting. The other main method is *divisive* clustering, which can be carried out by the `diana()` command.

12.4 You'll need to use the *pvclust* package to look at significance in hierarchical clustering. The `pvclust()` command requires the community data with samples as columns (which you have), and a dissimilarity metric:

```
> ant.pv = pvclust(ant, method.dist = "euclidean")
> ant.pv

Cluster method: average
Distance      : euclidean

Estimates on edges:

      au      bp   se.au   se.bp       v       c    pchi
1 0.614   0.481   0.029   0.005  -0.122   0.169   0.147
2 0.568   0.404   0.030   0.005   0.036   0.208   0.623
3 0.938   0.727   0.010   0.005  -1.071   0.468   0.185
4 0.609   0.414   0.030   0.005  -0.030   0.246   0.295
5 0.894   0.392   0.015   0.005  -0.486   0.761   0.658
6 0.731   0.431   0.025   0.005  -0.221   0.394   0.521
7 0.997   0.980   0.001   0.002  -2.429   0.373   0.244
8 0.761   0.793   0.026   0.004  -0.764  -0.054   0.565
9 1.000   1.000   0.000   0.000   0.000   0.000   0.000
```

The *p*-values you want are in the heading *au* (which stands for approximately unbiased) and high values (> 0.95) are significant. You can make a dendrogram plot containing the au values using the `plot()` command and can highlight significant clusters using the `pvrect()` command:

```
> plot(ant.pv)
> pvrect(ant.pv)
```

The `pvpick()` command will show you the significant clusters, giving edge and cluster membership information.

12.5 TRUE. In partitioning methods you decide how many clusters to create and the analysis proceeds to find the 'best' clusters based on that number of groups. R commands that require you to specify the number of clusters are: `kmeans()`, `pam()`, `fanny()` and `clara()`.

Chapter 13

13.1 Chi squared values are all positive and measure the magnitude of the difference between observed and expected values (standardised for sample size). This means that any particular value could represent a positive or negative association. Pearson residuals can be positive or negative and you therefore end up with a range of values with the smallest being the most negative associations and the largest being the most positive.

13.2 B. The aim is to sample a geographical area to identify the communities present. It does not matter if the sampling strategy is random or not, it is more important to cover the sampling area. You want to keep sampling units small to minimise 'accidental' overlap between community boundaries. You can collect abundance data but the analytical process will 'collapse' the data to presence-absence in calculating the co-occurrence.

13.3 TRUE. In a transect-based approach you are expecting the communities to change along the transect(s) – this is the entire point of the exercise! Your transects can lie along environmental gradients, where you expect the communities to change as conditions change, or you can use a successional gradient (which may encompass one or more environmental gradients).

13.4 You need to carry out a chi squared test on the contingency table. For 2 × 2 tables it is usual to apply Yates' correction. The significance works out to be $p = 0.053$, which is just above the threshold so is not significant. If you don't use Yates' correction the result comes out at $p = 0.036$, which would be taken as significant! If you use R then the chisq.test() command can be used, but first you need to make the data:

```
> td = matrix(c(150, 50, 40, 25), ncol = 2,
              dimnames = list(c("A+", "A-"), c("B+", "B-")))
> chisq.test(td)

    Pearson's Chi-squared test with Yates' continuity
    correction

data:  td
X-squared = 3.7426, df = 1, p-value = 0.05304
```

If you use Excel you would need to work out the marginal totals and then calculate the expected values, which turn out to be:

```
        B+     B-    Sum
A+    143.4  46.6    190
A-     56.6  18.4     75
Sum   200.0  65.0    265
```

In Yates' correction all expected values are reduced by 0.5 before carrying out the chi squared evaluation. You can then use the CHIDIST function to work out the significance with one degree of freedom.

13.5 FALSE. A species must have a significant positive association to be an indicator of a habitat but it must also have significant negative associations to other habitats. In addition, the species must be uncommon in those other habitats.

Chapter 14

14.1 If two axes are *orthogonal* it means that there is minimal correlation between them. The idea behind methods of ordination is to separate out the sites and species as much as you can along an axis so that you 'view' as much information as possible. If the second axis was highly correlated with the first, you would not gain much additional information. Using an orthogonal axis means that you separate the data 'more widely' and so get a clearer pattern.

14.2 First you need to rotate the *hsere.biol* and *hsere.ph* data so that the rows are the samples. Then you'll need the `rankindex()` command in the *vegan* package to compare the dissimilarity metrics (the Horn–Morisita metric is best). Finally you can run the `metaMDS()` command and extract the stress from it:

```
> hsere = t(hsere.biol)
> ph = t(hsere.ph)
> rankindex(ph, hsere, indices = c("bray", "euc", "gow", "horn"))
      bray          euc          gow         horn
0.8033467   0.7752816   0.7459648   0.8111865
> hsere.nmds = metaMDS(hsere, distance = "horn")
Square root transformation
Wisconsin double standardization
Run 0 stress 0.001497090
Run 1 stress 0.001589213
... procrustes: rmse 0.0004769718   max resid 0.0007185633
*** Solution reached

> hsere.nmds$stress
[1] 0.001497090
```

Your final stress may be slightly different as the command uses random starting points and a stable solution can be reached with subtly different values.

14.3 C. The `Condition()` instruction allows a variable to be specified that will be partialled-out of the analysis, i.e. the analysis is *conditioned* by that variable.

14.4 FALSE. When you are model-building the AIC values are important but it is low values you want to look for.

14.5 Start with the NMDS using the `metaMDS()` command in the *vegan* package. You can then run the `envfit()` command – using a formula will enable you to select the two variables for comparison. Neither has a significant fit (r^2: 0.2, 0.1 respectively) but the DO2 variable has the edge with the lower *p*-value:

```
> fw.nmds = metaMDS(fw.biol)
Square root transformation
Wisconsin double standardization
Run 0 stress 0.1741285
Run 1 stress 0.1741287
... procrustes: rmse 0.0005712724   max resid 0.001684685
*** Solution reached

> envfit(fw.nmds ~ DO2 + cond, fw.env)

***VECTORS

        NMDS1        NMDS2        r2   Pr(>r)
DO2    0.31018    0.95068   0.1957  0.200
cond   0.42659   -0.90444   0.1072  0.424
P values based on 999 permutations.
```

Your *p*-values may be slightly different as they are based on random permutations.

Appendix 2 Custom R commands in this book

These commands have been written especially to accompany the book. The commands can be found in the *CERE.RData* file, which you can download from the companion website. The file also contains the datasets used throughout the text. The commands are listed in alphabetical order and give basic details about useage – more details can be found on the companion website.

comm_grp

Group community data using grouping variable.
   This command can be used instead of `rowSum()` but essentially does the same thing.

```
comm_grp(comm, groups, data)
```

Parameters:

comm.   a community dataset as a `data.frame`, `matrix` or `table`.
groups   a grouping variable.
data   an optional `data.frame` where the grouping variable can be found.

Dapp

Bootstrap confidence intervals for Simpson's index.
   This is intended to be used with `apply()`; allowing bootstrap confidence intervals to be computed for multiple samples. The default is 1000 resamples and 95% CI.

```
Dapp(x)
```

Parameters:

x   a vector of values; a single sample from a community dataset.

dist_chi

Chi squared association for community analysis.
   Calculates pairwise chi squared statistics for a community dataset. The result has a class "distchi", which has `plot` and `summary` methods. The results include Pearson residuals as a dissimilarity matrix. The `plot` method produces an hierarchical cluster object, and plots the dendrogram.

```
dist_chi(data)
summary(x)
plot(x, method = "complete", ...)
```

Parameters:

Data   a community dataset (usually a `matrix` or `data.frame`) with columns as samples.
x   the result of a `distchi()` command.
method   the cluster joining algorithm to be used by `hclust()`, use one of "ward", "single", "complete" (the default), "average", "mcquitty", "median", or "centroid".
...   other parameters to pass to the plotting command.

diversity_comp

Diversity index comparison by permutation.

Computes significance of diversity index between two samples by permutation method (oecosimu). You can use any of the diversity indices used by the diversity() command in *vegan*.

```
diversity_comp(comm, index = "shannon", perm = 1999)
```

Parameters:

comm   a community dataset of exactly two samples.

index   a diversity index; use one of "shannon" (default), "simpson" or "invsimpson".

perm   the number of permutations, default = 1999.

eacc

Estimated species accumulation.

This command is intended to be used with apply() so that you can compute estimated species richness for multiple samples.

```
eacc(x)
```

Parameters:

x   a vector of community data.

entropy_plot

Plot results from Rényi or Tsallis calculations.

This command allows you to visualise the results from the renyi() or tsallis() commands. Usually you'll use it as an alternative to the usual *lattice* plotting command to overlay multiple results in a single plot window.

```
entropy_plot(H, type = "l", ylab = "Diversity", xlab = "Scale",
             legend = TRUE, col = palette(), lty = 1:6, cex = 1,
    pch = NULL, cex.legend = 1, ...)
```

Parameters:

H	the result from a renyi() or tsallis() command from the *vegan* package.
type	the type of plot, default "l" produces lines. Alternatives are "p", points, "b", both, "o" overplotted, "n" none.
xlab	labels for *x* and *y* axes.
ylab	
legend	Logical. The default, TRUE, displays a legend.
col	the default colours to use on the plot, the default uses the current colour palette. Colours are recycled as required.
lty	the line style(s) used, the default uses types 1:6 and these are recycled as needed.
cex	character expansion for plotted points, default = 1.
pch	plotting symbols, default is NULL.

cex.legend   character expansion for legend labels, default = 1.

...   other graphical parameters.

H_boot

Bootstrapping diversity indices.

Carries out a bootstrap resampling using the boot package. The results also compute the confidence intervals. You can specify any diversity index used by the `diversity()` command in *vegan* for a single sample.

```
H_boot(x, index = "shannon", R = 1000, ci = 95)
```

Parameters:

x	a vector of community data.
index	a diversity index; use one of "shannon" (default), "simpson" or "invsimpson".
R	the number of bootstrap resamples to use, defaults to 1000.
ci	the confidence interval to return, the default 95, returns 2.5% and 97.5% quantiles.

H_bss

Bootstrap significance between two diversity indices.

This command carries out a bootstrap analysis of the diversity indices of two samples. The result gives a *p*-value for the permutation as well as a calculated *z* -score. You can use any of the diversity indices used by the `diversity()` command in *vegan*. The result holds a class, "Hbss", which has `print` and `summary` methods.

```
H_bss(comm, R = 2000, index = "shannon")
print(x, ...)
summary(x)
```

Parameters:

comm	a community dataset containing exactly two samples.
x	the result of an H_bss() command.
R	the number of bootstrap resamples to use, defaults to 2000.
index	a diversity index; use one of "shannon" (default), "simpson" or "invsimpson".
...	other parameters to pass to print, such as digits.

H_ci

Bootstrap confidence intervals for a single sample.

This command carries out a bootstrap resampling to estimate confidence intervals for a single sample.

```
H_ci(x, index = "shannon", R = 1000, ci = 95)
```

Parameters:

x	a vector of community data.

index   a diversity index; use one of `"shannon"` (default), `"simpson"` or `"invsimpson"`.

R   the number of bootstrap resamples to use, defaults to 1000.

ci   the confidence interval to use, defaults to 95% (i.e. 2.5% and 97.5% quantiles).

H_CI

Confidence interval for Shannon or Simpson's *t*-test.

This command computes the confidence interval for the *t*-test used with Shannon or Simpson's indices. It is intended to be used with the `apply()` command so that you can plot error bars.

```
H_CI(x, index = "shannon", ci = 95)
```

Parameters:

x   a vector of community data.

index   a diversity index; use one of `"shannon"` (default) or `"simpson"`.

ci   the confidence interval to use the default is 95 and works out 95% CI.

H_sig

Special *t*-test for Shannon or Simpson's index.

This command carries out a version of the *t*-test for either Shannon or Simpson's indices. The result object has a class `"Hsig"`, which has `plot` and `summary` routines.

```
H_sig(x, y, index = "shannon", alpha = 0.05)
summary(x)
plot(x, ylab = "Index Value", xlab = "Samples", ...)
```

Parameters:

x, y   vectors of community data for comparison; these can be of different length. For `plot` and `summary` methods, x is the result of an `H_sig()` command.

index   a diversity index; use one of `"shannon"` (default) or `"simpson"`.

alpha   the significance level, the default 0.05 computes critical values for $p = 0.05$.

xlab, ylab   labels for the *x* and *y* axes.

...   additional graphical parameters to pass to `plot()`.

Happ

Bootstrap confidence intervals for the Shannon index.

This is intended to be used with `apply()`; allowing bootstrap confidence intervals to be computed for multiple samples. The default is 1000 resamples and 95% CI.

```
Happ(x)
```

Parameters:

x   a vector of values; a single sample from a community dataset.

pacc

Estimated species richness for accumulating samples.

This command is intended to be used with `lapply()` for use with a community dataset and a grouping variable. It calls the `poolaccum()` command from *vegan* and returns various estimators of species richness.

```
pacc(x, data)
```

Parameters:

x	a vector of index values.
data	the `data.frame` holding the community data.

plot_H

Point plot of diversity index and bootstrap confidence intervals.

This command carries out a bootstrap resampling of a community dataset and plots the results as a point plot that includes error bars showing the confidence intervals. You can use any diversity index used by the `diversity()` command in *vegan*.

```
plot_H(comm, index = "shannon",
       xlab = "Site Names", ylab = "Diversity Index",
       pch = 18, cex = 1.2, cex.axis = 0.75, ...)
```

Parameters:

comm	community dataset, usually a `matrix` or `data.frame`.
index	a diversity index; use one of `"shannon"` (default), `"simpson"` or `"invsimpson"`.
xlab, ylab	labels for x and y axes.
pch	plotting symbol, default = 18 (a filled diamond).
cex	character expansion, default = 1.2.
cex.axis	character expansion for axes, default = 0.75.
...	other graphical parameters to pass to `plot()`.

qstar

Plot Tsallis entropy profile using q*.

This command uses the `eventstar()` command from *vegan* to calculate q*. The profile is then plotted as three plots showing, evenness, diversity and effective species for the Tsallis scales selected.

```
qstar(x, q = seq(0, 2, 0.05), qmax = 5, type = "l",
      lcol = "red", llty = 1, llwd = 1, ...)
```

Parameters:

x	a vector of community data.
q	the scales to use for the Tsallis entropy (passed to `tsallis`).
qmax	the maximum q* value to use (passed to `eventstar`), default = 5.
type	the type of plot, default = `"l"` for lines.
lcol,	colour, type and width for the lines showing q* value and diversity values

```
llty,        at q*.
llwd
...          additional graphical parameters to pass to plot().
```

rad_aic

Compute AIC values from multiple RAD models.

This command takes the result of multiple RAD model-fitting (from radfit() in *vegan*) and computes the 'best' AIC value for each sample. The result shows the best AIC value and the corresponding model name for each sample.

```
rad_aic(x)
```

Parameters:

x
: the result of radfit() from *vegan*. The result must hold the class "radfit.frame", i.e. have been run on a community dataset with multiple samples.

rad.test

A significance test of differences in deviance of RAD models.

This command takes the result of a radfit() command (from *vegan*) and carries out a test of significance (ANOVA) of the various RAD models, examining their deviance. The result holds a class, "rad.htest" that has summary and plot methods.

```
rad_test(x, conf.level = 0.95, log = TRUE)
print(x, ...)
summary(x)
plot(x, which = "original", las = 1, ...)
```

Parameters:

x
: the result of a radfit() command from *vegan*. For summary and plot methods x should be the result of the rad_test() command.

conf.level
: the confidence level, the default is 0.95.

log
: Logical. If TRUE (the default) the log of the model deviance is used as the response variable in ANOVA.

which
: the type of plot, "original" (default) shows boxplot of model deviance, "post.hoc" displays Tukey post-hoc result.

las
: the orientation of the axis annotations, the default produces labels all horizontal.

...
: other parameters to be passed to plot() or print().

spec.rich

Species richness for a community dataset.

This command computes species richness for a community dataset.

```
spec.rich(data)
```

Parameters:

data
: a community dataset as a data.frame or matrix. Rows as samples.

species_assoc

Pairwise chi squared species association.

This command takes two species that you specify from a community dataset and carries out a pairwise chi squared association test.

```
species_assoc(spA, spB, data)
```

Parameters:

spA, spB	index values for the two species to be analysed, corresponding to their rowname attributes.
data	the data.frame containing the species for analysis; rows as species.

species_ind

Indicator species analysis by chi squared.

This command takes a community dataset and carries out a chi squared test for indicator species. The species can be arranged as the rows or columns. The samples are used as the groupings for the analysis, so you may need to group the dataset using rowSum() before analysis. The result shows only those species that have a significant association (positive or negative) for all samples/groups.

```
species_ind(x, rows = c("species", "samples"))
```

Parameters:

x	a community dataset, usually a data.frame or matrix.
rows	should the rows be treated as "species" (the default) or "samples"?

sr_est

Species richness estimation with accumulating sites.

This command uses accumulating species richness and a log model to determine the estimated species richness for a complete dataset. You can use a community dataset or the result of the specaccum() command from *vegan*. The result contains the estimated species richness and the log model. The result holds a class "srest", which has plot and summary methods.

```
sr_est(accum)
print(x, ...
summary(srest, digits = getOption("digits"))
plot(srest, col.points = "black", col.line = "black",
      lwd.line = 1, lwd.points = 1, ...)
```

Parameters:

accum	a vector of values corresponding to accumulating species richness, or a community dataset (rows as samples), or the result of a specaccum() command from *vegan*.
srest	the result of an sr_est() command.
digits	the number of significant figures to display.
col.points, lwd.points	the colour and width of the points used in the base plot.
col.line, lwd.line	the colour and width used for the best-fit line (produced via abline).
...	additional parameters to pass to plot() or print().

Bibliography

Anderberg, M.R. (1973) *Cluster Analysis for Applications*. Academic Press.

Anderson, M.J. (2001) A new method for non-parametric multivariate analysis of variance. *Australian Ecology* **26**, 32–46.

Anderson, M.J. (2006) Distance-based tests for homogeneity of multivariate dispersions. *Biometrics* **62**, 245–253.

Anderson, M.J., Ellingsen, K.E. and McArdle, B.H. (2006) Multivariate dispersion as a measure of *beta* diversity. *Ecology Letters* **9**, 683–693.

Anderson, M.J. and Millar, R.B. (2004) Spatial variation and effects of habitat on temperate reef fish assemblages in northeastern New Zealand. *Journal of Experimental Marine Biology and Ecology* **305**, 191–221.

Anderson, M.J. and Willis, T.J. (2003) Canonical analysis of principal coordinates: a useful method of constrained ordination for ecology. *Ecology* **84**, 511–525.

Anderson. E. and ten others (1999) *LAPACK Users' Guide*, third edition. SIAM. Available on-line at http://www.netlib.org/lapack/lug/lapack_lug.html.

Bauer, D.F. (1972) Constructing confidence sets using rank statistics. *Journal of the American Statistical Association* **67**, 687–690.

Benjamini, Y. and Hochberg, Y. (1995) Controlling the false discovery rate: a practical and powerful approach to multiple testing. *Journal of the Royal Statistical Society Series* B **57**, 289–300.

Benjamini, Y. and Yekutieli, D. (2001) The control of the false discovery rate in multiple testing under dependency. *Annals of Statistics* **29**, 1165–1188.

Beran, R. (1988) Balanced simultaneous confidence sets. *Journal of the American Statistical Association* **83**, 679–686.

Besag, J., Green, P.J., Higdon, D. and Mengersen, K. (1995) Bayesian computation and stochastic systems. *Statistical Science* **10**, 3–66.

Best, D.J. and Roberts, D.E. (1975) Algorithm AS 89: The upper tail probabilities of Spearman's *rho*. *Applied Statistics* **24**, 377–379.

Booth, J.G., Hall, P. and Wood, A.T.A. (1993) Balanced importance resampling for the bootstrap. *Annals of Statistics* **21**, 286–298.

Borg, I. and Groenen, P. (1997) *Modern Multidimensional Scaling. Theory and Applications*. Springer.

Cailliez, F. (1983) The analytical solution of the additive constant problem. *Psychometrika* **48**, 343–349.

Chambers, J.M., Cleveland, W.S., Kleiner, B. and Tukey, P.A. (1983) *Graphical Methods for Data Analysis*. Wadsworth & Brooks/Cole.

Chao, A. (1987) Estimating the population size for capture–recapture data with unequal catchability. *Biometrics* **43**, 783–791.

Chao, A., Chazdon, R. L., Colwell, R. K. and Shen, T. (2005) A new statistical approach for assessing similarity of species composition with incidence and abundance data. *Ecology Letters* **8**, 148–159.

Chase, J.M., Kraft, N.J.B., Smith, K.G., Vellend, M. and Inouye, B.D. (2011) Using null models to disentangle variation in community dissimilarity from variation in *alpha*-diversity. *Ecosphere* **2** [doi:10.1890/ES10-00117.1]

Clarke, K.R. (1993) Non-parametric multivariate analysis of changes in community structure. *Australian Journal of Ecology* **18**, 117–143.

Coleman, B.D, Mares, M.A., Willis, M.R. and Hsieh, Y. (1982) Randomness, area and species richness. *Ecology* **63**, 1121–1133.

Colwell, R.K. and Coddington, J.A. (1994) Estimating terrestrial biodiversity through extrapolation. *Philosophical Transactions of the Royal Society London* B **345**, 101–118.

Colwell, R.K., Mao, C.X. and Chang, J. (2004) Interpolating, extrapolating, and comparing incidence-based species accumulation curves. *Ecology* **85**, 2717–2727.

Cox, T.F. and Cox, M.A.A. (2001) *Multidimensional Scaling*. Chapman & Hall.

Crawley, M.J. (2002) *Statistical Computing: An Introduction to Data Analysis Using S-PLUS*. John Wiley & Sons.

Crist, T.O., Veech, J.A., Gering, J.C. and Summerville, K.S. (2003) Partitioning species diversity across landscapes and regions: a hierarchical analysis of α, β, and γ-diversity. *American Naturalist* **162**, 734–743.

Datta, A. and Rawat, G.S. (2003) Foraging patterns of sympatric hornbills during the nonbreeding season in Arunchal Pradesh, Northeast India. *Biotropica* **35**, 208–218.

Davison, A.C. and Hinkley, D.V. (1997) *Bootstrap Methods and Their Application*. Cambridge University Press.

Davison, A.C., Hinkley, D.V. and Schechtman, E. (1986) Efficient bootstrap simulation. *Biometrika* **73**, 555–566.

Dengler, J. (2009) Which function describes the species–area relationship best? A review and empirical evaluation. *Journal of Biogeography* **36**, 728–744.

Efron, B. and Tibshirani, R. (1993) *An Introduction to the Bootstrap*. Chapman & Hall.

Everitt, B. (1974) *Cluster Analysis*. Heinemann Educ. Books.

Excoffier, L., Smouse, P.E. and Quattro, J.M. (1992) Analysis of molecular variance inferred from metric distances among DNA haplotypes: Application to human mitochondrial DNA restriction data. *Genetics* **131**, 479–491.

Faith, D.P, Minchin, P.R. and Belbin, L. (1987) Compositional dissimilarity as a robust measure of ecological distance. *Vegetatio* **69**, 57–68.

Fisher, R.A., Corbet, A.S. and Williams, C.B. (1943) The relation between the number of species and the number of individuals in a random sample of animal population. *Journal of Animal Ecology* **12**, 42–58.

Forgy, E.W. (1965) Cluster analysis of multivariate data: efficiency vs interpretability of classifications. *Biometrics* **21**, 768–769.

Fox, J. (2008) *Applied Regression Analysis and Generalized Linear Models*, second edition. Sage.

Fox, J. and Monette, G. (1992) Generalized collinearity diagnostics. *Journal of the American Statistical Association* **87**, 178–183.

Fox, J. and Weisberg, S. (2011) *An R Companion to Applied Regression*, second edition. Sage.

Friendly, M. (2000) *Visualizing Categorical Data*. SAS Institute.

Fritsch, K.S. and Hsu, J.C. (1999) Multiple comparison of entropies with application to dinosaur biodiversity. *Biometrics* **55**, 4, 1300–1305.

Gabriel, K.R. (1971) The biplot graphical display of matrices with application to principal component analysis. *Biometrika* **58**, 453–467.

Gardener, M. (2012) *Statistics for Ecologists Using R and Excel: Data collection, exploration, analysis and presentation*. Pelagic Publishing.

Gilbert, G.S. and Sousa, W.P. (2002) Host specialization among wood-decay polypore fungi in a Caribbean mangrove forest. *Biotropica* **34**, 396–404.

Gleason, J.R. (1988) Algorithms for balanced bootstrap simulations. *American Statistician* **42**, 263–266.

Gordon, A.D. (1999) *Classification*, second edition. Chapman and Hall/CRC.

Gotelli, N.J. and Colwell, R.K. (2001) Quantifying biodiversity: procedures and pitfalls in measurement and comparison of species richness. *Ecology Letters* **4**, 379–391.

Gower, J.C. (1966) Some distance properties of latent root and vector methods used in multivariate analysis. *Biometrika* **53**, 325–328.

Gower, J.C. (1971) A general coefficient of similarity and some of its properties. *Biometrics* **27**, 623–637.

Gower, J.C. (1985) Properties of Euclidean and non-Euclidean distance matrices. *Linear Algebra and its Applications* **67**, 81–97.

Gower, J.C. and Hand, D.J. (1996) *Biplots*. Chapman & Hall.

Greenacre, M.J. (1984) *Theory and Applications of Correspondence Analysis*. Academic Press, London.

Gross, J. (2003) Variance inflation factors. *R News* **3**, 13–15.

Hall, P. (1989) Antithetic resampling for the bootstrap. *Biometrika* **73**, 713–724.

Hartigan, J.A. (1975) *Clustering Algorithms*. Wiley.

Hartigan, J.A. and Wong, M.A. (1979) A K-means clustering algorithm. *Applied Statistics* **28**, 100–108.

Heck, K.L., van Belle, G. and Simberloff, D. (1975) Explicit calculation of the rarefaction diversity measurement and the determination of sufficient sample size. *Ecology* **56**, 1459–1461.

Hill, M.O. (1973) Diversity and evenness: a unifying notation and its consequences. *Ecology* **54**, 427–473.

Hill, M.O. and Gauch, H.G. (1980) Detrended correspondence analysis: an improved ordination technique. *Vegetatio* **42**, 47–58.

Hinkley, D.V. (1988) Bootstrap methods (with Discussion). *Journal of the Royal Statistical Society B* **50**, 312–337, 355–370.

Hinkley, D.V. and Shi, S. (1989) Importance sampling and the nested bootstrap. *Biometrika* **76**, 435–446.

Hochberg, Y. (1988) A sharper Bonferroni procedure for multiple tests of significance. *Biometrika* **75**, 800–803.

Hoffmann, B.D. (2003) Responses of ant communities to experimental fire regimes on rangelands in the Victoria River District of the Northern Territory. *Australian Ecology* **28**, 182–195.

Hollander, M. and Wolfe, D.A. (1973) *Nonparametric Statistical Methods*. Wiley.

Holm, S. (1979) A simple sequentially rejective multiple test procedure. *Scandinavian Journal of Statistics* **6**, 65–70.

Hommel, G. (1988) A stagewise rejective multiple test procedure based on a modified Bonferroni test. *Biometrika* **75**, 383–386.

Hope, A.C.A. (1968) A simplified Monte Carlo significance test procedure. *Journal of the Royal Statistical. Society B* **30**, 582–598.

Hurlbert, S.H. (1971) The nonconcept of species diversity: a critique and alternative parameters. *Ecology* **52**, 577–586.

Hutcheson, K. (1970) A test for comparing diversities based on the Shannon formula. *Journal of Theoretical Biology* **29**, 151–154.

Johns M.V. (1988) Importance sampling for bootstrap confidence intervals. *Journal of the American Statistical Association* **83**, 709–714.

Johnson, N.L., Kotz, S. and Balakrishnan, N. (1995) *Continuous Univariate Distributions*. Wiley.

Jost, L. (2006) Entropy and diversity. *Oikos* **113**, 363–375.

Jost, L. (2007) Partitioning diversity into independent *alpha* and *beta* components. *Ecology* **88**, 2427–2439.

Kaufman, L. and Rousseeuw, P.J. (1990) *Finding Groups in Data: An Introduction to Cluster Analysis*. Wiley.

Kempton, R.A. and Taylor, L.R. (1974) Log-series and log-normal parameters as diversity discriminators for Lepidoptera. *Journal of Animal Ecology* **43**, 381–399.

Keylock, C.J. (2005) Simpson diversity and the Shannon–Wiener index as special cases of a generalized entropy. *Oikos* **109**, 203–207.

Kindt R., Van Damme P. and Simons A.J. (2006) Tree diversity in western Kenya: using diversity profiles to characterise richness and evenness. *Biodiversity and Conservation* **15**, 1253–1270.

Koleff, P., Gaston, K.J. and Lennon, J.J. (2003) Measuring *beta* diversity for presence–absence data. *J. Animal Ecol.* **72**, 367–382.

Krebs, C.J. (1999) *Ecological Methodology*. Addison Wesley Longman.

Kruskal, J.B. (1964a) Multidimensional scaling by optimizing goodness-of-fit to a nonmetric hypothesis. *Psychometrika* **29**, 1–28.

Kruskal, J.B. (1964b) Nonmetric multidimensional scaling: a numerical method. *Psychometrika* **29**, 115–129.

Krzanowski, W.J. and Marriott, F.H.C. (1994) *Multivariate Analysis. Part I. Distributions, Ordination and Inference*. Edward Arnold.

Lance, G.N. and Williams, W.T. (1966) A general theory of classifactory sorting strategies, I. Hierarchical systems. *Computer Journal* **9**, 373–380.

Lande, R. (1996) Statistics and partitioning of species diversity, and similarity among multiple communities. *Oikos* **76**, 5–13.

Legendre, P. and Anderson, M.J. (1999) Distance-based redundancy analysis: testing multispecies responses in multifactorial ecological experiments. *Ecological Monographs* **69**, 1–24.

Legendre, P. and Gallagher, E.D. (2001) Ecologically meaningful transformations for ordination of species data. *Oecologia* **129**, 271–280.

Legendre, P, and Legendre, L. (1998) *Numerical Ecology*, second English edition. Elsevier.

Legendre, P., Oksanen, J. and ter Braak, C.J.F. (2011) Testing the significance of canonical axes in redundancy analysis. *Methods in Ecology and Evolution* **2**, 269–277.

Lloyd, S.P. (1957, 1982) Least squares quantization in PCM. Technical Note, Bell Laboratories. Published in 1982 in *IEEE Transactions on Information Theory* **28**, 128–137.

MacQueen, J. (1967) Some methods for classification and analysis of multivariate observations. In *Proceedings of the Fifth Berkeley Symposium on Mathematical Statistics and Probability*, eds L. M. Le Cam & J. Neyman, **1**, pp. 281–297. University of California Press.

Mantel, N. (1967) The detection of disease clustering and a generalized regression approach. *Cancer Research* **27**, 209–220.

Mardia, K.V., Kent, J.T. and Bibby, J.M. (1979) *Multivariate Analysis*. Academic Press.

McArdle, B.H. and Anderson, M. J. (2001) Fitting multivariate models to community data: A comment on distance-based redundancy analysis. *Ecology* **82**, 290–297.

McCune, B. (1997) Influence of noisy environmental data on canonical correspondence analysis. *Ecology* **78**, 2617–2623.

McCune, B. and Grace, J.B. (2002) *Analysis of Ecological Communities*. MjM Software Design.

McQuitty, L.L. (1966) Similarity analysis by reciprocal pairs for discrete and continuous data. *Educational and Psychological Measurement* **26**, 825–831.

Mendes, R.S., Evangelista, L.R., Thomaz, S.M., Agostinho, A.A. and Gomes, L.C. (2008) A unified index to measure ecological diversity and species rarity. *Ecography* **31**, 450–456.

Mielke, P.W. and Berry, K.J. (2001) *Permutation Methods: A Distance Function Approach*. Springer Series in Statistics. Springer.

Miller, R.G. (1981) *Simultaneous Statistical Inference*. Springer.

Minchin, P.R. (1987) An evaluation of relative robustness of techniques for ecological ordinations. *Vegetatio* **69**, 89–107.

Mountford, M. D. (1962) An index of similarity and its application to classification problems. In: P.W.Murphy (ed.), *Progress in Soil Zoology*, 43–50. Butterworths.

Noreen, E.W. (1989) *Computer Intensive Methods for Testing Hypotheses*. Wiley.

Oksanen, J. (1983) Ordination of boreal heath-like vegetation with principal component analysis, correspondence analysis and multidimensional scaling. *Vegetatio* **52**, 181–189.

Oksanen, J. and Minchin, P.R. (1997) Instability of ordination results under changes in input data order: explanations and remedies. *Journal of Vegetation Science* **8**, 447–454.

Palmer, M.W. (1986) Pattern in corticolous bryophyte communities of the North Carolina Piedmont: Do mosses see the forest or the trees? *The Bryologist* **89**, 59–65.

Palmer, M.W. (1993) Putting things in even better order: the advantages of canonical correspondence analysis. *Ecology* **74**, 2215–2230.

Palmer, M.W. (1990) The estimation of species richness by extrapolation. *Ecology* **71**, 1195–1198.

Patefield, W.M. (1981) Algorithm AS159. An efficient method of generating r × c tables with given row and column totals. *Applied Statistics* **30**, 91–97.

Patil, G.P. and Taillie, C. (1982) Diversity as a concept and its measurement. *Journal of the American Statistical Association* **77**, 548–567.

Pereira, I.M. (2003) Use-history effects on structure and flora of Caatinga. *Biotropica* **35**, 154–165.

Peres-Neto, P.R. and Jackson, D.A. (2001) How well do multivariate data sets match? The advantages of a Procrustean superimposition approach over the Mantel test. *Oecologia* **129**, 169–178.

Pielou, E.C. (1975) *Ecological Diversity*. Wiley.

Preston, F.W. (1948) The commonness and rarity of species. *Ecology* **29**, 254–283.

Ripley, B.D. (1996) *Pattern Recognition and Neural Networks*. Cambridge University Press.

Rogers, J.A. and Hsu, J.C. (2001) Multiple comparisons of biodiversity. *Biometrical Journal* **43**, 617–625.

Rousseeuw, P.J. (1987) Silhouettes: A graphical aid to the interpretation and validation of cluster analysis. *Journal of Computational and Applied Mathematics* **20**, 53–65.

Sammon, J.W. (1969) A non-linear mapping for data structure analysis. *IEEE Transactions on Computing* **C-18** 401–409.

Sarkar, S. (1998) Some probability inequalities for ordered MTP2 random variables: a proof of Simes conjecture. *Annals of Statistics* **26**, 494–504.

Sarkar, S. and Chang, C.K. (1997) Simes' method for multiple hypothesis testing with positively dependent test statistics. *Journal of the American Statistical Association* **92**, 1601–1608.

Scherer, R. (2010) Simultaneous Confidence Intervals for Biodiversity Indices with Application to Overdispersed Multinomial Count Data. http://www.biostat.uni-hannover.de/research/thesis/MSc-Scherer20100525.pdf

Seber, G.A.F. (1984). *Multivariate Observations*. Wiley.

Shaffer, J.P. (1995) Multiple hypothesis testing. *Annual Review of Psychology* **46**, 561–576.

Shimodaira, H. (2002) An approximately unbiased test of phylogenetic tree selection, *Systematic Biology* **51**, 492–508.

Shimodaira, H. (2004) Approximately unbiased tests of regions using multistep-multiscale bootstrap resampling, *Annals of Statistics* **32**, 2616–2641.

Shimwell, D.W. (1971) *The Description and Classification of Vegetation*. Sidgwick & Jackson.

Sibson, R. (1972) Order invariant methods for data analysis. *Journal of the Royal Statistical Society B* **34**, 311–349.

Smith, B.T, Boyle, J.M., Dongarra, J.J., Garbow, B.S., Ikebe,Y., Klema, V. and Moler, C.B. (1976) *Matrix Eigensystems Routines – EISPACK Guide*. Springer-Verlag Lecture Notes in Computer Science **6**.

Smith, E.P and van Belle, G. (1984) Nonparametric estimation of species richness. *Biometrics* **40**, 119–129.

Sneath, P.H.A. and Sokal, R.R. (1973) *Numerical Taxonomy*. Freeman.

Struyf, A., Hubert, M. and Rousseeuw, P.J. (1996) Clustering in an object-oriented environment. *Journal of Statistical Software* **1**.

Struyf, A., Hubert, M. and Rousseeuw, P.J. (1997) Integrating robust clustering techniques in S-PLUS, *Computational Statistics and Data Analysis* **26**, 17–37.

Suzuki, R. and Shimodaira, H. (2004) An application of multiscale bootstrap resampling to hierarchical clustering of microarray data: How accurate are these clusters? *The Fifteenth International Conference on Genome Informatics 2004*, P034.

Ter Braak, C.J.F. (1986) Canonical correspondence analysis: A new eigenvector technique for multivariate direct gradient analysis. *Ecology* **67**, 1167–1179.

Torgerson, W.S. (1958) *Theory and Methods of Scaling*. Wiley.

Tóthmérész, B. (1995) Comparison of different methods for diversity ordering. *Journal of Vegetation Science* **6**, 283–290.

Tsallis, C. (1988) Possible generalization of Boltzmann–Gibbs statistics. *Journal of Statistical Physics* **52**, 479–487.

Ugland, K.I., Gray, J.S. and Ellingsen, K.E. (2003) The species-accumulation curve and estimation of species richness. *Journal of Animal Ecology* **72**, 888–897.

Van Sickle, J. and Hughes, R.M. (2000) Classification strengths of ecoregions, catchments, and geographic clusters of aquatic vertebrates in Oregon. *Journal of the North American Benthological Society*. **19**, 370–384.

Venables, W.N. and B.D. Ripley (2002) *Modern Applied Statistics with S*. Springer-Verlag.

Westfall, P.H. and Young, S.S. (1993) *Resampling-based Multiple Testing: Examples and Methods for p-Value Adjustment*. Wiley.

Whittaker, R.H. (1960) Vegetation of Siskiyou mountains, Oregon and California. *Ecological Monographs* **30**, 279–338.

Whittaker, R.H. (1965) Dominance and diversity in plant communities. *Science* **147**, 250–260.

Wilkinson, J.H. (1965) *The Algebraic Eigenvalue Problem*. Clarendon Press.

Williamson, M. and Gaston, K.J. (2005) The lognormal distribution is not an appropriate null hypothesis for the species–abundance distribution. *Journal of Animal Ecology* **74**, 409–422.

Wilson, J.B. (1991) Methods for fitting dominance/diversity curves. *Journal of Vegetation Science* **2**, 35–46.

Wolda, H. (1981) Similarity indices, sample size and diversity. *Oecologia* **50**, 296–302.

Wright, S.P. (1992) Adjusted P-values for simultaneous inference. *Biometrics* **48**, 1005–1013.

Yandell, B.S. (1997) *Practical Data Analysis for Designed Experiments*. Chapman & Hall.

Zapala, M.A. and Schork, N.J. (2006) Multivariate regression analysis of distance matrices for testing associations between gene expression patterns and related variable. *Proceedings of the National Academy of Sciences, USA* **103**, 19430–19435.

Index

9 781907 807619